Measurement Systems: Applications and Design

Edited by **Miles Eron**

Clanrye International

New Jersey

Published by Clanrye International,
55 Van Reypen Street,
Jersey City, NJ 07306, USA
www.clanryeinternational.com

Measurement Systems: Applications and Design
Edited by Miles Eron

International Standard Book Number: 978-1-63240-347-6 (Hardback)

Printed in China.

Contents

Preface

This book provides an introduction to measurement systems, their designs and various applications. Measurement is an interdisciplinary experimental science. Measurement systems synergistically merge science, manufacturing and arithmetical methods to give basic data for study, design and improvement, management of processes and operations, and make possible safe and economic performance of systems. In the past few years, measuring methods have extended and developed through wide research activities and hardware enhancements.

Significant researches are present in this book. Intensive efforts have been employed by authors to make this book an outstanding discourse. This book contains the enlightening chapters which have been written on the basis of significant researches done by the experts.

Finally, I would also like to thank all the members involved in this book for being a team and meeting all the deadlines for the submission of their respective works. I would also like to thank my friends and family for being supportive in my efforts.

Editor

1

Measurement: System, Uncertainty and Response

Md. Zahurul Haq*
Bangladesh University of Engineering & Technology (BUET)
Bangladesh

1. Introduction

A measuring instrument transforms a *measurand*, i.e., a physical variable to be measured, to provide comprehensible output. An instrument can be aptly analysed and synthesized as a *system* which is a set of interconnected components functioning as a unit, while *response* is a measure of such unit's fidelity to its purpose. A real instrument's response is neither perfect nor identical even under static and replicate-conditions. So appropriate approach to quantify the deviation from the *true value* is required. It is now widely recognized that measurement results should be expressed in terms of *estimated value* and an associated *uncertainty value* obtained by proper analysis. System response analysis is further complicated in dynamic measurements as an instrument does not respond instantaneously to an input that varies in time. This obviously creates a measurement problem, and if these effects are not accounted for, dynamic errors are introduced. Therefore, the performance of a measuring instrument is to be specified in terms of both static and dynamic performance parameters.

2. Measurement system

Measurement is the act of assigning a specific value to a measurand. Mass, distance, time, temperature, force, and other physical quantities, as well as the properties of matters, materials, and devices, must be measured and described in common terminology. Measuring instruments are designed to generate a fixed and reproducible magnitude of the measurand which is expressed by a number (the magnitude ratio) followed by the matching unit, e.g., a length of 2.5 m. So measurement provides quantitative information on the actual state of the measurand that otherwise could only be estimated. *ISO/IEC Guide 98:2008 Uncertainty of Measurement – Guide to the Expression of Uncertainty in Measurement (GUM)* (ISO, 2008) reports the following scope of applications of measurements:

- To maintain quality control and quality assurance in production;
- To comply with and enforcing laws and regulations;
- To conduct basic/applied research and development, in science and engineering;
- To develop, maintain and compare international and national physical reference standards, reference materials, and also to achieve traceability to national standards.

---

*http://teacher.buet.ac.bd/zahurul/

According to *ISO/IEC Guide 99:2007 International Vocabulary of Metrology – Basic and General Concepts and Associated Terms (VIM)* (ISO, 2007), *measurement* is defined as:

> *process of experimentally obtaining one or more quantity values that can reasonably be attributed to a quantity.*

Thus measurement is an experimental science and most experiments are classified into following four categories (Dunn, 2010):

1. *Variational experiments.* These are carried out with an objective to establish the mathematical relations between the experiment's variables.
2. *Validation experiments.* These are carried out to validate a specific hypothesis.
3. *Pedagogical experiments.* These are aimed to demonstrate something that is already known.
4. *Exploration experiments.* These are conducted to explore an idea or possible theory.

2.1 General measurement system

It is sensible to have a generalized description of both the operation and performance of a measuring instrument without recourse to any specific physical hardware. A measuring instrument can be described in terms of its functional elements (Fig. 1); and these elements form the bridge between the *input* to the measurement system and the system *output*, a quantity that is used to infer the value of the measurand. Most measurement systems fall within the general framework consisting of three functional stages (Holman, 2001):

1. *Sensor-transducer stage. Sensor* is directly affected by the measurand, while *transducer* transduces the sensed information to provide an output quantity having a specified relation to the input quantity. Examples of sensors-transducer include thermocouple, strain gauge, manometer, load-cell, etc. There are three basic phenomenon in effect in any sensor operation:
   i) The change (or the absolute value) in the measurand causes an equivalent change in the sensor property, e.g., displacement, voltage, resistance, capacitance, inductance, magnetic flux, etc.
   ii) The change in the sensor property is converted into a more usable form, e.g., temperature change results in the change in generated voltage by a thermocouple.
   iii) The exposure of the sensor to the effects of the measurement environment may lead to some exchange of energy to cause *loading effect*; e.g., a thermometer when inserted into a cup of tea takes some heat from it to cause a difference between the true value and the indicated value.
2. *Signal-conditioning stage.* Transduced signal is modified by one or more basic operations, such as amplification, filtering, differentiation, integration, averaging, etc. for further processing, i.e., display, storage or use in feed-back control systems, etc.
3. *Output stage.* It provides the information sought in a form comprehensible to one of the human senses or to a controller (Beckwith et al., 2007). Output may be *analogue* or *digital*, it may be displayed (using LCD or seven-segment display) or saved (using data-loggers) or may be transmitted to a computer or controller (using data-acquisition system) for further use.

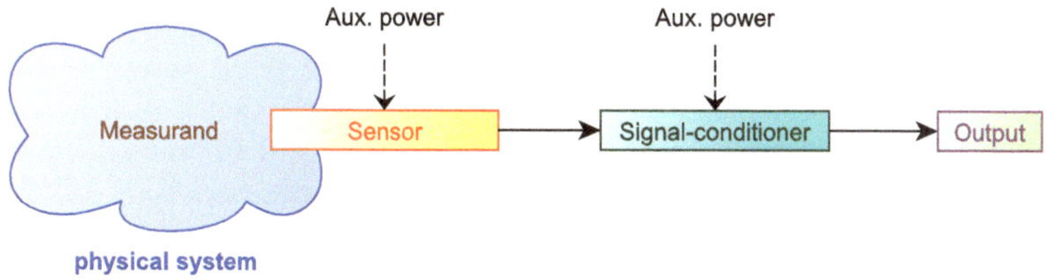

Fig. 1. Functional elements of an instrument or a measurement system.

A glass-bulb mercury thermometer may be analysed to illustrate the above concepts. Here, mercury acts as the sensor whose volume changes with change in its temperature. The transducer is the thermometer-bulb where the change in mercury volume leads to mercury displacement because of the bulb's fixed volume. The stem of the thermometer is the signal-conditioner which physically amplifies the mercury displacement and the graduated scale on the stem offers the required temperature indication. So a pre-established relationship between the input (temperature) and output scale is utilized which is obtained by *calibration*.

2.2 Instrument's performance characteristics and calibration

Measuring instrument transforms the measurand into suitable output that is functionally related to the input and the relationship is established by calibration. *Calibration* is defined in ISO (2007) as:

> *operation that, under specified conditions, in a first step, establishes a relation between the quantity values with measurement uncertainties provided by measurement standards and corresponding indications with associated measurement uncertainties and, in a second step, uses this information to establish a relation for obtaining a measurement result from an indication.*

Calibration provides the opportunity to check the instrument against a known *standard* and subsequently to reduce errors in measurements. There is a hierarchy of standards which arranges themselves in order of decreasing accuracy with the *primary standards* being the most accurate (Doebelin, 2004). Tables 1 and 2, respectively, list hierarchy of standards and errors associated with various levels of temperature standards.

Primary standard	Maintained as absolute unit standard
Transfer standard	Used to calibrate local standards
Local standard	Used to calibrate working standards
Working standard	Used to calibrate local instruments

Table 1. Hierarchy of Standards (Figliola & Beasley, 2011).

Level	Method	Uncertainty [°C]
Primary	Fixed thermodynamic points	0
Transfer	Platinum RTD	±0.005
Working	Platinum RTD	±0.05
Local	Thermocouple	±0.5

Table 2. Examples of Temperature Standards (Figliola & Beasley, 2011).

Calibration is the process of comparison of the output of a measuring system to the values of a range of known inputs and the results may be expressed by a statement, calibration function, calibration diagram, calibration curve, or calibration table. In some cases, it may consist of an additive or multiplicative correction of the indication with associated measurement uncertainty. Over the time, it is possible for the indicated values to drift and it makes recalibration necessary at regular intervals. In general, calibration of measuring instruments needs to be *traceable* to national standardizing laboratory. Hence, *traceability* is defined by ISO (2007) as:

> *property of a measurement result whereby the result can be related to a reference through a documented unbroken chain of calibrations, each contributing to the measurement uncertainty.*

When the measurand maintains a steady value or slowly varies with time, system performance can be described in terms of *static characteristics*. Systems dealing with rapidly varying measurand require additional performance parameters termed *dynamic characteristics*. System performing satisfactorily during static calibration may not provide correct results during dynamic conditions. So relationship between the dynamic input and output must be examined and dynamic performance parameters are required to provide satisfactory results.

3. Static characteristics and specifications

When reporting the measurement result of a measurand, it is necessary that some quantitative indication of the quality of the result be given so that those who use it can assess its reliability (ISO, 2008). Without such an indication, measurement results cannot be compared, either among themselves or with reference values given in a specification or standard. *ASME Power Test Codes (PTC) 19.1: Test Uncertainty* (ASME, 2005) cites the following objectives of uncertainty analysis:

- To facilitate communication regarding measurement and test results;
- To foster an understanding of potential error sources in a measurement system and the effects of those potential error sources on test results;
- To guide the decision-making process for selecting appropriate and cost-effective measurement systems and methodologies;
- To reduce the risk of making erroneous decisions; and
- To document uncertainty for assessing compliance with agreements.

3.1 Measurement errors and uncertainties

Every measurement has error, which results in a difference between the measured value and the true value. The difference between the measured value and the true value is the *total error*. Hence, *accuracy* is defined as the degree of conformity of an indicated value to a recognized accepted standard, or ideal or true value (ISO, 2007). Since the true value is unknown, total error cannot be known and therefore only its expected values can be estimated.

ASME (2005) quantifies the following two components of total error:

1. *Random error*, ϵ, is the portion of the measurement error that varies randomly in repeated measurements throughout the conduct of a test (ISO, 2007). Random errors may arise from uncontrolled test conditions and nonrepeatabilities in the measurement system, measurement methods, environmental conditions, data reduction techniques, etc.
2. *Systematic error*, β, is the component of measurement error that in replicate measurements remains constant or varies in a particular manner (ISO, 2007). These errors may arise from imperfect calibration corrections, measurement methods, data reduction techniques, etc.

Random errors are identified and quantified through repeated measurements while trying to keep the conditions constant and by statistical analysis of the results. Systematic errors can be revealed when the conditions are varied, whether deliberately or unintentionally. Figliola & Beasley (2011) reported the following four methodologies to reduce systematic errors:

1. *Calibration,* by checking the output(s) for known input(s).
2. *Concomitant method,* by using different methods of estimating the same thing and comparing the results.
3. *Inter-laboratory comparison,* by comparing the results of similar measurements.
4. *Experience.*

Errors in the measurement process are classified into four groups (ASME, 2005; Figliola & Beasley, 2011) and examples of the error sources are reported in Table 3.

Error group	Error source
1. Calibration error	Standard or reference value errors Instrument or system errors Calibration process errors Calibration curve fit
2. Loading error	Interaction between the instrument and test media Interaction between test article and test facility
3. Data-acquisition error	Measurement system operating conditions Sensor-transducer stage (instrument error) Signal conditioning stage (instrument error) Output stage (instrument error) Process operating conditions Sensor installation effects Environmental effects Spatial variation error Temporal variation error
4. Data-reduction error	Calibration curve fit Truncation error

Table 3. Error classifications and sources.

It is now widely recognized that, when all of the known or suspected components of error have been evaluated and the appropriate corrections have been applied, there still remains an uncertainty about the correctness of the stated result, i.e., a doubt about the quality of the result of the measurement. The word *uncertainty* means doubt, and thus in its broadest sense *uncertainty of measurement* means doubt about the validity of the result of a measurement. According to ISO (2007), *uncertainty* is defined as:

> *non-negative parameter characterizing the dispersion of the quantity values being attributed to a measurand, based on the information used.*

There are two widely accepted professional documents on uncertainty analysis:

1. ASME Power Test Codes (PTC) 19.1: *Test Uncertainty* (ASME, 2005), and

2. ISO/IEC Guide 98:2008 *Uncertainty of Measurement – Guide to the Expression of Uncertainty in Measurement (GUM)* (ISO, 2008).

These two documents differ in some terminology and how errors are catalogued. For example, ASME (2005) refers to random and systematic error terms to classify errors by how they manifest themselves in the measurement; ISO (2008) refers to Type A and Type B errors. Type A uncertainties have data with which standard deviation can be calculated, while Type B uncertainties do not have data to calculate a standard deviation and must be estimated by other means. These differences are real but the final result of an uncertainty analysis by either method will yield a similar value (Figliola & Beasley, 2011).

3.2 Analysis of measurement data

Any single measurement of a parameter, x, is influenced by different elemental random error sources. In successive measurements of the parameter, the values of these elemental random error sources change resulting in the evident random scatter. Measurement scatters common in science and engineering are, in general, described by Normal or Gaussian distribution which predicts that the scatter in the measured data-set will be distributed symmetrically about some central tendency (Fig. 2). If an infinite number of measurements of a parameter were to be taken following the defined test process, the resulting population of measurements could be described statistically in terms of the *population mean*, μ, and the *population standard deviation*, σ; and the true value of x would be μ (Doebelin, 2004).

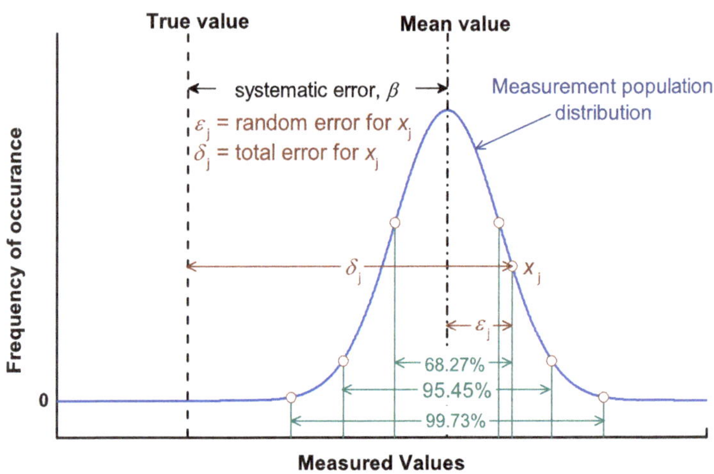

Fig. 2. Illustration of measurement errors.

In practice, only finite-sized data are available. So measured data can only provide an *estimate* of the true value (Figliola & Beasley, 2011). If the measured variable is described by discrete data of size N, the *mean value*, $\overline{x}$, of the measured value, x_j is given by:

$$\overline{x} = \frac{1}{N}\sum_{j=1}^{N} x_j \tag{1}$$

, and the *standard deviation* is given by

$$S_x = \sqrt{\frac{1}{N-1}\sum_{j=1}^{N}(x_j - \overline{x})^2} \tag{2}$$

In computing the sample average, $\overline{x}$, all N data are independent. However, the standard deviation uses the result of the previous calculation for $\overline{x}$. Hence, number of degrees of freedom is reduced by one and in calculating the standard deviation, $(N-1)$ is used instead of N. It may also be noted that, for infinite number of measurements $(N \to \infty)$: $\overline{x} \to \mu$ and $S_x \to \sigma$. For finite data sets, the *standard deviation of the mean*, $S_{\overline{x}}$, can be estimated from a single finite data set as:

$$S_{\overline{x}} = \frac{S_x}{\sqrt{N}} \tag{3}$$

For a normal distribution of x about some sample mean value, $\overline{x}$, it can be stated statistically

$$x_j = \overline{x} \pm u_x = \overline{x} \pm k(\nu,\gamma)S_x = \overline{x} \pm t(\nu,\gamma)S_x \quad (P\%) \tag{4}$$

where the u_x is the *uncertainty interval* or *precision index* associated with the estimate of x. The value of u_x depends on a constant known as *coverage factor* that depends on the probability distribution function, the confidence level, γ (which is expressed by probability, $P\%$) and the amount of data, N. For finite data set, $k(\nu,\gamma) = t(\nu,\gamma)$. The Student's t value for a given probability, $P\%$ and degrees of freedom in data, $\nu = N-1$, can be obtained from Table 4, which is a short tabulation of the *Student's t distribution*. The estimate of the *true mean value*, x_t, based on the finite data-set can be estimated (Figliola & Beasley, 2011) using:

$$x_t = \overline{x} \pm k(\nu\gamma)S_{\overline{x}} \quad (P\%) \tag{5}$$

ν	t_{50}	t_{90}	t_{95}	t_{99}
1	1.000	6.314	12.706	63.657
2	0.816	2.920	4.303	9.925
5	0.727	2.015	2.571	4.032
10	0.700	1.812	2.228	3.169
20	0.687	1.725	2.086	2.845
30	0.683	1.697	2.042	2.750
∞	0.674	1.645	1.960	2.576

Table 4. *Student's t distribution* (Figliola & Beasley, 2011).

The uncertainty interval, u_x, assumes a set of measured values with only random error present. Furthermore, the set of measured values is assumed to have unbound significant digits and to have been obtained with a measuring system having infinite resolution. When finite resolution exits and truncation of digits occurs, the uncertainty interval may be larger than that predicted by the consideration of the random errors only. The uncertainty interval can never be less that the resolution limits or truncation limits of the measured values.

To illustrate the above concepts, consider the data of Table 5, where 20 values of an arbitrary measurement is presented.

9.78	10.29	9.68	10.30	9.69	9.63	10.35	9.88	10.29	9.72
9.85	10.25	9.75	10.24	9.89	10.31	9.94	10.35	10.27	9.65

Table 5. Sample of random variable, x.

- Sample mean, $\overline{x} = \frac{1}{N}\sum_{j=1}^{N} x_j = 10.0$
- Standard deviation, $S_x = \sqrt{\frac{1}{N-1}\sum_{j=1}^{N}(x_j - \overline{x})^2} = 0.28$
- If a 21st data points are to be taken:

$$\begin{aligned} x_{j=21} &= 10.00 \pm (1.729 \times 0.28) = 10.00 \pm 0.48 \qquad (90\%) \\ &= 10.00 \pm (2.093 \times 0.28) = 10.00 \pm 0.59 \qquad (95\%) \end{aligned}$$

  So, there is a 95% probability that the 21st value will be between 9.41 and 10.59.
- The true value, x_t, is estimated by the sample mean value, $\overline{x}$ and the standard deviation of the mean, $S_{\overline{x}} = S_x/\sqrt{N} = 0.0626$:

$$\begin{aligned} x_t &= 10.00 \pm 0.11 \qquad (90\%) \\ &= 10.00 \pm 0.13 \qquad (95\%) \end{aligned}$$

3.3 Uncertainty analysis: error propagation

In calibration experiments, one measures the desired results directly. In nearly all other measurements, results are obtained through functional relationship with measured values. So, it is necessary to compute the uncertainty in the results from the estimates of the uncertainty in the measurements. This computation process is called the *propagation of uncertainty*.

Consider a result, R, which is determined through some functional relationship between independent variables, $x_1, x_2, \cdots, x_N$, defined by

$$R = R(x_1, x_2, \cdots, x_N) \tag{6}$$

where N is the number of independent variables and each variable contains some measure of uncertainty to affect the final result. The best estimate of the true mean value, R_t, would be stated as:

$$R_t = \overline{R} \pm u_R \qquad (P\%) \tag{7}$$

where the sample mean, $\overline{R}$ is found from

$$\overline{R} = R(\overline{x}_1, \overline{x}_2, \cdots, \overline{x}_N) \tag{8}$$

and the uncertainty in $\overline{R}$ is found from

$$u_R = \sqrt{\left(\frac{\partial R}{\partial x_1} u_1\right)^2 + \left(\frac{\partial R}{\partial x_2} u_2\right)^2 + \cdots + \left(\frac{\partial R}{\partial x_N} u_N\right)^2} \tag{9}$$

where, $u_1, u_2, \cdots, u_N$ are the uncertainties associated with $x_1, x_2, \cdots, x_N$.

To illustrate the concepts presented in the present section, let us consider the calculation of electrical power, P from

$$P = VI$$

, where, voltage, V and current, I are measured as

$$V = 100\text{ V} \pm 5\text{ V}$$
$$I = 10\text{ A} \pm 0.1\text{ A}$$

Hence, the nominal value of power is 1000 W. By taking the worst possible variations in voltage and current, we would calculate

$$P_{max} = (100+5)(10+0.1) = 1060.5\text{ W}$$
$$P_{min} = (100-5)(10-0.1) = 940.5\text{ W}$$

Thus, using the simple calculations, the uncertainty in the power is +6.05%, −5.95%. However, it is quite unlikely that the power would be in error by these amount because the variation of voltage reading would probably not be corresponding with the current reading. Hence, it is quite unlikely these two independent parameters would achieve maximum or minimum values simultaneously. Hence, using Eq. 9:

$$\frac{\partial R}{\partial x_1} = \frac{\partial P}{\partial V} = I = 10\text{ A} \qquad u_1 = u_V = 5\text{ V}$$

$$\frac{\partial R}{\partial x_2} = \frac{\partial P}{\partial I} = V = 100\text{ V} \qquad u_2 = u_I = 0.1\text{ A}$$

$$u_R = \sqrt{(10\times 5)^2 + (100\times 0.1)^2} = \sqrt{2500+100} = 51.0\text{ W} = 5.1\%$$

Hence, estimated power, $P = 1000\text{ W} \pm 5.1\%$.

If we ignore the uncertainty contribution due to the current reading ($u_I \to 0$), the uncertainty in power measurement would be 5%. Hence, very little is gained by trying to reduce the 'small' uncertainties. Any improvement in result should be achieved by improving the instrumentation or technique connected with relatively large uncertainties.

4. Specification of dynamic systems and response

A measuring instrument requires a certain amount of time to achieve complete response. In case of a dynamic measurement, actual response undergoes some attenuation, delay and distortion. A measurement system should be capable of yielding the true value(s) from the time varying input value(s). Study of dynamic characteristics of such system requires considerable analyses and the responses are conveniently stated in terms of step response, harmonic response and frequency response of the instrument.

4.1 Modelling of measuring instruments

The response of a measurement system, i.e., output, $x(t)$, when subjected to an input forcing function, $f(t)$, may be expressed by a linear ordinary differential equation with constant coefficients of the form (Doebelin, 2004):

$$a_n \frac{d^n x}{dt^n} + a_{n-1}\frac{d^{n-1}x}{dt^{n-1}} + \cdots + \underbrace{a_2\frac{d^2x}{dt^2} + \underbrace{a_1\frac{dx}{dt} + \overbrace{a_0 x = f(t)}^{0^{th}\ order}}_{1^{st}\ order}}_{2^{nd}\ order} \tag{10}$$

where, $f(t) \triangleq$ Input quantity imposed on the system,
$x(t) \triangleq$ Output or the response of the system,
a's $\triangleq$ Physical system parameters, assumed constants.

Hence, the *order* of the system is designated by the order of the differential equation. Once the governing equation of the instrument is established, its response can be obtained if the input (measurand) is a known function of time. Dynamic response of a system can be estimated by solving Eq. 10 with proper initial and boundary conditions. While the general model of Eq. 10 is adequate for handling any linear measurement system, certain special cases (e.g., *zero-*, *first-* and *second-order* systems) occur so frequently in practice that warrant special attention. Furthermore, complex systems may profitably be considered in terms of the combinations of simple cases and the response can be inferred from the observations made from these cases.

4.1.1 Zero-order instrument

The simplest case of Eq. 10 occurs if all the values of a's other than a_0 are assumed to be zero, and Eq. 10 is then transformed into a simple algebraic equation:

$$a_0 x = f(t) :\Longrightarrow x(t) = \Bbbk\, f(t) \tag{11}$$

The constant, $\Bbbk \triangleq 1/a_0$ is called the *static sensitivity* of the system, and it represents the scaling between the input and the output. For any-order system, static sensitivity always has the same physical interpretation, i.e., the amount of output per unit input when the input is static and under such condition all the derivative terms of Eq. 10 are zero.

A practical example of a zero-order instrument is a displacement measuring potentiometer where a strip of resistance material is excited with a voltage, V_s, and provided with a sliding contact responding to displacement. If the resistance is linearly distributed along the length, L, the output voltage, $e_o(t)$, may be written as a function of displacement, $l(t)$, as:

$$e_o(t) = \frac{V_s}{L} l(t) = \Bbbk\, l(t) \tag{12}$$

where, $\Bbbk \triangleq V_s/L$ (Volts per unit length).

4.1.2 First-order instrument

If in Eq. 10, all a's other than a_1 and a_o are taken as zero, we get:

$$a_1 \frac{dx}{dt} + a_0 x = f(t) :\Longrightarrow \tau \frac{dx}{dt} + x = \Bbbk\, f(t) \tag{13}$$

where, $\Bbbk \triangleq 1/a_o \triangleq$ static sensitivity,
$\tau \triangleq a_1/a_o \triangleq$ time-constant.

The *time constant* has the dimension of time, while the *static sensitivity* has the dimension of output divided by input. When time constant of a system is very small, the effect of the derivative term in Eq. 13 becomes negligible and the governing equation approaches to that of a zero-order system. It is experimentally observed that an instrument with small value of the time-constant is associated with good dynamic response.

A practical example of a first-order instrument is a mercury-in-glass thermometer. If a thermometer, initially at room temperature, is inserted into a hot fluid at temperature, T_∞, then the convective heat gain from the hot fluid will result in the increase of internal energy of mercury at the same rate. The thermometer takes a while to give the correct reading, and the temperature reading, $T(t)$, can be obtained using the following energy-balance equation:

$$\dot{Q}_{in} = hA\left[T_\infty - T(t)\right] = mC\frac{dT(t)}{dt} :\Longrightarrow \tau\frac{dT(t)}{dt} + T(t) = T_\infty \qquad (14)$$

Hence, $\tau \triangleq mC/hA$ and $\Bbbk = 1$; and h is the convective heat transfer coefficient, A is the surface area over which heat is transferred, m is the mass of mercury, and C is its specific heat.

4.1.3 Second-order instrument

A second-order instrument is one that follows the equation:

$$a_2\frac{d^2x}{dt^2} + a_1\frac{dx}{dt} + a_o x = f(t) :\Longrightarrow \frac{1}{\omega_n^2}\frac{d^2x}{dt^2} + 2\frac{\zeta}{\omega_n}\frac{dx}{dt} + x = \Bbbk\, f(t) \qquad (15)$$

$$\begin{aligned} &\Bbbk \triangleq 1/a_o \triangleq \text{static sensitivity,} \\ \text{where, } &\omega_n \triangleq \sqrt{\tfrac{a_o}{a_2}} \triangleq \text{undamped natural frequency,} \\ &\zeta \triangleq \tfrac{a_1}{2\sqrt{a_o a_2}} \triangleq \text{dimensionless damping ratio.} \end{aligned}$$

A practical example of a second-order instrument is a weight measuring spring balance as shown in Fig. 3. Hence, k is the spring constant of the ideal spring and b is the friction constant where damper friction force is linearly proportional to the velocity of mass, $\dot{x}(t)$. Summing the forces acting on the mass and utilizing the Newton's second law of motion yields,

$$f(t) - kx - b\frac{dx}{dt} = m\frac{d^2x}{dt^2} :\Longrightarrow \frac{1}{\omega_n^2}\frac{d^2x}{dt^2} + 2\frac{\zeta}{\omega_n}\frac{dx}{dt} + x = \Bbbk\, f(t) \qquad (16)$$

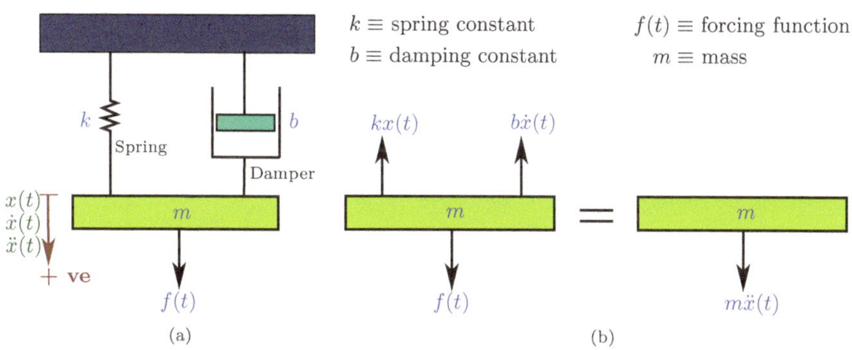

Fig. 3. (a) Spring-mass-damper system, (b) Free-body diagram.

Final form of Eq. 15, could represent, among other things, an electrical resistor-inductor-capacitor (RLC) circuit. The variables of the electric circuit behave exactly as the analogous variables of the equivalent mechanical system. The spring-mass-damper system and analogous RLC circuit are illustrated in Fig. 4. Both of these systems share the same governing equation and therefore have analogous response when subjected to input forcing.

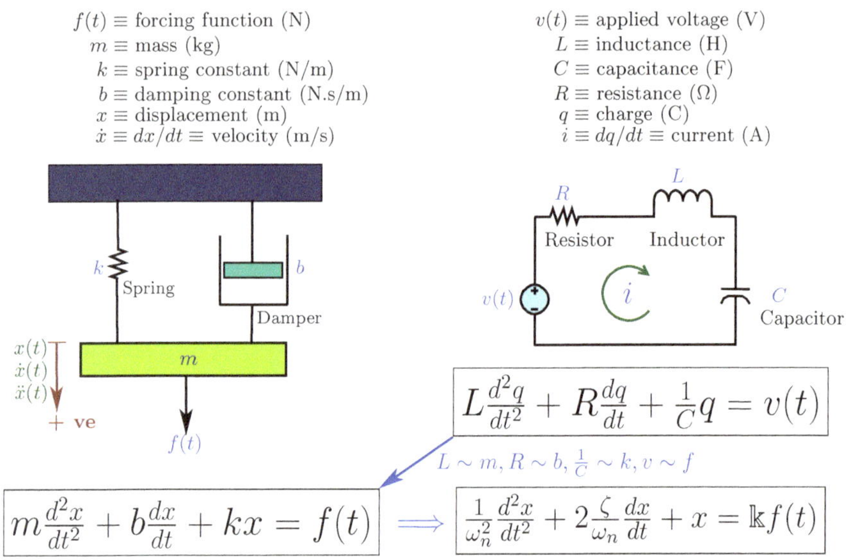

Fig. 4. Spring-mass-damper system and analogous RLC circuit.

4.2 Modelling of response of measuring instruments

Once the governing equation of a system is established, its dynamic response to forcing element can be obtained if the forcing can be represented as a function of time. The fundamental difficulty in this approach lies in the fact that the quantities to be measured usually do not follow some simple mathematical functions. However, it is fruitful to study a system's response to common forcing inputs, e.g., step and harmonic inputs. These forcing functions have been found to be very useful in the theoretical and experimental analyses of measurement system's response. Mathematically, the *step function* of magnitude, A, can be expressed as:

$$f(t) = \begin{cases} 0 & \text{at } t = 0 \\ A & \text{for } t > 0 \end{cases} \tag{17}$$

, and the *harmonic function* of circular frequency, ω, can be expressed as:

$$f(t) = \begin{cases} 0 & \text{at } t = 0 \\ A\sin\omega t & \text{for } t > 0 \end{cases} \tag{18}$$

Graphical representation of these two functions are shown in Fig. 5.

4.2.1 Response of zero-order instruments

The response of a zero-order instrument is governed by a simple algebraic equation (Eq. 11). So the output, $x(t)$, follows the input, $f(t)$, perfectly with no distortion or time lag. Hence, zero-order instrument represents *ideal* dynamic performance, and is thus a standard against which less perfect or dynamic instruments are compared. Typical responses of a zero-order instrument are shown in Fig. 6.

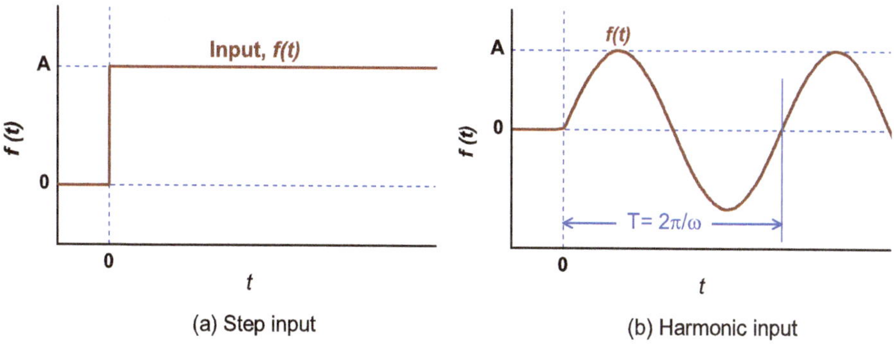

Fig. 5. Step and harmonic inputs.

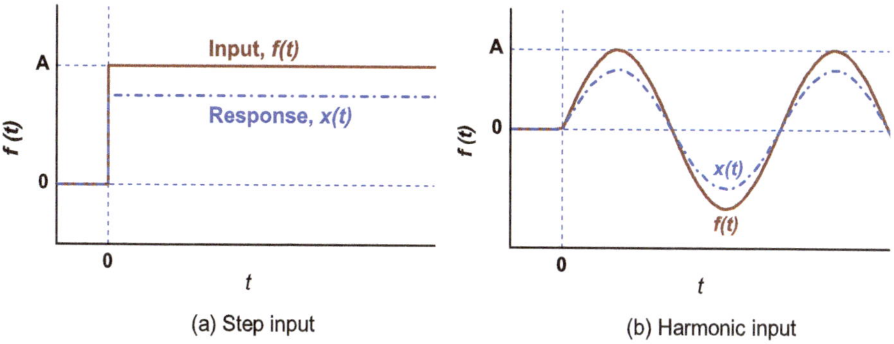

Fig. 6. Zero-order instrument's response for step and harmonic inputs (for $\Bbbk = 0.75$).

Zero-order system concept is used to analyse real system's response in static calibration. When dynamic signals are involved, higher order differential equations are required to estimate the dynamic response as most of the real systems possess inertial/storage capability and are subjected to some viscous resistance and dissipation (Figliola & Beasley, 2011).

4.2.2 Response of first-order instruments

Response for Step Input: If the governing equation for first-order system (Eq. 13) is solved for a step input function and with initial condition $x|_{t=0} = x_0$, the solution is:

$$x(t) = \underbrace{(x_o - A\Bbbk)\exp(-t/\tau)}_{\text{transient response}} + \underbrace{A\Bbbk}_{\text{steady-state response}} \tag{19}$$

For large values of time, $x(t \to \infty) = A\Bbbk \triangleq x_\infty$. Hence, $x_\infty \triangleq$ *steady-state response* of the system. Equation 19 can be rearranged to give non-dimensional response, $M(t)$, as:

$$M(t) = \frac{x(t) - x_o}{x_\infty - x_o} = 1.0 - \exp(-t/\tau) \tag{20}$$

, where, $M(t)$ is the ratio between output and input amplitudes. Non-dimensional Eq. 20 is valid for all first-order system's response to step-input forcing; and plot of Eq. 20 is shown in

Fig. 7. It is observed that, the response approaches the steady-state value monotonically and the response reaches 63.2% of its steady-state value when $t = \tau$. When elapsed time is 2τ, 3τ, and 4τ, the response is 86.5%, 95% and 98%, respectively. It is also seen that, the slope of the response curve (Fig. 7) at the origin is 1.0, i.e., if the initial response rate were maintained, the response would be completed in one time constant. However, the final value (steady-state response) is achieved only after infinite time (Fig. 7). In practice two response specifications are used to describe the first-order system response:

1. *Settling time*, t_s. It is defined as the time required for the system to reach and stay within the 2% of the final value. Hence, for first-order system, $t_s \simeq 4\tau$.
2. *Rise time*, t_r. It is defined as the time required for the response to go from a small percentage (10%) to a large percentage (90%) of the step input. Hence, for first-order system, $t_r \simeq 2.2\tau$.

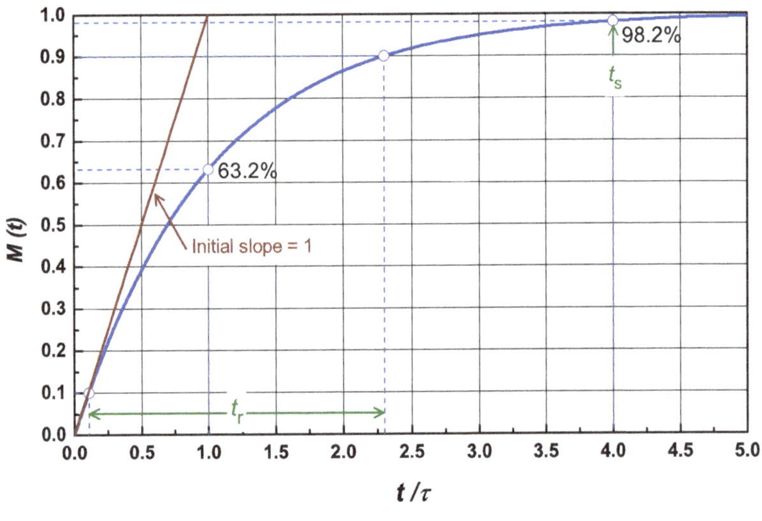

Fig. 7. First-order system response to a unit step-input.

Response for Harmonic Input: If the governing equation for first-order system (Eq. 13) is solved for harmonic input and $x|_{t=0} = 0$, the solution is:

$$\frac{x(t)}{A\Bbbk} = \underbrace{\frac{\omega\tau}{1+(\omega\tau)^2}\exp(-t/\tau)}_{\text{transient response}} + \underbrace{\frac{1}{\sqrt{1+(\omega\tau)^2}}\sin(\omega t+\phi)}_{\text{steady-state response}} \tag{21}$$

where, $\phi \triangleq \tan^{-1}(-\omega\tau) \triangleq$ *phase lag*. Hence, *time delay*, Δt, is related to phase lag as:

$$\Delta t = \frac{\phi}{\omega} \tag{22}$$

For $\omega\tau >> 1$, response is attenuated and time/phase is lagged from input, and for $\omega\tau << 1$, the transient response becomes very small and response follows the input with small attenuation and time/phase lag. Ideal response (without attenuation and phase lag) is obtained when the system time constant is significantly smaller than the forcing element period, T. As $t \to \infty$, the first term on the right side of Eq. 21 vanishes and leaves only the steady-state solution:

$$x(t)|_s = \frac{A\mathbb{k}}{\sqrt{1+(\omega\tau)^2}}\sin(\omega t+\phi) = \frac{A\mathbb{k}}{\sqrt{1+(\omega\tau)^2}}\sin\omega(t+\Delta t) = f(t)\times G_a\angle\phi \tag{23}$$

Hence, $G_a \triangleq \mathbb{k}/\sqrt{1+(\omega\tau)^2} \triangleq$ *steady-state gain*. Equation 23 indicates that the attenuated steady-state response is also a sine wave with a frequency equal to the input signal frequency, ω, and it lags behind the input by phase angle, ϕ.

To illustrate the above concepts, the response of a thermocouple subjected to a harmonic temperature variation is plotted in Figs. 8 and 9; and summary results are also reported in Table. 6. Initial transient response is clearly visible and steady-state response is achieved only after the expiration of 4τ time. The response is of same frequency but attenuated in magnitude and lags behinds the input signal. However, once the steady-state is achieved, knowing the values of the gain, G_a, and phase lag, ϕ, it is possible to estimate the correct input value, and an example is also shown in Fig. 8. The effect of time constants on dynamic response is illustrated in Fig. 9 where response from three thermocouple systems with time constants of 1, 5 and 50 s are plotted. Figure 9 provides a meaningful insight into the effects of time constant as it shows that systems with small time constants follow the input with less attenuation and time delay, and vice versa. However, when average value of a measurand is desired (e.g., in chemical reactant mixing chamber), system with higher time constants are suitable and near average result is observed as shown in Fig. 9 for a thermocouple with τ = 50 s.

Unit	τ [s]	τ/T	ϕ [deg]	Δt [s]	G_a
01	01	0.04	-14.0	-0.98	0.97
02	05	0.2	-51.3	-3.58	0.62
03	50	2.0	-85.4	-5.96	0.08

Table 6. Thermocouple's response to harmonic temperature variation.

4.2.3 Response of second-order instruments

Response for Step Input: For a second-order system subjected to a step input, the solution (response) is readily available (Dunn, 2010) and may be written as:

$$\frac{x(t)}{\mathbb{k}A} = \begin{cases} 1-\cos(\omega_n t) & \text{for } \zeta = 0 \\ 1-\frac{1}{\sqrt{1-\zeta^2}}\exp(-\zeta\omega_n t)\sin[\omega_n t\sqrt{1-\zeta^2}+\cos^{-1}(-\zeta)] & \text{for } 0<\zeta<1 \\ 1-(1+\omega_n t)\exp(-\omega_n t) & \text{for } \zeta = 1 \\ 1-\exp(-\zeta\omega_n t)\left[\cosh(\omega_n t\sqrt{\zeta^2-1})+\frac{\zeta}{\sqrt{\zeta^2-1}}\sinh(\omega_n t\sqrt{\zeta^2-1})\right] & \text{for } \zeta>1 \end{cases} \tag{24}$$

The nature of the response of the second-order system for a step input depends on the value of the damping ratio, ζ, as depicted in Fig. 10, and four types of response may be identified:

1. *Harmonic oscillation* ($\zeta = 0$). Response oscillates at the natural frequency, ω_n, and the oscillations are undamped.

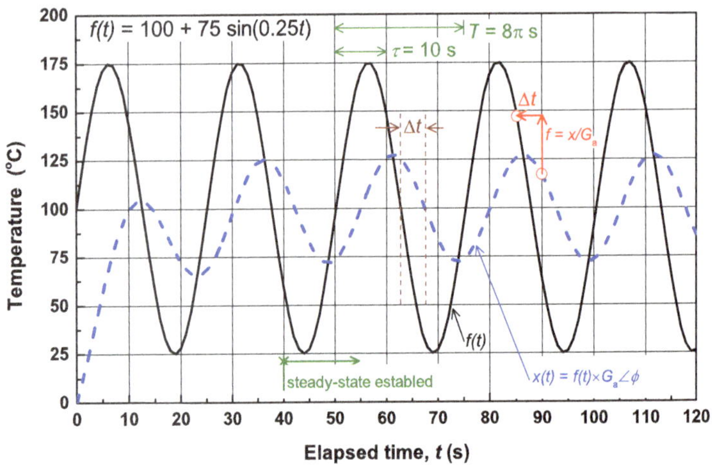

Fig. 8. Response of thermocouple for harmonic temperature input.

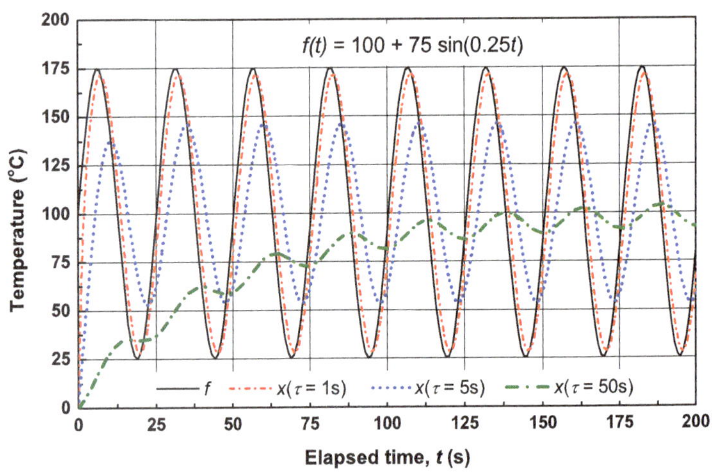

Fig. 9. Effect of time constant on thermocouple's response to harmonic temperature input.

2. *Underdamped response* ($0 < \zeta < 1$). Response overshoots the steady-state value initially, and then eventually decays to the steady-state value. The smaller the value of ζ, the larger the overshoot.
3. *Critically damped response* ($\zeta = 1$). An exponential rise in response occurs to approach the steady-state value without any overshoot, and the response is as fast as possible without overshoot.
4. *Overdamped response* ($\zeta > 1$). Response approaches the steady-state value without overshoot, but at a slower rate. Hence, excessive time is required to complete a measurement and therefore frequency at which the measurement is possible is limited. So, little attention is focused on the overdamped systems for dynamic measurements.

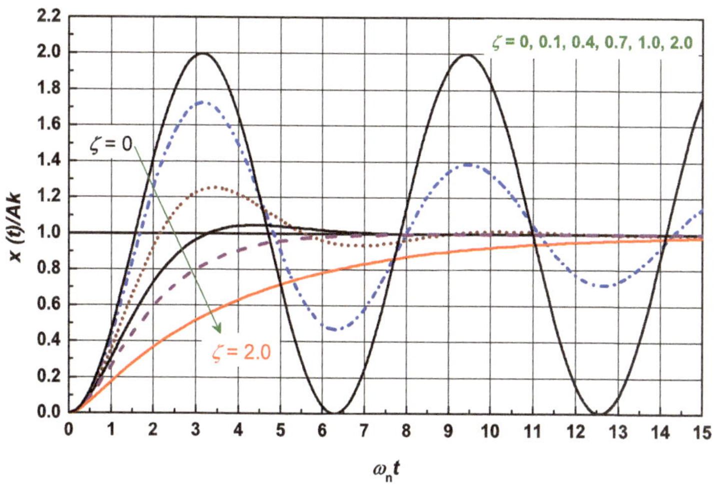

Fig. 10. Transient response of a second-order system for a step-input at $t = 0$.

For underdamped system, as ζ decreases, response becomes increasingly oscillatory. The transient response oscillates about the steady-value and occurs with a period, T_d, given by:

$$T_d \triangleq \frac{2\pi}{\omega_d} \quad : \quad \omega_d \triangleq \omega_n\sqrt{1-\zeta^2} \tag{25}$$

, where, $\omega_d \triangleq$ *ringing frequency*. Standard performance parameters in terms of step response is shown in Fig. 11. It is observed that, the duration of the transient response is controlled by $\zeta\omega_n$. In fact, its influence is equivalent to that of a time constant in a first-order system, such that we could define a second-order time constant as $\tau_e \triangleq 1/\zeta\omega_n$. The system settles to steady state value, $x_\infty = \mathbb{k}A$, more quickly when it is designed with large $\zeta\omega_n$, i.e., small τ_e. Nevertheless, for all systems with $\zeta > 0$, the response will eventually indicate the steady value as $t \to \infty$ (Figliola & Beasley, 2011). Like a first-order system, the swiftness of the response may be described by *settle time*, t_s and *rise time*, t_r. For second-order system, $t_s \simeq 4\tau_e$ and $t_r\omega_n \cong 2.16\zeta + 0.60$ for $0.3 \leqq \zeta \leqq 0.8$ (Dorf & Bishop, 1998).

Equation 24 also reveals that for second-order system, whenever ω_n appears, it appears as the product of $\omega_n t$. This means that, if we say, double ω_n, the same range of the response will occur in exactly one half the time. Thus ω_n is a direct and proportional indicator of response speed. For a fixed value of ω_n, the speed of response is determined by ζ (Doebelin, 1998).

4.2.4 Response in time-domain and postscript

The response of a zero-order system is an ideal one and response follows the input without attenuation and phase lag. In practice, no measurement system is ideal and therefore does not possess zero-order characteristics. First-order system follows the input signal with some attenuation and phase lag, but no oscillations are present. Second-order system responses are further complicated as in addition to output attenuation and phase lag, it may exhibit oscillation. The total response of a non-zero-order dynamic system due to a harmonic input is the sum of transient response, which is independent of the frequency of the input, and a

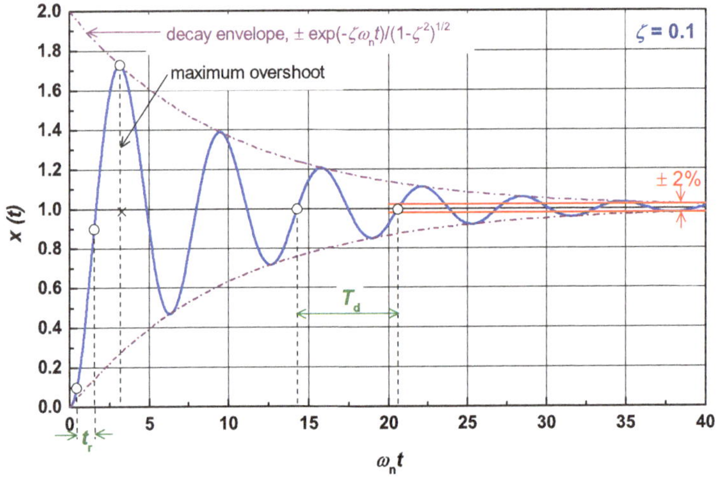

Fig. 11. Second-order underdamped response specifications.

response which depends on the frequency of the input (Dorf & Bishop, 1998). For a stable system, the transient response decays exponentially and eventually becomes insignificant compared to the response that is specific to the input. When this occurs, the system has reached the steady-state response (Kelly, 2003). Time-domain analysis is often complicated because of the initial transient-response. However, in a measurement system, we are interested in the steady-state response where the system responds to the input without being affected by the transient effect of the system itself.

4.3 Frequency response of instruments

A very practical and convenient alternative to time-domain modelling to capture the dynamic response of linear system is through its *frequency response* which is defined as the steady-state response of the system to a sinusoidal input. The sinusoidal is a unique input signal, and the resultant output signal for a linear system as well as signal throughout the system, is sinusoidal in the steady-state; it differs from the input wave-form only in amplitude and phase angle (Dorf & Bishop, 1998). Moreover, almost all types of functions can be described through Fourier analysis in terms of the sums of sine and cosine functions. So, if a linear system's response for sinusoidal input is determined, then its response to more complicated input can be described by linearly superimposing the outputs determined for each of the sinusoidal-input that were identified by Fourier analysis. In frequency response analysis, sinusoidal input signal is varied over a range of frequencies, and for each frequency, there is a gain and phase angle that give the characteristic response at that frequency. The results are plotted in a pair of graphs known as *Bode diagram* which consists of two plots:

1. Logarithmic gain, $L(\omega) \triangleq 20\log_{10} G_a(\omega)$ vs. $\log_{10}(\omega)$, and
2. Phase angle, $\phi(\omega)$ vs. $\log_{10}(\omega)$

The vertical scale of the amplitude Bode diagram is in decibels (dB), where a non-dimensional frequency parameter such as frequency ratio, (ω/ω_n), is often used on the horizontal axis.

Laplace transformation (LT) is the key to frequency-domain modelling. It is a mathematical tool to transform linear differential equations into an easier-to-manipulate algebraic form. In frequency domain, the differential equations are easily solved and the solutions are converted back into time-domain to provide system response (Fig. 12). Some of the common Laplace transform (LT) pairs are reported in Table 7.

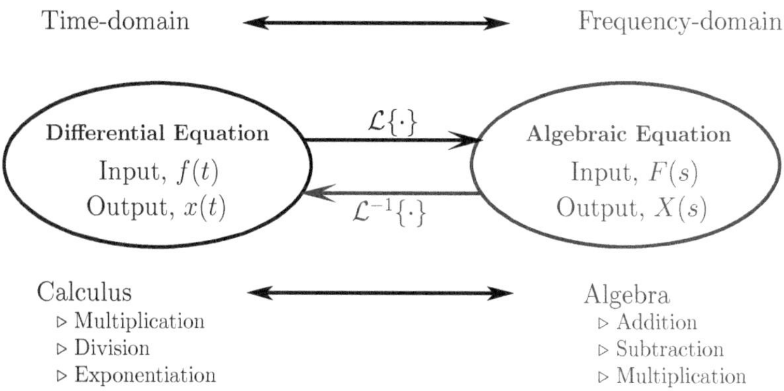

Fig. 12. Laplace transform (LT) takes a function from time-domain to frequency-domain. Inverse LT takes a function from frequency-domain to time-domain.

$f(t)$	$F(s)$
$\delta(t)$, unit impulse	1
Step function, A	A/s
At; At^n	$\frac{A}{s^2}$; $\frac{An}{s^{n+1}}$
Ae^{-at}	$\frac{A}{s+a}$
$A\sin(\omega t)$	$\frac{A\omega}{s^2+\omega^2}$
$\cos(\omega t)$	$\frac{s}{s^2+\omega^2}$
$f'(t)$	$sF(s) - f(0)$
$f''(t)$	$s^2F(s) - sf(0) - f'(0)$
$\frac{\omega_n}{\sqrt{1-\zeta^2}} e^{-\zeta\omega_n t} \sin \omega_n \sqrt{1-\zeta^2}\, t$, $\zeta < 1$	$\frac{\omega_n^2}{s^2+2\zeta\omega_n s+\omega_n^2}$
$c_1 f_1(t) + c_2 f_2(t)$	$c_1 F_1(s) + c_2 F_2(s)$

Table 7. Laplace transform (LT) pairs of some common functions.

4.3.1 Laplace transform (LT) and Transfer function (TF) methods

Transfer function (TF) is widely used in frequency-domain analysis and it establishes the size and timing relationship between the output and the input. Transfer function of a linear system, $G(s)$, is defined as the ratio of the Laplace transform (LT) of the output variable, $X(s) \triangleq \mathcal{L}\{x(t)\}$, to the LT of the input variable, $F(s) \triangleq \mathcal{L}\{f(t)\}$, *with all the initial conditions are assumed to be zero.* Hence,

$$G(s) \triangleq \frac{X(s)}{F(s)} \tag{26}$$

The Laplace operator, $s \triangleq \sigma + j\omega$, is a complex variable. Hence, $j \equiv \sqrt{-1}$, and ω translates into a sinusoid in time-domain, σ translates into an exponential term, $\exp(\sigma t)$. For steady-state sinusoidal input, $\sigma = 0$, and system response can be evaluated by setting $s = j\omega$. So,

$$F(s) \longrightarrow \boxed{G(s)} \longrightarrow X(s) \quad :\Longrightarrow \quad x(t) = f(t) \times G_a \angle\phi$$

$$\text{Hence, } \begin{array}{lll} G(j\omega) & \triangleq G(s)|_{s=j\omega} & \triangleq \text{steady-state transfer function,} \\ G_a(\omega) & \triangleq |G(j\omega)| & \triangleq \text{gain frequency response,} \\ \phi(\omega) & \triangleq \angle G(j\omega) & \triangleq \text{phase frequency response,} \\ G_a(\omega)\angle\phi(\omega) & \triangleq G_a\angle\phi & \triangleq \text{frequency response.} \end{array}$$

For example, transfer function of a first-order system is evaluated below:

- time-domain equation: $\tau \frac{dx}{dt} + x = \Bbbk\, f(t)$
- s-domain equation: $\tau s X(s) + X(s) = \Bbbk\, F(s)$

$$\blacktriangleright\ F(s) \longrightarrow \boxed{G(s) = \frac{\Bbbk}{1 + s\tau}} \longrightarrow X(s)$$

- $G_a = |G(s \Leftarrow j\omega)| = \left|\frac{\Bbbk}{1+j\omega\tau}\right| = \frac{\Bbbk}{\sqrt{1+(\omega\tau)^2}}$
- $\phi = \angle G(j\omega) = \tan^{-1}(-\omega\tau)$
- For input, $f(t) = A\sin(\omega t)$, steady-state response, $x(t) = \frac{\Bbbk A}{\sqrt{1+(\omega\tau)^2}} \sin(\omega t + \phi)$.

Expressions of $G(s)$, G_a and ϕ for zero-, first- and second-order systems are reported in Table 8.

	Zero-order	First-order	Second-order
Differential Equation	$x(t) = \Bbbk\, f(t)$	$\tau \frac{dx}{dt} + x = \Bbbk\, f(t)$	$\frac{1}{\omega_n^2}\frac{d^2x}{dt^2} + 2\frac{\zeta}{\omega_n}\frac{dx}{dt} + x = \Bbbk f(t)$
Transfer function, $G(s)$	$\Bbbk$	$\frac{\Bbbk}{\tau s+1}$	$\frac{\Bbbk}{\frac{1}{\omega_n^2}s^2+2\frac{\zeta}{\omega_n}s+1}$
Gain, G_a	$\Bbbk$	$\frac{\Bbbk}{\sqrt{1+(\omega\tau)^2}}$	$\frac{\Bbbk}{\sqrt{\left[1-(\frac{\omega}{\omega_n})^2\right]^2+4\zeta^2(\frac{\omega}{\omega_n})^2}}$
Phase angle, ϕ	0	$\tan^{-1}(-\omega\tau)$	$\tan^{-1}\left[-\frac{2\zeta(\frac{\omega}{\omega_n})}{\left[1-(\frac{\omega}{\omega_n})^2\right]}\right]$

Table 8. Transfer function, gain and phase angle of zero-, first- and second-order systems.

4.3.2 Bode-diagram and selection of instrumentation parameters

Ideal system response of a zero-order system is shown in Fig. 13, where both the values of $L(\omega)$ and $\phi(\omega)$ are zero. In case of a first-order system response, as shown in Fig. 13, when the measurement system responds with values of $L(\omega) \simeq 0$, system transfers all or nearly all of the input signal amplitude to the output and with very little time delay. At larger values of $\omega\tau$, measurement system will essentially filter out frequency information of the input by responding with smaller amplitudes and larger time delays. So, any combination of $\omega\tau$ will produce the same results: if one wants to measure signals with high frequency content, then a system with an appropriately small τ is required and vice versa (Figliola & Beasley, 2011).

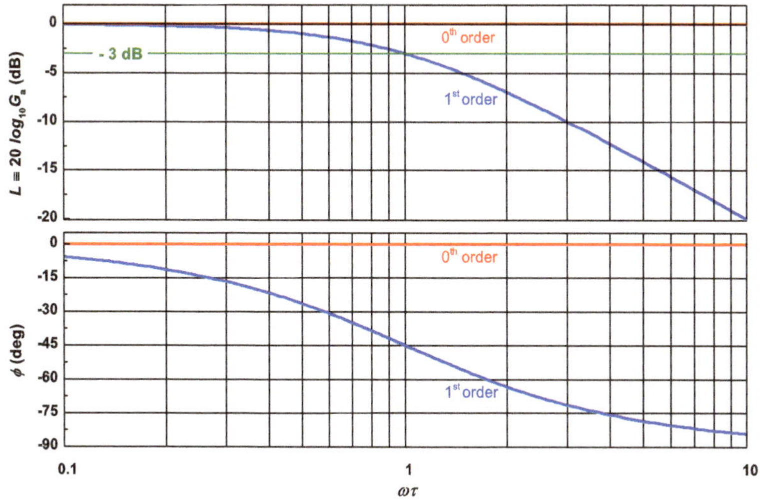

Fig. 13. Bode-diagram for zero- and first-order systems.

In a second-order system response, as shown in Fig. 14, at low values of ω/ω_n, values of $L(\omega)$ and $\phi(\omega)$ remain near to zero. This indicates that information concerning the input signal of frequency, ω, will be passed through to the output with little attenuation and phase lag. This region of frequency response curves is called the *transmission band*. The actual extent of the frequency range for near unity gain depends on the system damping ratio, ζ. The transmission band of a system is typically defined as $-3\text{ dB} \leqq L(\omega) \leqq 3\text{ dB}$. At large values of ω/ω_n, the system will attenuate the amplitude information of the input signal and a large phase shift occurs (Figliola & Beasley, 2011). For a system with $\zeta \to 0$, $L(\omega)$ will be very high and $\phi(\omega) \to -180^o$ in the vicinity of $\omega/\omega_n = 1$. This behaviour is a characteristics of

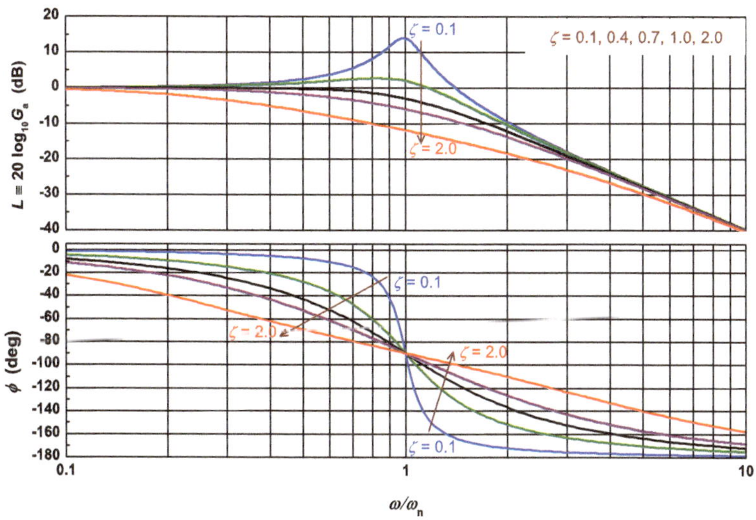

Fig. 14. Bode-diagram for second-order systems.

system resonance. However, real systems possess some damping to limit the abruptness and magnitude of resonance, but underdamped systems will experience some resonance.

For second-order systems, response shown in Fig. 14, following observations can be made:

- System has a good *linearity* for low values of ζ and up to a frequency ratio, ω/ω_n, of about 0.3 as the amplitude gain is very nearly unity ($G_a \simeq 1$) with $L(\omega) \simeq 0$.
- For large values of ζ, the amplitude is reduced substantially.
- The phase shift characteristics are a strong function of ω/ω_n for all frequencies.
- As a general rule of thumb, the choice of $\zeta = 0.707$ is optimal since it results in the best combination of amplitude linearity and phase linearity over the widest range of frequencies (Holman, 2001).

The universal curves (Figs. 13 and 14) may be used as guidance in the selection of measurement instruments and system components. As the values of $L(\omega)$ and $\phi(\omega)$ deviate from zero, these curves take on rather steep slopes, and in these regions, small errors in the predictions of τ and the deviation of the real system from ideal one may lead to significant errors in measurements (Figliola & Beasley, 2011).

5. References

ASME (2005). ASME PTC 19.1-2005: Test Uncertainty. Amarical Society of Mechanical Engineers (ASME), New York, USA.

Beckwith, T., Marangoni, R. & Lienhard, J. (2007). *Mechanical Measurements*, 6th edn, Pearson Prentice Hall.

Doebelin, E. O. (1998). *System Dynamics: Modeling, Analysis, Simulation, Design*, Marcel Dekker, Inc. NY, USA.

Doebelin, E. O. (2004). *Measurement Systems: Applications and Design*, 5th edn, McGraw-Hill.

Dorf, R. & Bishop, R. (1998). *Modern Control Systems*, 8th edn, Pearson Prentice Hall.

Dunn, P. F. (2010). *Measurement and Data Analysis for Engineering and Science*, 2nd edn, CRC Press.

Figliola, R. & Beasley, D. (2011). *Theory and Design for Mechanical Measurements*, 5th edn, John Wiley.

Holman, J. (2001). *Experimental Methods for Engineers*, 7th edn, McGraw-Hill.

ISO (2007). ISO/IEC Guide 99:2007 International Vocabulary of Metrology – Basic and General Concepts and Associated Terms (VIM). International Organization for Standardization (ISO), Geneva, Switzerland.

ISO (2008). ISO/IEC Guide 98:2008 Uncertainty of Measurement – Guide to the Expression of Uncertainty in Measurement (GUM). International Organization for Standardization (ISO), Geneva, Switzerland.

Kelly, S. G. (2003). *System Dynamics*, Cengage Learning.

2

Measurement Systems for Electrical Machine Monitoring

Mario Vrazic, Ivan Gasparac and Marinko Kovacic
University of Zagreb, Faculty of Electrical Engineering and Computing, Croatia

1. Introduction

Monitoring systems on electrical machines, nowadays largely present on new electrical rotating (synchronous generators) and non rotating machines (transformers) have to endure harsh conditions such as vibrations and electromagnetic disturbances. Also, especially for rotating machines, rotor physical quantities (temperature, strain, current...), have to be transferred to the standstill part of the measurement system without galvanic contact.

Therefore, monitoring systems should be, and usually are, designed for a specific purpose and for a specific machine. Some essential parts are the same but most of the monitoring system should be custom made especially if one takes into consideration that there is a small serial production of large electrical machines in the world and that most of the large units are unique.

Modern monitoring of big transformers covers permanent on-line monitoring of electrical, magnetic and mechanic quantities. It is specific when compared to rotational electrical machines because it also usually includes monitoring of dissolved gasses in transformer oil. This information, with proper interpretation, represents an important element for transformer state determination. However, transformer monitoring systems are not the focus of this chapter.

This chapter deals with all problems, obstacles and solutions applied to two measurement systems installed on a 35 MVA hydro-generator and a 250 MVA turbo-generator.

Large electric machines justify investment in measurement systems for several reasons. Usually, measurement systems for bearings are installed. In such systems vibrations and oil parameters (pressure and temperature) are monitored to help maintenance. Measurements of stator and less often rotor currents are also present as parts of different systems. For example, a differential stator current measurement is present for protection purposes and enables detection of current leakage (short circuit between phase windings). Monitoring systems that incorporate all necessary measurements and, of course, interpretations of those measurements are almost non existing and very rare. This chapter will concentrate on synchronous generators.

2. The monitoring system

The monitoring system is basically a data acquisition system with on line data processing and possible data interpretation. A standard data acquisition system consists of a sensor, signal transfer, A/D conversion, signal conditioning and data storage.

The sensor is usually analogue, like a Pt1000 probe for temperature measurement. The temperature is measured as a resistance over two, three or four wires. If resistance is measured over four wires, the current flows through two wires and the other two wires are used for voltage measurement. This system configuration is very sensitive to electromagnetic disturbances thus making measurement inherently inaccurate. In order to avoid that and to make measurement more accurate signal amplifiers and converters are designed. They enable steady current regardless temperature dependence of connecting wires. But still, voltage measurement is subjected to electromagnetic disturbances since the measured voltage has an mV range. Of course, all of these disturbances and most of their influence can be avoided by proper wiring. For example, bifilar wire configuration and shielded cables significantly reduce electromagnetic disturbances if the shield is properly grounded.

So, such configuration, for example the one installed in Plomin thermo power plant, has a sensor (a Pt1000 probe), a rather large amount of wire (15 m), a signal converter and amplifier (placed in the measurement cabinet), a ±10 VDC signal path (a 2 m cable), an acquisition system, communication (90 m Ethernet FTP cable) and finally a measurement computer.

The reasons for such configuration will be presented later in this chapter. Of course, if one has the opportunity and technical possibility, one should install a more advanced temperature measurement. For example, there are digital temperature probes on the market that are basically small chips with temperature probes and all the necessary signal conversion. All they need is power and communication. They obtain them both through a three wire system that is completely independent of electromagnetic disturbances. Such a system has a serial connection and can have up to 127 probes in one system.

This description shows one review of the acquisition system that, of course, can grow to an enormous size. For example, in CERN in Switzerland, there is a measurement system with over 175 000 channels.

The monitoring system incorporates data acquisition and enables instant data processing and interpretation. For instance, before information enters the PC, the analogue signal goes through the A/D conversion and scaling. With this, for example, a signal of 5.4 mV becomes 24.5 °C. After that, the information about the exact temperature enters a PC application that operates with it. The temperature, maybe, does not represent the physical quantity, so vibrations will be used from now on.

Vibrations are measured with accelerometers. They measure vibration accelerations and such information enters the PC. But for a reasonable vibration analysis different quantities should be obtained from vibration acceleration such as vibration velocity and vibration displacement. Furthermore, for all the three quantities (acceleration, velocity and displacement) a base harmonic and some higher harmonics should be known. All of this can be done within data acquisition unit but then this unit will be very expensive. So, the general idea is that such advance data processing is done within application in the monitoring PC.

The monitoring system can now have access to all the necessary data and can conclude upon the electrical machine condition. Of course, this conclusion can be based on simple or complex algorithms. Simple algorithms will include allowed limits, alarm limits and shutdown limits for some or most of the physical quantities. Complex algorithms include monitoring of some physical quantity trend lines and from that info concludes on further actions. Those conclusions can lead to a well-timed maintenance action that is more efficient and cheaper than waiting for a fault to occur and lead to an unplanned stopping of electrical machine and process for which this machine is necessary.

In the following text three measurement systems and one monitoring system will be presented.

The general idea of building a measurement system, and further on, a monitoring system on synchronous generators started with a desire to build a user P-Q diagram of a synchronous aggregate. In order to adapt FEM and analytical models to a real generator a lot of measurement of different physical quantities should be done. This demanded a measurement system to be made. Such a measurement system should be able to measure magnetic inductances B (T), temperatures T (°C), acceleration of vibrations a (m/s^2), stator and rotor voltages U (V) and currents I (A).

So the whole idea was tested on synchronous hydro-generator and synchronous turbo-generator.

3. Measurements on the hydro-generator

When research for a user P-Q diagram started, the access was approved to Vinodol hydropower plant where three generators (Fig. 1) are installed. Their rated data is: S_n=35 MVA, U_n=10.5 kV, f=50 Hz, n=500 rpm, $cos\varphi$=0.9.

Fig. 1. Hydro-generators in hydro power plant Vinodol - 35MVA each

A larger measurement system was installed on the first generator. This enables a reduction of measurement systems installed on the other two generators. This text will describe the larger measurement system.

The research demanded a magnetic induction measurement, so 22 Hall probes were installed in winding end part and in the air gap. Since calibrated Hall probes are very expensive it was decided to buy only three calibrated Hall probes with their calibration certificates. The rest of hall probes were cheap sensors (their cost was approx. $5 per probe).

Hall probes have 4 connections. Two of them need a steady current flow and on the other two voltage can be measured. The voltage corresponds to the magnetic induction. Both current (1 mA or 5 mA) and voltage (range of mV) have rather small values. To satisfy those conditions current sources were designed. The current range can be selected with a switch within the current source. The Hall probe voltage is amplified with a signal amplifier to a ±10 VDC signal range and as such prepared for data acquisition. Of course, there are still two problems: calibration and electromagnetic disturbances.

But before calibration, all Hall probes were placed on an electronic board (Fig. 2) that has two possible connection places. If there is enough room for probe placement, then Fig. 2 option a can be used, but if there is not enough place (air gap, cooling channel) then a minor board can be cut out from the bigger one (Fig. 2 option b) and wires can be connected to it.

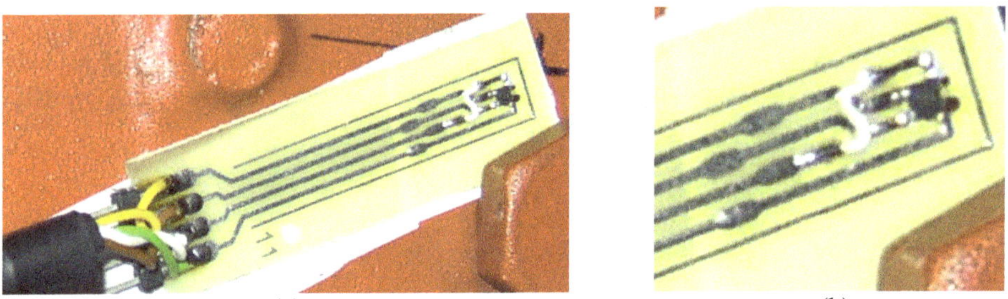

(a) (b)

Fig. 2. Hall probe board

The calibration problem was resolved when it was done in the laboratory. A three phase choke was used (Fig. 3) for that purpose. The air gap on the choke can be changed from 0 to 32 mm.

The calibrated Hall probe (the expensive one) was used as a reference probe and all the necessary calibrations (with AC and DC signals) for all the other Hall probes were made with it: the calibration constant, the DC offset and the thermal dependence. Several Hall probes were discarded because of their lack of linearity but most of the cheap hall probes were successfully calibrated.

Still, there was a problem of electromagnetic disturbances since the voltage signal from Hall probes is very small (range of mV). The distance from a probe to signal amplifiers is approx. 15 meters and that also contributes to the problem. This issue was minimized by using a 4 wire shielded cable. So there is one cable for each probe. Since those synchronous generators are air cooled there was no problem with the cable entrance through generator housing. Standard cable glands were used. All the cables were shielded on one side, near signal amplifiers. Such configuration ensures electromagnetic disturbance discharge thus minimizing their influence on the measured signal.

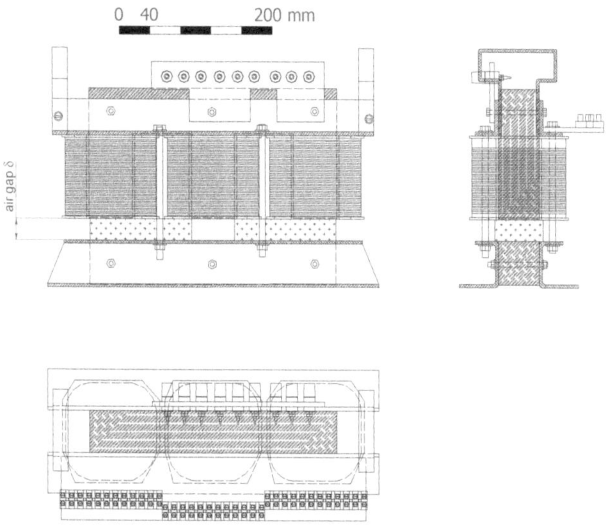

Fig. 3. Choke for Hall probe calibration

Similar measurement was made for temperature. Pt1000 probes and 2 wire measurement were used. Signal converters and amplifiers in one device were used to obtain a ±10 VDC signal for the data acquisition system. Two wire shielded cables were used for each signal and were grounded as the ones from Hall probes.

But a different problem was identified here. The problem was heat transfer from the surface to the probe. So before mounting Pt1000 probes onto the generator, the probes were conditioned. They were aligned on a thin pertinax base to ease mounting on the generator and as well as to assure the wanted relative position. Each probe was glued with super glue to pertinax (Fig. 4). After that isolation foil was glued over the probe. This foil assures electrical isolation but does not influence thermal transition. Cable wires were soldered on the Pt probe and the cable was glued on pertinax. When mounting on the generator, thermal paste was applied between the surface and the probe to ease thermal conductivity.

After that everything was suffused with thermal resistant glue.
In order to be able to obtain the exact wanted position of probes within the generator, Hall (A1, B1-B4, C1-C4, D1-D4, E1-E4, F1-F3, G1-G2) and Pt1000 (GTS1-GTS14) probes were placed on thin boards in the exact position (Fig. 5). The boards were then placed at the exact position within the generator (Fig. 6, Fig. 7 and Fig. 8). All probe positions were carefully recorded and later used when 2D and 3D FEM models were confirmed by measurements (Fig. 9).

One of the instruments developed during the measurement system design was an instrument for measuring the load angle (δ). During the designing period several devices were designed and commissioned. The method of load angle measurement is based on time measurement between zero crossing of the generator voltage and the rotor position represented by positive edge of proximity switch impulse. At first, this operation was carried out with a laboratory device and then after the production of a more professional version, the third, compact and universal model was produced.

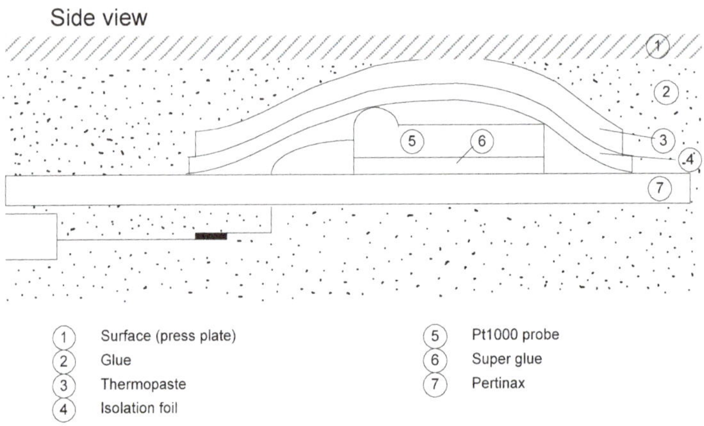

Fig. 4. Mounting scheme of Pt1000 probes

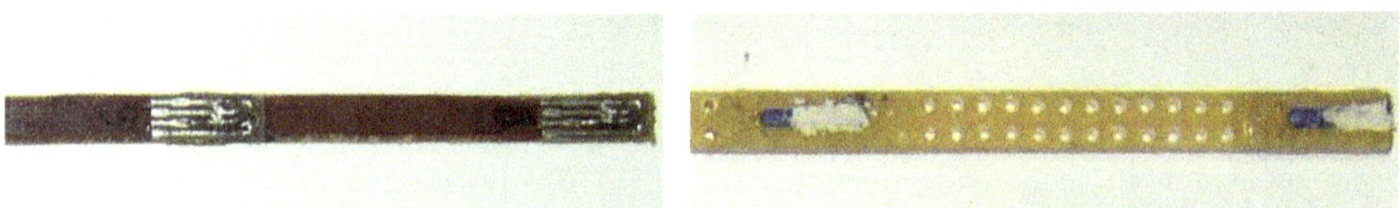

(a) (b)

Fig. 5. Prepared probes: a) Hall probes; b) Pt1000 probes

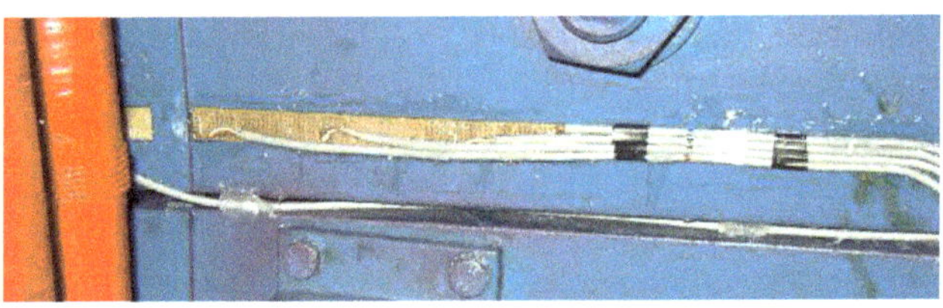

Fig. 6. Built-in Hall probes

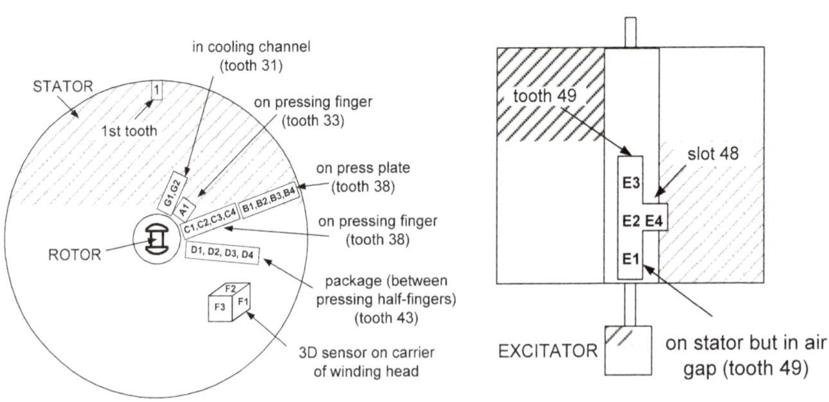

Fig. 7. Placement of Hall probes in generator

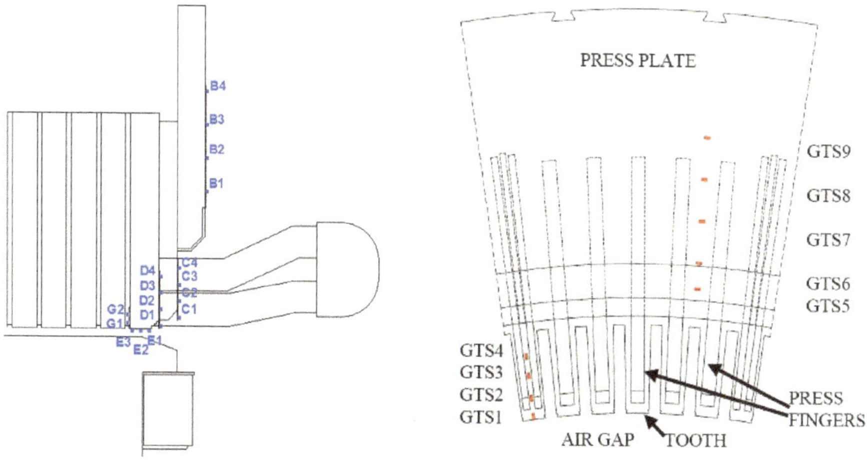

Fig. 8. Placement of Hall and Pt1000 probes

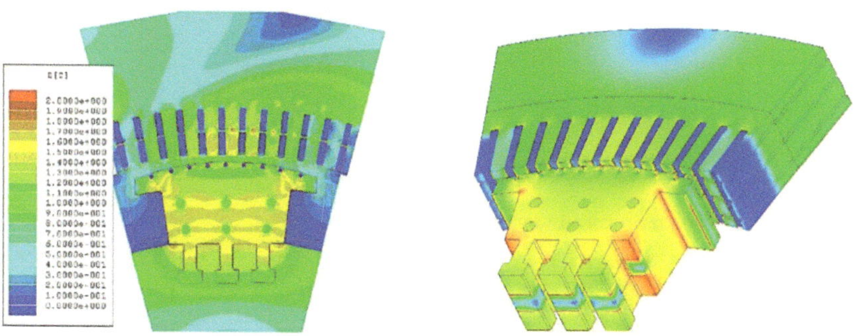

Fig. 9. 2D and 3D FEM induction calculations for no load operation

Also, as previously mentioned, two vibration probes (accelerometers) were installed on the stator package (Fig. 10).

Fig. 10. Vibration accelerometer installed on the stator package

After all of that, the measurement system needed to be designed and built. Since, at the time, only 16 channel data acquisition systems were available to us, 4 different data acquisition systems (Iotech) were used for complex measurements. All together there were a total of 51 signals (22 Hall probes, 14 Pt1000 probes, 3 armature currents, 3 armature voltages, 1 excitation current, 1 excitation voltage, 1 impulse per rotation, 1 load angle, 3 grid voltages, 2 vibration accelerations). All data acquisition systems had trigger possibility on one channel so this was used for synchronous start of data acquisition. Since there were so many signals there are lots of cables present. Most measurements were made with a temporary measurement system (Fig. 11). In order to be sure that electromagnetic

Fig. 11. Temporary measurement system

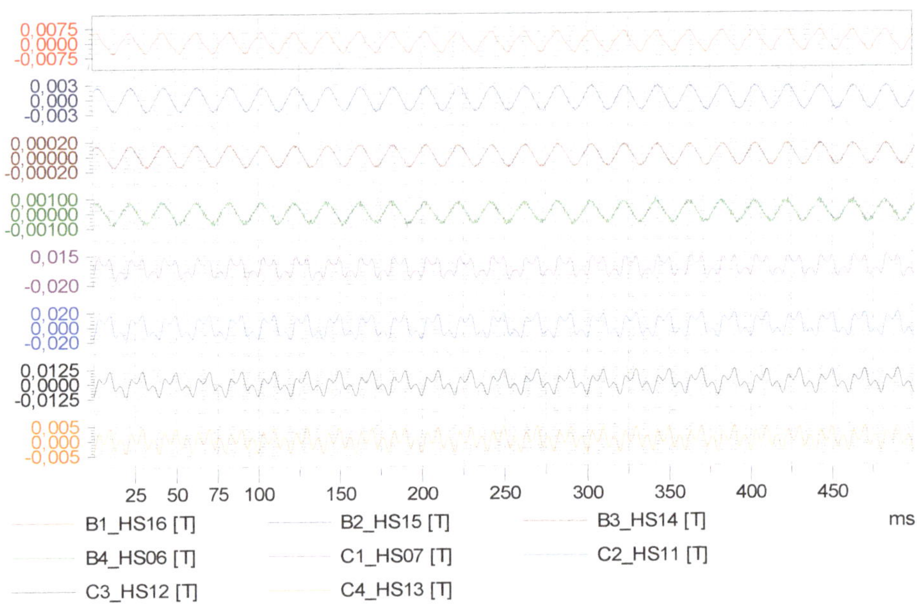

Fig. 12. Hall probe measurements – signals of different probes

disturbances are not present, or at least are acceptable, each signal was checked several times and compared to the theory. For example, each magnetic induction signal was checked separately and together with the rest of them. If two signals are the same and their probes are not in the position that can give the same signals (for example the same radial position on the press plate) this is the indication that one signal is copied to the other channel. Also, most of signals must have some angular difference and this is the other way to check them (Fig. 12).

After several measurements with the temporary measurement system there was a 3 year delay with this power plant due to circumstances beyond our influence. So, the measurement and monitoring system development for this power plant was halted. But all designs have been done and the equipment has been bought and installed in the electrical cabinet (Fig. 13). This cabinet (measurement cabinet) will be installed in the power plant during the first half of next year. Data acquisition will be realised with National instruments Compact RIO product. The main interface will be touch panel. The plan is to place a PC near this cabinet and connect the system to the Internet.

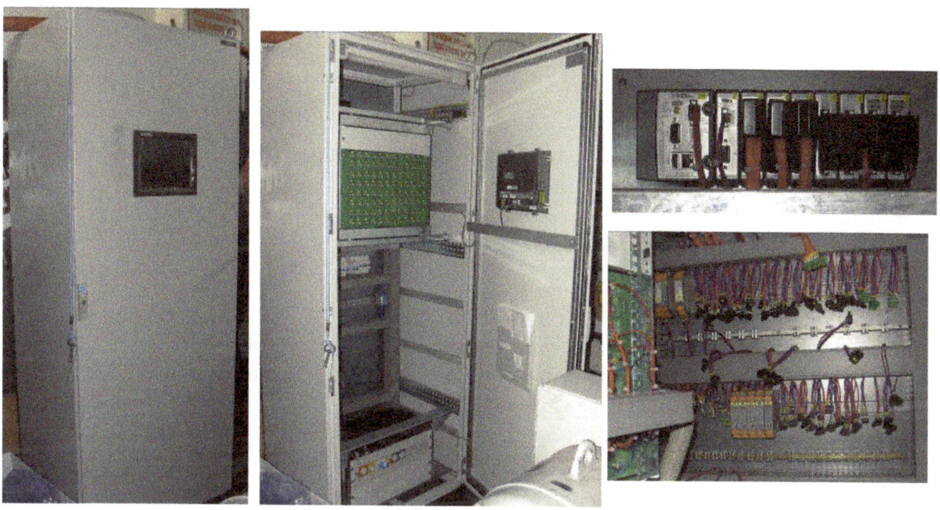

Fig. 13. Measurement cabinet for measurement systems installed on all three generators in hydro power plant Vinodol

4. Measurements on the turbo-generator

A measurement system was also installed on one turbo-generator (Fig. 14). This generator is installed in Plomin 2 thermo power plant. Its rated data are 250 MVA, 3000 rpm.

A similar set of probes was installed in this generator as it was in the generator in HPP Vinodol. In total, 55 probes were installed in the generator: 30 Hall probes, 14 Pt1000 probes and 11 accelerometers. All armature and excitation voltages and currents are measured together with one grid voltage. All those currents and voltages are measured in such a way that a malfunction in the measurement system does not influence systems in the power plant. For instance, armature currents are measured with current transformers (5A/333mV)

Fig. 14. Turbo-generator 250 MVA in Plomin 2 thermo power plant

placed around the wires from the secondary windings of the major current measurement transformers (2000A/5A) of the generator. Voltages are also measured on the secondary windings of major voltage measurement transformers. This connection was made over 250mA fuses. Major voltage transformers have the ability to burn those fuses (in case of a measurement system malfunction) without jeopardizing their primary function.

This generator is cooled with hydrogen. Since this is a highly explosive coolant all signals were checked for energy level they import in the generator. The energy level must be lower than the energy that is high enough for sparking and that may ignite hydrogen. This is the first step to ensure that the measurement system can do no harm to the generator. The next step was to shut down all the measurements and cut off the power in the measurement cabinet in case of hydrogen leakage. So, if sensors detect hydrogen outside of generator, the main contactor used for powering the measurement cabinet switches off. The third step was installation of a hydrogen detection probe within the measurement cabinet since signal cables can lead hydrogen from the generator to the measurement cabinet in case of cable glands malfunction. The fourth step was to design special cable glands to ensure absence of hydrogen leakage. Cable glands (Fig. 15) were designed, tested, installed and they have been functioning flawlessly for two years.

Signals from the generator were conducted through a signal cable. This cable has 16 twisted pair wires. Each twisted pair is shielded and the whole cable is shielded, too. This double shielding ensures better protection from electromagnetic disturbances. The measurement system is similar to the one in HPP Vinodol. The difference occurs in the data acquisition system.

After conditioning all the signals go to one device (NIcDAQ9188) and not to four different data acquisition systems. The conditions and customer demanded that the PC should be positioned in the control room approximately 90m away from the measurement cabinet (Fig. 16) so Ethernet cable FTP cat.6 was used for that connection.

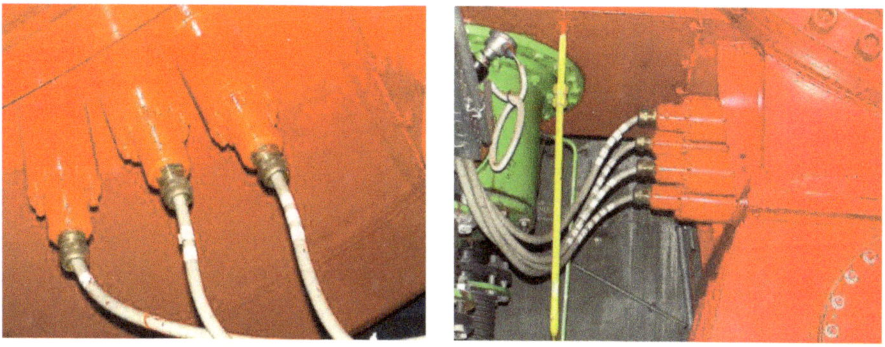

Fig. 15. Signal cable glands

Fig. 16. Measurement cabinet with NIcDAQ9188 data acquisition system

Unlike the measurement system in HPP Vinodol, where measurement and monitoring systems should be finished, everything is finished in TPP Plomin2. A lot of work was done in order to produce the application in LabVIEW for measurement, signal processing and data storage.

The front panel application was developed in LabVIEW and consists of several tabs. All of the measurements are divided between multiple tabs. The first three tabs are used for graphical view of the measured signals. It is possible to observe waveforms of all of the acquired signals (magnetic field flux density, vibration accelerations and electrical signals such as stator currents, voltages, field current, voltage etc.). The user is also able to read the exact value of the waveform using a cursor tool and move over the waveform using zoom in/out and drag tools (Fig. 17).

After successful acquisition of the all the signals the application computes all the necessary quantities such as rms, peak-to-peak, base harmonics of the acceleration, velocity and

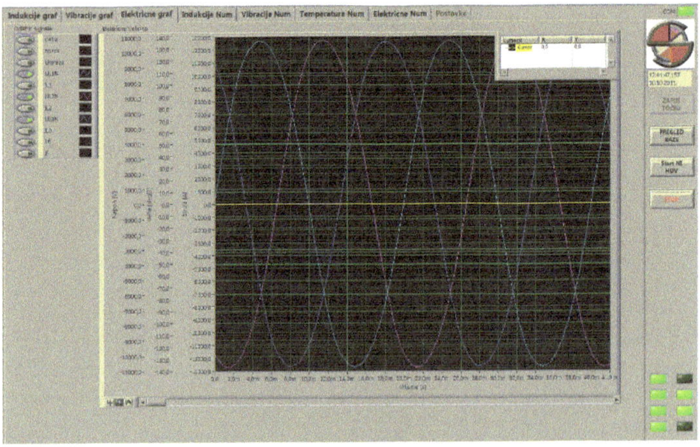

Fig. 17. Front panel of the application, showing real time waveforms of the 3-phase system voltage

displacement for vibration signals; dc-offset, rms value, base harmonics for magnetic flux density signals; mean value for temperature signals; full 3 phase power analysis for electrical signals. The signals that can also be measured on the synchronous machine are field current, voltage and load angle of the machine. The computed quantities are also divided in four separate tabs (magnetic flux density, vibrations, temperature and electrical), (Fig. 18).

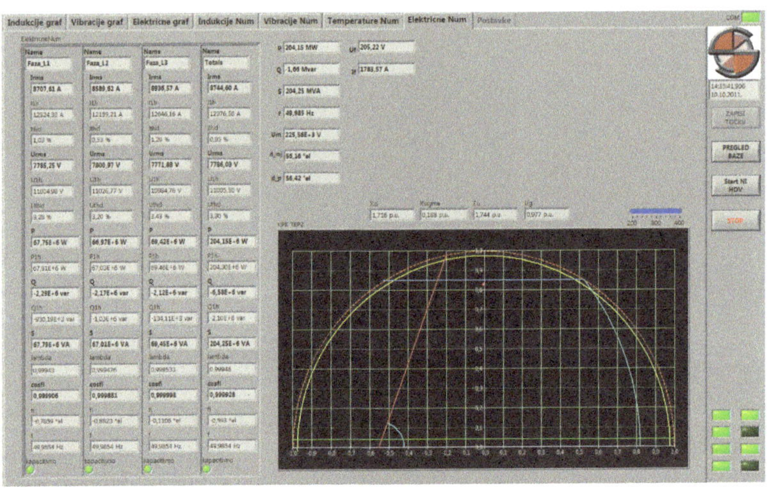

Fig. 18. Front panel of the application, showing real time electrical measurements and capability diagram

After a successful calculation, all of the calculated quantities are logged into the historical database (Fig. 19). A user can access the database from the application GUI, and observe variations or trends of the selected quantities over a desired time period. Trend analysis is especially valuable tool in vibration monitoring. Based on the trend of the calculated vibrations user can predict a maintenance of the machine.

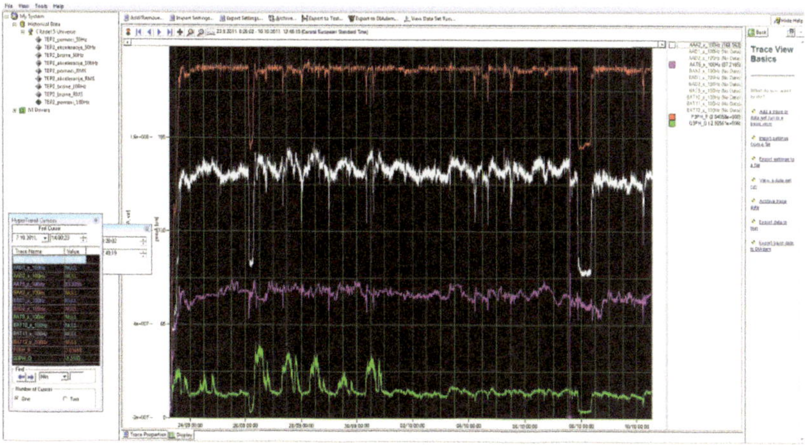

Fig. 19. Historical data view of computed variables (red-active power, green- reactive, white- tangential end winding vibration, violet-axial end winding vibration)

The described method of monitoring incorporates measurement of the signal waveforms. All of the data logged into the database are just physical quantities calculated from those waveforms. There are cases when a user demands to log the waveform data for further, more detailed, analysis. A continuous waveform logging requires a large storage capacity as all of the waveforms are sampled at rather a high frequency (more than 10 kHz). To lower the demands on storage capacity the application must be able to decide which of the waveforms to store. The stationary waveforms are no of interest for most of the users so the application must detect only a change in the signal and based on that change it has to trigger storing of the waveform. When designing such an application one has to be careful that it records a certain amount of signal before the actual trigger event. Furthermore, in this type of measurement the acquisition system and the data processing have to be able to withstand the continuous sampling of the signal to ensure that none of the samples are lost due to the buffer overrun or the slow data processing. Fig. 20 shows an armature current transient properly recorded with pre-buffer time set to 5 s and total transient time set to 20 s.

5. Measurements on the rotor

Modern synchronous generator monitoring and controlling systems require measurement of various physical quantities. One of those is the temperature of the machine. It is relatively easy to measure temperature of the stationary part of the machine, but the difficulties arise with the measurement of the rotor temperature. Conventional temperature measurement requires slip-rings for each sensor. The measurement system which uses digital wireless data transfer and digital temperature sensors will be described in short.

The DS18B20 digital sensors were chosen for the temperature measurement. The accuracy of the temperature sensors is ±0.5 °C. Sensors use a 1-wire protocol for communication with the microcontroller. Bluetooth communication was selected because of the high operating frequency 2.4 GHz, low power requirement and wide usage in various IT equipments. A special embedded device was developed to convert data from temperature sensors. The measurement system consists of several specific parts:

- Bluetooth communication module
- Microcontroller
- Battery charge management
- 14-bit ADC
- Li-ion battery

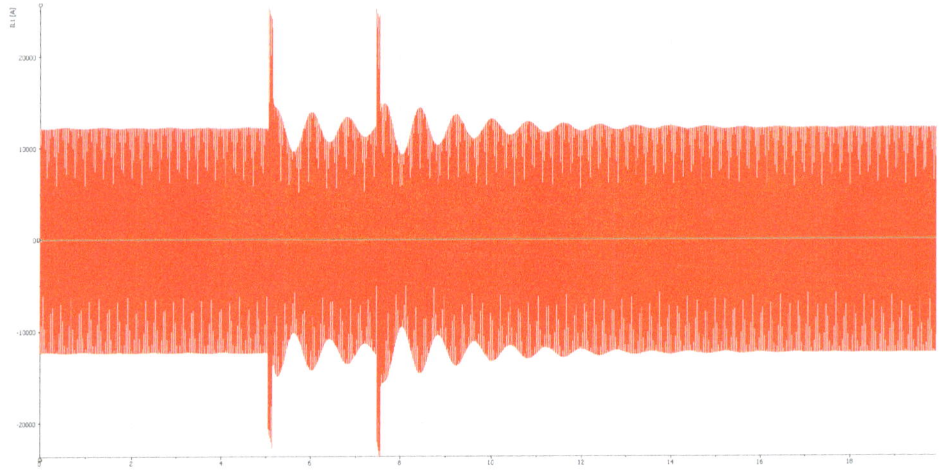

Fig. 20. TEP2 armature current transient caused by the short circuit on the power-line (5 s of pre-buffer time and 20 s of total signal time)

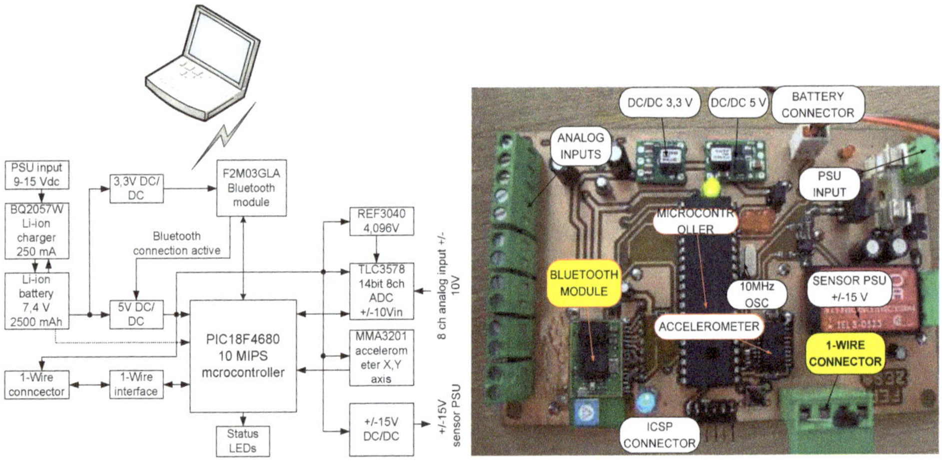

Fig. 21. Block diagram and PCB of the wireless measurement system

The heart of the measurement system is a microcontroller which routes data from the sensors to the Bluetooth interface. Each DS18B20 has a unique 64-bit serial code, which allows multiple DS18B20 to function on the same 1-Wire bus. A parallel connection of the sensors simplifies wire routing over the machine rotor. Routines for searching, addressing and reading 1-Wire temperature sensors are integrated in the embedded code of the microcontroller.

The Bluetooth module is connected with the microcontroller via a USART interface so the system acts as a virtual serial port on the host computer. It is advisable that the microcontroller has a pre-programmed bootloader, a part of software which enables updating of firmware in the rotation through the Bluetooth link.

Because the system operates on the rotating part of the machine it had to be battery supplied. An absence of the alternating magnetic field on the rotor side of the synchronous generator excluded the usage of coils for parasite battery charge. Instead, batteries can be charged via a DC/DC converter connected to the synchronous machine excitation. This procedure has to be performed with caution due to the high voltage spikes generated by the excitation system.

To save the battery, only the Bluetooth module has to be powered while the connection is inactive while all other circuits (microcontroller, sensor supply) should be powered after a successful Bluetooth connection to the host system is established.

The measurement system has additional features intended for future use, such as a 2-axsis accelerometer, an 8 channel 14-bit analogue to digital converter and a ±15V sensor power supply.

Data from all the sensors had to be monitored and logged on to the host computer. The GUI application sends commands to the measurement system via the Bluetooth and waits for its response. The time base of the measurement is generated by the host system. The application searches for sensors on the 1-Wire bus and monitors the temperature of each sensor. Temperatures of each sensor together with time base are logged in a textual file which is used for further analysis. The GUI application can also read the battery voltage and battery charge status via the analogue input of the microcontroller. After a successful sensor search, the application enters in the temperature polling mode. A small database which joins names to the unique 64-bit address of a specific sensor can be incorporated into the application. This way, it is easier to monitor temperature of each sensor during an experiment.

It can be useful if the application has an ability to render 3D temperature distribution over the surface. The 3D representation greatly improves visualization of temperature distribution, especially locations of the hot-spots. The 3D model of the salient, shown in Fig. 22 (right), has been drawn and imported in LABView using "3D Sensor Mapping Tool". Sensors were positioned on the same coordinates as on the real salient pole. Temperature distribution over the surface between sensors is calculated using interpolation.

The measurement system was mounted on the 400 kVA, 1000 rpm synchronous generator, Fig. 23. Because of the Bluetooth transceiver it had to be placed in a non metal case. It was decided that the measurement system would be mounted on a shaft using a custom made holder. It was mounted in such a way that centrifugal force pushes the plastic case of the measurement system towards the metal holder and provides reliable fixation. Balancing weights were mounted on the opposite side of the shaft to reduce vibrations.

After a successful installation of the measurement system prototype on the laboratory synchronous generator, various experiments have been conducted. Fig. 24 shows an example of a measurement recorded by the measurement system. The excitation winding was first heated in standstill and afterwards in the rotation.

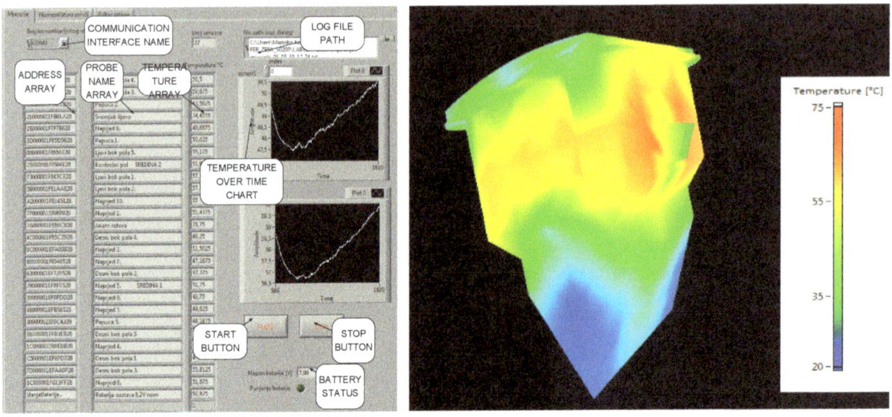

Fig. 22. GUI application and 3D temperature field representation

Fig. 23. Mounting of the temperature probes (left); measurement system in rotation (right)

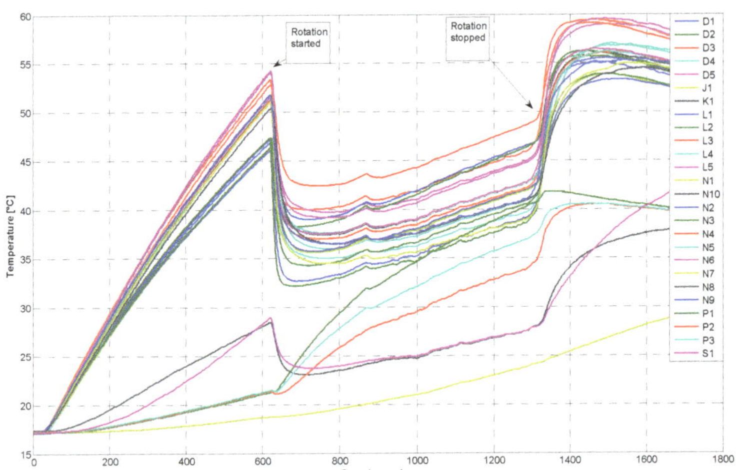

Fig. 24. Temperatures from 27 sensors placed on synchronous generator rotor (2 s sampling time)

6. Conclusion

Measurement systems described in this chapter show measurement of most of necessary physical quantities needed for proper rotating electric machine monitoring. Of course, measurement is not everything. A good monitoring system should be based on expert knowledge so that the monitoring system can give advice for maintenance. For large units such as synchronous generators monitoring systems are different for each machine. During the creation of such monitoring systems electric machine manufacturer should at least be present, because he did, during the design process, lots of modelling (analytically, with FEM etc.) and knows most of limitations of the produced electric machine. It would be the best if the manufacturer also produces such a monitoring system and has a permanent access to measurements. In such a way the manufacturer can improve his product, correct some flaws and optimize it.

7. References

Mohan, N. *Electric Drives: An Integrative Approach*, Mnpere, ISBN 978-0971529250, Minneapolis, USA, 2003.

Tavner, P.; Ran, L.; Penman, J.; Sedding, H. *Condition Monitoring of Rotating Electrical Machines*, The Institute of Engineering and Technology, ISBN 978-0-86341-739-9, Stevenage Herts, United Kingdom, 2008.

Vas, P; *Parameter estimation, condition monitoring, and diagnosis of electrical machines*, Oxford science publication, ISBN 0-19-859375-9, Oxford University Press Inc., New York, USA, 2001

Babichev, S.A.; Zakharov, P.A.; Kryukov O.V. Automated monitoring system for drive motors of gas-compressor units, *Automation and remote control*, Vol.72, No.1, January 2011, pp. 175-180, ISSN: 0005-1179, DOI: 10.1134/S0005117911010176

Despalatovic M., Jadric M., Terzic B. Real-time power angle determination of salient-pole synchronous machine based on air gap measurements, *Electric power systems research*, Vol.78, No.11, pp. 1873-1880, November 2008, ISSN: 0378-7796, DOI: 10.1016/j.epsr.2008.03.013

Yang, W.; Tavner, P.J.; Wilkinson, M.R. Condition monitoring and fault diagnosis of a wind turbine synchronous generator drive train, *Renewable Power Generation*, IET, Vol.3, No.1, pp. 1-11, March 2009, ISSN: 1752-1416, DOI : 10.1049/iet-rpg:20080006

Yang W., Tavner P.J., Crabtree C.J., Wilkinson M. Cost-Effective Condition Monitoring for Wind Turbines, *IEEE Transactions on industrial electronics*, Vol.57, No.1, pp. 263-271, January 2010, ISSN: 0278-0046, DOI 10.1109/TIE.2009.2032202

Rangnekar, S.; Nema, R.K.; Raman, P.; PC based data acquisition and monitoring system for synchronous machines, *Proceedings of IEEE/IAS International Conference on Industrial Automation and Control*, 1995, pp. 195 - 197, DOI 10.1109/IACC.1995.465843

Rahimian, M.M.; Butler-Purry, K.; Modelling of synchronous machines with damper windings for condition monitoring, *Proceedings of Electric Machines and Drives Conference*, 2009., pp. 577 - 584, ISBN 978 - 1 - 4244 - 4252 - 2, DOI 10. 1109 / IEMDC. 2009. 5075264

Batzel, T.D. Observer-Based Monitoring of Synchronous Generator Winding Health, Proceedings of Power Systems Conference and Exposition, 2006., pp. 1150 - 1155, DOI 10.1109/PSCE.2006.296470

Hanic, Z.; Kovacic, M.; Vrazic, M. Influence of Mounting Temperature Probes on the Excitation Winding on its Temperature Field, *Proceedings of the XIX International Conference on Electrical Machines (ICEM 2010)*, ISBN 978-1-4244-4175-4, pp 1-6, Roma, Italy, September 2010.,

Hanic, Z.; Vrazic, M.; Stipetic, S. Some problems related to surface temperature measurement of synchronous generator excitation winding in rotation, *Proceedings of 14th International Power Electronics and Motion Control Conference (EPE-PEMC 2010)*, ISBN 978-1-4244-7854-5, pp T7-15 - T7-20, Skopje, Macedonia, September 2010.

Kovacic, M.; Vrazic, M.; Gasparac, I. Bluetooth wireless communication and 1-Wire digital temperature sensors in synchronous machine rotor temperature measurement, *Proceedings of 14th International Power Electronics and Motion Control Conference (EPE-PEMC 2010)*, ISBN 978-1-4244-7854-5, pp T7-25 - T7-28, Skopje, Macedonia, September 2010.

Stipetic, S.; Hanic, Z.; Vrazic, M. Application of IR Thermography to Measurements on Synchronous Hydro-generator in Rotation, *Proceedings of the 10th edition of the Quantitive Infrared Thermography (QIRT 10)*, ISBN 978-2-9809199-1-6, pp 1-8, Québec City, Canada, July 27-30, 2010

Vrazic, M.; Stipetic, S.; Kutija, M. Methodology of verifying IR temperature measurement on synchronous generator in rotation, *Proceedings of EDPE 2009 - 15th international conference on electrical drives and power electronics*, ISBN 978-953-6037-56-8, pp 1-5, Dubrovnik, Croatia, October 12-14, 2009

Vrazic, M.; Gasparac, I.; Pavlica, M. Ensuring synchronous generator special standstill tests, *Proceedings of 2008 International Conference on Electrical Machines*, pp. 1-5, Vilamoura, Portugal, 6-9 September, 2008.

Vrazic, M.; Gasparac, I.; Pavlica, M. Some Problems of Synchronous Hydro-Generator Temperature Measurement, *Proceedings of 2008 International Conference on Electrical Machines*, pp. 1-4, Vilamoura, Portugal, 6-9 September, 2008.

Maljkovic, Z.; Gasparac, I.; Vrazic, M. Electromechanical oscillations in bulb type hydropower unit, *Proceedings of 13th International Conference on Electrical Drives and Power Electronics*, ISBN 953-6037-43-2, pp. 7-15, Dubrovnik, Croatia, 26-28. September, 2005

Gasparac, I.; Maljkovic, Z.; Vrazic, M. Vibrations measurement during hydrogenerator testing, *Proceedings of XVII IMEKO WORLD CONGRESS - Metrology in the 3rd Milenium*, ISBM 953-7124-00-2, pp. 924-928, June 22-27, 2003, Dubrovnik, Croatia

3

Internal Combustion Engine Indicating Measurements

André V. Bueno[1], José A. Velásquez[2] and Luiz F. Milanez[3]
[1]Federal University of Ceara, UFC,
[2]Federal University of Technology - Parana, UTFPR,
[3]University of Campinas, UNICAMP,
Brazil

1. Introduction

Engine indicating includes the measurement of instantaneous in-cylinder pressure, the determination of the top dead centre (TDC) and the measurement of the instantaneous crank angle (c.a.). These measurements are fundamental for engine combustion diagnosis and for indicated work calculation.

In engine combustion diagnosis, the apparent heat release rate and the combustion reaction extent are the most useful quantities obtainable from engine indicating data. The apparent heat release rate is calculated by computing the amount of fuel chemical energy release necessary to obtain the experimentally observed pressure, while the combustion reaction extent is evaluated through the released fraction of the total fuel chemical energy. Heat release analysis is often complemented using optical techniques and its utilization as a diagnostic tool covers a wide range of objectives, including the development of new combustion systems, the analysis of alternative fuel burning, the validation of mathematical models for engine simulation, the investigation of combustion chamber insulation effects and the study of new injection strategies.

Taking into account the importance of indicating measurements for engine research and development, this chapter addresses the description of a modern engine indicating measurement system and includes recommendations about good-practice procedures and cares that a researcher must observe in order to perform accurate experiments and to obtain valuable information from measured data.

1.1 A word on the evolution of engine indicating measurement systems

The measurement of the working fluid pressure of heat engines was a topic of interest for engineers since the advent of the steam engine, for which the Watt's indicator was developed. When the internal combustion engine became the most widespread heat machine, its analyses and improvements also demanded the measuring of in-cylinder pressure data and early measurements were accomplished utilizing several configurations of mechanical indicators (Amann, 1985). However, as the operating speeds increased with the development of the engines, the frequency response of such mechanical indicators became deficient, making them obsolete in the mid-sixties of the last century.

High-speed electronic transducers, capable of converting the deflection of a low inertia diaphragm into an electrical signal were designed to meet the demand for pressure measuring instruments with superior characteristics. Early versions of such devices already had adequate frequency response to the phenomena occurring in the engine combustion chamber, having been built using extensometers (Draper & Li, 1949) and piezoelectric crystals (Kistler, 1956) as sensing elements. Nevertheless, these new electronic pressure transducers were initially coupled to analog data acquisition systems made of a signal conditioning amplifier, a cathode-ray oscilloscope and a photo camera that was used to record the pressure signal from the oscilloscope screen (Brown, 1967). This data generating procedure was cumbersome and had a number of uncertainties related to the film exposure time as well as to the oscilloscope trace (Benson & Pick, 1974).

At the end of the 1960s, complex analog systems capable of carrying out a fully electronic processing of the transducer signal became available. These devices were firstly used for specific applications, such as determining the indicated power (Alyea, 1969), indicated mean effective pressure (Brown, 1973), as well as studying knock and misfire in spark ignition engines (Alwood et al., 1970). The user interface was through a voltmeter, which displayed a voltage proportional to the indicated mean effective pressure or, alternatively, by means of an electromechanical counter that showed the number of cycles in which knock or misfire occurred.

In the mid-seventies, when analog-to-digital converters were included in the instrumentation, multipurpose and of-less-complexity experimental sets became available (Benson & Pick, 1974; Ficher & Macey, 1975; Marzouk & Watson, 1976). From this time, the signal from instrumentation amplifiers has been digitalized and stored in a computer, allowing further manipulation via software. Therefore, greater storage capacity and flexibility in data analysis were assured, maintaining an acceptable level of accuracy.

Nowadays these are typical characteristics of a modern engine indicating system, as the one schematically represented in Figure 1, which includes components suitable for measurements in diesel engines. In this scheme, the fuel injection line pressure is measured with a strain-gauge based sensor connected to an instrumentation amplifier, while the signal generated by a piezoelectric pressure sensor can be conditioned following two measurement procedures, which are addressed in detail in the present chapter. In the first procedure, the in-cylinder pressure is obtained by employing a charge amplifier for conditioning the transducer signal, while in the second procedure a current-to-voltage converter is used to measure the in-cylinder pressure derivative.

Pressure data are indexed with the angle position of the crankshaft, with reference to the compression TDC. Usually the crankshaft angle position is determined with an optical angle encoder, which provides both – one pulse per revolution in a channel used to establish the TDC angle reference, and 720 pulses per revolution in a second channel to determinate the instantaneous relative angle position. External pulse multipliers may also be available to improve the relative angle position resolution up to 3600 pulses per revolution. Each angle position pulse triggers a high speed data acquisition system, which should be able to simultaneously acquire the signals provided by the conditioning amplifiers, collect the acquired data among multiple cycles for cycle averaging, and save it in a storage computer.

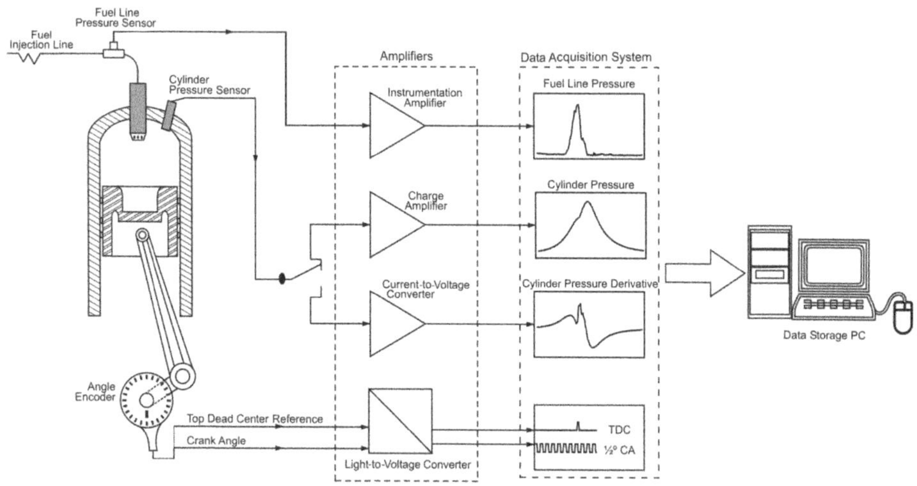

Fig. 1. A typical engine indicating measurement system.

2. The transducer for in-cylinder pressure measurement

The analysis of the processes occurring in the cylinder of internal combustion engines requires pressure transducers with high specifications regarding linearity, frequency response and resistance to thermal solicitations. Studies comparing transducers available at the end of 1960 decade (see for example Brown, 1967) found that those having piezoelectric crystals as measuring elements exhibited better tolerance to thermal solicitations than those based on strain gauges. By this reason, piezoelectric transducers eventually spread to measure the in-cylinder pressure, while sensors based on strain gauges (metal or piezo-resistive) were preferably used in measurements where the thermal solicitations are modest, such as pressure measuring in the fuel injection line and in the intake manifold.

Piezoelectric transducers are capable of maintaining high characteristics in frequency response and linearity over a wide range of pressures. On the other hand, the main reported drawbacks to their use include instability of the baseline and low intensity of its output signal. These drawbacks and their control will be discussed in Section 2.4, where the techniques for conditioning the pressure transducer signal will be presented.

2.1 Operating principle of the piezoelectric pressure transducer

Figure 2 illustrates the principle of operation of a piezoelectric pressure transducer. The pressure change rate (*dP/dt*) experienced by the transducer diaphragm is transmitted to a piezoelectric crystal through intermediate elements, causing its deformation at a rate $d\varepsilon/dt$.

Due to the piezoelectric effect, this deformation polarizes charge q in the transducer electrode originating an electric current i, which constitutes the transducer output signal:

$$i = -\frac{dq}{dt} = -G_s \frac{dP}{dt} \tag{1}$$

where G_s is the transducer sensitivity (gain).

During the measurement of in-cylinder pressure, the transducer is exposed to a transient heat flow that causes continuous changes in its temperature. These temperature changes modify the sensitivity of the piezoelectric element and impose thermal stresses in the diaphragm and in the sensor housing, generating spurious forces that act on the quartz element and cause additional distortion of the signal provided by the transducer. The inaccuracy due to these effects receives the name of temperature drift (Marzouk & Watson, 1976).

It is usual to separate the temperature drift into two components. The first one corresponds to changes in heat flow that occur within each cycle, and is named short-term drift or thermal shock. The second component represents the slow change in the transducer temperature resulting from variations in the engine operating conditions and is known as load-change drift or long-term drift.

Normally, the long-term drift only promotes a slow instability in the baseline for transducer signal, whose consequences and control depend on the circuitry chosen for its polarization.

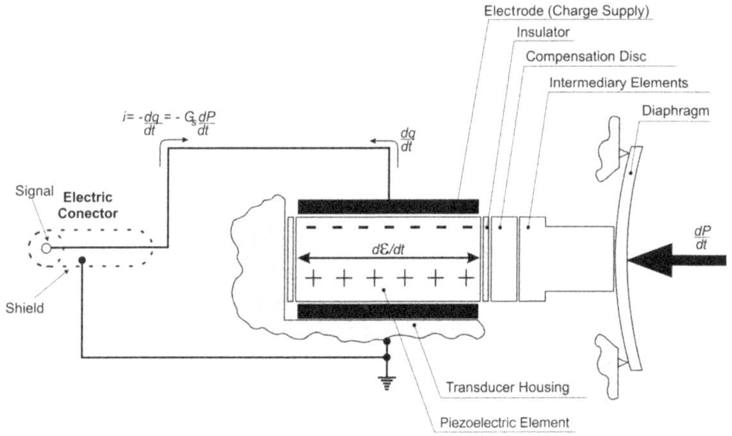

Fig. 2. The piezoelectric pressure transducer.

The effects of the short-term drift, in turn, depend on the intensity with which this phenomenon occurs. A severe short-term, such as that resulting from operating conditions that can permanently damage the transducer, may produce pressure values below the atmospheric pressure at the end of the expansion process. At more restrained levels, however, the presence of short-term drift cannot be readily identified, causing the pressure reading to be higher than the actual in-cylinder pressure during combustion, and lower throughout the remainder of the expansion. Although modern piezoelectric sensors are designed to minimize the effects of short-term drift, it must be taken into account that the intensity of the latter strongly depends on the thermal load at the transducer location. This thermal load is influenced by the intense flows during the gas exchange process, by the approximation of the fuel jet in a diesel engine or by the flame front approximation in a spark-ignition engine. Thus, checking the extent of the thermal shock at the transducer location in each particular case is a good practice for an accurate measuring.

2.2 Choice of the transducer mounting location

The installation of the piezoelectric pressure transducer must be preceded by the calibration of the complete measuring chain formed of the piezoelectric transducer, the signal conditioning amplifier and the data acquisition system. It is possible to use a dead-weight calibrator for this task in accordance with the recommendations made by Lancaster et al. (1975). The choice of the location where the transducer will be mounted should prioritize well-cooled regions of the head and avoid these where thermal stresses can induce deformations in the transducer housing. The diaphragm of the sensor should be positioned as recommended by the manufacturer (usually, with a gap of 1.5 to 3.0 mm from the inner surface of the head). Water-cooled transducers feature superior gain (higher signal/noise ratio), linearity and thermal resistance respective to the miniature uncooled transducers, and should be the first choice for use in cases where there is enough space in the head. Channels connecting the combustion chamber with the cavity where the sensor diaphragm is located may behave as acoustic resonators, generating pressure oscillations and their consequent measurement inaccuracies, which can invalidate the evaluation of both, the indicated thermodynamic parameters and the combustion energy release. So, the use of these channels (as typically occurs when the sensor is embedded in a spark plug) is recommended only for the identification of abnormal combustion in spark ignition engines.

In-cylinder pressure measurements in direct injection (DI) diesel engines demand extra care due to the high compression ratio and to the shape of the combustion chamber. In these engines, when the piston is near the TDC, about 90% of the working fluid mass is inside the piston bowl and in the region above this cavity. The pressure of this mass portion is representative of the cylinder averaged pressure. The remaining mass occupies the crevice regions between piston and head as well between piston and sleeve, and its pressure may exhibit oscillations of up 10 bar amplitude, which are due to the turbulent in-cylinder flow and to acoustic phenomena resulting from combustion. Thus, the transducer must be located at a point from which the pressure of the mass above the piston bowl can be accessed. Finally, it is important to point out that when choosing the transducer mounting point the fuel jet impingement on the transducer diaphragm should also be avoided.

2.3 Validation of the transducer mounting location

In order to illustrate the procedure for validating the transducer mounting location we will consider as example the case of a fast, direct injection diesel engine with three valves per cylinder, in which indicating measurements were conducted using an uncooled mini-transducer AVL GM 12 D, mounted above the piston bowl in the glow plug place, as shown in Figure 3.

The method proposed by Randolph (1990) allows checking the occurrence of short-term drift by comparing the cycle-to-cycle variability of the in-cylinder pressure readings at specified instants along the working cycle. Certain variability is normal due to the random nature of the combustion process, which causes one cycle to be slightly different from other under unchanged operating conditions. This cycle-to-cycle variability leads to changes in the thermal load acting on the transducer and, when short-term drift occurs, it leads also to changes in transducer sensibility, thus enlarging the scattering of pressure readings.

In order to apply the Randolph test, consider two points along the cycle, designated as C1 and B2. The first point is the beginning of the exhaust process, thus characterizing a moment when the transducer is under the influence of combustion thermal loads. The second point is located in the compression stroke, thus soon after the gas exchange, during which the transducer is cooled. So, if short-term drift was occurring, it should produce pressure readings at C1 more scattered than those at B2 (Randolph, 1990; Roth el al., 2001). Figure 4 shows the pressure deviations from the sample mean-value, for 56 consecutive cycles. According to the description given above, the short-term drift would cause the scattering of points along the x-axis be greater than that along the y-axis. However, the points in this figure are distributed equitably with respect to the axes, thus implying that short-term drift is trivial in this case. The scattering of points with respect to the diagonal of the graph allows judging about the repeatability of the experiment. The behaviour observed in Figure 4 attests for a good repeatability and is similar to that reported by Roth et al. (2001) for transducers akin to the one used in this example.

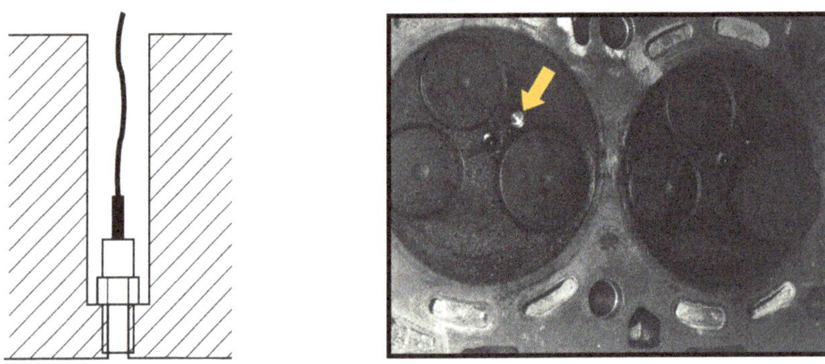

Fig. 3. Pressure transducer mounting location.

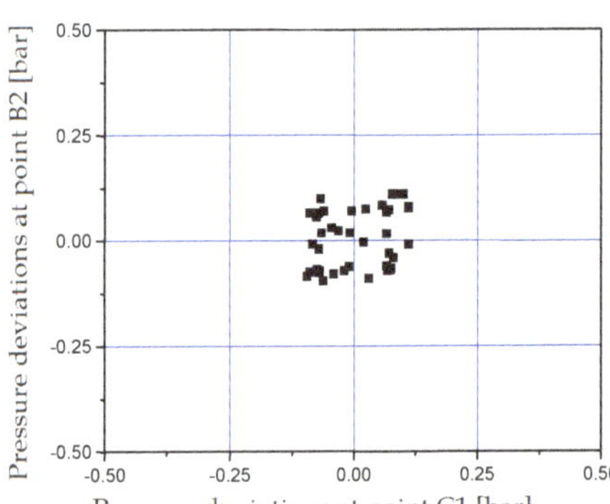

Fig. 4. Deviation of pressure readings respective to sample mean-value. Point C1: 145 c.a. degrees after compression TDC. Point B2: 80 c.a. degrees before compression TDC.

In cases of significant short-term drift, it is recommendable to mount the transducer via an adapter, which avoids its direct contact with the cylinder gas, thus eliminating local heating of the transducer components and, mainly, of its diaphragm. Another solution consists in installing the transducer recessed by means of a measuring channel. However, such mounting procedure may lead to inaccuracies caused by an oscillating flow in the channel. The extent of such inaccuracies has been studied in detail by Hountalas & Anestis (1998).

2.4 Strategies for processing the piezoelectric transducer signal

The strong influence of the pressure derivative on the energy release rate was recognized by several authors (Marzouk & Watson, 1976; Woschni, 1965; Krieger & Borman, 1966). Despite this, indicating measurements usually aim obtaining the in-cylinder pressure, even when combustion diagnosis is the major objective. As a rule, this task is carried out using a piezoelectric transducer polarized by a charge amplifier (Brown, 1976; Marzouk & Watson, 1976; Benson & Pick, 1974; Lancaster et al., 1975; Lapuerta et al., 2000; Benson & Whitehouse, 1983). The preference for this practice can be attributed to the legacy of the mechanical indicators, which were created especially for pressure measuring, as well as to the importance of knowing the in-cylinder pressure for calculating the indicated work or for characterizing the thermodynamic state of the working fluid. Thus, obtaining pressure derivative data for heat release analysis was eventually relegated to the numerical manipulation of available pressure data. As it will be discussed below, this procedure amplifies the inaccuracies in the pressure data, originating pressure derivative curves with oscillations that are eventually transferred to the results of energy release rate calculation.

The numerical instabilities inherent to the differentiation operation motivated the development of experimental arrangements in which this operation was carried out by means of electronic circuits. A reference to such an approach can be found in the work of Marzouk & Watson (1976) who processed the analog signal from a charge amplifier by an electronic differentiator. While avoiding the numerical differentiation of pressure data, these authors did not obtain satisfactory results for the pressure variation rate, underestimating its peak value during the premixed combustion. This was due to the deficient frequency response of the adopted differentiation circuit, which severely damaged the performance processing of those data that make up the high frequencies region of the spectral distribution (see Section 4).

The drawbacks mentioned above can be overcome by employing an alternative technique of polarization of the piezoelectric sensor – the direct conversion of the current signal supplied by the transducer into an analog voltage level proportional to the time rate of change of the in-cylinder pressure. The circuitry presented in Section 2.4.2 is based on this technique.

2.4.1 Signal processing with a charge amplifier

Figure 5 shows a simplified schematic of the experimental setup commonly used to obtain in-cylinder pressure data. A shielded cable with high insulation resistance conducts the charges polarized by the transducer to the inlet of the charge amplifier. This device is based on an integrator circuit, which provides an output voltage proportional to the time integration of the electric current applied at its input during a time interval Δt, taken from the instant at which it was started (or reset) until the desired instant of measurement. Thus, the pressure variation during the interval Δt is given by

$$\left(P - P_{ref}\right) = \frac{v_{ch} \cdot G_a}{G_s} \tag{2}$$

where P and P_{ref} are values of the pressure acting on the transducer diaphragm at the end and at the beginning of the time interval Δt, respectively; G_a is the gain of the charge amplifier and v_{ch} is the charge amplifier output voltage.

Since the charge produced by the pressure transducer is pretty small, reaching just tens of picocoulombs per bar, the association between a piezoelectric transducer and a charge amplifier results extremely sensitive to non-idealities of electronic circuits, mainly to current leakages occurring through the insulation resistance of the measurement system or even within the integrator circuit. Such current leakages cause a slow and steady decline in the level of the output voltage of the charge amplifier, thus yielding pressure values lesser than the actual ones. In order to control this inaccuracy it is necessary to have high input impedance in the charge amplifier, while taking especial care for maintaining the electrical contacts clean and for operating the data conditioning system in a low-humidity environment.

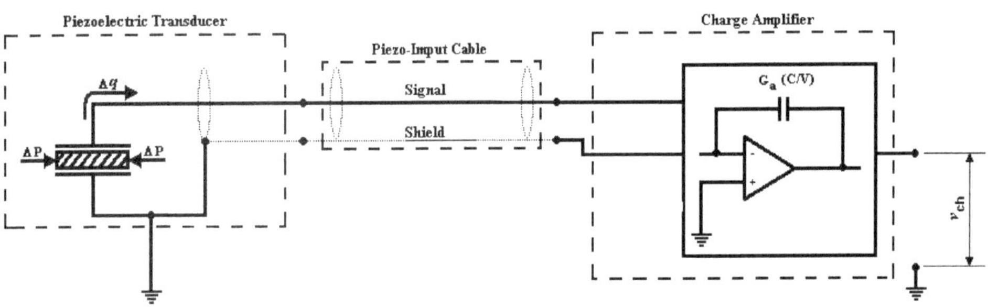

Fig. 5. Transducer signal conditioning through a charge amplifier.

Long-term drift also promotes an incessant and slow displacement of the data baseline as the integrator circuit processes spurious current induced by changes in the transducer temperature, which in its turn result from changes in engine operation conditions.

Therefore, even when properly using a good-quality measuring system, the combined action of circuitry non-idealities and long-term drift leads to instability of the pressure baseline, which is a drawback inherent to the use of a charge amplifier and is reported in the literature as pressure baseline floating (Marzouk & Watson, 1976; Benson & Pick, 1974; Lancaster et al., 1975; Benson & Whitehouse, 1983). This baseline floating can produce pressure deviations of tens of bars during a long-duration measurement, making it mandatory to reset periodically the amplifier in order to avoid its saturation.

The deviations mentioned above as well the fact that pressure data provided by a charge amplifier are related to an initial measurement time make it necessary to correct the baseline at each cycle, as well as the pressure readings. This task can be accomplished by determining the pressure baseline that leads to a null value of the apparent heat release rate over the first 80 c.a. degrees after the intake valve closing. This criterion for correcting the pressure readings is based on the recommendations given by Lapuerta et al. (2000).

2.4.2 Signal processing with a current-to-voltage converter

The most evident experimental setup for converting the current polarized by a piezoelectric transducer consists in conducting it to the ground through an electrical resistance, thus originating a voltage drop equal to the product of the value of this resistance by the current. Such procedure would result, however, in the formation of an RC circuit between the transducer inherent capacitance and the measurement resistance, which, in turn, would cause attenuation in the voltage drop and a phase delay with respect to the polarized current. In order to avoid this problem Bueno et al. (2009, 2010, 2011) proposed to use a circuit that converts current into voltage (Franco, 2001), consisting of an operational amplifier and a negative feedback resistor, as shown in Figure 6. This circuit operates stably in a range of gains typical for engine indicating measurements, and may undergo minor changes related to balancing techniques, depending of which operational amplifier was chosen for its construction. The virtually zero value of the input impedance of the current-to-voltage converter, given by the ratio of R_A and the open-loop gain of the operational amplifier A_1, makes it immune to inaccuracies arising from the inherent capacitance of the piezoelectric transducer. Besides that, in order to isolate the converter with respect to the impedance of the instrument to which it is connected, a voltage follower amplifier was used. Thus, the pressure rate of change dP/dt is given by:

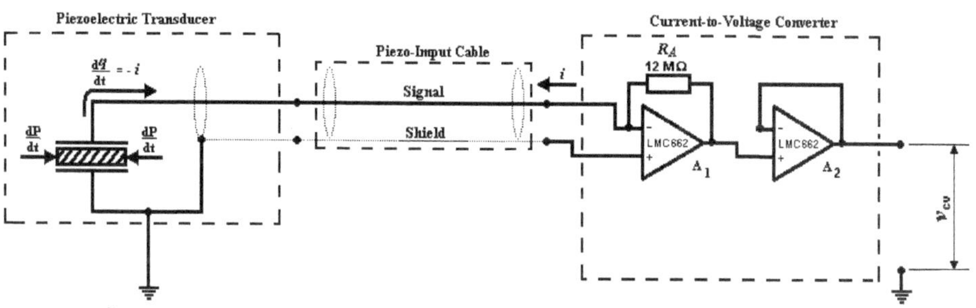

Fig. 6. Transducer signal conditioning through a current-to-voltage converter.

$$\frac{dP}{dt} = \frac{v_{cv}}{G_s \cdot R_A} \tag{3}$$

where v_{cv} is the output voltage of the current-to-voltage converter and R_A is the gain adjusting resistance of the current-to-voltage converter. Taking the crankshaft angle speed as a constant, the following expression for the pressure derivative with respect to crank angle can be obtained:

$$\frac{dP}{d\theta} = \frac{v_{cv}}{6 \cdot G_s \cdot R_A \cdot RPM} \tag{4}$$

in this equation RPM represents the engine operating speed, in revolutions per minute.

The in-cylinder pressure can be determined by numerically integrating the data obtained for its derivative. A fourth order Runge-Kutta algorithm is recommended for this task (Teukolsky, 1996) and the pressure baseline can be determined by applying the same correction technique recommended in Section 2.4.1.

The use of a current-to-voltage converter eliminates the need of special care for insulation resistance and leakage currents, as in this case the charges polarized by the transducer unrestrictedly flow to the ground. In addition, the long-term drift makes the pressure derivative baseline to deviate from its original position returning, however, after the engine reaches a steady state. Therefore, this technique also eliminates the need for resetting the transducer polarization circuit due to the large displacements of the data baseline. Further details involving the conditioning of the pressure derivative signal using a current-to-voltage converter can be found in references Bueno et al. (2009) and Bueno et al. (2011).

3. Crank angle measurements and TDC position identification

In modern engine indicating systems the instantaneous position of the crankshaft is determined with the aid of an optical angular encoder, whose operating principle is based on photoelectric scanning of a sequence of thin opaque lines. These lines are etched on a transparent disk that rotates together with the crankshaft and are arranged in the radial direction at equal angular intervals, forming the so-named incremental track (see Figure 7).

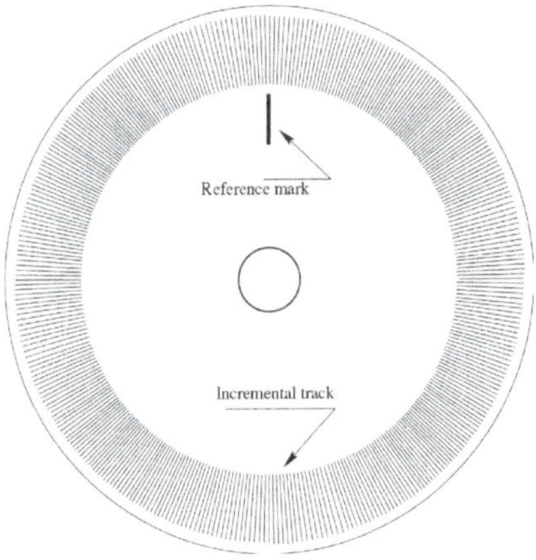

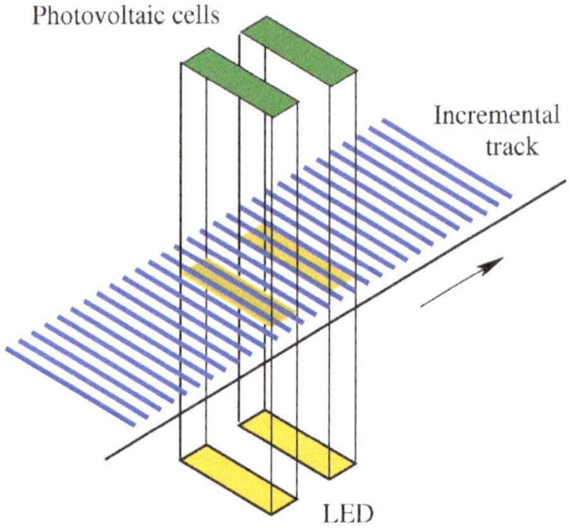

Fig. 7. The optical angle encoder.

During operation of the optical angular encoder a light beam emitted by a LED falls perpendicular to the disk plane on the incremental track. As the disk is rotating, this light beam will be reflected if meets one of the etched lines, but it will pass across the disk while falling on the gap between two lines. Then, a sequence of light pulses synchronized with the crankshaft angular position will pass across the disk and will be conducted through an optical fiber to a light-pulse converter, where photovoltaic cells transform this light signal into an electrical one. Therefore, the encoder outputs a pulse string in response to the amount of rotational displacement of the crankshaft. A separate counter counts the number of output pulses to determine the crankshaft angle position.

In engine indicating measurements the angular position of the crankshaft must be determined in relation to a reference, which is generally also used as reference to identify the position of the TDC. Such reference is generated with the aid of a second incremental track that has a single mark, which is used to reset the counting of the pulses.

Special attention should be given to the correct identification of the TDC position, as small errors in this measurement lead to significant errors in the evaluation of indicated work as well as combustion heat release rate (Krieger & Borman, 1966; Pischinger & Glaser, 1985; Lapuerta et al., 2000). In order to achieve adequate precision in determining the TDC position it is recommended to perform a dynamic measurement, running the engine motored and unfired or, alternatively, preventing combustion only in the cylinder where the measurement is going on, while the other cylinders function fired to keep the engine running.

When performing dynamic measurement, the inaccuracies that would be generated by the bearing clearances are eliminated. Such measurement is usually carried out with a capacitive proximity sensor, or in the absence of such a sensor, the TDC position can be inferred from the motored-engine pressure data.

The capacitive proximity sensor uses two conductive objects separated by a dielectric material. A voltage difference applied to the conductive objects generates an imbalance of electrical charges between them, originating an electric field in the dielectric material. When this voltage is alternated the electrical charges move continuously, going from one of the conductive objects to the other and generating an alternating electric current, which is the output signal of the sensor. The amount of current flow is determined by the capacitance, and the capacitance depends on the proximity of the conductive objects. Closer objects cause greater current than more distant ones.

In the capacitive sensors used to determine the TDC position, one of the conductive objects is the sensor probe itself, while the piston plays the role of the second conductive object (Figure 8). The sensor is mounted in the head in such a way that when the engine is running the piston will move closer or away from the sensor, but without actually touching it. Thus, the sensor will produce a signal with amplitude, which is inversely proportional to the distance between the TDC sensor tip and the piston top. The exact TDC position will correspond to the maximum amplitude of the TDC sensor signal, which can be determined with great accuracy because of the high degree of symmetry of the signal.

When using the motored-engine pressure data for identifying the TDC position, the main arising difficulty is that the peak pressure precedes the actual TDC position, which corresponds to the minimum volume. This occurs due to heat transfer and mass losses, and the angle interval between these events is named loss angle (Figure 9). Several enough accurate methods have been proposed for determining the loss angle (Pinchon, 1984; Stas, 1996), and usually manufacturers of indicating equipment include in their manual recommendations to estimate loss angle values, which depend on the kind of the engine (spark ignition or diesel) and compression ratio.

The advantage of the direct measurement of TDC position, compared with its determination from the motored-engine pressure curve, is that there is no need for a correction involving the loss angle, which always increases the uncertainty.

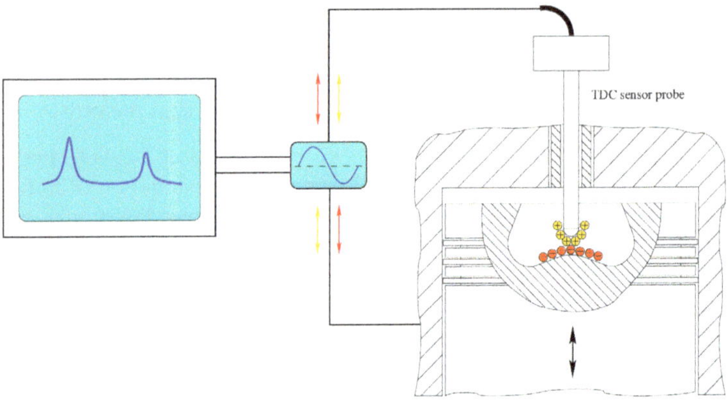

Fig. 8. The capacitive TDC sensor.

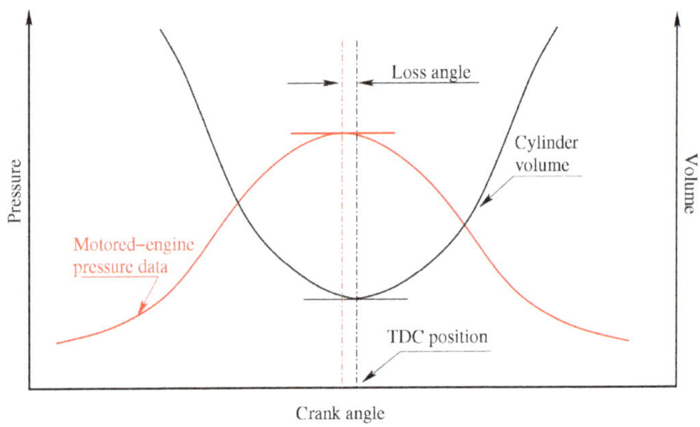

Fig. 9. Definition of loss angle.

4. Uncertainty sources and data treatment

Cylinder pressure data measured by a transducer can be understood as being the sum of two components: *(i)* a smooth component, corresponding to the instantaneous average pressure through the entire cylinder volume; and *(ii)* a spurious component, originated by both the turbulent flow of gas inside the cylinder and the acoustic pulsations associated to combustion. According to the hypothesis of the single-zone combustion model (Krieger & Borman, 1966), the first component must be used to characterize the thermodynamic state of the gases in the cylinder, constituting the information of interest for heat release analysis. The second component is small when compared to the first one and its extent is influenced by the transducer mounting location in the combustion chamber. In addition to these two components of the measured pressure, the experimental data also include noise generated by the measurement system during the conditioning of the transducer output signal. Thus, for accurate heat release analysis it is necessary to isolate the volume-averaged pressure component from both the flow and the combustion driven spurious component as well as from the measurement noise.

Typically, the assessment of the smooth volume-averaged pressure component has been accomplished by means of numerical treatment of experimental data, which is usually carried out after data averaging over several consecutive cycles. A diversity of methods can be used for this numerical treatment, including FFT filters, windowing filters, smoothing splines and Wiebe function regression (Zhong et al., 2004; Payri et al., 2011; Ding et al., 2011). However, the choice of a proper data treatment methodology requires, at least, a basic knowledge about the behaviour of each one of the above mentioned components of the experimental data. In order to assess this behaviour Bueno et al. (2009) proposed to estimate the smooth component corresponding to the average pressure through the entire cylinder volume and, then, to employ it to analyze the spurious component of the experimental data. Sections 4.1 through 4.3 show how this task is accomplished.

4.1 The volume-averaged smooth data component

Smooth component curves (referred here as *reference curves*) for both in-cylinder pressure and in-cylinder pressure derivative data can be estimated by means of a computer routine based on the single-zone combustion model. In this routine, the combustion heat release rate is modelled through Wiebe functions (Wiebe, 1970) whose adjusting constants are obtained by optimizing curve-fitting to heat release rate data calculated from measured values. Figure 10 shows typical reference curves and experimental data obtained with each one of the signal conditioning procedures described in Sections 2.4.1 and 2.4.2. The curves shown in Figure 10 are related to a direct injection diesel engine and they show that the premixed burning causes a steep slope in the pressure curve region subsequent to ignition (circled area in Figure 10a), whilst in the pressure derivative curve it is reflected as a prominent peak (circled area in Figure 10b). By comparing these two representations, it becomes evident that the premixed combustion is more visible in the pressure derivative curve, a fact that led to the choice of the pressure derivative as the basic representation for the analysis of the data given by the indicator system.

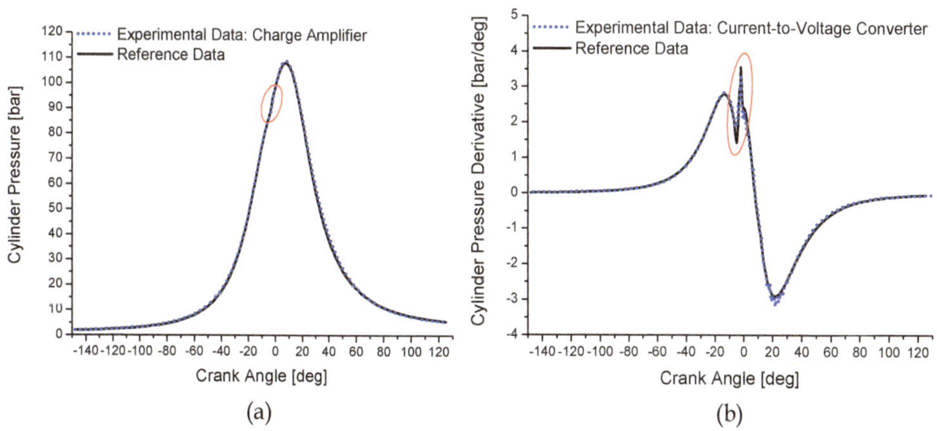

Fig. 10. Typical reference curves for pressure and pressure derivative data (D.I. diesel engine running at 2600 rpm and 80% of full load).

Frequency domain representations such as Lomb periodograms (Lomb, 1976) constitute a suitable tool for the analysis of experimental data. In these periodograms the spectral power is normalized with respect to the variance of the analyzed data, and its value represents the degree of participation of the signal associated to a given frequency in the data composition. In Figure 11, spectral distributions of two pressure derivative curves obtained through numerical simulation are compared. The first curve, named *Reference Data*, was calculated using two Wiebe functions, corresponding to premixed and diffusive stages of combustion, respectively. The second curve, named *Diffusive Combustion Data*, was obtained suppressing the premixed combustion and only one Wiebe function representing diffusive combustion was used. As can be seen in Figure 11, these curves exhibit similar spectral power distribution in the low-frequency region (up to 1000 Hz); however, in the region between 1000 and 5000 Hz the suppression of premixed combustion caused considerable attenuation of the pressure derivative data. Thus, it may be concluded that the contribution of the premixed combustion is located in this frequency range (from 1000 to 5000 Hz) and that the utilization of low-pass filters or numerical treatment to smooth the transducer signal at frequencies below 5000 Hz may cause distortion or even the elimination of the influence of premixed combustion on experimental data. It may also be observed that the effects of both the compression process and the diffusive combustion play a major role in the spectral distribution of the data of interest, occupying the frequency range with highest spectral power values (below 1000 Hz).

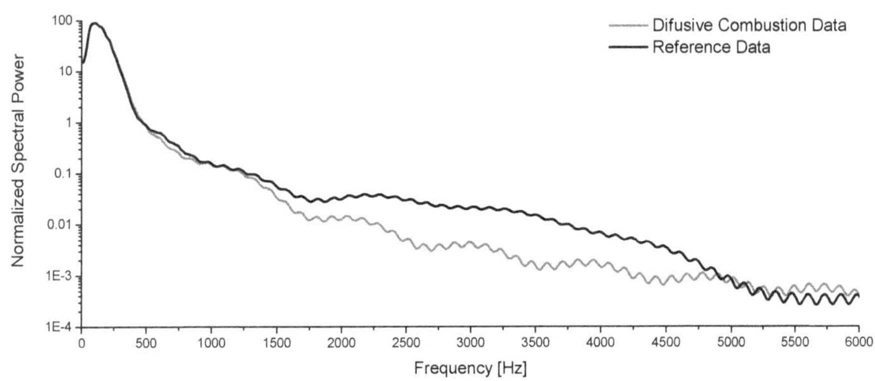

Fig. 11. Spectral distribution for simulated in-cylinder pressure derivative data (D.I. diesel engine running at 2600 rpm and 80% of full load).

4.2 Measurement system noise and the flow driven spurious component

Uncertainties associated to the data acquisition system are dominated by the truncation error of the analog-to-digital converter, acting as a source of white noise in the acquired experimental signal. Therefore, the measurement uncertainty due to this noise (*MU*) can be estimated in terms of the analog-to-digital converter accuracy, which corresponds to the least-significant bit (*LSB*) of the used measurement system.

When a transducer is polarized by a charge amplifier, the influence of this uncertainty on the pressure derivative curve is indirect, because the measured parameter in this case is the in-cylinder pressure. Nevertheless, it is possible to estimate the measurement uncertainty imposed by the quantization noise to the pressure derivative (MU_{dP}) using finite differences approximation of fourth order. Doing so and taking into account that data reported in Figures 10 and 11 refer to measurements made with a system whose *LSB* value is 0.131 bar, results

$$MU_{dP} = \pm\sqrt{\left(\left(\frac{LSB}{12\Delta\theta}\right)^2 + \left(\frac{-8LSB}{12\Delta\theta}\right)^2 + \left(\frac{8LSB}{12\Delta\theta}\right)^2 + \left(\frac{-LSB}{12\Delta\theta}\right)^2\right)} = \pm 0.25 \ \textbf{bar/}\deg \quad (5)$$

On the other hand, when the direct polarization of the transducer is used, the pressure derivative is accessed directly and the measurement uncertainty imposed by the quantization noise to the pressure derivative is given by the corresponding *LSB* value. In the case of the data reported in Figures 10 and 11 this value is 54.694 bar/s, then

$$MU_{dP} = \pm\ 54.694\ \textbf{bar/s} = \pm 3.5\times 10^{-3}\ \textbf{bar} / \deg \quad (6)$$

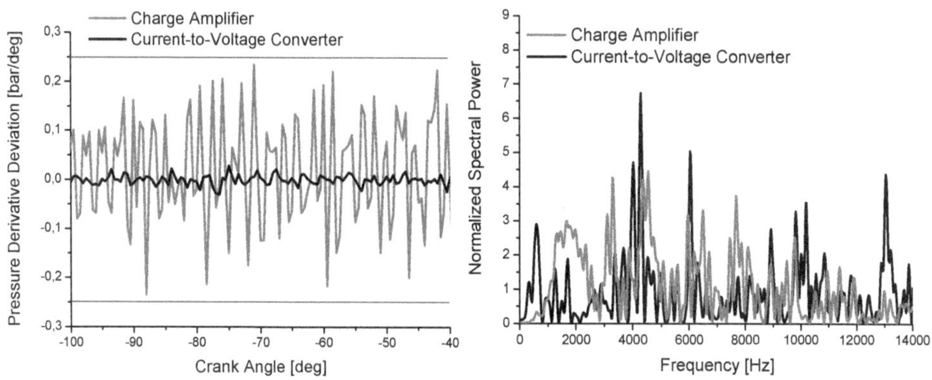

(a) (b)

Fig. 12. Deviations due to the measurement system noise and to the in-cylinder flow (D.I. diesel engine running at 2600 rpm and 80% of full load).

It must be noticed that the utilization of a current-to-voltage converter allowed reducing the measurement uncertainty imposed by the quantization noise to only 1.4% of the value obtained using a charge amplifier. As it will be discussed later, this fact has important consequences for obtaining smooth heat release rate diagrams. Additionally, when signal conditioning is carried out using a charge amplifier, the refinement of the encoder resolution increases the measurement uncertainty, whilst for the case of the current-to-voltage converter the measurement uncertainty does not depend on the encoder resolution.

Figure 12a shows experimental data deviation from the reference pressure derivative curve during the compression process, when combustion does not occur and, therefore, the spurious components of the experimental data can be attributed only to the in-cylinder flow and to the data acquisition system. It may be observed in this figure that the utilization of a charge amplifier imposes oscillations, which are within the limits of the quantization noise (±0.25 bar/deg); whilst for the case of the current-to-voltage converter these oscillations exceed the corresponding limits (±3.5×10-3 bar/deg). Therefore, it is possible to conclude that when a charge amplifier is used the effects of the noise generated by the data acquisition system predominate over the spurious component generated by in-cylinder flow, whilst in the case of the current-to-voltage converter this spurious component is the major responsible for the oscillations observed along the compression process.

The spectral distributions corresponding to the deviation curves shown in Figure 12a can be found in Figure 12b. These distributions characterize both the in-cylinder flow as well as the quantization error, as sources of white noise (random noise), having a considerable part of its spectral composition in the same frequency range occupied by the data of interest. Such behaviour makes it difficult to use low-pass filters or numerical smoothing without loss of useful signal, demanding especial care during the choice of the smoothing parameters, which vary with engine running condition (Zhong et al., 2004). Due to the high quantization noise in the data obtained with the charge amplifier, the remaining analysis of the experimental data composition was based on the current-to-voltage converter approach.

4.3 The combustion-driven spurious component

The ignition in diesel engines exhibits an eminently random character, being strongly influenced by the interaction of chemical and hydrodynamic phenomena occurring along the ignition delay (Maunoury et al., 2002). Because of this, a diesel engine operating in steady state presents cycle-to-cycle variations in the amount of fuel available for premixed combustion as well as in the locations where ignition occurs by first. In addition, the natural vibration modes of the gas contained in the cylinder are excited by the high burning rates that occur in the beginning of the combustion (Schmillen & Schneider, 1985), generating pressure oscillations of large amplitude that are rapidly dampened due to the high area/volume ratio of the cylinder cavity. Due to cycle-to-cycle combustion variability, different modes are excited in each cycle, causing the oscillations to vary in phase, characteristic frequencies and amplitude from one cycle to other.

Figure 13a shows the typical behavior of experimental pressure derivative data, where the cycle-to-cycle variability associated to the premixed combustion and to the following acoustic oscillations contrasts with the repeatability inherent to both the compression process and the diffusive combustion. In the periodograms presented in Figure 13b, the combustion-generated variability appears in the frequencies above 1000 Hz. This result is in accordance with the observations of Strahle (1977) and Strahle et al. (1977a, 1977b), who attributed the pressure data spectrum above 1000 Hz to the random phenomena associated to combustion.

In Figure 14a the deviation of the experimental data respective to the reference pressure derivative curve is shown along the combustion process for two cycles. As can be noticed, after 25 c.a. degrees the amplitude of the oscillations is dampened to a constant level slightly higher than that observed during compression, indicating that combustion amplifies the influence of the turbulence on experimental data. The spectral analysis of this deviation is shown in Figure 14b, where a periodic signal whose frequencies should correspond to the cylinder cavity characteristic frequencies can be seen (9.0 kHz for cycle 2; and 2.0, 6.0 and 11.0 kHz for cycle 13). Figure 14 confirms that the oscillations vary in phase, in characteristic frequencies and in amplitude from one cycle to other. This behavior may be used to remove the combustion-driven oscillations by data overlapping.

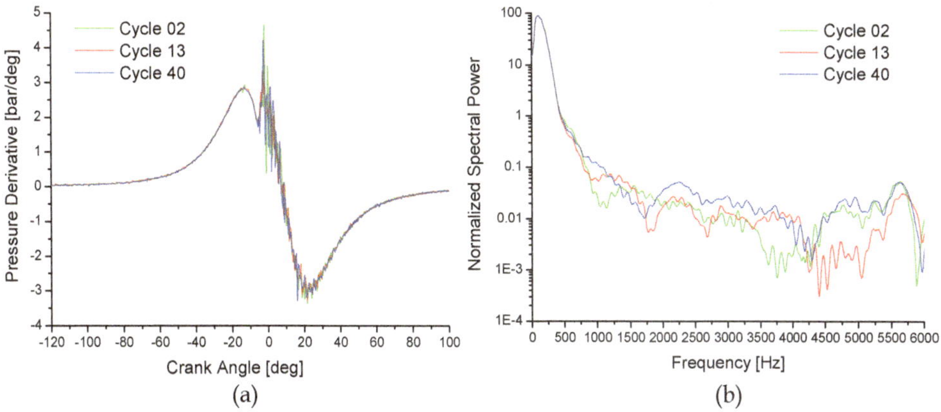

Fig. 13. Pressure derivative experimental data and its spectral composition (D.I. diesel engine running at 2600 rpm and 80% of full load).

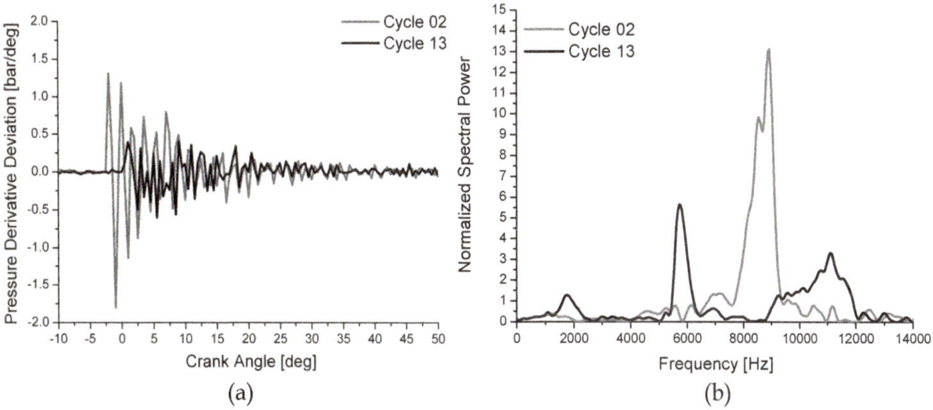

Fig. 14. Combustion-driven spurious component of the experimental pressure derivative data (D.I. diesel engine running at 2600 rpm and 80% of full load).

4.4 Cycle-averaging as a data smoothing method

Cycle-averaging over a significant number of successive cycles is an effective method of data treatment, easing by overlapping the effects of cycle-to-cycle variations, combustion driven oscillations and measurement noise. This can be seen in Figure 15, which shows in its upper part the spectral composition of the spurious components of experimental data, whilst in its lower part the spectral composition of single-cycle pressure derivative data is compared to that of mean data (averaged over a sequence of 56 cycles). Data corresponding to cycles with the lowest noise level (13th cycle) and with the highest noise level (2nd cycle) are shown in the left and in the right parts of this figure, respectively.

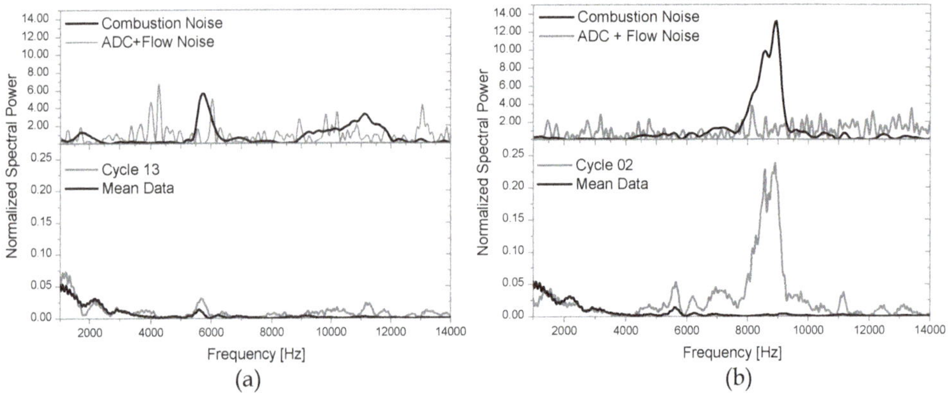

Fig. 15. Noise in individual cycles and its reduction by cycle averaging (D.I. diesel engine running at 2600 rpm and 80% of full load).

Figures 16 and 17 show data of cylinder pressure and its derivative for the same engine operating conditions, where 95% confidence limits are displayed. Data of Figure 16 were obtained using a charge amplifier while for data of Figure 17 a current-to-voltage converter was used. In both cases data were averaged over a sequence of 56 cycles. Analyzing these figures it can be concluded that both transducer signal-conditioning procedures resulted in similar values for the cycle averaged cylinder pressure and its confidence limits, so that they are equivalent when the cylinder pressure is the parameter of interest. However, the studied procedures gave different results when evaluating pressure derivative, which is the data of greater interest for heat release analysis. The pressure derivative confidence interval obtained with the current-to-voltage converter was smaller than that obtained using the charge amplifier and subsequent numerical derivation, the former reaching only about 1/50 of the latter during most of the compression process as well as during late combustion.

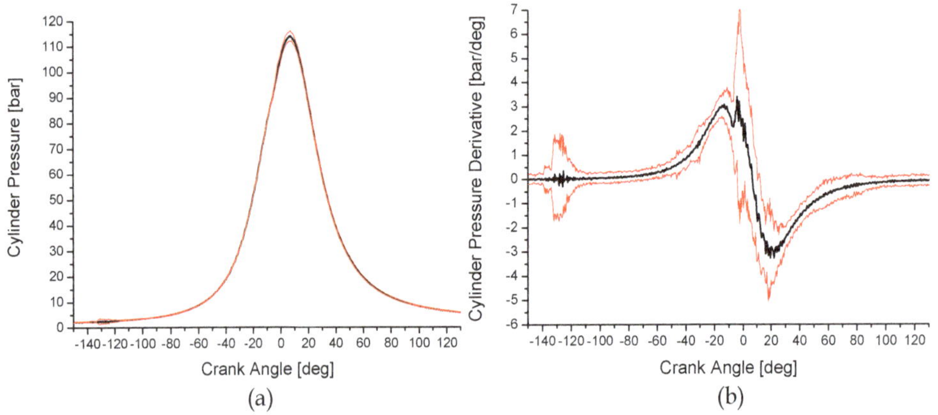

Fig. 16. Indicating data obtained using a charge amplifier (D.I. diesel engine running at 2900 rpm and full load).

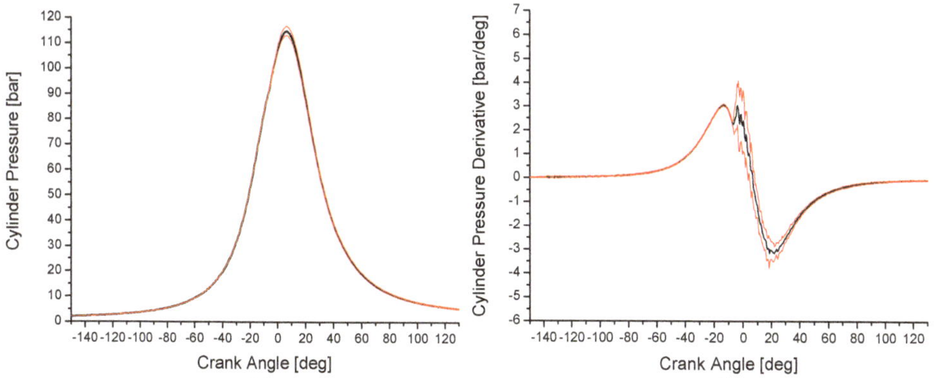

(a) (b)

Fig. 17. Indicating data obtained using a current-to-voltage converter (D.I. diesel engine running at 2900 rpm and full load).

Considering the extension of the confidence intervals of pressure derivative data, it is expected that heat release results calculated from the charge amplifier data exhibit a more irregular and oscillatory pattern than those calculated from the current-to-voltage converter data. It is worth mentioning that large oscillations in the calculated heat release rate curve can result in negative values during the compression as well as during late combustion, which makes no physical sense. These oscillations make it difficult to determine the pressure data baseline, which in turn increases the uncertainty in the assessment of the burned fraction of fuel mass. Data shown in Figure 18a confirm that heat release curves

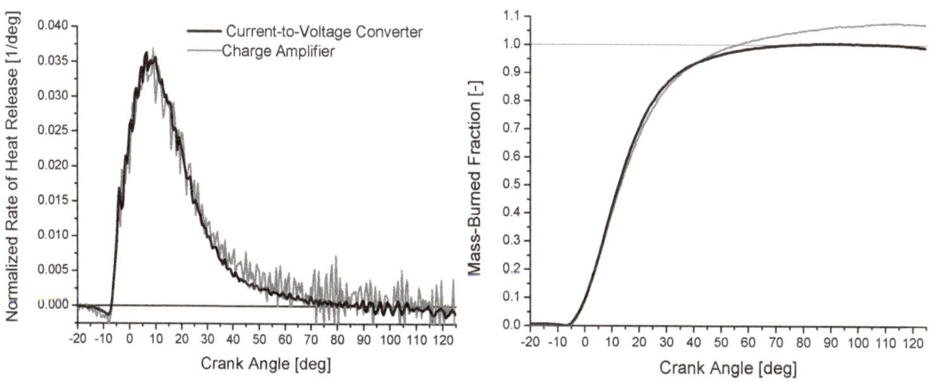

(a) (b)

Fig. 18. Normalized heat release rate and fuel mass burned fraction (D.I. diesel engine running at 2900 rpm and full load).

calculated from charge amplifier data exhibit a higher noise level than those calculated from current-to-voltage converter data, while Figure 18b gives evidence of the differences in the calculated burned fraction of fuel mass, which are caused by this higher noise level.

5. References

Alwood, H. I. S.; Harrow, G. A. & Rose, L. J. (1970) A Multichannel Electronic Gating and Counting System for the Study of Cyclic Dispersion, Knock and Weak Mixture Combustion in Spark Ignition Engines. *SAE Paper 700063.*

Alyea, J. W. (1969) The Development and Evaluation of an Electronic Indicated Horsepower Meter. *SAE Paper 690181.*

Amann, A. C. (1985). Classical Combustion Diagnostics for Engine Research. *SAE Paper 850395.*

Benson, R. S. & Pick, R. (1974) Recent Advances in Internal Combustion Engine Instrumentation with Particular Reference to High-Speed Data Acquisition and Automated Test Bed. *SAE Paper 740695.*

Benson, R. S. & Whitehouse, N. D. (1983) *Internal Combustion Engines, 1st Ed.* Pergamon Press.

Brown, W. L. (1967) Methods for Evaluating Requirements and Errors in Cylinder Pressure Measurement. *SAE Paper 670008.*

Brown, W. L. (1973) The Caterpillar imep Meter and Engine Friction. *SAE Paper 730150.*

Bueno, A. V.; Velásquez, J. A. & Milanez, L. F. (2009) A New Engine Indicating Measurement Procedure for Combustion Heat Release Analysis. *Applied Thermal Engineering*, Vol. 29, pp. 1657–1675.

Bueno, A. V. ; Velásquez, J. A. & Milanez, L. F. (2010) Heat Release and Engine Performance Effects of Soybean Oil Ethyl Ester Blending into Diesel Fuel. *Energy*, Vol. 36, pp. 3907-3916.

Bueno, A. V., Velásquez, J. A. & Milanez, L. F. (2011) Notes on 'A Methodology for Combustion Detection in Diesel Engines Through In-cylinder Pressure Derivative Signal'. *Mechanical Systems and Signal Processing*, Vol. 25, pp. 3209-3210.

Ding, Y. ; Stapersmal, D. ; Knoll, H. & Grimmelius, H. T. (2011) A New Method to Smooth the In-cylinder Pressure Signal for Combustion Analysis in Diesel Engines. *Proceedings of the Institution of Mechanical Engineers, Part A: Journal of Power and Energy*, Vol. 225, pp. 309-318.

Draper, S. C. & Li, T. Y. (1949) New High-Speed Indicator of Strain-Gauge Type. *Journal of Aerospacial Science*, Vol. 16, pp. 593-610.

Ficher, R. V. & Macey, J. P. (1975) Digital Data Acquisition with Emphasis on Measuring Pressure Synchronously with Crank Angle. *SAE Paper 750028.*

Franco, S. (2001) *Design with Operational Amplifiers and Analog Integrated Circuits, 3rd Ed.* McGraw-Hill.

Hountalas, D. T. & Anestis, A. (1998) Effect of Pressure Transducer Position on Measured Cylinder Pressure Diagram of High Speed Diesel Engines. *Energy Conversion and Management*, Vol 39, pp. 589-607.

Kistler (1956). *SLM Pressure Indicator*, DEMA Bull. 14.

Krieger, R. B. & Borman, G. L. (1966) The Computation of Apparent Heat Release for Internal Combustion Engines, *ASME paper 66-WA/DGP-4.*

Lancaster, D. R. ; Krieger, R. B. & Lienesch, J. H. (1975) Measurement and Analysis of Engine Pressure Data. *SAE Paper 750026.*

Lapuerta, M. ; Armas, O. & Bermúdez, V. (2000) Sensitivity of Diesel Engine Thermodynamic Cycle Calculation to Measurement Errors and Estimated Parameters. *Applied Thermal Engineering,* Vol. 20, pp. 843-861.

Lomb, N. R. (1976) Least-Squares Frequency Analysis of Unequally Spaced Data. *Astrophysics and Space Science,* Vol. 39, pp. 447-462.

Marzouk, M. & Watson, N. (1976) Some Problems in Diesel Engine Research with Special Reference to Computer Control and Data Acquisition. *Proceedings of the Institution of Mechanical Engineers,* Vol. 190, pp. 137-151.

Maunoury, B. ; Duverger, T. ; Mokaddem, K. & Lacas, F. (2002) Phenomenological Analysis of Injection, Auto-ignition and Combustion in a Small D.I. Diesel Engine. *SAE Paper 2002-01-1161.*

Payri, F. ; Olmeda, P. ; Guardiola, C. & Martín, J. (2011) Adaptive Determination of Cut-off Frequencies for Filtering the In-cylinder Pressure in Diesel Engines Combustion Analysis. *Applied Thermal Engineering,* Vol. 31, pp. 2869-2876.

Pinchon, P. (1984) Calage Thermodynamique du Point Mort Haut des Moteurs à Piston. *Revue de l'institut du Pétrole,* Vol 39(1).

Pischinger, R. & Glaser, J. (1985) Problems of Pressure Indication in Internal Combustion Engines. *Proceedings of COMODIA 1985,* Tokyo, Japan.

Randolph, A. L. (1990) Methods of Processing Cylinder-Pressure Transducer Signals to Maximize Data Accuracy. *SAE Paper 900170.*

Roth, K. J. ; Sobiesiak, A. ; Robertson, L. & Yates, S. (2001) In-Cylinder Pressure Measurements with Optical Fiber and Piezoelectric Transducers. *SAE Paper 2002-01-0745.*

Schmillen, K. & Schneider, M. (1985) Combustion Chamber Pressure Oscillations as a Source of Diesel Engine Noise. *Proceedings of COMODIA 1985,* Tokyo, Japan.

Stas, M. J. (1996) Thermodynamic Determination of T.D.C. in Piston Combustion Engines. *SAE Paper 960610.*

Strahle, W. C. (1977) Combustion Randomness and Diesel Engine Noise: Theory and Initial Experiments. *Combustion and Flame,* Vol. 28, pp. 279-290.

Strahle, W. C. ; Handley, J. C. & Varma, M. S. (1977a) Cetane Rating and Load Effects on Combustion Noise in Diesel Engines. *Combustion Science and Technology,* Vol. 17, pp. 51-61, 1977.

Strahle, W. C. ; Muthukrishnan, M. & Handley, J. C. (1977b) Turbulent Combustion and Diesel Engine Noise. *Proceedings of the Combustion Institute,* Vol. 16, pp. 337-346.

Teukolsky, S. A. ; Vetterling, W. T. & Flannery, B. P. (1966) *Numerical Recipes in C: The Art of Scientific Computing.* William H Press.

Wiebe, J. J. (1970) *Brennverlauf und Kreisprozes von Verbrennungsmotoren,* VEB-Verlag Technik Berlin, Germany.

Woschni, G. (1965) Computer Programs to Determine the Relationship between Pressure, Flow, Heat Release and Thermal Load in Diesel Engines. *SAE Paper 650450.*

Zhong, L. ; Henein, N. A. & Bryzik, W. (2004) Effect of Smoothing the Pressure Trace on the Interpretation of Experimental data for Combustion in Diesel Engines. *SAE Paper 2004-01-0931.*

4

Experimental System for Determining the Magnetic Losses of Super Paramagnetic Materials; Planning, Realization and Testing

Miloš Beković and Anton Hamler
University of Maribor, Faculty of Electrical Engineering and Computer Science, Slovenia

1. Introduction

In recent decades we have witnessed the development of nanotechnology throughout all fields of science including the field of magnetic materials. When it comes to nanotechnology is usual for at least one of the dimensions to be in nm scale and therefore the nano-materials or nano-equipment exhibit certain unique characteristics. The area of magnetic materials also followed the developments within this field, with the focus on magnetic nanoparticles, which are typically super-paramagnetic. Within the context of nanotechnology, it is necessary to observe particles in comparison with solid magnetic materials because the unique characteristics enable some completely new applications.

Within this section, the magnetic fluid is exposed, as a representative of nano-materials, because of its unique features of super-paramagnetic properties and liquid state. The merger of these two properties is enabled because of the magnetic nanoparticles, as discussed in details in the second chapter. At this point several technical and biomedical applications can be listed and the benefits arising from the unique properties of magnetic fluids. Example applications in engineering mechanics regarding rotary sealing have been described by the authors (Jibin et al., 1992), whilst the biomedical use of these materials enabled new methods in medical diagnostics and therapeutic methods (Kumar, 2009). In this context the method of medical hyperthermia has been introduced, which is becoming a promising alternative treatment for cancerous tissues in contrast to the use of conventional methods (Barba et al., 2010; Pavel & Stancu, 2009; Pollert, et al., 2007). The aforementioned method exploits the phenomenon of warming ferrofluids within alternating magnetic field with the aim of the thermally destroying cancer cells. This paper presents a physical mechanism responsible for the heating of magnetic fluid within a magnetic field, as well as the realization of an experimental system heating evaluation.

If modern medicine actually succeeds its usage of magnetic fluid for hyperthermia treatment then one of the important parameters is certainly its heating power. In reality, this is about the magnetic losses of magnetic materials in alternating magnetic field but in the context of a heating element for application within medical hyperthermia regarding heating power. Therefore, the precise characterization of magnetic losses is crucial and due to the specific properties of this material differing from conventional approaches, for determining any losses. This requires a measurement system that can determine these losses, this being the essence of this article. The following shows the design of such a system and its realization, as well as the actual testing on a magnetic fluid sample.

2. Magnetic losses

Magnetic fluids consist of stable colloidal dispersions of magnetic nanoparticles within a carrier liquid, where the Brownian motion keeps the particles permanently suspended, whilst adsorbed long-chain species on the particle's surfaces prevent their agglomeration (Rosensweig, 1997). Their physical properties are governed by the usage of carrier liquid whilst the magnetic properties are dependent on the types, sizes and concentrations of magnetic nanoparticles.

If such a fluid is exposed to an alternating magnetic field, its magnetic particles tend to align in the direction of the magnetic field, also known as magnetic relaxation. Each particle alignment is accompanied by energy consumption from the magnetic field, which is outwardly manifested as magnetic losses. Since these fluids usually contain a great number of particles and the rotations are frequent, as in the case of high-frequency fields, then the operation of the dissipation mechanisms elevates the temperature of the material. On the other hand, the amount of heat released within one magnetization cycle is equal to the area of the hysteresis loop (Carrey, et al., 2011, Beković, 2010).

These two properties of the fluid within the alternating magnetic field serve as the basis for determining of its heating power. The calorimetric method, where the determination of losses is based on the elevated temperature according to the equation below, is where the specific absorption rate (SAR) is determined as

$$\mathrm{SAR} = \frac{\rho C_{\mathrm{s}}}{m_{\mathrm{Fe}}}\left(\frac{\Delta T}{\Delta t}\right)_{t=0}, \tag{1}$$

where ρ is the density of the fluid, C_s is the sample specific heat-capacity, m_{Fe} is the total content of the maghemite ($g_{\mathrm{Fe}}/cm^3_{\mathrm{sample}}$), and $\Delta T/\Delta t$ is the temperature change, respectively the initial slope of the measured temperature rise curve. In order to determine any loss by this method, in addition to the constant parameters of the liquid, measured temperature response is required, which varies depending on the amplitude and frequency of applied magnetic field.

The second method provides specific heating power (SHP) or magnetic losses P_{h} by determining the surface of the BH hysteresis loop, as given by the equation

$$\mathrm{SHP} = \frac{f}{\rho}\int_0^T H(t)\frac{\mathrm{d}B(t)}{\mathrm{d}t}\mathrm{d}t\,, \tag{2}$$

where the $H(t)$ is the magnetic field strength, $B(t)$ is the measured magnetic induction and f the frequency of of magnetic field. Both these constants can be used for calculating the specific heating power p in W/mm³ according to (3), where φ is mass density of the magnetic particles.

$$p = \mathrm{SAR}\rho\varphi = \mathrm{SHP}\rho \tag{3}$$

Both presented approaches provide an equivalent loss, as demonstrated in (3), whilst equations (1) and (2) serve as the base for designing a measurement system.

3. Experimental system

The basic idea for designing this experimental system was determining the magnetic losses of the magnetic fluid exposed to the magnetic field. The primary need for the realization of such a system is the generation of alternating magnetic fields with the possibility of changing the amplitude and frequencies of the magnetic fields, and the container for the measured sample. The specific needs of a measuring system for determining losses using the chosen method where determined according to equations (1) and (2). In the case of calorimetric methods, the temperature of the sample must be measured in order to determine the derivative ($\Delta T/\Delta t$), whilst in case of the magnetic measurements method the time-dependent magnetic field strength and magnetic induction must be evaluate in order to determine the integral in (2). Other constants appearing in equations (1) and (2) are within the domain of the fluid's manufacturer of and are usually known.

3.1 The planning

The construction of the system, the design of which can be seen in the figure below (Fig. 1), was compiled from the results carefully planning its electrical and thermal properties. The figure only shows the basic elements of the system, but thereafter it is represented by the result of analyse which has resulted in a decision in the design and selected materials.

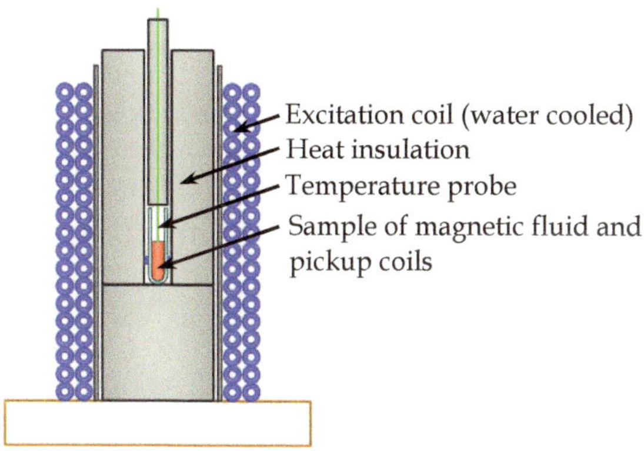

Fig. 1. The scheme of the key elements of the measurement system for the characterization of magnetic fluid heating power.

The measurement system was incorporated within a measurement scheme as seen in Fig. 2. which shows it's crucial elements. It is evident that a function generator is used for power amplifier control, by generating a sinus waveform of various amplitudes and frequencies. The power amplifier (Amplifier research) supplies an LC circuit, where L is the inductance of the excitation coil, and C serves to create resonance conditions at a selected frequency. The temperature measurements are performed using two sensors, fibre optic and resistance. The measurements are performed using an oscilloscope, whilst control and data acquisition is done via the *LabView* software package; communication being performed via standard GPIB and RS 232 busses.

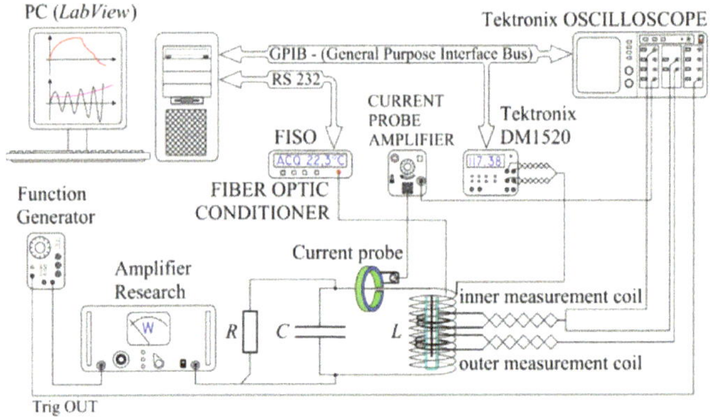

Fig. 2. Measurement scheme.

3.1.1 Magnetic field properties

The authors of some articles on the topic of magnetic fluid-loss characterization use those measurement systems where the measured sample is exposed to locally-different values of the magnetic fields, which means that losses are defined as the average value of a given field. In order to avoid this anomaly, the homogeneity of the field must be ensured; it can be obtained via the relatively long axial-length of the excitation coil.

This study tested different lengths or coils by increasing the number of coil turns, as show on the left side of Fig. 3. Turns of the coil were added, as seen, until the desired field deviations of less than 0.5% for amplitude H had been achieved within the area housing the sample magnetic fluid. The finite-element method (FEM) software *Vector Fields OPERA* was used and a static FEM analysis was performed for each number of turns.

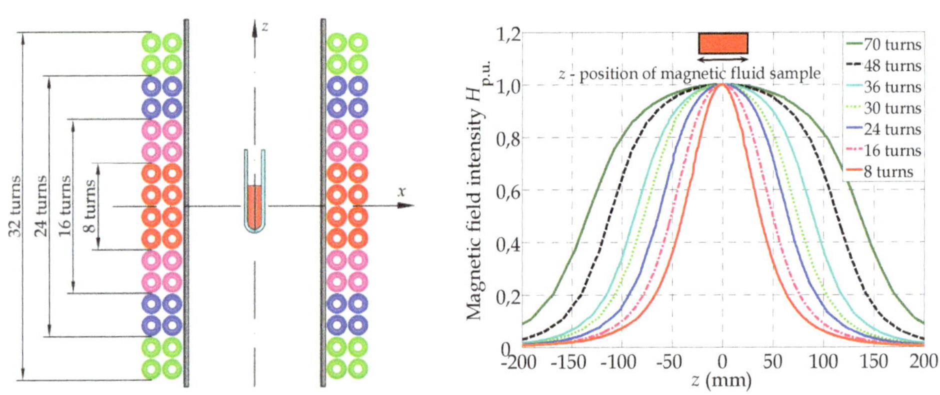

Fig. 3. Left side shows the principle of adding image excitation coil turns, right side-the calculated results of magnetic field strength along the z-axis for different numbers of turns.

Partial results for the FEM analyse for the seven cases of different excitation coil turns are presented at the right side of Fig. 3, showing the amplitudes of the magnetic field strengths along the *z*-axis. It can be seen that, in the case of the least coil turns, *H* has a normal distribution. Thus it means that the sample was not exposed to the same magnetic field, both in terms of the direction and the amplitude. In this case, the deviation from the maximum value was up to 25 % less, so it was necessary to add the coil-turns in order to achieve homogeneity of the magnetic field. The graph shows the result of increasing number of turns where it can be seen that adding field-coil turns reduces the deviation of the magnetic field within the position of the measured sample. A deviation of 0 %, would, in theory be achieved with an endless coil, therefore, a tolerance less than 0.5 % in the *z*-direction was acceptable, which was achieved with 70 turns, whilst the discrepancies in the *x* and *y*-directions were negligible.

Fig. 4 shows three examples of magnetic FEM analysis, which show the magnetic field strengths at the *zx*-plate in the centre of the measurement system. These results show that, in the case of a smaller number of turns, the magnetic field in the region of the sample is clearly non-homogeneous; in addition to archive the same value of *H* higher values of excitation current is required and it is limited by the amplifier.

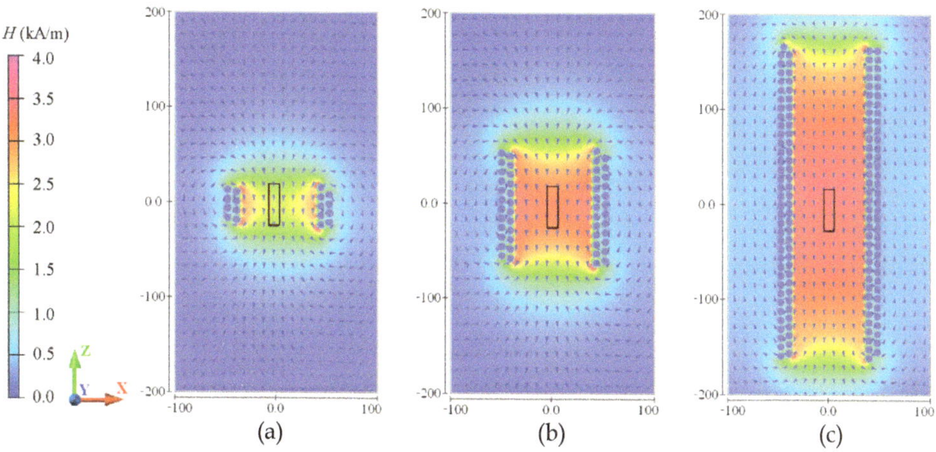

Fig. 4. Results of FEM analysis; magnetic field intensity within the measurement system for different numbers of excitation coil turns; a) 8 turns, b) 24 turns and c) 70 turns.

3.1.2 Thermal field properties

Besides magnetic calculation during the design process of the measurement system, a thermal analysis was preformed to ensure heat insulation within the system and the temperature stability of the sample in case of outside temperature disturbances. Several papers on magnetic fluid heating power characterization have not focused on the thermal isolation of their systems, which could result in improper temperature measurements; essential in calorimetric method. The resulting temperature increase of a sample can also be the consequence of heat transfer from the heated excitation coil. Therefore, sufficient thermal insulation must be included within the system, in such manner that any generated temperature increase is merely a result of the desired heating mechanism, which is the power-loss of the magnetic fluid sample.

For this part of the paper, an excitation coil with 70 turns was selected and thermal analysis for different examples performed. Fig. 5 displays part of an *xz*-cross-section of the measurement system, and the used materials. It also displays the distances or thicknesses of the materials. The first influence looked at was the impact of warming the excitation coil on the temperature of the magnetic fluid sample. The heating of the excitation coil was taken into consideration due to the Joule losses and magnetic fluid for heating caused by the aforementioned heating mechanisms. The maximum coil-current was set as given by the maximum output of power supply. By knowledge of the coil resistance, the maximum loss of excitation coils was determined together with heating power of the measuring sample. The second influence examined was the influence of the cooling water within the measuring system, since the plan was to produce a water-cooled coil. Final influence observed was the presence of a glass vacuum tube, as visible in Fig. 5 and marked within area-I and the presence of polystyrene as an insulating material on the same figure marked within area-II. Four different examples of the FEM analysis are collected on the Table. I. For this research, none of the dimensions where changed, as marked in Fig . 5 but only the presence of the materials and the water cooling of excitation coil.

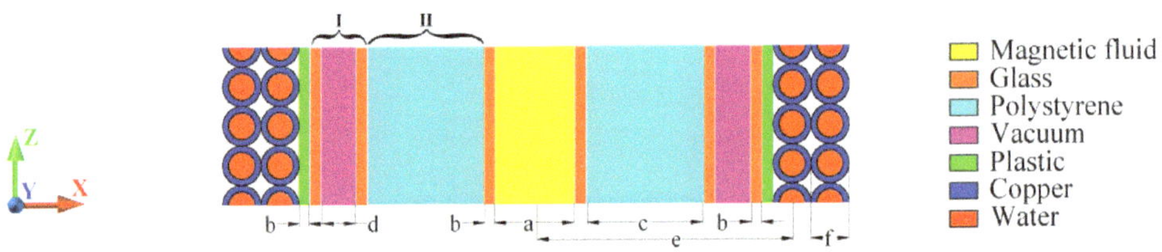

Fig. 5. The realized measurement system on the left and 3D model for thermal FEM analysis on the right.

Ex.	Heat: Mag. fluid (W/mm³)	Heat: Excit. coil (W/mm³)	Cooling Water T (°C)	Polystyrene	Vacuum tube	results: T_{max} (°C)	results: T_{avg} (°C)
A	1e-4	1e-7	no	no	no	152	147
B	1e-4	1e-7	no	no	yes	148	139
C	1e-4	1e-7	no	yes	yes	129	121
D	1e-4	1e-7	yes	yes	yes	110	108

Table 1. Examples of thermal analyses.

The result for one thermal analysis for example C can be seen in Fig. 6, which shows a) the temperature distribution in the measuring system on the xz-cross-section, b) the temperature derivative and c) the heat flux.

The results for all four FEM examples where the calculated temperature field distributions throughout the volume; in this case, the most interesting were displayed by the variables along the x-axis. The straight line was thus at the height $z = 0$ and x between 60 and - 60, as demonstrated in Fig. 4.

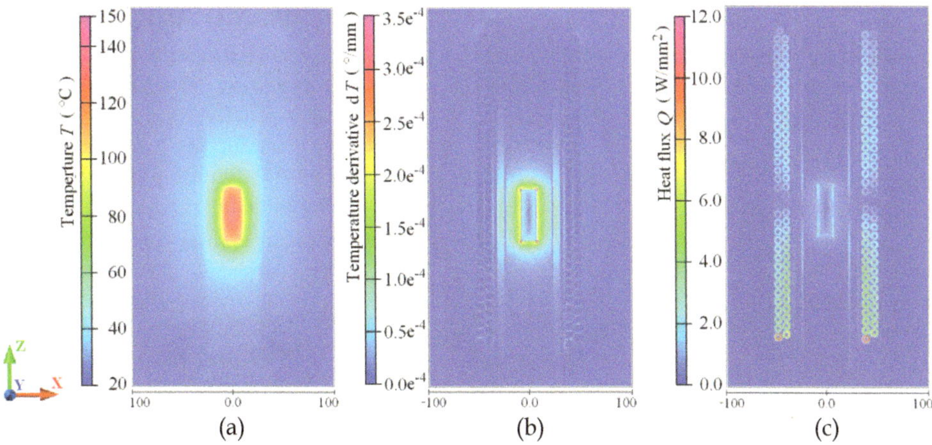

Fig. 6. Results of FEM analysis; a) the temperature distribution in the system
b) the temperature derivative and c) the heat flux within the system; results obtained for ex- C.

Figure 7 shows the results for temperature evaluated along the aforementioned line, for all four cases. It can be seen that the sample's temperature was generally higher; less of the insulation was included in the calculation. Notably in the last case, when the cooling water with the same temperature as the surroundings is included, the sample fluid is heated only because of attributed heating power, hence the lowest temperature. The impact of the ambient was minimized. When observing in the course of the heat flux (right picture) for all four cases along the same line it appeared, that in the position $40 < z < 50$, where the coil was located, there was a source of heat-flow directed to the outside and inside of the system, thus causing unnecessary heating of the sample. When a layer of insulation was added between the coil and the sample, heating was reduced (ex. C).

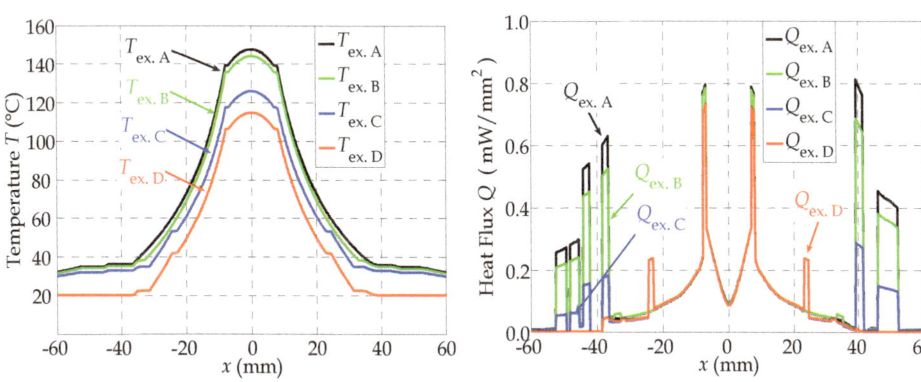

Fig. 7. Results of FEM analysis; temperature along the x-axis for examples A, B, C and D and the right picture, the heat flux along the x-axis for the same examples.

3.2 The realization

The result of this planning was a realization of the measurement system shown at the left of Fig. 8. The right picture is a 3D model and reveals more details, since it is partly cut out. This model was also used for all the FEM calculations presented in the previous subchapter.

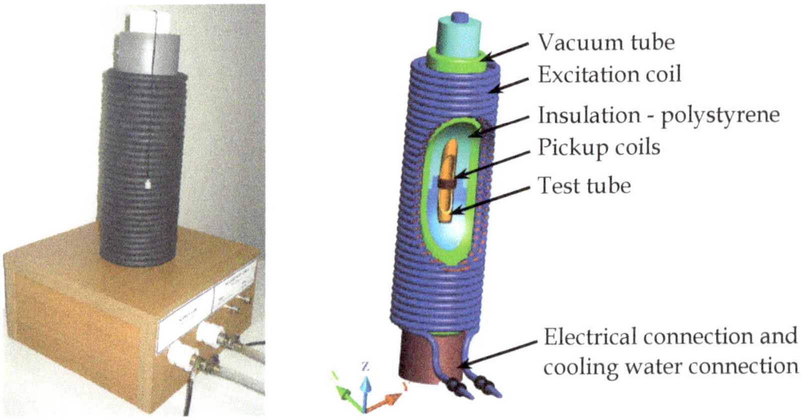

Fig. 8. The realized measurement system left side and 3D model for thermal FEM analysis right side.

3.3 Measurements of magnetic field parameters

So far the focus has been on the thermal insulation of the measurement system for better implementation when characterizing fluids, according to calorimetric method (3), whilst these sections have shown the principle of magnetic field variables measurement when determining the losses based on the determination of the hysteresis loop's surface (2). From the equation, it is clear that it is necessary to measure the magnetic field strength $H(t)$ and magnetic flux density $B(t)$. For this measurement a system of three measuring coils has been created, which bond to each other, as shown in Fig. 9. The top view displays the magnetic fluid sample, the middle shows around the sample containing a glass test tube for sample storage, and further outwards the two pickup coils for measuring the induced voltages e_1 and e_2. These two voltages are used in determination of magnetic field strength and magnetic flux density according to (5) and (6).

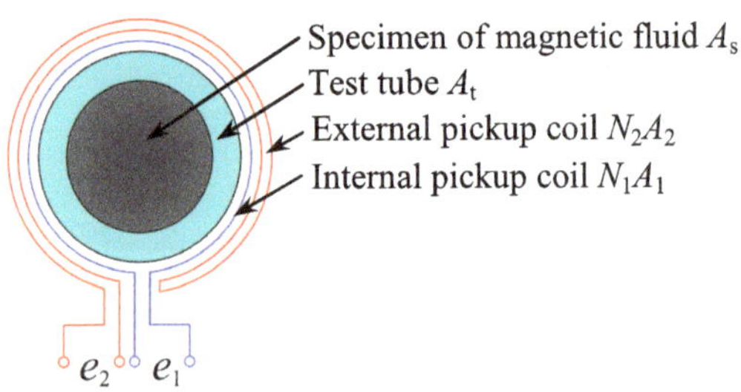

Fig. 9. Principle of measuring B(t) and H(t) by means of two pickup coils.

First let's look at the magnetic flux density $B(t)$, as defined by equation (4), where N_1 is the number of inner pickup coil turns, A_s is the cross-section of the sample, and ϕ_1 the magnetic flux in the sample.

$$B(t) = \frac{\phi_1}{N_1 A_s} \tag{4}$$

The inner measuring tube coil tightly embraces the vial so that the magnetic flux can be replaced by the integral of a measured induced voltage, as shown in (5).

$$B(t) = \frac{1}{N_1 A_s} \int e_1(t)\, dt \tag{5}$$

Measurement of the magnetic field strength is a greater challenge, since the magnetic flux ϕ_2 should be measured close to the sample, whilst excluding sample's magnet flux. The external pickup coil marked in Fig. 9 is used for this measurement and it is evidently wound in such a manner that embraces the sample.

$$H(t) = \frac{1}{\mu_0} \frac{\phi_2}{N_2 A_2} = \frac{1}{\mu_0} \frac{1}{N_2 A_2} \int e_2(t)\, dt \tag{6}$$

Equation (6) contains N_2 and A_2 which are constants and represent the number of coil turns and its cross-section. It was determined experimentally in such a manner, that the magnetic flux linkages where measured whilst exposing the coil to the known density of the magnetic field.

4. Testing

This section verifies the operation of the measuring system on a sample of magnetic fluid. Firstly, the calorimetric method is described and then those methods based on the magnetic field quantities measurements.

The magnetic fluid sample used in this work is commercially available; therefore some tests were performed to determine the basic structural properties of the fluid. Chemical analysis revealed particles of maghemite γ-Fe_2O_3 dispersed in mineral oil. Transmission electron microscopy revealed the mean diameter of particles to be 10.9 nm. Magnetic measurement of its magnetization curve was performed and modified according to the Langevine theory of paramagnetism (Pshenichnikov) the magnetization of particles, close to saturation, is described by

$$M = M_s \left(1 - \frac{k_B T}{\mu_0 M_{s\,\text{bulk}} \pi / 6 d_{core}^3} \frac{1}{H} \right), \tag{7}$$

The volume fraction of the magnetic particles in the used sample of magnetic fluid was determined from a ration of M_s/M_s bulk, where $M_{s\,\text{bulk}}$ was 400 kA/m and it is 10.57 %.

4.1 Magnetic fluid characterization SAR

Calorimetric measurements were carried out using a calorimeter, which is an instrument that allows the heat-effects within it to be determined by direct measurement of temperature, as proposed in (1). The sample of magnetic fluid was exposed to an ac magnetic field. The measurement principle for the calorimetric measurements required a stable temperature of the sample, before the magnetic field could be switched on. For this reason, the measurement could only begin when the temperature of the sample was constant, usually at the value of cooling water. When this condition had been fulfilled, the excitation coil was turned on and constant frequency and regulated current set up the magnetic field at the desirable value. The measurement lasted until the temperature steady state had been reached or if the other heat production equalled the heat dissipation.

Several tests, for an 8 ml sample of magnetic fluid, of different magnetic field strengths were performed and the results for temperature rise are presented in Fig. 10. The measurements were performed at a frequency of 400 kHz and constant field amplitudes, that were regulated to achieve a discrepancy of ± 10 A/m around the selected values.

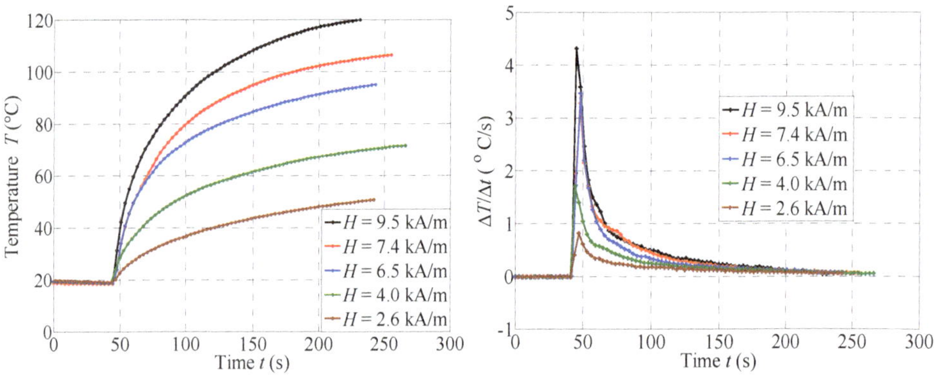

Fig. 10. Calorimetric measurements of temperature rise for different values of magnetic field strengths at frequency of the supply field of 400 kHz, and the temperature derivatives.

This approach is not new regarding magnetic fluid determination, but it serves as a verification method for determining losses through hysteresis loops or through complex susceptibility, as presented below.

4.2 Magnetic fluid characterization SPL

This section explores the methods for determining the losses of magnetic fluids at constant temperature. For this, equation (3) was used to calculate the losses from the area of the hysteresis loop and, in the second case, out of a complex susceptibility, as explained later, both based on the measurement of time courses $H(t)$ from equation (6) and $B(t)$ from (5). In the analysis, magnetic polarization was used instead of magnetic flux density, where

$$B = J + \mu_0 H \,. \tag{8}$$

Another approach for determining losses is based on (9) according to Carrey, where the parameter of the complex susceptibility χ'' represents the essential component

$$P_\chi = \frac{1}{\rho} \pi \mu_0 H^2 \chi'' f \,. \tag{9}$$

This section shows the equivalences of these methods for the sampling the magnetic fluid and point out the advantages of these methods against the calorimetric method. In this case, the characterization losses are executed under such conditions as would be applied during actual treatment with hyperthermia.

The relationship between relative permeability and susceptibility was determined using this expression

$$\mu_r = \frac{B}{\mu_0 H} = \frac{J + \mu_0 H}{\mu_0 H} = \frac{J}{\mu_0 H} + 1 = \chi + 1 . \quad (10)$$

The responses of the material to the applied magnetic field were studied using the complex susceptibility approach. In this case the susceptibility was treated as a real χ' component, which is within the phase with the magnetic field and imaginary component χ'' that lags behind the 90°, the latter is also related to the energy losses or absorbed energy by the sample from the ac field respectively. The terms for both were written (11), where δ was the angle between the measured signals H(t) and $J(t)$ or $B(t)$

$$\chi = \frac{J}{\mu_0 H} \quad \Rightarrow \quad \chi' = \frac{\hat{J}}{\mu_0 \hat{H}} \cos\delta \, , \, \chi'' = \frac{\hat{J}}{\mu_0 \hat{H}} \sin\delta \, , \quad (11)$$

The characterization was carried out in such a manner that different values for power amplifier currents were set, and thus the magnetic field strength whilst measuring the voltages of the two pickup coils. Such an example occurs in the lower two graphs showing the courses of the induced voltages e_1 and e_2 for the change of gain over seventeen steps.

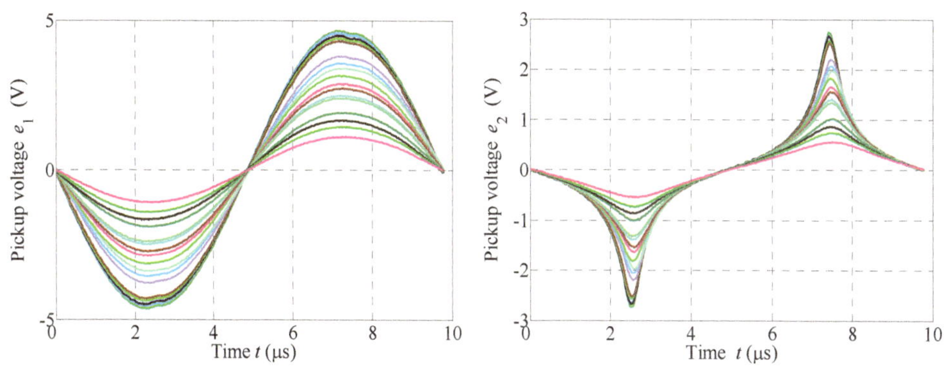

Fig. 11. Measured induced voltages e_1 and e_2 for various examples of magnetic fields between 5 and 15 kA/m.

Here it should be noted that the measurements must be carried out within a very short period of time, since a long exposure of the sample to the high value of the field quickly rises the temperature, as is clear from the heat curves in Fig. 10. Magnetic fluid at higher temperatures behaves differently, since it's the closer to Curie's temperature which also decreases its heating power because of the narrower hysteresis loop. From the measured induced voltage the desired variables $H(t)$ and $B(t)$ can be calculated, which are plotted in Fig 12 for the same example.

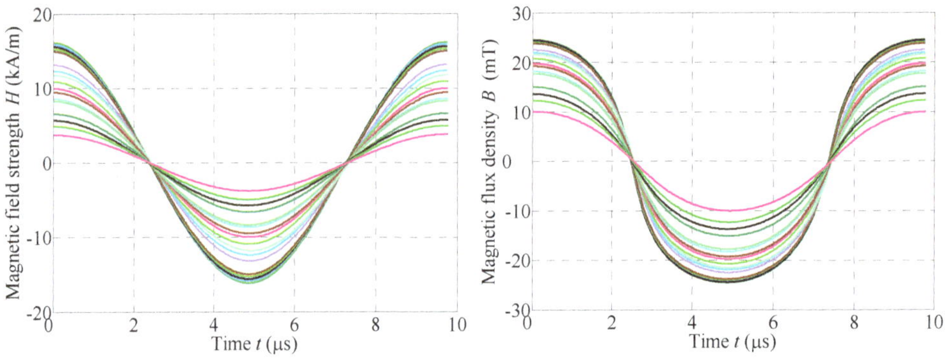

Fig. 12. The time courses of the calculated values $H(t)$ and $B(t)$ for the same example.

The measured signals provide the determination of losses using both methods, namely by determining the hysteresis loop area using equation (2) or through the determination of the complex susceptibility and losses calculated by equation (9). Fig. 13 below shows the two approaches, the left images are the outlined *JH* hysteresis loops for determining the losses by the first method. Determination of the surface hysteresis is no novelty, but an interesting approach for the evaluation of the dynamic hysteresis loop areas of super paramagnetic materials, and the determination of loss by this method.

The right side figure shows the magnetic filed $\mu_0 H$ box and the response of materials to the field of magnetic polarization $J(t)$ (you could also use $B(t)$). The goal was to determine the phase shifts of signals δ to be able to determine the χ'' according to (11).

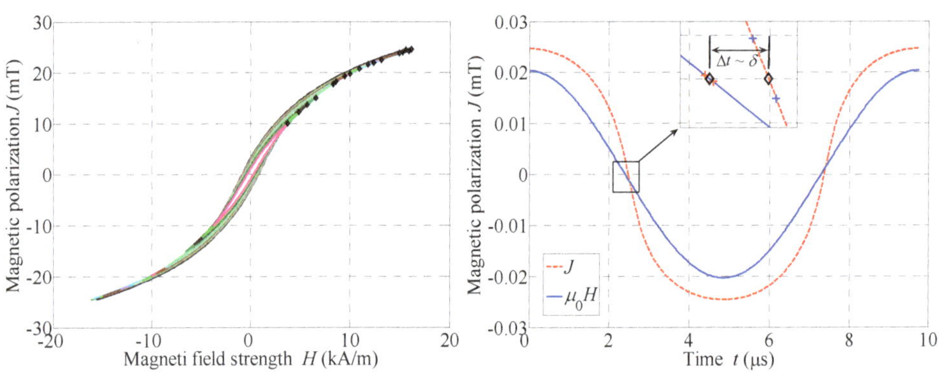

Fig. 13. Hysteresis loop and the time course of signals for determining magnetic losses.

If the plot of the phase shifts in delta is now done for all seventeen cases, it is found that only the latter is not constant but decreases linearly, as shown in Fig. 14. According to the formula for the losses determined from a complex susceptibility P_χ (9) the desired characteristic $P = f(H)$ can be calculated, which is for example of the 100 kHz frequency given in the same graph.

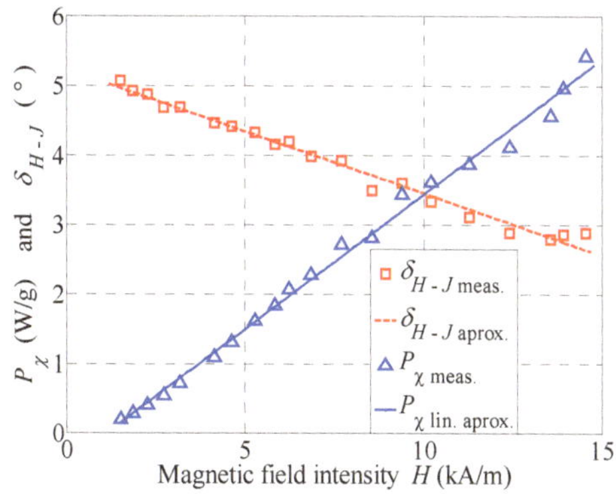

Fig. 14. The result of the phase shift δ measurements for different magnetic field strengths and associated loss of magnetic fluid sample at 100 kHz.

When the losses of P_h were plotted in the same graph as the loss of P_χ it was evident, that both analyses of the measured signals offered the same results. The results are summarized in Fig. 15 for four different frequencies of the applied field and it was discovered that the losses expectedly increased when increasing the frequencies.

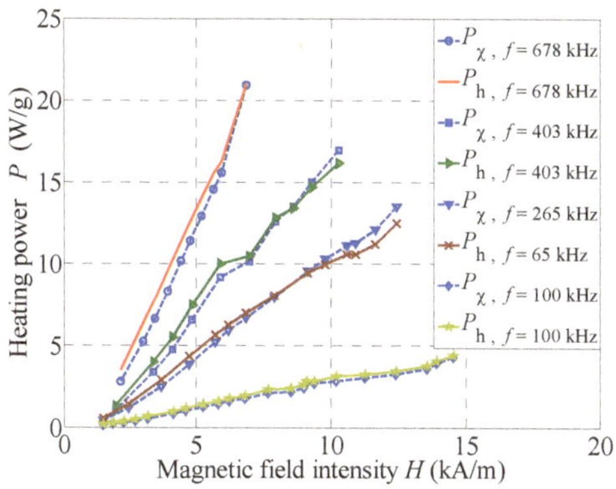

Fig. 15. A comparison of heating powers P_χ and P_h for four different frequencies, and the amplitudes of magnetic field intensity.

5. Conclusion

In conclusion we can point out specific guidelines that apply to measuring systems for determining the losses of magnetic fluids. In this context, the losses are a desired property, because it means greater heat output and thus higher quality material for usage in medical hyperthermia.

In order to build such a measurement system it is necessary firstly to provide a source magnetic field that must be able to change both the amplitude and frequency of the magnetic field. In achieving this objective, we have tried to approximate as closely as possible those fields parameters that would actually be best suited for implementing this treatment.

Secondly, we need a measurement system which is sufficiently precise and will allow for a clear characterization of magnetic fluid losses. This article presents three methods, where each of them exposed their weakness, and the possibility of erroneous measurements.

In the calorimetric method, the emphasis is on adiabatic changes in temperature, at least at the start of the measurement, which determines the maximum slope of warming. For FEM analysis, we have demonstrated the importance of certain measures for improving the methodology.

The method of magnetic measurement emphasizes the accuracies of the measuring coil parameters, since only with precisely defined parameters can accurate measurements be performed. We here shown the equivalence of both methods, where we see that the determination of the hysteresis loop area and determination of the angle between the polarization and the magnetic field brings almost the same result for determining the losses, depending on the accuracies of certain constants.

6. References

Barba, P., Dughiero, F., & Seini, E. (2010). Magnetic field synthesis in the design of inductors for magnetic fluid hyperthermia. *IEEE Transaction on Magnetics*, Vol. 46, No. 8, pp. 2931-2943, 0018-9464.

Beković, M., & Hamler, A. (2010). Determination of the heating effect of magnetic fluid in alternating magnetic field. *IEEE Transaction on Magnetics*, Vol. 46, No.2, pp. 552-555, 0018-9464.

Carrey, J., Mehdaoui, B., & Respaud, M. (2011). Simple models for dynamic hysteresis loop calculations of magnetic single-domain nanoparticles: Application to magnetic hyperthermia optimization. *Journal of Applied Physics* 109, 083921-1-17, 0021-8979.

Jeong, J. R., Lee, S. J., Kim, J. D., & Shin, S. C. (2004), Magnetic properties of Fe_3O_4 nanoparticles encapsulated with poly (D,L Lactide-Co-Glycolide). *IEEE Transaction on Magnetics*, Vol. 40, No.4, pp. 3015-3017, 0018-9464.

Jibin, Z., & Yongping L., (1992). Numerical calculations for ferrofluid seals. *IEEE Transaction on Magnetics* vol. 28, No.6, pp. 3367 - 3371, 00189464.

Kim, D. H., Thai, Y. T., Nikels, D. E., & Brazel, C. S., (2009). Heating of aqueous dispersions containing MnFe2O4 nanoparticles by radio-frequency magnetic field induction, *IEEE Transaction on Magnetics*, Vol. 45, No.1, pp. 64-70, 0018-9464.

Kumar, C (Ed(s)). (2009). *Magnetic Nanoparticles*, nanomaterials for life science Volume 4, Wiley-vch Verlag GmbH&Co.

Pavel, M., & Stancu, A. (2009). Ferromagnetic nanoparticles dose based on tumor size in magnetic fluid hyperthermia cancer therapy. *IEEE Transaction on Magnetics*, Vol. 45, No.11, pp. 5251-5254, 0018-9464.

Pshenichnikov A.F., Mekhonoshin V.V., & Lebedev A.V., "Magneto-granulometric analysis of concentrated ferrocolloids", J. Magn. Magn. Mater. 161, pp. 94-102, 1996.

Pollert, E., Knižek, K., Maryško, M., Kašpar, P., Vasseur, S., & Duguet, E., (2007), New T_C-tuned magnetic nanoparticles for self-controlled hyperthermia. *Journal of Magnetism and Magnetic Materials*, 316, pp. 122-125, 0304-8853.

Rosensweig, R.E., "Ferrohydrodynamics", Dover publications, inc, Mineola New York 1997.

5

Study on Wireless Torque Measurement Using SAW Sensors

Chih-Jer Lin*, Chii Ruey Lin, Shen-Kai Yu, Guo-Xing Liu, Chih-Wei Hung and Hai-Pin Lin
National Taipei University of Technology, Taiwan, R.O.C.

1. Introduction

An acoustic wave is a vibration in an elastic medium that propagates in space and time, thus transferring the energy supplied by an excitation source along the medium in the form of oscillation or vibration. Acoustic wave propagation entails elastic deformation of the medium along the propagation axis or in other axes as well. In contrast to electromagnetic waves, acoustic waves do require a medium to propagate, and their propagation speeds depend on the mechanical properties of the wave-supporting material. Virtually any material is capable of supporting acoustic wave propagation, including silicon. Nevertheless, the piezoelectric properties of certain materials facilitate the wave propagation. Thus, for improving the electromechanical energy conversion, piezoelectric materials are usually chosen as the acoustic layer of many acoustic-wave resonators. Also known as sound speed, the acoustic-wave phase velocities are several times slower than those of the electromagnetic wave traveling in the same medium(Auld,B. A.,1990). There exist two types of acoustic waves, surface acoustic waves (SAW) and bulk acoustic waves (BAW). A surface acoustic wave is a type of mechanical wave motion which travels along the surface of a solid material. As shown in Fig. 1, the surface particles of an isotropic solid move in ellipses in planes normal to the surface and parallel to the direction of the wave propagation. The particle displacement is significant at a depth of about one wavelength. This motion decreases at the surface at thinner depths and increases at greater depths. As the size of the ellipses is smaller, its eccentricity changes for particles in the deeper material. Surface acoustic waves were discovered in 1885 by Lord Rayleigh (Rayleigh,1885); therefore, they are often named after him: Rayleigh waves. Rayleigh showed that SAWs could explain one component of the seismic signal due to an earthquake, a phenomenon not previously understood. The velocity of acoustic waves is typically 3000 m/s, which is much lower than the velocity of the electromagnetic waves.

On the other hand, bulk acoustic waves are longitudinal, shear-mode, or combination of both. Longitudinal waves travel through the medium parallel to the same axis of the oscillations or vibrations of the particles in the medium; that is, in the same or opposite direction as the motion of the wave as shown in Fig. 2. Longitudinal mode waves are confined in a resonant cavity, thus displaying a particular standing-wave pattern. All other

*Corresponding Author

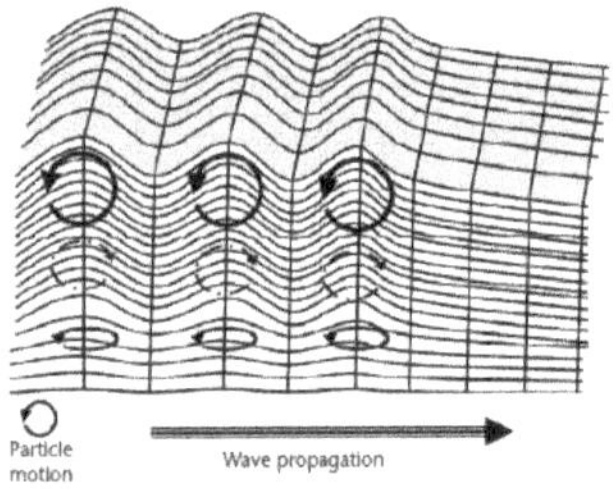

Fig. 1. Rayleigh wave propagation. (Auld ,1990).

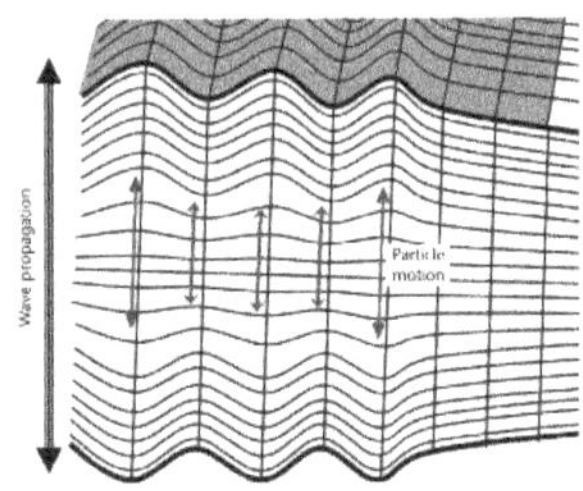

Fig. 2. The longitudinal-mode waves (Auld ,1990).

wavelengths experience destructive interference and are suppressed. While longitudinal modes have a pattern with their nodes located axially along the length of the cavity, transverse modes, with nodes located perpendicular to the axis of the cavity, may also exist. A transverse or shear-mode wave propagates and transfers its energy in the direction perpendicular to the oscillations occurring in the medium.

Although the existence of the Surface Acoustic Wave (SAW) was first discussed in 1885 by Lord Rayleigh, it did not receive engineering interest for a long time. The first SAW device was actually made in 1965 by White and Voltmer, who found out how to launch a SAW in a piezoelectric substrate by an electrical signal. White and Voltmer suggested that SAWs can be excited and detected efficiently by using an interdigital transducer (IDT) placed on a piezoelectric substrate(R.M. White and F.W. Voltmer,1965). Using IDT is a very convenient way on a piezoelectric substrate for excitation and detection of an acoustic wave, because very fine IDTs can be mass-produced by using photolithography, which has been well developed for semiconductor device fabrication, and proper design of the IDT enables the construction of transversal filters with outstanding performance. Starting around 1970, SAW devices were developed for band-pass filters, pulse compression radar and oscillators for TV sets and professional radio. Since 1980, the rise of mobile radio caused a dramatic increase in demand for filters, particularly for today's cellular telephones.

All acoustic wave devices and sensors use a piezoelectric material, such as quartz crystal, to generate the acoustic wave. Piezoelectricity, which was discovered by Pierre and Jacque Curie in 1880, means "pressure electricity" due to the Greek word "piezo" for pressure. The Curie brothers were able to demonstrate the generation of electric charge in response to applied pressure or stress; this is so called the *direct piezoelectric effect*, as shown in Fig. 3. Piezoelectricity is received its name in 1881 from Wilhelm Hankel, and remained largely a curiosity until 1921, when Walter Cady discovered the quartz resonator for stabilizing

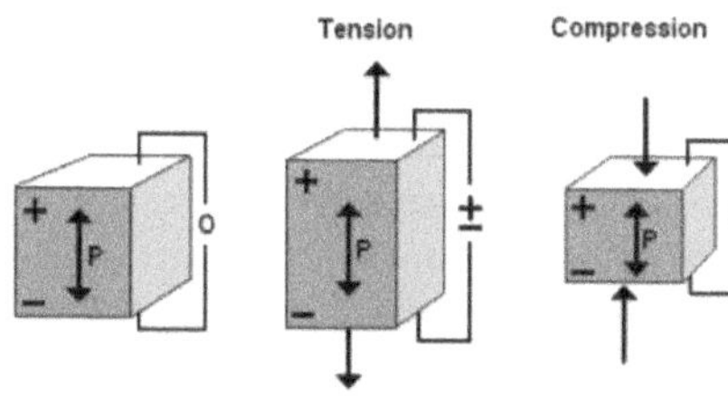

Fig. 3. The direct Piezoelectric Effect.

electronic oscillators. The piezoelectric effect has been understood as the linear electromechanical interaction between the mechanical and the electrical state in crystalline materials with no inversion symmetry (Gautschi, G., 2002). For precision positioning purpose, the *converse piezo effect* is usually applied to produce precise displacements in response to the applied electric field, as shown in Fig. 4. The phenomenon is reciprocal, but not linear with hysteresis.

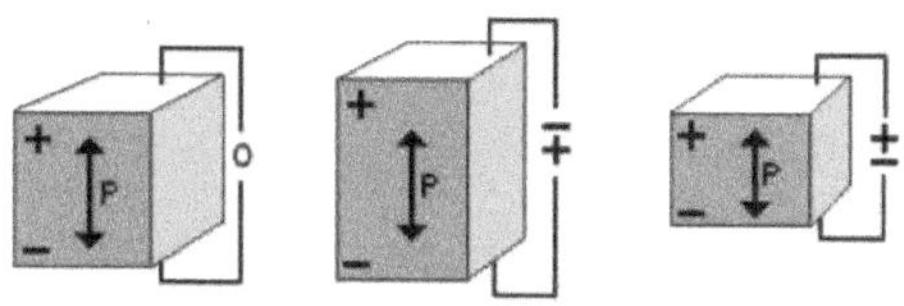

Fig. 4. The reverse Piezoelectric Effect.

Therefore, an oscillating electric field is used in piezoelectric acoustic wave sensors to create a mechanical wave, which propagates through the substrate and is then converted back to an electric field for measurement. Based on this principle, SAW devices have found diverse applications for measuring physical quantities such as temperature, pressure, torque, acceleration, tire-road friction, humidity, etc.. There are several methods to measure forces and torques. The force to be measured is often converted into a strain on a flexible element. The change of strain is subsequently measured by a sensor, for example, a piezoresistive, a capacitive or a resonant sensor. Unfortunately, the metallic resistance strain gauge is relatively insensitive such that the actual output voltage is usually only several mV of analog voltage before amplification and the gauges must not be significantly overstrained. The measurement range and overloading capabilities for strain gages are seriously restricted. In general, measurement instrumentation needs smaller sensing devices of lower power consumption and with large measurement range and overload capabilities. Digital microelectronics with greater compatibility is highly desirable. Noncontact and wireless operation is sometimes needed; in addition, batteryless devices are also necessary in some application cases.

SAW devices became important for sensor purposes in recent years. SAW sensors are applied as well as wired sensor elements in active circuits and as remote passive devices. The SAW sensors are passive elements, which do not need power supply, and can be accessed wirelessly, enabling remote monitoring in harsh environment. They can work in the frequency range of 10 MHz to several GHz. In addition, SAW sensors have the advantages such as compact structure, outstanding stability, high sensitivity, low cost, fast real time response, extremely small size (lightweight), and their fabrication is compatible with CMOS and micro-electro-mechanical (MEMS) integrated circuits technology. The SAW sensors are used for identification of moving object and parts (so called ID tags) and wireless measuring of temperature, pressure, stress, strain, torque, acceleration, tire-road friction, and humidity. The SAW sensors are well suited for measuring pressure in car and truck tires and tire-road friction monitoring. Their characteristics offer advantages over technologies such as capacitive and piezoresistive sensors, which require operating power and are not wireless. SAW has been used in electronics for many years, notably in quartz resonators which provide high Q-value as a result of the low acoustic losses. Exploiting the delay lines give a long delay in a small space with low acoustic velocities. SAW resonant sensors offer many benefits, such as improved sensitivity and accuracy, and reduced power consumption, among others. The resonance frequencies of the resonator-based sensors will be changed due to the action of the external excitation; therefore, we can use this characteristic of SAW device to detect the external force or torque via designing a interrogation (read-out) electronic circuit. Therefore, in this study we focus on designing an embedded system to measure the interrogation resonance frequency via a RF antenna set. In addition, this research's purpose is to establish a torque measurement based on a wireless SAW measurement system. Finally, we illustrate how to measure the torque wirelessly using the relation between the frequency shift and the torque, which is applied to a rotary rod.

2. Wirelessly torque measurement

2.1 Literature reviews for wireless torque measurement

A **torque sensor** or **torquemeter** is a device for measuring and recording the torque on a rotating system, such as a bicycle crank, rotor, gearbox, crankshaft, transmission, or an engine. The torque measurement can be separated into two categories, which are static torque and dynamic torque measurements. For static torque measure, using strain gage applied to a shaft or axle as torque sensors or torque transducers is the most common way. Static torque measurement is relatively easy with respect to dynamic measurement, because the measured shaft or axle is static. However, dynamic torque is not easy to measure because the shaft is rotating. Therefore, for a dynamic torque measuring system, the torque effect is generally transferred into some electric- or magnetic-type signals transmitting via wireless technologies from the rotary shaft being measured to the static system. Dynamic torquemeters is usually used to measure the torque being transmitted between the two machines which are connected. There are several varieties of torquemeter couplings, which are used for continuous on-line torque monitoring. Torquemeter designs are faced with the task of detecting a physical change due to torsion in the coupling while the shaft is rotating. Therefore, the dynamic torque measurement is necessary to perform through non-contacting means and get the torque information to a stationary output device via wireless device. Over the years, many methods have been devised to measure the torsional effects exhibited by the coupling; from the past literatures or present commercial products, we can find the following types of non-contact torque measurement, which are strain gage type torquemeters using twist angle or phase shift (torsional deflection) measurement, magnetoelastic torque sensors, and SAW torque sensors. We will make a literature survey as the following.

2.1.1 Strain gage integrated with a wireless transfer system

In the 1930's, the first in-line torque transducer was used on the liner Queen Mary, where employed the phase displacement principle to measure torque. It needed a length of 4 meters to adequately measure the angle of twist with around ±5% accuracy because of crude electronics. The modern torquemeters use the twist between a pair of toothed flanges, which is used to generate sinusoidal signals in magnetic pickups in the form of internally toothed rings and circumferential coils, to measure phase displacement. Fig. 5 describes the modern torque transducer using the phase displacement principle. Then, the resultant phase change of the two signals is measured by digital electronics to compute the applied torque. Phase displacement transducers are best suited to high speed and high temperature applications and they can achieve the accuracies of ±0.1%.

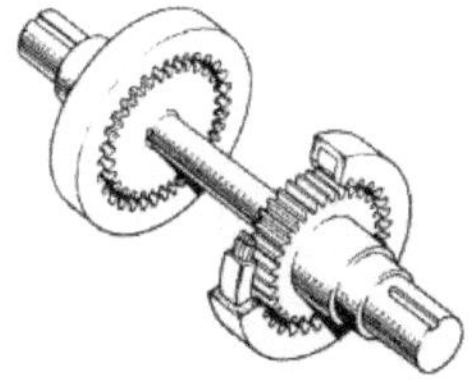

Fig. 5. The torquemeters using phase shift with 2 channels pick-up rings (Corcoran, joe and D'Ercole, Steve, 2000).

Different from the phase shift measurement, another way to measure torque uses the angle of twist resulting from the applied torque, which is measured using two angular position sensors and a differential transformer circuit. The measurement system consists of a differential transformer measuring system, two concentric cylinders fixed to a shaft either side of a torsion section and two concentric coils attached to the housing, as shown in Fig. 6. Both cylinders have circumferential slots, which rotate inside the coils. An alternating current flows through a primary coil and when the slots start to overlap due to shaft twist, a torque-proportional EMF is induced in a secondary coil. However, its performance is worse than the phase displacement transducer, but its cost is generally lower than the phase displacement transducer.

The strain gage is also called load cell, which is delivered by Load Delvin in 1856; the principle of the strain gage is using the relation between the resistance and the strain for metal materials. When the metal martial suffers pull or tension, its resistance will be increased due to the strain. Otherwise, the resistance decreases when the martial suffers pressure. According to the relation between the strain of the metal and the applid torque, the strain gages can be used to detect torque. However, the strain gage needs the electrical source to produce the variation of voltage for measurement so that the torque sensor using strain gage needs an external power supply. On one hand, a means to power the strain gauge bridge is necessary, as well as a means to receive the signal from the rotating shaft. On the other hand, the non-contact measurement is accomplished using slip rings, wireless telemetry, or rotary transformers. The first rotating strain gauge torque transducer employed a system of slip rings to make the electrical connections from the casing to the rotating shaft. For the slip ring torque transducers, there are two problems which should be noted. The first one is that the slip rings are carrying only millivolt signals from the strain

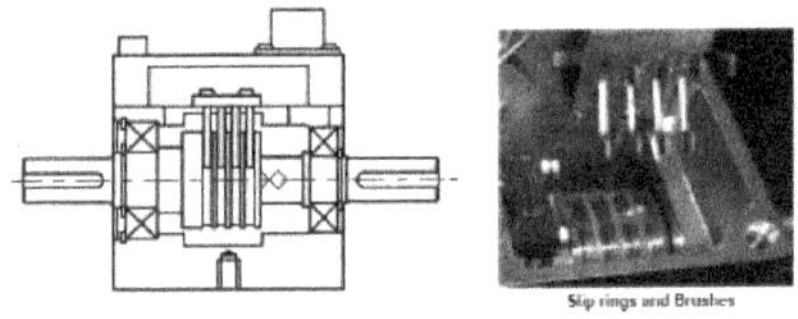

Fig. 6. The torquemeters using angle of twist resulting from the applied torque (David Schrand).

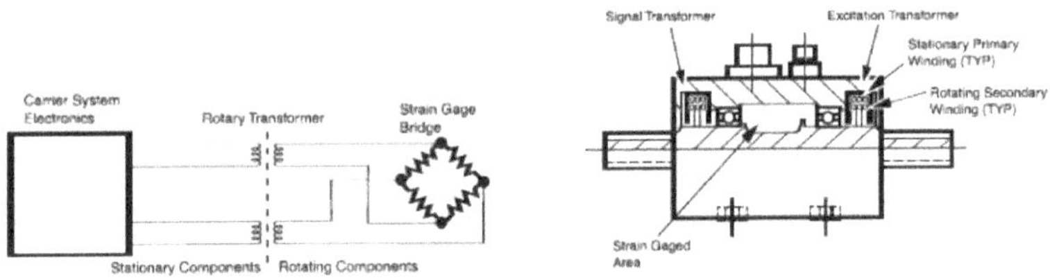

Fig. 7. The strain-gage type torquemeter with RTS (PCB load & torque).

gauges; therefore, the materials for both the slip rings and the brushes have to be carefully selected. The second is that slip ring torque transducers should be applied for slower speed and short-term test applications due to the lifetime of brushes. The rotary transformer system (RTS), which uses induction principles to transfer signal or power to the measurement system, became popular in the 1980's. RTS uses high-performance, high-temperature electronic components to be as a telemetry system but it only uses kHz carrier frequencies, not the MHz of radio systems. Several companies use the rotary transformer system with strain gauges, and with typical inaccuracies of ± 0.2%, and speeds up to 50,000rpm it can be a very cost-effective torque transducer.

For novel types of torque transducers, conditioning electronics and an A/D converter may be integrated into a microchip which is fixed to the rotating shaft; then, stator electronics read the digital signals and convert those signals to a high-level analog output signal. In this case, the wirelessly strain-gage type torque sensor, which needs a a bettery to support the power of the system, is not a passive device and it needs a wireless communication system to transfer the measurement data via RF. Wireless communication technologies are convent for remote control, such as ZigBee, Bluetooth or wirefire networks. ZigBee is a new communication technology in the wireless field and ZigBee comes from the bee which communicates with other bees about position of pollen via shape of dance. In the Fig.8 describes a measurement system using the strain gage integrated with Zigbee. The torque sensor integrated with ZigBee wireless communication sensor can detect the torque of shaft to prevent breaking due to the fact that the torque exceeds the material limit. The torque sensor systems with ZigBee involves of two parts, measuring unit (strain gages) and receiver unit (microprocessor). The measuring unit is used to measure torque from strain gages and transmit digital signal to receiver unit. Then, the MCU is used to perform A/D conversion and signal processing tasks. A simple MCU is enough perference for measuring task and transmit signals. After that the Zigbee chip transfers the digital signal into RF signal and transmits it to the receiver.

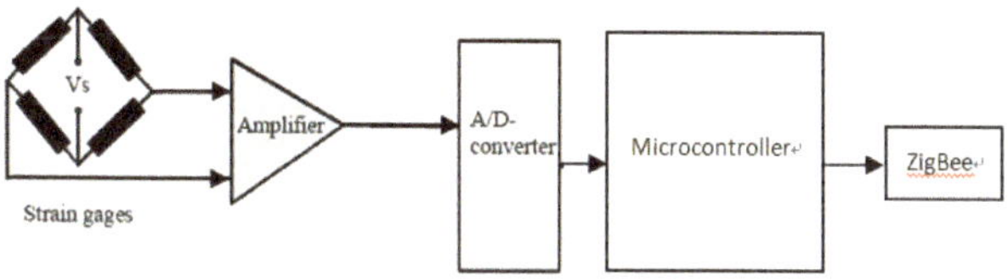

Fig. 8. The block diagram of ZigBee torque measuring system.

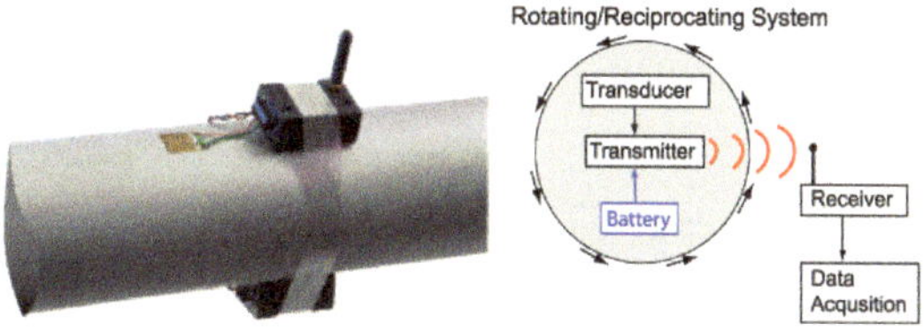

Fig. 9. The block diagram of wireless measuring unit (Binsfeld Engineering Inc.).

2.1.2 Magnetoelastic torque sensors

Besides of strain-gage type torquemeter, one way to achieve this non-contact measure uses the magnetic characteristics of the shaft with a series of permanent magnetic devices. The magnetic characteristics will vary according to the applied torque, and thus can be measured using magnetoelastic torque sensors, which are generally used for in-vehicle applications on racecars, automobiles, and aircraft. The United States Patent, the inventor Bunyer, Scott L(Bunyer, Scott L, 2007) in his patent " Magneto-Elastic Resonator Torque Sensor " , the torque sensor consists of a substrate and a magneto-elastic sensing component formed from or on the substrate. The magneto-elastic sensing component and the substrate together form a magneto-elastic torque sensor, which when subject to a stress associated with a torque, shifts a characteristic frequency thereof linearly in response to the torque, thereby inducing a pathway by which magneto-elastic energy is coupled to excite vibrations in a basal plane of the magneto-elastic sensor, thereby generating torque-based information based on resonator frequency thereof. Figs. 10-11 illustrate the torque sensing system using magneto-elastic resonator. A few years ago, a magneto-elastic device was developed in the USA, which uses a magneto-elastic sleeve fixed to a stainless steel shaft. This permanent magnet generates a magnetic field proportional to torque. The resultant magnetic field is less dense that the earth's field, hence internal shielding is needed for the hall-effect probes, which is fine for applications whose accuracy is not critical. The ABB Company proposed the magneto-elastic torque sensor, which is called Torductor®-S (ABB company, 2007). Since the sensor is part of the load-carrying shaft and Torductor®-S gives a true non-contact and rugged torque sensor without any moving parts. Therefore, the measured torque is the true transmitted torque, which enables Torductor®-S to combine high accuracy with high overload capacity and fast response at all times. A high output signal ensures integrity against electrical or magnetic interference from the surroundings. The magneto-strictive transducer, which is similar to the magneto-elastic torque sensor, is mechanically simpler than the magneto-elastic unit and relies on a pre-magnetized shaft proportionally changing its magnetic field when torque is applied. The magneto-strictive transducer is a low-cost transducer; however, zero drift with time and temperature and the effect of adjacent magnetic fields can be a problem for. These two devices can provide an accuracy of around ±1%.

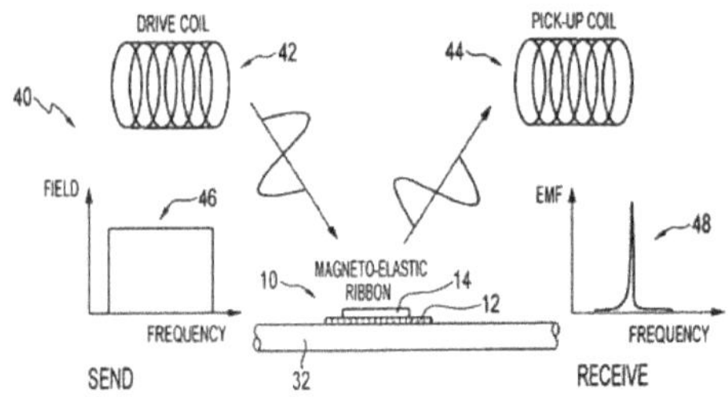

Fig. 10. The torque sensing system using magneto-elastic resonator (Bunyer, Scott L, 2007).

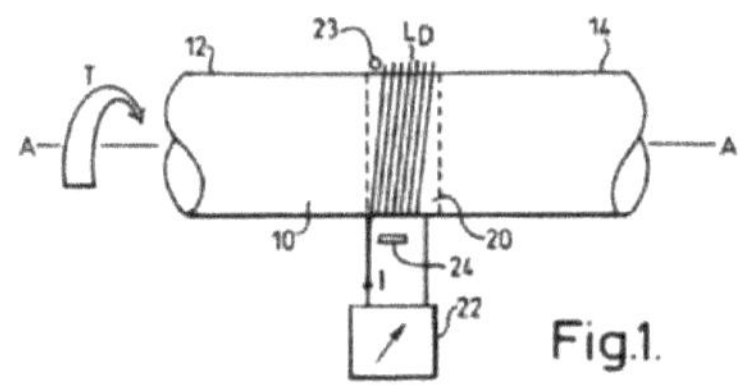

Fig. 11. The torque sensing system using magneto-elastic resonator (Bunyer, Scott L, 2007).

2.1.3 SAW torque sensors

Although the existence of the Surface Acoustic Wave (SAW) was first discussed in 1885 by Lord Rayleigh, it did not receive engineering interest for a long time. Until 1965, R.M. White suggested that SAWs can be excited and detected efficiently by using an interdigital transducer (IDT) placed on a piezoelectric substrate(R.M. White and F.W. Voltmer, 1965). This is because very fine IDTs can be mass-produced by using photolithography, which has been well developed for semiconductor device fabrication, and proper design of the IDT enables the construction of transversal filters with outstanding performance. Acoustic devices are robust in respect to temperature and mechanical stress discussed by Seifert et al. (F. Seifert, A. Pohl, R. Steindl, L. Reindl, M.J. Vellekoop, and B. Jakoby, 1999). They are reliable more than the lifetime of other electronic devices. Application of SAW devices to non-contact torque measurement was first suggested and patented in 1991 by A. Lonsdale and B. Lonsdale (A. Lonsdale, B. Lonsdale, 1991). Non-contact torque sensors based on SAW reflective delay lines were also introduced in (U. Wolft et al.; R. Grossmann et al. 1996). Since then some reasearchers fabricated and applied torque sensors based on SAW resonators to a number of industrial customers (A. Lonsdale, 2001; P. Merewood, 2000). In 2002, non-contact torque sensors based on SAW resonators application was proposed by Beckley et al. (J. Beckley, V. Kalinin, M .Lee, and K. Voliansky, 2002). Although the operation of both types of sensor was successfully demonstrated in the abovementioned publications, they did not cover such important aspects as limitations on the accuracy of the sensors resulting from their non-contact interrogation and the temperature stabilization of their sensitivity to torque. SAW resonators have been a commercial success on radio frequency applications, especially for filter and oscillator implementations. Their impact has made possible considerable reductions in size and power of the chipsets of mobile devices(Fujitsu Media Devices Ltd., 2006; Epcos AG, 2008). More modest and important applications of SAW resonators are the measurements in mass detector and pressure sensor devices with application in bio-particle detection(Martin, F., 2004; Talbi, A., 2006).

In recent years some instrument manufacturers of force and torque measurement devices have investigated instead of using resistance strain gauges. For example, one leading manufacturer of weighing machines now uses metallic and quartz resonant tuning fork technologies for industrial applications and the others have established some applications using surface acoustic wave (SAW) technology, optical technology, and magneto-elastic technology. Further commercial developments are taking place to enhance device manufacturability and improve device sensitivity and robustness in operation. Measurement on stiffer structures at much lower strain levels than before is possible now. Reviewing the world patents, we can find many torque measure patents using Piezo or SAW devices. In the United States Patent US 2003/0000309 A1, A. Lonsdale and B. Lonsdale claimed the patent of "Torque measurement", in which they proposed a method and apparatus for measuring the torque transmitted by a member comprises a SAW device secured to the member such that mechanical stress in the member due to torque transmitted thereby induces bending of the SAW device (A. Lonsdale and B. Lonsdale, 2003), as shown in Fig. 12. In another United States Patent, LEC, RYSZARD MARIAN, Magee et al. claimed the "Torque Measuring Piezoelectric Device and Method" using non-contact measurement of torque applied to a torque-bearing member such as a shaft. The method involves the use of a piezoelectric transducer mechanically coupled to the shaft for rotation therewith, and having electrical characteristics responsive to applied torque(Steven J. Magee, 2004). Electrical signal characteristics are changed by the torque-dependent transducer characteristics. In 2007, in another United States Patent, Steven J. Magee claimed that the sensor systems and methods are disclosed herein, including a sensor chip, upon which at least two surface acoustic wave (SAW) sensing elements are centrally located on a first side of the sensor chip. The SAW sensing elements occupy a common area on the first side of the sensor chip. An etched diaphragm is located centrally on the second side of the sensor chip opposite the first side in association with the two SAW sensing elements in order to concentrate the mechanical strain of the sensor system or sensor device in the etched diagram, thereby providing high strength, high sensitivity and ease of manufacturing.

Low propagation velocity enabled the use of SAW devices for time delays and filtering in radar systems and television but their application burgeoned within the mobile phone market. Honeywell SAW sensors use small piezo-electric quartz die upon which two or three single-port resonators, with natural resonant frequencies around 434 MHz, are fabricated in aluminum using standard photo-lithographic techniques, as shown in Fig. 13. (Magee et al., Honeywell, 2007). A SAW resonator is excited by a short radio-frequency (RF) burst. The centrally placed interdigital transducer (IDT) converts the electrical input signal to a mechanical wave through the piezo-electric effect. The waves propagate from IDT to reflectors and back until a forced resonance exists as a standing wave. After the transmit signal is switched off, the resonator continues to oscillate but at a frequency modified by any applied mechanical and/or thermal strain. The decaying oscillation is converted back to an electrical signal via the piezo-electric effect and retransmitted to the SAW interrogation board where the frequencies are analyzed and converted to engineering parameters. A more recent development is the use of SAW devices attached to the shaft and remotely interrogated. The strain on these tiny devices as the shaft flexes can be read remotely and output without the need for attached electronics on the shaft. The probable first use in volume will be in the automotive field as, of May 2009, Schott announced it has a SAW sensor package viable for in vehicle uses.

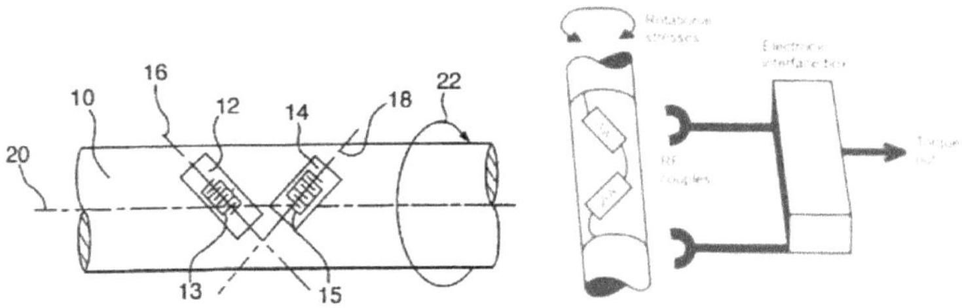

Fig. 12. The block diagram of a SAW torque sensor system. (A. Lonsdale and B. Lonsdale, 2003).

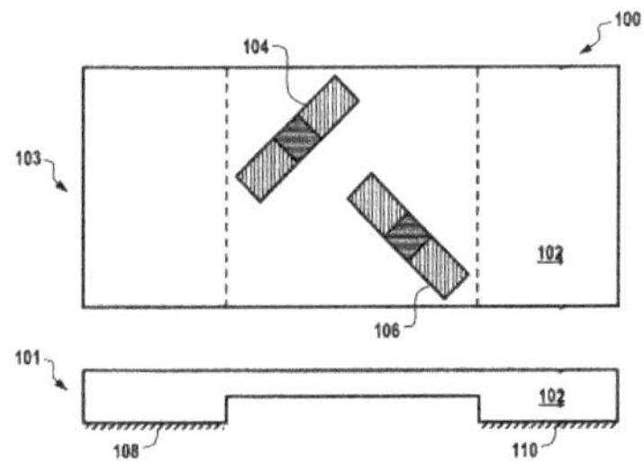

Fig. 13. The diagram of a SAW torque sensor system. (Magee et al., Honeywell, 2007).

A British company, Sensor Technology Ltd., has developed a Torqsense® digital RWT310/320 series transducers using surface acoustic wave (SAW) technology. Torqsense® is a registered trademark used for Electric or Electronic Transducers for providing electric or electronic sensors for sensing torque and owned by Lonsdale, Bryan, Lonsdale, Anthony. Based on a revolutionary new, patented, technology using SAW strain sensing elements, Torqsense® sensors offer a superior solution to non-contact rotary torque measurement in industrial applications. The SAW torque devices are strain sensitive elements with very high frequency modulating and it has the ability to convert an electrical signal into an acoustic signal with the same frequency. Because of a reduction in propagation velocity of about five orders of magnitude, the SAW device has a much smaller wavelength, which allows manipulation of an RF signal in a very small package. Therefore, we can consider a SAW device as a frequency dependent strain gauge, which can be used to measure the change in resonant frequency caused by the strain in the shaft. For dynamic torque measurement, the SAW device needs an RF couple to transmit the signal from the shaft to a fixed pick-up. It is a relatively low-cost device offering a ± 0.25% accuracy giving a total system accuracy of around ± 0.35% (patent Magee, 2007).

2.2 Types and applications of SAW devices

Wirelessly interrogated passive SAW sensors can employ one of the two basic types of SAW devices, one-port resonators(F. Jerems et al, 2001; A. Lonsdale, B. Lonsdale, 1991; A. Lonsdale, 2001; P. Merewood, 2000) and reflective delay lines (F. Schmidt, G. Scholl; A. Pohl, F. Seifert, 1997; U. Wolft et al.). Obviously, ordinary delay lines and two-port resonators can also be used as passive sensing elements, although they are less preferable because of the larger number of RF coupling devices are required (F. Schmidt, G. Scholl). If active components are allowed on the shaft then both the delay line and the resonator can be used in a feedback loop of a SAW oscillator, wirelessly coupled to an interrogation unit and wirelessly powered by an RF signal. This is an approach inherited from a traditional wired SAW sensor design.

2.2.1 Reflective delay lines of SAW devices

Most SAW passive sensors are designed using a reflective delay line. The surface acoustic wave in a reflective delay line propagates towards reflectors distributed in a characteristic barcode-like pattern and is partially reflected at each reflector. Fig. 14 shows the SAW delay line sensor and the principle of the sensor is drawn in Fig.15. Fig. 16 shows reflectors lined of same track and different tracks.

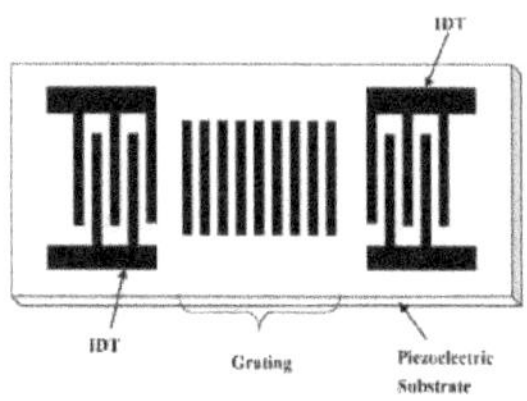

Fig. 14. The SAW delay line sensor.

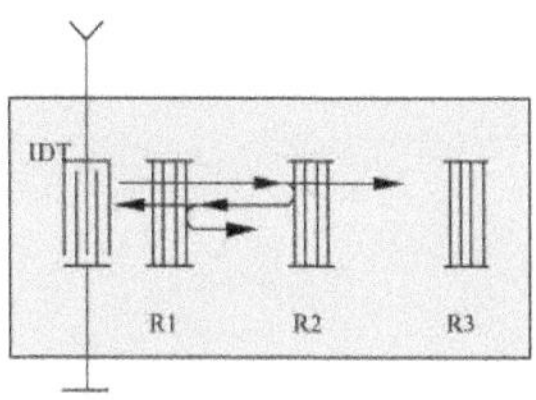

Fig. 15. The reflectors lined of same track (Weidong Cheng et al., 2001).

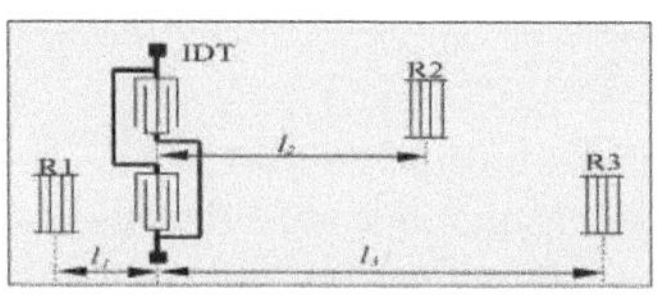

Fig. 16. The reflectors lined of different tracks(Weidong Cheng et al., 2001).

2.2.2 Resonators of SAW devices

SAW resonators exploit SAW propagation and electromechanical transduction to implement electronic circuits as filters, oscillators, and sensors. A SAW resonator is basically a resonant cavity in which a first transducer electrode converts the electric signal into a lateral mechanical wave. The resulting SAW propagates on the piezoelectric to reach the second electrode, where it is transduced back into the electrical domain. When arriving at the second electrode, and typically aided by one or more reflector electrodes, the acoustic wave bounces back in the direction of the first electrode, and the electromechanical conversion is repeated indefinitely, as depicted in Fig. 17. Thus, the acoustic wave is trapped in the cavity formed by the resonator electrodes.

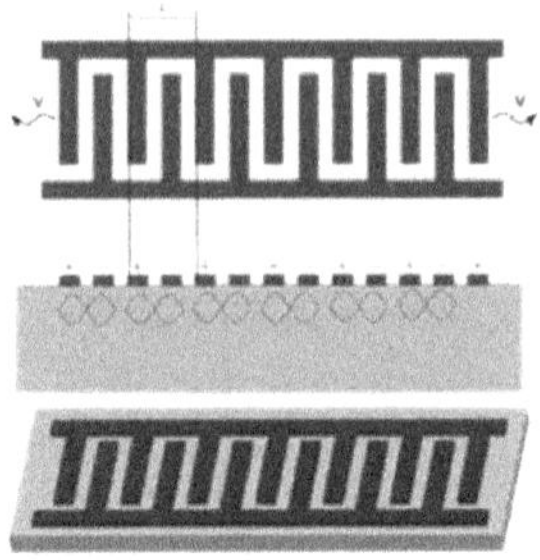

Fig. 17. The surface acoustic wave (SAW) resonator (Humberto Campanella, 2010).

The available photolithography resolution limits the dimensions of the IDTs and the piezoelectric layer determines the maximum operating frequency of the fundamental mode of the resonator. However, as well as other electromechanical resonators, SAW devices can be operated at over-tone modes to bypass near-to-fundamental bulk or other resonance modes. Typical frequencies for SAW-resonator-based applications are in the UHF band below 1 GHz, although high-performance commercial devices in the 1.5-GHz GPS and 1.9-GHz PCS bands are available so far(Epcos AG, 2002; TriQuint Semiconductor, 2009). The fundamental frequency of the SAW resonator mainly depends on the pitch of the IDTs, which is chosen to be equal to the SAW wavelength λ, and the sound speed of the piezoelectric layer v:

$$f_0 = \frac{v}{\lambda} \tag{1}$$

SAW resonators are found in two types of port configurations one-port and two-port resonators. One-port resonators are two-terminal devices, and they find application in oscillator circuits like VCOs or Colpitts oscillators. Two-port resonators behave more like narrow band-pass filters.

Fig. 18 shows a typical configuration One-port SAW resonators, The One-port SAW resonators have a single IDT generating and receiving the SAW, and two grating reflectors, which reflect the SAW and generate a standing wave between the two reflectors. The IDT and reflectors are fabricated on quartz crystal substrate or another piezoelectric material and patterned by photolithographic processes(Ken-Ya Hashimoto, 2000). As Fig. 19 shown, its equivalent circuit near resonance, where C_M and L_M are the dyamical capacitance and inductance, respectively, corresponding to the contributions of elasticity and inertia. On other hand, C_O is the static capacitance of the IDTs, and R_M is the motional resistance corresponding to the contribution of damping. On the other hand, two-port SAW resonators exhibit two IDTs, one of them generating the SAW, and the second one picking it. As with one-port devices, two grating reflectors aid the SAW to be reflected and confined between the IDTs. The generic connection is connected to the first and second IDTs as shown in Fig. 20 for two-port SAW resonator and Fig. 21 shows the resonator equivalent circuit.

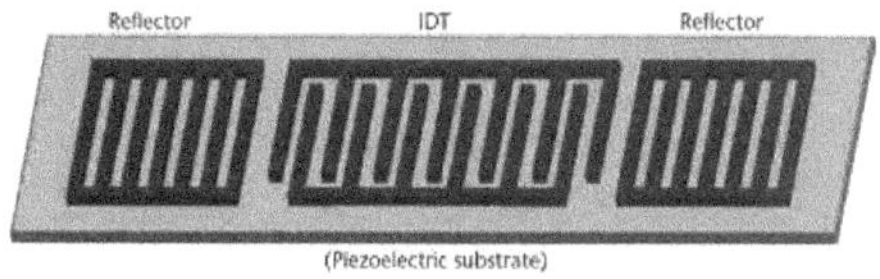

Fig. 18. The one-port SAW resonator (Humberto Campanella, 2010).

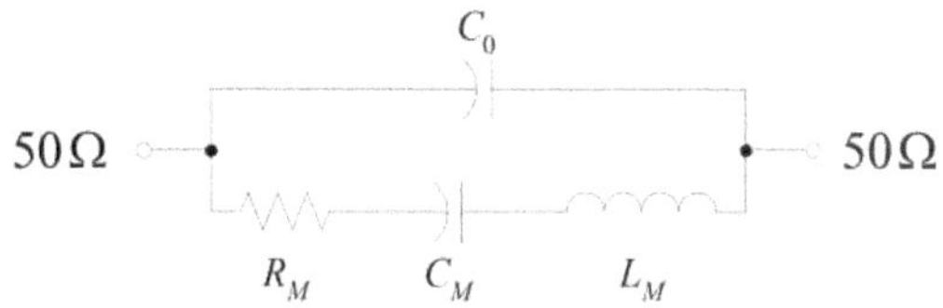

Fig. 19. The equivalent circuit for the one-port SAWR (Ken-Ya Hashimoto, 2000).

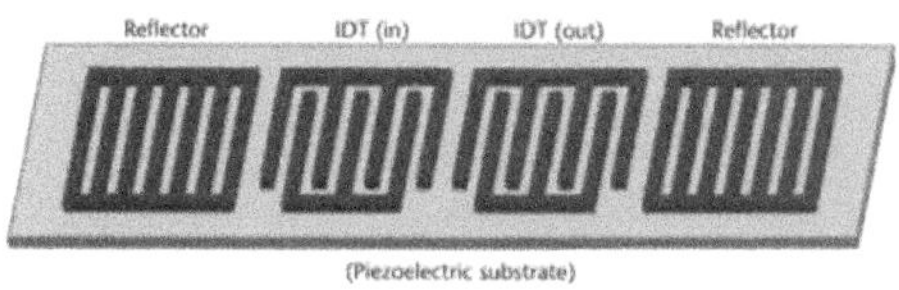

Fig. 20. The two-port SAW resonator (Humberto Campanella, 2010).

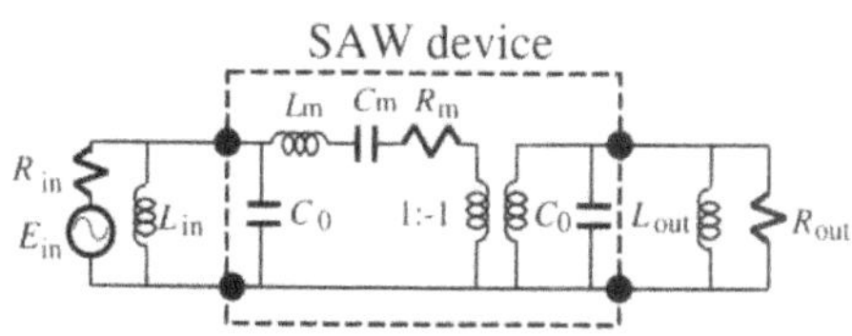

Fig. 21. The two-port SAW resonator equivalent circuit (Ken-Ya Hashimoto, 2000).

2.2.3 Characteristics of one-port SAW resonators

In this study, we using one-port SAW resonators device. Fig. 18 shows typical configuration one-port SAW resonators. Fig. 22 shows frequency response for resonator and in this study we just measure the resonance frequency ω_r or antiresonance frequency ω_a.

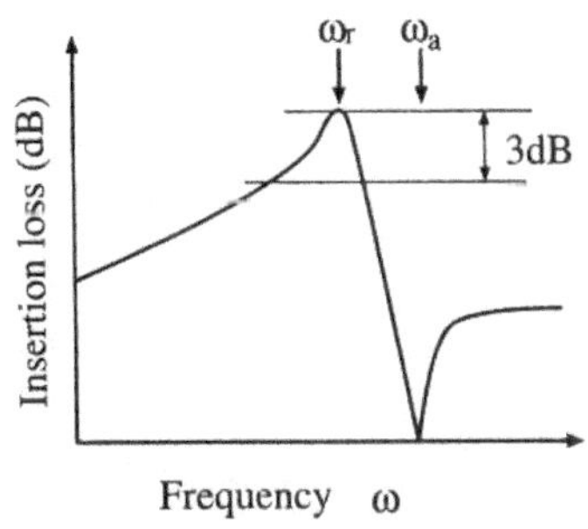

Fig. 22. The frequency response for a resonator (Ken-Ya Hashimoto, 2000).

From the equivalent circuit of one-port SAW device in Fig. 19, we have

$$\omega_r = \frac{1}{\sqrt{L_M C_M}}, \tag{2a}$$

$$\omega_a = \frac{1}{\sqrt{\frac{L_M C_M C_O}{C_M + C_O}}}, \tag{2b}$$

where C_M and L_M are the dyamical capacitance and inductance, respectivly. On the other hand, C_O is the static capacitance of the IDTs, and R_M is the motional resistance corresponding to the contribution of damping. Fig. 23 shows typical electrical resonance characteristics of one-port SAW resonator. As shown in Fig. 22, the conductance G takes a maximum at the frequency above the resonance frequency ω_r; in addition, there exists the anti-resonance frequency ω_a where the resistance G^{-1} takes a maximum.

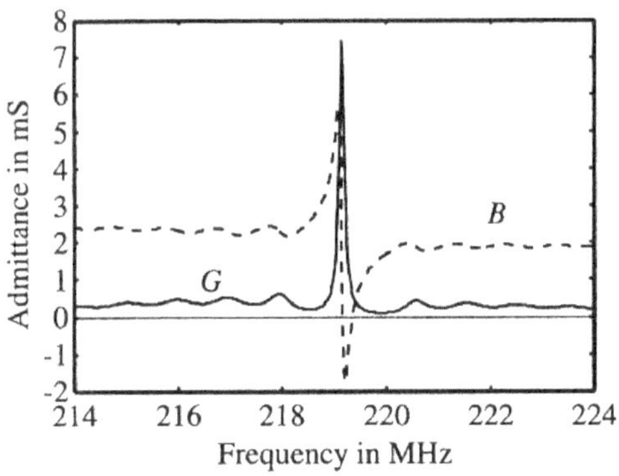

Fig. 23. The electrical resonance characteristics of one-port SAW resonator (Ken-Ya Hashimoto, 2000).

As mentioned in Ref. (Ken-Ya Hashimoto, 2000), C_M and L_M can be obtained from the measured C_O, ω_r and ω_a. R_M can be determined by G^{-1} at $\omega = \omega_r$. The capacitance ratio γ is frequently used as a measure of the resonator performance, and is for:

$$\gamma = \frac{C_O}{C_M} = \frac{1}{(\omega_a / \omega_r)^2 - 1}, \tag{3}$$

which corresponds to the inverse of the effective electromechanical coupling factor. The quality factor Q at the resonance frequency is called the resonance quality factor, which is denoted by Q_r. This is also an important measure, which is shown by:

$$Q_r = \frac{\omega_r L_M}{R_M} = \frac{1}{\omega_r C_M R_M}, \tag{4}$$

Both resonance frequency ω_r and antiresonance frequency ω_a will be shift by temperature change and effect of applied stress. Effect of temperature can have a considerable affect on propagation velocity for surface acoustic wave device, since many of the material constants involved are themselves temperature dependent. The temperature dependence of surface wave attenuation for quartz in particular has been studied over various temperature and frequency range(M. R. Daniel and J. De Klerk, 1970). Different piezoelectric materials, propagation directions, and cuts all show different temperature dependencies over different range. It is thus possible to engineer that a SAW device has specific temperature characteristics, which has led to the development of several SAW based temperature sensors(D. Hauden et al., 1981; J. Neumeister et al., 1990; M. Viens and J. D. N. Cheeke, 1990). In effect of applied stress also contributes to acoustic wave attenuation and velocity change, and has been studied extensively by Slobodnik(A. J. Slobodnik, 1972). In this study, the attenuation is due to the generation of surface acoustic waves in the gas in contact with the SAW device. In other words, the shear vertical component of the wave causes periodic compression and rarefaction of the gas, resulting in a coupling of acoustic energy from the SAW device into the air.

2.2.4 Applications of torque sensors based on SAW resonators

Although the existence of the Surface Acoustic Wave (SAW) based on torque measurement was first discussed in 1996 by A.Lonsdale (A. Lonsdale et al.). Torque (radial strain) transducers are one of the more common devices used by development engineers. Knowledge of torque and rotational speed can be used to indicate power, from which the efficiency of gearboxes, transmissions, electrical machines and many other systems can readily be assessed. To apply the SAW element principle, two devices are used in a half bridge, analogous to the classic resistive strain gauging configuration; one positioned so as to be sensitive to the principal compressive strain and the other positioned to observe the principal tensile strain. In the absence of bending moments and axial forces, the principal stress planes lie perpendicular to one another at 45° to the plane about which the tensional moment is applied. This is illustrated in Fig. 24.

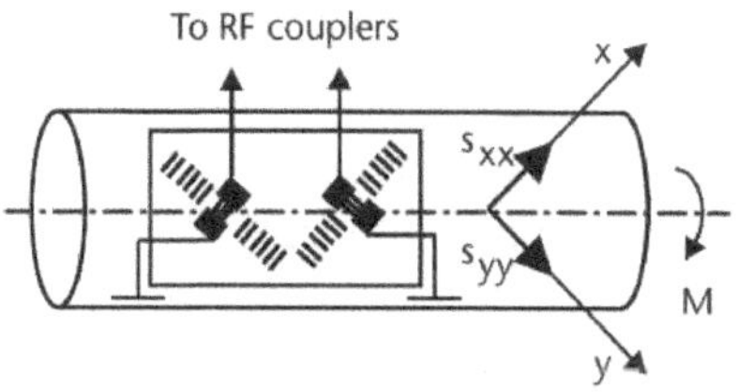

Fig. 24. Torque sensing element based on SAW resonators(Stephen Beedy et al., 2004).

In this SAW sensors applied rely on the fact that the torque M applied to the shaft creates two principal components of strain, $S_{XX} = -S_{YY} = S$. As a result, one of the SAW devices is under tension and the other one is under compression, causing the opposite change of resonant frequency in the devices. The resonators have the same or better performance for the same size of substrate and are less demanding in terms of the receiver bandwidth and sensitivity. Resonator Q factors are about 10,000. The torque sensor interrogation system can employ continuous frequency tracking of reflected frequencies from the two SAW resonators.

This applied can achieve temperature compensation and eliminate sensitivity to shaft bending. From the technique described it is apparent that the output signal will be in the frequency domain as shown in Fig. 25.

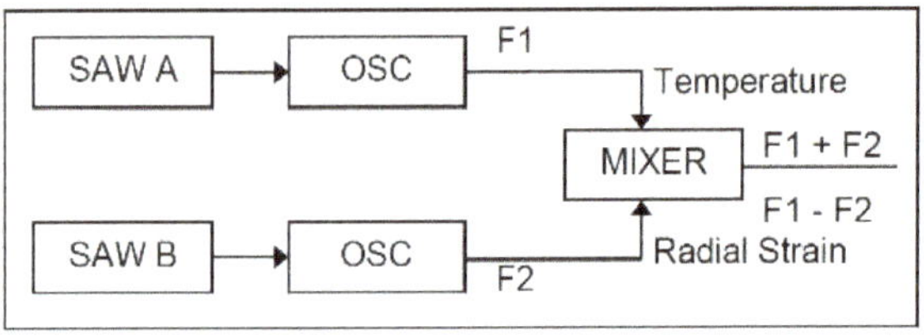

Fig. 25. The schematic of frequency domain for SAW signals(A. Lonsdale, 1993).

3. Wirelessly measurement architecture based on SAW device

An acoustic wave is a disturbance in an elastic medium that propagates in space and time, thus transferring the energy supplied by an excitation source along the medium in the form of oscillation or vibration. There exist two types of acoustic waves: surface acoustic waves (SAW) and bulk acoustic waves (BAW). In this study, two one-port SAWRs (manufactured by *f*-tech Corporation) with the resonant frequency of 433.42 MHz and 433.92 MHz are used to be torque sensing devices for this study. The specifications for the SAWR (433.92 MHz) are descibled in Table 1. Fig. 26 shows the wafer of the SAW resonators using in the experiments and Figs. 27 and 28 show the die size of the SAWR and its microimage.

Fig. 26. The Wafer of f-tech SAWRs (433.92MHz).

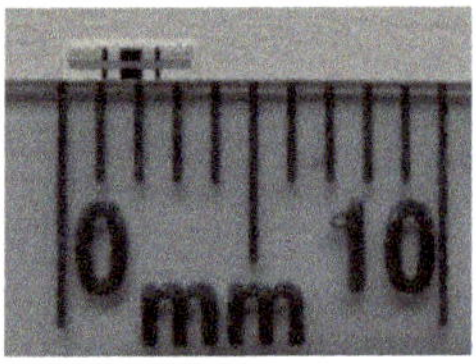

Fig. 27. The photo of the SAWR's die size.

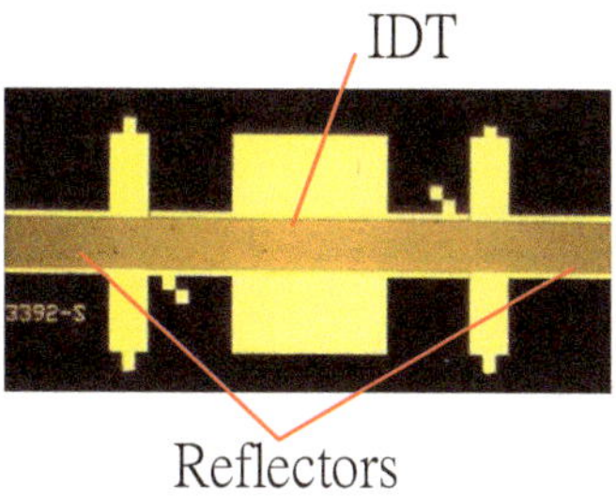

Fig. 28. The microimage of the SAWR's die.

Item	Unit	Min.	Typ.	Max.
Center Frequency, f_r	MHz	433.845	433.92	433.995
Insertion Loss, IL	dB	-	1.5	2.0
Unloaded quality factor, Q_U		-	12000	-
Loaded quality factor, Q_L		-	1880	-
RF Equivalent RLC Model				
Motional Resistance, R_M	Ω	-	18.5	-
Motional Capacitance, C_M	fF	-	1.653	-
Motional Inductance, L_M	μH	-	81.35	-
Shunt Static Capacitance, C_0	pF	-	2.4	-
Turnover Temperature, T_0	°C	-	30	-
Frequency Temperature Coefficient, FTC	ppm/k^2	-	0.032	-

Table 1. 433.92 MHz SAWR specifications (www.f-tech.com.tw).

In this study, we used two one-port SAWRs to obtain the torque signal via measuring the frequency shift of SAWs, Δf, which is relate to the strain ε and the fundamental requency of the SAWR f_0. For each SAWR, the relation between the frequency shift and the strain can be described as follows.

$$f_0 = \frac{v_0}{4d}$$

where f_0 is the fundamental frequency of the SAWR, which mainly depends on the pitch of the IDT, d, and v_0 is the speed of SAW.

Obtaining the variation of the fundamental frequency according the the variation of the speed and pitch for the SAW, we have:

$$f_0 + \Delta f = \frac{v_0 + \Delta v}{4(d + \Delta d)} = \frac{v_0 + \Delta v}{4d(1+\varepsilon)} \tag{5}$$

where ε is the strain. Neglect the variation of the SAW speed, becasue it is very small. Then, we have

$$f_0 + \Delta f \approx \frac{4df_0}{4d(1+\varepsilon)} = \frac{f_0}{1+\varepsilon} \tag{6}$$

Using Taylar expansion to consider the approximation, we can obtain the following.

$$\frac{1}{1+\varepsilon} = 1 - \varepsilon + \frac{1}{2}\varepsilon^2 - \frac{1}{3!}\varepsilon^3 + \ldots \cong 1 - \varepsilon \tag{7}$$

then the frequency shift can be approximated as follows

$$\Delta f = f_0 \cdot \left(\frac{1}{1+\varepsilon} - 1 \right) = f_0 \frac{(1-\varepsilon-1)}{1+\varepsilon} \approx -\varepsilon f_0 \tag{8}$$

Therefore, we can obtain the relation between the frequency shift and the strain as follows

$$\Delta f = -\varepsilon \cdot f_0 \tag{9}$$

where Δf is the shift of the resonant frequency, ε is the strain and f_0 is the resonant frequency. In the implementation, two one-port SAW resonators oriented at $\pm 45^o$ to the shaft axis, which are shown in Fig. 29. Fig. 30 shows the locations of the two SAWR on the shaft; we also attach the strain gage with two directions of $\pm 45^o$ on the opposit side of the SAWR to obtain the actual strain as shown in Fig. 30. We use the network analyzer (Agilent E5071A) to measure the S11 value of this two SAWRs for obtaining the fundamental frequency of these two SAWRs as shown in Fig. 31. Then, we can find the relation between the frequency shift and the applied torque as shown in Fig. 32. Fig. 32 shows that the resonant frequency shifts for SAWR1 and SAWR2 on the location of $\pm 45^o$ will move in the opposite directions as soon as torque is applied to the shaft, because the stress's direction is inverse due to the applied torque as σ_x and σ_y described in Fig. 29. Therefore, we can obtain the value of torque measuring the difference between the two frequencies and this arrangement can compensate as mestioned in the past literature (A. Lonsdale, 1993). From the frequency shifts of the SAWR1 and SAWR2 measured in Fig. 32, we can obtain the actual strain which is computed according to Eq. (9).

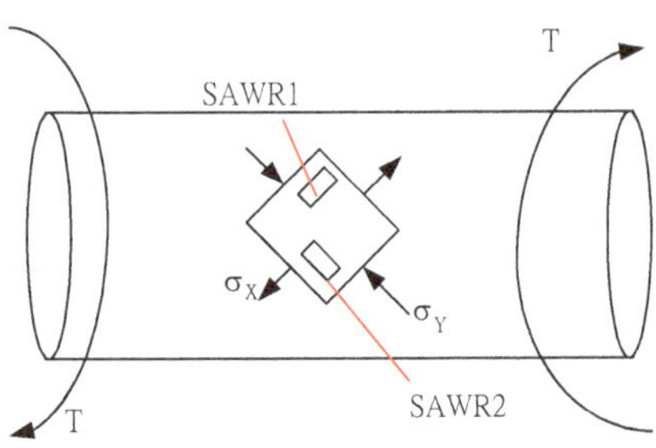

Fig. 29. The relation between the stress and the applied torque for the SAWRs.

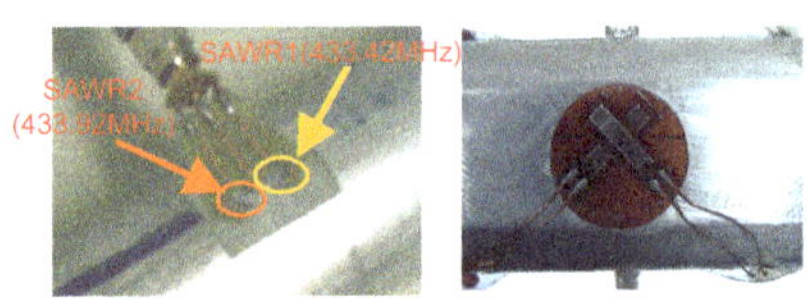

Fig. 30. The setup of the SAWRs and the strain gage on the shaft.

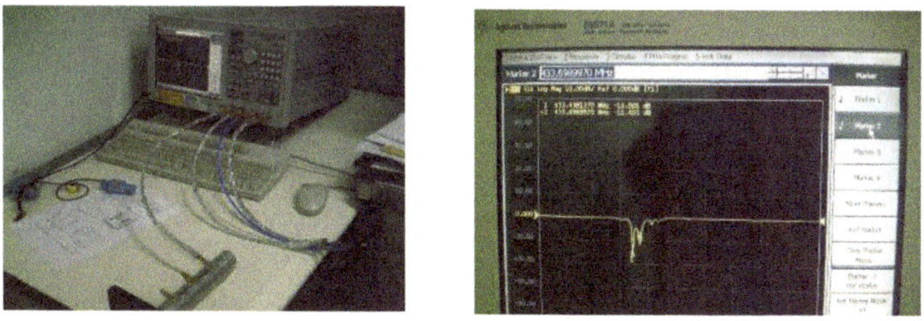

Fig. 31. Torque measurement using the network analyzer (Agilent E5071A).

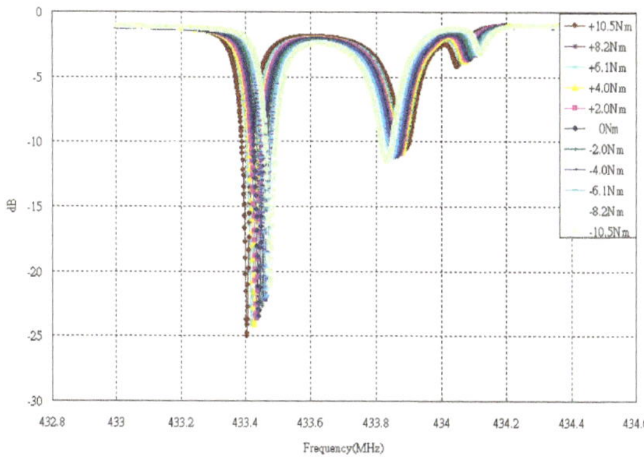

Fig. 32. The relation between the S11 value and the frequency of the two SAWRs.

Fig. 33 shows the relation between the applied torque and the strain, which is measured using the SAWRs. The theoretical torque can be computed as follows.

$$\tau_{\max} = \frac{Tc}{J} = \frac{T \times r}{\frac{\pi}{2}(r)^4}$$

$$\tau_{\max} = \frac{\sigma}{2} \tag{10}$$

$$\varepsilon = \frac{\sigma}{\mathrm{E}} = \frac{2\tau_{\max}}{\mathrm{E}}$$

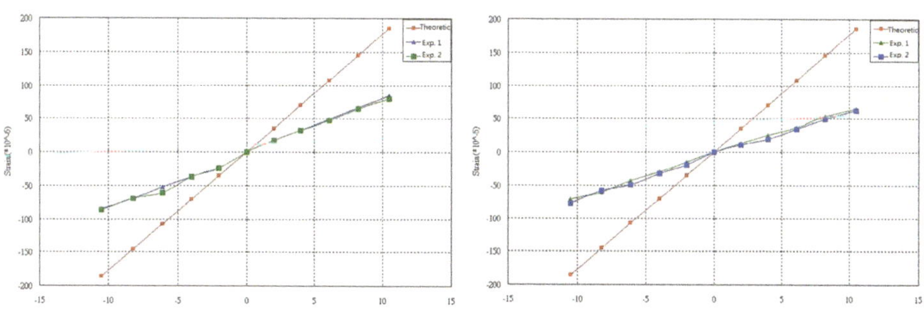

Fig. 33. The relation between the applied torque and the strain using two SAWRs.

where E is Young's modulus of the material, σ is the stress, ε is the strain, r is the radius of the torsion bar and T is the applied torque. Although the SAWRs response only 40% strain which is compared from the strain gage in Fig. 34, the relation between the torque and the strain is linear; therefore, we can measure the torque applied on the shaft according to the relation obtained in Fig. 34 for the SAWRs.

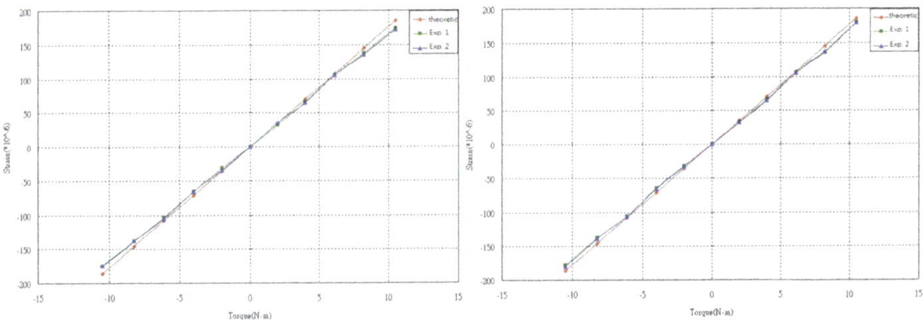

Fig. 34. The relation between the applied torque and the strain using the strain gage.

If the measurement of the resonant frequency in the torque sensor is performed by means of continuous tracking of S11 minimum at the input of the rotary coupler then a remarkably low standard deviation of the frequency measurement error of 0.3-0.4 ppm can be achieved at less than 1 ms update period. However the continuous tracking requires one automatic frequency control (AFC) loop per resonator. In this study, the RSSI value is used to obtain the resonant frequency of SAWR instead of measuring S11 minimum. Fig. 35 shows the fabricated torque sensor, which consists of the four microstrip lines and two SAWRs. The connection between the microstrip and the SAWR is established by the IC wire bonding machine. The parameters of the microstrip line are designed optimally by the software of Advanced Design System 2006A. Moreover, the wireless signals are transmitted by four coupling antennas, which are described in Fig. 36. The proposed wireless torque sensory system is described in Fig. 37.

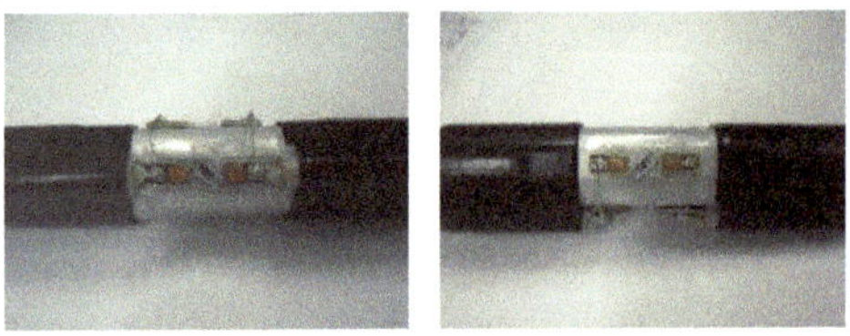

Fig. 35. The rotational shaft equipped with two SAWRs.

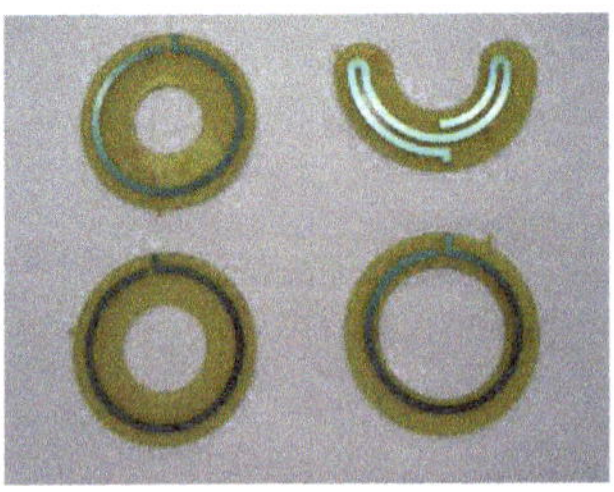

Fig. 36. The coupling antenna for the sensory system.

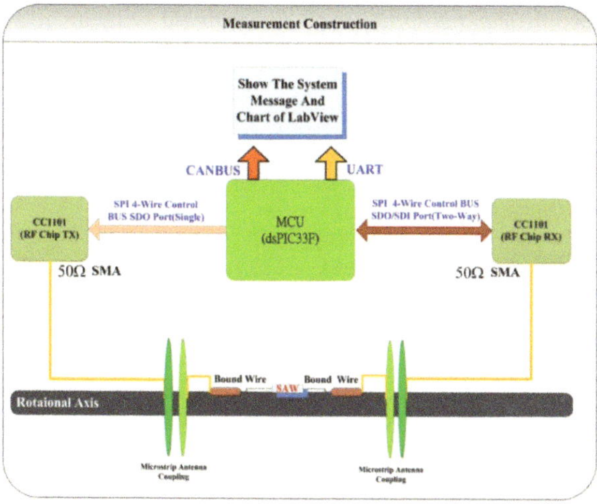

Fig. 37. The block diagram of the proposed wireless torque sensory method.

A wireless passive torque measurement system using SAW sensors is presented as shown in Fig. 38. To measure the torque via wireless sensing, a dsPIC architecture based on two SAW sensors via wireless RF signals is proposed in this paper. When the torque is applied to the shaft, the two SAWRs will change their resonant frequencies in opposite directions in the frequency domain. According to the past literature, measuring the difference between the two frequencies one can obtain the value of torque and achieve partial temperature compensation, which are performed by means of continuous tracking of minimumal S11 at the input of the rotary coupler. Different from the past method using the frequency with the minimal S11, this study tries to obtain the maximum of S21 to measure the value of torque via continuous tracking of output signal. An dSPIC33F microchip is used to control the RF chips (C1100) to achieve the transmitting and receiving task. The frequency with maximum received signal strength indication (RSSI), where the value of S21 is maximal, will be influenced by the applied torque. The measuring system mainly uses the RSSI signal with respect to the frequency to find the SAW center frequency value, and then the torque values can be obtain according to the frequency shifts. Therefore, the frequency, which has the maximal RSSI value, can be used to obtain the applied torque of the rotation shaft according to the frequency shift. In addition, to achieve the wireless transmission, two sets of coupling-type antenna are applied on the shaft. Finally, the torque signal is passed through CANBUS to integrate with the other vehicles electron systems. In this study, a wireless torque sensing system has been established to measure the torque of the rotaional shaft via the RF antenna.

To illustrate the process of the proposed wireless torque measuring system, a single electromagnetic RF interrogation system is studied to produce RF interrogation signals to the SAW device. According to the measurement of the frequency shift, the environmental influence can be measured by the specified relation via experimental results. A dsPIC33 chip is used to control the RF chip (C1100) to achieve the transmitting and receiving between 433.1MHz and 433.85 MHz, and the interval scanning frequency is 0.5 KHz. According to the response of the RSSI value with repect to the scanning frequency, the torque value can be obtained according the look-up table obtained by the experimental results in Fig. 34.

After all, the microcontroller (dsPIC33) itself is equipped with a controller area network (CAN) interface, which is responsible for the calculation of the sensor data and providing the communication with other electronic systems(dsPIC33FJ256GP710A user guid). According to above system, the CC1101 chip is a low- power consumption RF chip developed by Chipcon. The main operating frequency is designed at 315, 433, 868, and 915MHz. The receive chip of RF has an amplifier to enhance a signal, and then change the signal to IF (Intermediate Frequency). When the CC1101 plays an IF medium frequency receiver, the I/O signals are transferred to digital signals using A/D converter. On the other hand, the transmit chip of RF includes a LC VCO (LC Voltage-controlled oscillator) and a LO (Local Oscillator). Fig. 38 describes the block diagram of the proposed architecture using CC1100 and dsPIC to measure the frequency shift of the SAW devices. In the measurement system, the MCU's performance is the critical component, which dominates system bandwidth, sensitivity, and flexibility. Therefore, we choose the dsPIC33F to improve our measuring system. The dsPIC33F is a 16-bit DSC chip, which is developed by to integrate 16bits MCU and DSP with high speed computation. The dsPIC provides 84 commands, 16 registers and DMA as shown in Fig. 39. The wireless signals are transmitted by four coupling antennas, which are described in Fig. 37. Fig. 38 shows the flow chart of the proposed measurement system, which microchip controls two RF chips to perform transmission and receiving. Then, the signals are transmitted through the SAW via coupling antennas; at the same time, the dsPIC33 can obtain the RSSI with respect to frequency for SAW devices. According to the property of the SAW, the frequency intensity of RSSI is larger than others when the signal's frequency approaches the resonant frequency. After

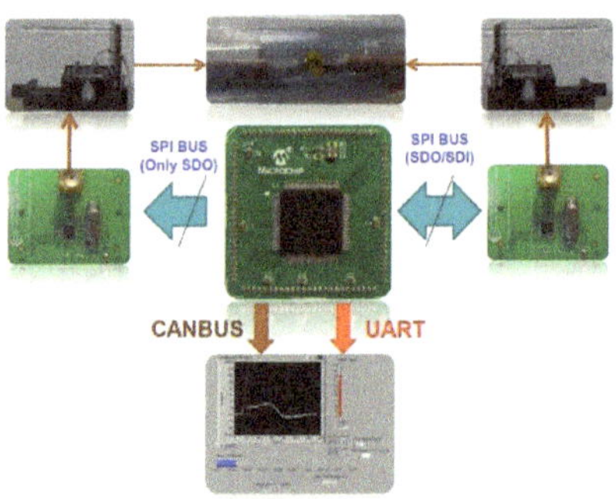

Fig. 38. The block diagram of the proposed measurement system.

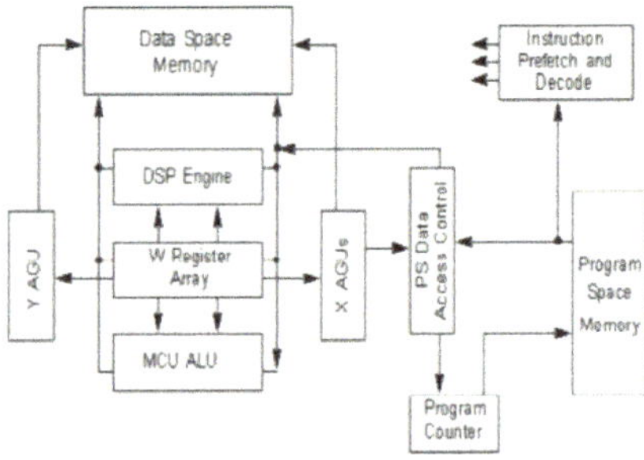

Fig. 39. The architecture of dsPIC33.

dsPIC33 receives the RSSI analysis for the signal in the scanning frequency domain, we can obtain a shift value of frequency with comparison from the initial resonant one for each signal. To control the RF transmit and receive modules of CC1100, the kernel of measurement system is programed in the microchip dsPIC33F which possesses UART and CANBUS transferring functions. The dsPIC33 controller is connected with CC1101 RF modules, which include transmitting and receiving task as shown in Fig. 38.

The dsPIC33 MCU uses four lines of SPI to communicate with CC1101 module such as sweep scanning time. In addition, the RSSI value of CC1101 RF receiving module uses 8-bit digital signal to communicate with the dsPIC33, which controls two RF chips for transmitting and receiving via the SPI signals. For the RF chip which performs transmitting task, dsPIC33 is only set the frequency for transmitting in the register of CC1100. When the RF chip which performs receiving task, the dsPIC33 sets the frequency for receiving in the register of CC1100 and receives the RSSI value in DMA at the same time via SPI signals. Then, the RSSI value is transferred via UART to the host PC to describe the measurement results. The flow-chart of the above process is describled in Fig. 40.

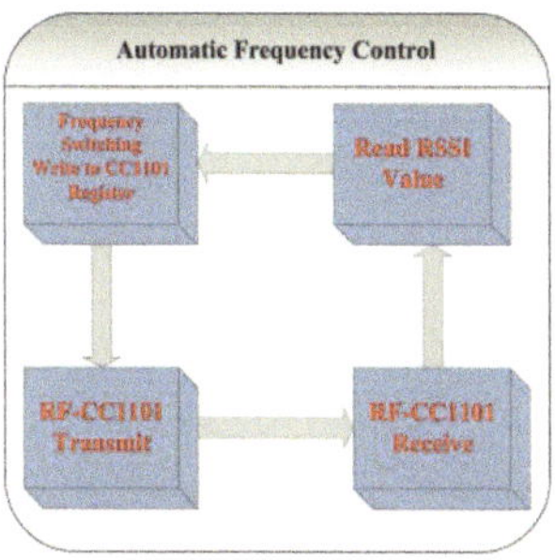

Fig. 40. The flow chart of the AFC Loop code in the dsPIC33F.

4. Experimental results and discussions

In this implementation, the SAW sensors with wireless RF coupling antennas as shown in Fig. 41, are attached on a shaft; the dsPIC33 microchip is used to achieve the transmitting

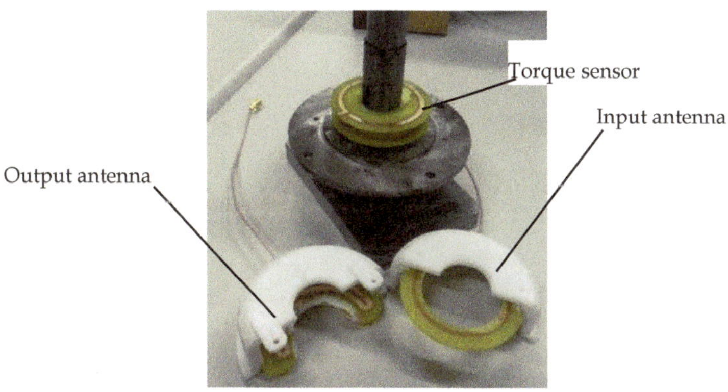

Fig. 41. The wireless torque sensory system.

and receiving RF signal tasks via the RF chips (CC1100). The program embedded in dsPIC33 microchip is coded using MPLAB software, which is the development environment based on C++ language. The microchip dsPIC33 is used to set the transmit frequency of C1100 from 433.1MHz to 435MHz and the interval between the scanning frequency is from 0.5 to 10 KHz for the specified bandwidth. On the receiver side of CC1100, the dsPIC33 makes the RF receiver chip's frequency to match the frequency of transmitting synchronously and then it read the DMA to obtain the intensity of the RSSI. Therefore, the frequency with the maximal RSSI value is obtained and transmitted to PC using CANBUS network. The flow chart of the above processes is described in Fig. 42. On the PC side, a Labview program is developed to display the frequency response as shown in Fig. 43. In this case, we have developed a simple virtual instrument whose purpose is similar to the network analyzer.

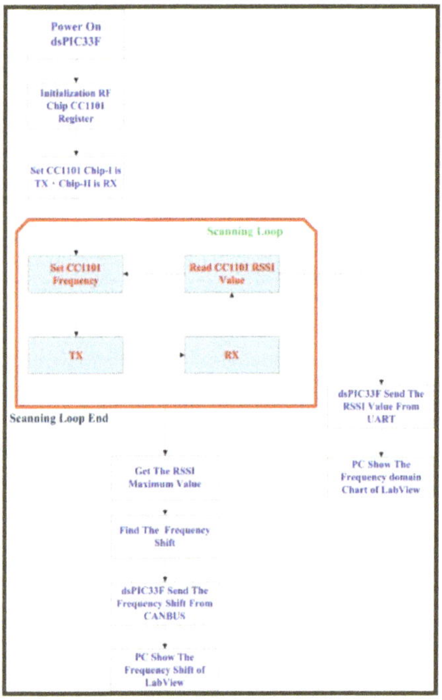

Fig. 42. The flow chart of the dSPIC 33F program.

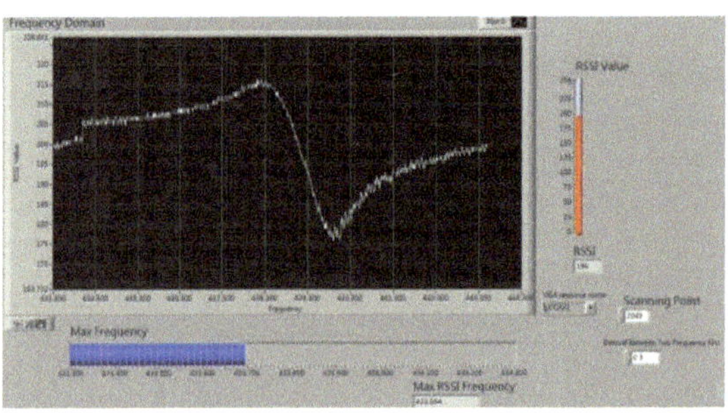

Fig. 43. The RSSI response with respect to frequency in LabView interface.

Fig. 44 shows the relation between the torque and the frequency shift. We can establish a linear mapping relation between the torque and the frequency shift via using the experimental results in Fig. 44. Then, the torque can be obtained in real-time. For this wireless torque sensing system, the performance index for the sensory system is important for feasible applications. On one hand, we use dsPIC33F with a wide bandwidth to increase torsion sensor performance. On the other hand, different frequency scanning interval introduces the different measurement performance as shown in Fig. 45. The developed sensory system's bandwidth is described in Table 2. From the experimental results, we can summarize the following comments. (1) The smaller scanning interval will cause the lower system bandwidth, but the measurement system is more stable. (2) The less number of scanning points can improve the system performance with bad stability.

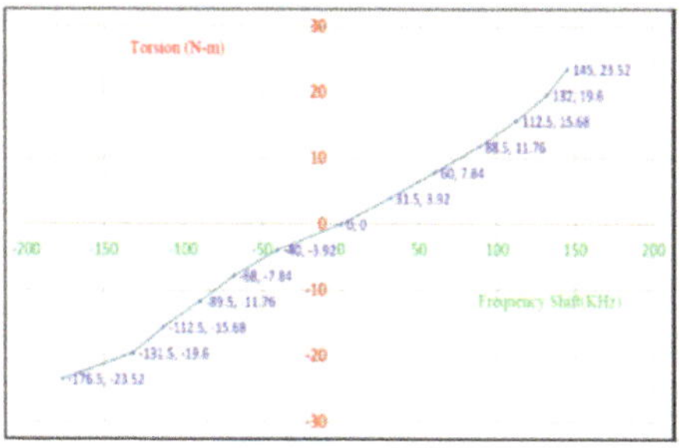

Fig. 44. The experimental results between the torque and frequency shift.

Fig. 45. Number of frequency scanning points v.s. the system bandwidth.

No. of scanning points	System bandwidth (Hz)	Scanning interval (KHz)	Perturbation (KHz)
2000	15	0.5	1
1000	38.43	1	2
500	64.11	2	4
200	191.35	5	10
100	383.1	10	20

Table 2. The bandwidth of the developed torque sensory system.

5. Conclusion

In this paper, we study SAW devices to establish a wireless torque measurement system, which uses the analysis of RSSI to obtain the frequency shift of SAW using embedded microchip dsPIC33F. From the experimental results, a prototype of virtual instrument based on microchip integrated with LabView is used to measure the frequency response of the SAW devices; that is, we developed a simple PC-based Spectrum Analyzer. The measurement through sweeping varies frequency to obtain intensity of RSSI signals can establish the relation between the torque and the frequency shift via the proposed measurement system using UART and CANBUS. The experimental results validated the proposed measurement system.

6. Acknowledgment

This work was supported by the National Science Council under NSC grant NSC 100-2221-E-027-031.

7. References

A. J. Budreau and P. H. Carr, "Temperature Dependence of the Attenuation of Microwave Frequency Elastic Surface Waves in Quartz," *Appl. Phys. Lett.*, 1971, 18, 239.

A. J. Slobodnik, Jr., "Attenuation of microwave acoustic surface waves due to gas loading," Appl. Phys., 1972, 43, 2565.

A. Lonsdale, "Dynamic rotary torque measurement using surface acoustic waves", 2001, *Sensors*, vol. 18, pp. 5 1-55

A. Lonsdale, "DYNAMIC ROTARY TORQUE MEASUREMENT USING SURFACE ACOUSTIC WAVES," 2001, Technical Director Sensor Technology Ltd

A. Lonsdale, B. Lonsdale, "Method and apparatus for measuring strain", Int. patent public. No. WO 91/13832, 19 Sept. 1991, Int. Applic. No. PCT/GB91/00328, Int. filing date: 4 March 1991, Priority: 9004822.4, 3 March 1990, GB .

A. Pohl, F. Seifert, "Wirelessly interrogable suacoustic wave sensors for vehicular applications", 1997, *IEEE Trans. Instrum. Meas.*, vol. 46, pp. 1031-1 038.

A.Lonsdale, Transence Technologies, and N.Schofield, MST Ltd. "An Intergrated Low Cost Sensor For The Direct Torque Control of Brushless DC MOTORS", 1996, Machines for automotive applications, Vol. 4, pp. 6/1-7

Auld, B. A., 2nd, "Acoustic Fields and Waves in Solids", 1990, *Krieger Publishing Company*, Vols. I and II, Malabar.

Bunyer, Scott L et al., "Magneto-Elastic Resonator Torque Sensor", Aug. 28, 2007, United States Patent, Patent NO. US 7,261,005 B2.

Corcoran, Joe and D'Ercole, Steve, "A New Development in Continuous Torque Monitoring Couplings," 2000, Proceedings of ASME DETC 2000 Power Transmission and Gearing Conference.

D. Hauden, G. Jaillet, and R. Coquerel, "Temperature Sensor Using SAW Delay Line", *IEEE Ultrason. Symp. (Proc.)*, 1981, 148.

D. Hauden, M. Michel, G. Bardeche, and J. J. Gagnepain, " Temperature Effects on Quartz-Crystal Surface-Wave Oscillators", *Appl. Phys. Lett.*, 1977, 31, 315.

dsPIC33FJ256GP710A user guide, Microchip Inc.

Epcos AG, "SAW Frontend Module GSM 850/EGSM/DCS/PCS/WCDMA 2100," 2008, Surface Acoustic Wave Components Division.

F. Jerems, M. Pro& D. Rachui and V. Kalinin, "A new generation non-torsion bar torque sensors for electromechanical power steereing applications", 2001, Proc. Int. Congress on Electronic Systemsfor Vehicles, pp. 27-28 September, Baden Baden.

F. Schmidt, G. Scholl, *"Wireless SAW identification and sensor systems "*, *Advances in Surface Acoustic Wave Technology, Systems and Applications,* vol. 2, C. C. W. Ruppel and T. A. Fjeldly, Eds. Singapore, World Scientific.

F. Seifert, A. Pohl, R. Steindl, L. Reindl, M.J. Vellekoop, and B. Jakoby, " Wirelessly interrogable acoustic sensors." 1999 IEEE INTERNATIONAL FREQUENCY CONTROL SYMPOSIUM., pp. 1013 - 1018.

Gautschi, G, "Piezoelectric Sensorics, Force, Strain, Pressure, Acceleration and Acoustic Emission Sensors", 2002, *Materials and Amplifiers*

http://www.fujitsu.com., Fujitsu Media Devices Ltd., Cellular SAW Duplexer, FAR-D5GA-881M50-D1AA, 2006,

http://www.honeywell.com/sensing, Honeywell Inc.

http://www.pcb.com/Linked_Documents/Force-Torque/Catalog/Sections/FTQ200G_0107_Torque.pdf, Torque Sensor PCB Piezotronics, Inc

http://www.sendev.com/catalog/pdf/torque-measurement.pdf, David Schrand, Sensor Developments Inc

http://www.sensors.co.uk/torqsense/RWT310-230, sensor technology

http://www.triquint.com., "881.5/1960 MHz SAW Filter," 2009, TriQuint Semiconductor.

http://www05.abb.com/global/scot/scot264.nsf/veritydisplay/eef8846525950f25c1257380003623c6/$File/3BSE022227R0101_-001.pdf, Torductor®-S, 2007, ABB Company.

Humberto Campanella "Acoustic Wave and Electromechanical Resonators"2010 ARTECH HOUSE

Humberto Campanella, "Acoustic Wave and Electromechanical Resonators", 2010 ARTECH HOUSE Inc..M. R. Daniel and J. de Klerk, "TEMPERATURE-DEPENDENT ATTENUATION OF ULTRASONIC SURFACE WAVES IN QUARTZ", 1970,Appl. Phys. Lett., Vol. 16, pp. 30.

J. Beckley, V. Kalinin, M. Lee, and K. Voliansky," Non-Contact Torque Sensors Based on SAW Resonators," 2002 IEEE International Frequency Control Symposium and PDA Exhibition.

J. Neumeister, R. Thum, and E. Lüder, "A SAW delay line oscillator as a high-resolution temperature sensor", *Sens. Actuators A*, 1990, A21-23, 670.

Ken-Ya Hashimoto, "Surface Acoustic Wave in Telecommunications Modeling and Simulation" 2000, Springer, Inc..

M. Viens and J. D. N. Cheeke, "Highly Sensitive Temperature Sensor Using SAW Oscillator", *Sens. Actuators A*, 1990, A24, 209.

Magee et al, "Surface Acoustic Wave Sensor Methods and Systems," Jan. 23, 2007, United States Patent, Patent NO. US 7,165,455 B2 .

Magee et al., "Torque Measuring Piezoelectric Device and Method", Jan. 20, 2004, United States Patent, Patent NO. US 6,679,123 B2 .

Martin, F., "Pulse Mode Shear Horizontal-Surface Acoustic Wave (SH-SAW) System for Liquid Based Sensing Applications," 2004, *Biosensors and Bioelectr.*, Vol. 19, pp. 627–632.

P. Merewood, "A novel ZnO/Si thin-film SAW torque measurement microsystem," Sept., 2000, Int. Forum on Wave Electron and Its Applications, pp. 14-18, St.-Petersburg.

R. Grossmann, J. Michel, T. Sachs, E. Schrufer, "Measurement of mechanical quantities using quartz sensors," March 1996, Proc. Europ. Freq. Time Forum, pp. 376-381, 5-7.

R.M. White and F.W. Voltmer, "Direct Piezoelectric Coupling to Surface Elastic Wave", 1965, Appl. Phy. Lett., Vol. 17 pp.314-316.

Rayleigh, J.W.S., "On waves propagated along the plane surface of an elastic solid", 1885, Proceedings of the London Mathematical Society, Vol. 17, No. 1, pp. 4-11.

Stephen Beeby, Graham Ensell, Michael Kraft, Neil White, " MEMS Mechanical Sensors", 2004, Artech House, Inc.

Talbi, A., et al., "ZnO/Quartz Structure Potentiality for Surface Acoustic Wave Pressure Sensor," *Sens. Actuators A: Phys.*, Vol. 128, 2006, pp. 78–83.

U. Wolft; F. Schmidt, G. Scholl, V. Magory, "Radio Accessible SAW sensors for non-contact measurement of torque and temperature", 1996, Proc. IEEE Ultrason. Symp.

Weidong Cheng, Yonggui Dong, and Guanping Feng, "A Multi - Resolution Wireless Force Sensing System Based upon a Passive SAW Device," Sept. 2001, IEEE Trans. Ultrason. Freq. Contr. Vol. 48, NO.5, pp. 1438 - 1441.

6

Non Contact Measurement System with Electromagnets for Vibration Tests on Bladed Disks

Christian Maria Firrone and Teresa Berruti
Department of Mechanics, Polytechnic University of Turin (POLITO), Turin, Italy

1. Introduction

In vibration testing one of the main concerns in the measurement of the forced response of vibrating structures is the coupling between the test article and the exciter. When the structural behaviour of the structure is pretty linear a hammer test allows getting a first idea of the resonant frequencies and mode shapes without perturbing the dynamic response. Unfortunately, real applications of structures whose behaviour is linear are rare. Usually, structures are made of components assembled together by means of joints whose behaviour may result highly nonlinear. Depending on the amount of excitation, joints can dramatically change the dynamic behaviour of the whole system and the modelling of this type of constraint is therefore crucial for a correct design. For this reason, tests are performed to characterize the contact parameters of joints focusing the attention in the material used in the application or extending the study by considering the whole components coupled by a particular geometry of joint. The characterization of joints in dynamics requires the knowledge of the amplitude of the exciting force which is usually set as a harmonic waveform whose frequency varies within a range of interest in the so called stepped-sine test which overcomes two limits of the hammer test. In fact, the stepped-sine test controls the force amplitude at each value of frequency in a given range by means of a feedback control which activates the measurement only when the target force is reached. Moreover a realistic value of force amplitude must be reached in order to explore the nonlinear behaviour of the structure. For this reason electromagnetic shakers are used by connecting the exciter to the test article by means of a stinger. Different strategies are used to detune the dynamics of the test article itself from the dynamics of the exciter , they are based on the design of the stinger or on the constraint of the shaker. Unfortunately, these strategies are not useful in particular applications where a small perturbation may introduce large changes in the dynamics of structures or in general it may interfere with the measurement of the physical phenomena. In turbomachinery, this is the case when the nonlinear friction damping of the blade root attachment to the disk is studied. The amount of damping introduced by the contact is very small, nonetheless it is still of the same order of magnitude of the structural damping of the blade and it can be increased by a proper design of the blade root geometry (dovetail, fir-tree, flat, crowned). A contact excitation of the blade like the one required by the electromagnetic shaker would perturb the measurement of the friction damping, for this reason it is preferred to excite directly the base of the platform where the blade is attached. Another example is the study of the forced response of rotors excited by a multiple source of excitation, which is the application of the study presented in this chapter. Two main features of the bladed disk design concerning the dynamic response are veryimportant. 1) The rotor is subjected to an excitation pattern where each blade is loaded with a series of pulses within a complete disk rotation. The number and the intensity of pulses depend on the architecture of the engine preceding the rotor (number of combustion chambers, stages, stator vanes) and the wide spectral content is usually characterized by several harmonic components whose excitation frequencies are mainly a multiple of the rotor angular speed. The harmonic index which defines the multiplicity is called *Engine Order* (*EO*). 2) A rotating disk is nominally characterized by cyclic symmetry properties (Thomas, 1979) where one fundamental sector constituted by the blade and its associated disk sector is repeated for a number of times around the rotation axis. The periodical repetition of geometric and material features (*tuned* configuration) generates a particular dynamic behaviour of the disk where natural frequencies, and corresponding modes, can be grouped into families featuring the same deformed shape of the vibrating blades and disk (Castanier & Pierre, 2006).

The main objective at design stage is to avoid any matching of the rotor natural frequencies with the excitation frequency of each harmonic component of the load spectrum which shortens the life of the component due to HCF failure of blades. Unfortunately, bladed disk dynamics may show unexpected response due to small differences between sectors nominally identical which determine a *mistuned* configuration of the system (Castanier & Pierre, 2006, Kenyon & Griffin, 2003). In worst cases, a distorted and localized forced response occurs in particular for high modal density regions where one or more blades vibration amplitude is higher than the tuned response amplitude. Causes of mistuning are multiple: irregularities associated to the material properties constituting the blades, asymmetry generated during assembly or during service due to wear assessment, different contact condition at the blade root joints, snubbers, shrouds in case of non-integral bladed disks, Foreign Object Damage (FOD) or modifications introduced during maintenance and repair processes. Another source of mistuning comes from the so-called underplatform dampers (UPDs) which are metal parts located within cavities between contiguous blades. UPDs' aim is to limit the vibration amplitude of blades in case of excitation at resonance by means of friction damping at the contact surfaces. Mistuning is introduced at the contacts which couple directly the blades subjected to wear and change of the contact position during service. Interesting studies were carried out where UPDs are seen not only as a further source of mistuning to take into account but also as a mean to mitigate negative effects of inherent mistuning (Pierre et al., 2002, Petrov et al., 2010) even by distributing a number of UPDs smaller than the number of the blades (Avalos & Mignolet, 2008) or by systematically using UPDs having different mass (Gotting et al., 2004).

Fimite Element Modelling, mistuning identification and numerical results obtained by taking into account mistuning in the design must be validated with tests on dummy or real disks. Test campaigns are performed on test rigs which can be classified in two groups having a basic dual property: the first group has rotating disks with fixed excitation source, on the contrary the second group supports a static rotor excited by a travelling excitation. Both test rigs have advantages and limitations which must be addressed with respect to the objective of the test campaign.

Rigs with rotating test articles are generally more complex than static rigs, first of all due to the design of the disk containment which must fulfil safety requirements in case of blade release (usually they are built in a pit or within a protective bunker). Moreover, different levels of complexity are found in the comprehension of the mistuning phenomenon if tests are performed on real engines and cold flow rigs since fluid-structure interaction must be considered. In detail, correlations between mistuning and aero-coupling, damping and elastic effects of gas flow which decreases the magnification factor of the mistuned response (Beirov et al., 2008) and the influence of mistuning on flutter instabilities onset can be studied. On the contrary, fluid-dynamics effects can be avoided in vacuum chamber (spin test rig). For both types of rotating rigs, however, gyroscopic and centrifugal force effects (e.g., blades untwisting, dependence of UPDs contact pressure on disk velocity) should be taken into account. The excitation provided by the gas flow must be substituted in vacuum rigs with an alternative source. Broad applications of non-contact excitation provided by magnetic fields are found in literature by means of permanent magnets or electromagnets whose attractive force is exerted on ferritic bladed disks. Permanent magnets are compact, not expensive exciters which can be easily bought at different material grade of magnetization working in some cases at a temperature up to 200°C. One blade is excited by the magnets which are equally spaced around the circumferential coordinate and is loaded during one disk rotation with a number of pulses equal to the number of magnets. The excitation frequency is therefore strictly proportional to the rotation frequency by the number of magnets which determines the engine order of the force pattern. Electromagnets allows to release the excitation frequency from the rotation speed by using an AC feeding (Prchlik et al., 2009) in spite of a more complexity of the excitation system: more room is required to place the EMs, heat must be dissipated (which can be a critical issue in vacuum), an amplifier system and a signal generator must be provided. Hybrid solution can be also found in literature as in (Rice et al., 2007) where magnetic excitation is used as an extra source of excitation in engine tests with gas flow in order to explore flutter onset at different engine orders for the same rotation speed i.e., for the same aerodynamic load, air mass flow and centrifugal effects. Moreover in (Szwedowicz et al., 2007) a really cheap exciter is constituted by an air jet simply generated by the pressure drop between room condition and the vacuum chamber. In this case only one exciter can be provided since it is not possible to control a given time delay between more pipes which are necessary if an engine order type excitation is required. On the contrary, a narrow, single pulse virtually contains all the engine order indices which can be calculated by Fourier series decomposition of the force profile along the spatial domain of the disk circumferential coordinate. A contacting excitation system is used in (Szwedowicz et al., 2006, Gilbert et al., 2010) by means of piezoelectric actuators applied to the root of the blades. Piezo-actuators need electric feeding supplied by an end-shaft slip-ring connected to the shaft. Since they are contacting devices, on one hand this technique may cause non-intentional mistuning which should be taken into account. On the other hand, a precise force control is achievable through the calibration of the piezo-devices. However, the dynamic perturbation induced by the exciters can be limited if they are glued to the disk (disk mistuning) which usually does not generate critical mistuning phenomena as perturbed blades do (blade mistuning).

The determination of the magnification factor requires the measurement of each blade response typically obtained by means of strain gauges which are again contacting devices which can cause a non-intentional blade mistuning (Beirov et al., 2009). Tip timing techniques (Heath & Imregun, 1998, Heinz et al., 2010) proximity probes and rotating laser Doppler vibrometer (LDV) purposely designed (Sever, 2004) or already available in commerce are valid equipments which do not interfere with the original bladed disk dynamics. The first two techniques measure the disk response with respect to a stationary reference system while the third technique allows measuring the disk response on a reference system rotating with the disk. Therefore, attention must be paid to correlate input and output data of the rig with simulation results since the transformation of the coordinate system from static to rotating and vice versa is required. A not negligible aspect involving LDV measurement is the necessity of visual access to the rotating bladed disk which may be in conflict with safety requirements.

The second group of test rigs carries a static rotor and a rotating excitation. The travelling wave is in this case generated by contacting (piezo-actuators (Kruse & Pierre, part II, 1997), shakers (Strehlau & Kuhhorn, 2010)) or non-contacting systems (electromagnets with AC feeding (Kruse & Pierre, part I, 1997), speakers (Judge et al., 2003, Jones & Cross, 2010)). Whatever the case, a control system must be purposely developed in order to activate the exciters with a given phase shift in time to mimic the engine order force pattern. Since the bladed disk does not rotate, the dynamic response is directly measured on the reference system fixed to the rotor by using the same devices already described. In particular, the rotating LDV is substituted with a scanning LDV which measures, in its basic configuration, a set of points previously defined on the plane including the rotor (Strehlau & Kuhhorn, 2010, Berruti et al., 2010). Compared to the case of rotating rotors in this case aerodynamic coupling effects are avoided as well as complications given by data transmission. Such test rigs allow to easily collect measurements of the dynamics of dummy or real bladed disks directly related to mistuning and are therefore suitable for validating mistuning identification methods and reduction techniques with precision.

When non-linear phenomena are present as friction damping occurring at joints, the control of the force input is of primary importance since the forced response is strongly influenced by the amount of the local displacements of the structure at the contact surfaces. The force pattern can be measured during tests or estimated by means of a calibration process. Forces generated by air jets, for instance, can be measured with manometer probe measuring the pressure field before testing (Chang & Wickert, 2002) or with probes embedded in the surface of the blades (Petrov, 2010) or estimated as in (Szwedowicz et al., 2007) by calculating analytically the peak force applied to the blade as the product of the velocity at the end of the pipe and the mass flow hitting the blade. In this case the calculation relies on some simplifications about the area of the impacting gas and the distribution of the pressure on the blade airfoil. A similar approach can be used for oil jets whose higher density is suitable to excite very large blades. Much lower force amplitudes can be reached by speakers which can be calibrated by recording the spectra of a white noise by means of a microphone (Judge et al., 2003, Jones & Cross, 2010). The dynamics of the speaker can be recorded and used as a calibration curve in order to guarantee the same intensity at each frequency, while the value can be estimated by measuring the pressure on the microphone multiplied by the area of the speaker. Attention must be paid to the air gap between each blade and speaker since a small difference can cause an unacceptable error on the time delay between the harmonic force profiles which actually load the blades.

Force measurement using the magnetic field as a source of excitation can be performed by placing a force transducer below the magnet. In the case of permanent magnets, since it is usually a single piece, compact device, it can be considered as an additional mass which must be taken into account in order to verify if the frequency upper limit of the force transducer drops to values within the tested bandwidth. In the same way, force transducers can be used with electromagnets. Unfortunately, electromagnets are more complex devices than the permanent magnets, since components like conducting wires, packed plates, and insulating materials are assembled together and may generate a non-linear dynamic behaviour of the exciter itself. Similar to piezo-actuators (Kruse & Pierre, part II, 1997), they must be supported by an adequate power amplifier system in order to generate the required force amplitude in particular for high frequencies if non-linear dynamics characterization induced by slipping phenomena is addressed.

In this chapter, the design, manufacture and calibration of a travelling wave excitation system for bladed disks is described and tested. The typical application of the system is the Frequency Response Function (FRF) measurement by means of the stepped-sine technique for the characterization of the complex phenomenon of mistuning and non-linear dynamic response in presence of friction contacts such as blade root joints, shrouds or underplatform dampers. The measurements provide accurate results to be compared with numerical prediction.

The essential requirements of a travelling wave excitation system for such kind of tests in presence of mistuning are:

- the absence of contact between blade and exciter
- the same excitation force amplitude on each blade modulated with a given time lag which is determined by the engine order index.

If the test setup involves the characterization of nonlinearities due to friction the requisites listed above are not enough, since the excitation system should be able to provide also:

- a high excitation force amplitude in order to produce slipping between the friction contacts,
- an accurate measurement of the force amplitude of the rotating force pattern.

The excitation produced by an array of electromagnets (EMs) is non-contact and is then suitable for this kind of measurements. Moreover, any travelling wave force can be generated on the disk by changing the phase delay of the waveform of each current signal flowing on the EMs. The system hardware and software are purposely designed to satisfy all the requirements listed above. The system set up requires an accurate calibration process to generate the travelling force applied to the structure by each EM. The calibration consists of two steps: the first step is based on an instrumented EM which measures the force exerted on the structure, the second step is a tuning process in order to regulate all the other EMs for the generation of the same force amplitude on each blade.

In the first part of the chapter the hardware and software of the excitation system setup are described, together with a summary of the theoretical background that is necessary for the EM design. In the second part the system set up and calibration is presented by focusing on the proposed tuning method that allows the calibration of the EMs on site i.e., under the disk in their final position. In the last part of the chapter an example of application on a dummy bladed disk is presented.

2. The excitation system

The excitation system sketched in Figure 1a) is based on an electromagnetic unit, that is the electromagnet + the aluminium support (EM) repeated cyclically in order to excite a cyclic structure like a blade array where one EM is placed under each blade (Berruti et al, 2010, 2011). Each EM is mounted with a certain gap with respect to the blade so that the blade is not in contact with the exciter. The force is generated by the magnetic induction flowing through the EM, the air gap and the ferromagnetic material of the blade. The EMs are mounted on an aluminium circular plate (Figure 1-b). The aluminium is chosen in order to not interfere with the magnetic flux flowing in the EMs. Each EM is fed with a voltage with a given amplitude and phase. The signal generator system National Instruments cRIO 9074 is connected to the audio amplifiers Crown CDi 2000 (2 channels each, 800 W power supply for each channel). Each EM is connected to one channel. A LabVIEW software is purposely developed in order to generate a number of voltage harmonic signals equal to the number of EMs. The voltage signals are amplified in order to feed the EMs with a given amplitude and phase. The resulting excitation system has the following features:

- by feeding each EM with a voltage with a different phase from the neighboring it is possible to reproduce an engine order (EO) type excitation typical of the excitation of turbine disks in the engine;
- by controlling the force amplitude it is possible to keep it constant during the step-sine test;
- by tuning the EMs with respect to a reference one it is possible to reach the same excitation force amplitude on each EM.

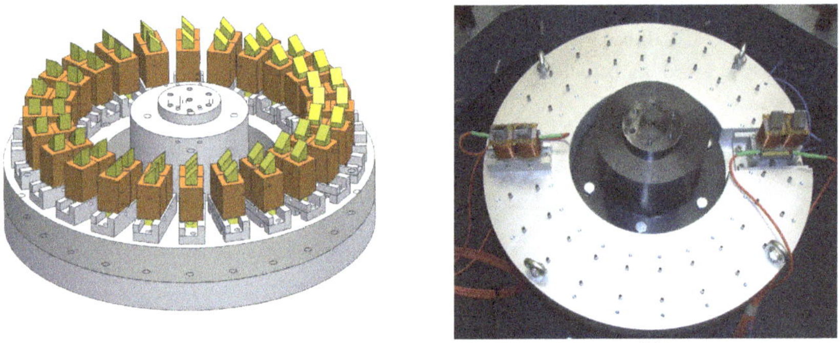

(a) (b)

Fig. 1. a) The cyclical excitation system; b) The aluminium circular plate with two EMs.

2.1 The electromagnet unit design

The electromagnet design strictly depends on the type of application. The general design criteria are hereinafter addressed. The electromagnet is designed in order to satisfy two essential criteria related to i) the geometry of the cyclic structure (bladed disk) excited by the EM and ii) the maximum force and maximum frequency of the planned travelling wave excitation, in detail:

- the EM size must be limited according to the room between two adjacent blades of the disk under test. As an example, the bladed disk shown later in the application allows a maximum square cross area for a commercial EM core S=16x16mm². The U-shaped core outer dimensions are therefore automatically determined: 64X48x16 mm.

- the EM must produce a force amplitude up to 10N in a mechanical frequency range from 100Hz to 600Hz given an average air gap l_{air}=2mm minimum. The last parameter is set in order to consider the variation of the gap during vibration negligible. For this reason l_{air} will be considered constant during the design process.

Figure 2 shows the designed EM unit which must face the blade with a gap that can be adjusted according to the application. Each electromagnet is made of two coils (1) of wire wrapped around a U-shaped core of ferromagnetic packed plates. Two prismatic ferromagnetic extensions (2) are glued at the two ends of the U shaped core in order to face the blade at the stagger angle α. This solution was preferred instead of producing a single piece EM core in order to change the prism in case of bladed disks with different stagger angles. The prismatic extensions are made up of SMC, i.e. a sintered ferromagnetic material which strongly limits the eddy currents.

Fig. 2. Detail of the EM unit.

2.2 The theoretical model of the magnetic force

The attractive force on the blade can be estimated by means of a simple model of the magnetic circuit of the electromagnet facing the blade. Electromagnet and the equivalent magnetic circuit are sketched in Figure 3 a) and b) respectively. The two coils are modelled as two generators of magnetomotive force Ni, where N is the number of turns for each coil and i is the current. The part of the circuit corresponding to the air and the part of the circuit corresponding to the ferromagnetic materials are characterized by the values of the reluctance $\Re$:

$$\Re = \frac{l}{\mu S} \tag{1}$$

where l and S are respectively the length and the cross area of the part of the circuit characterized by $\Re$, μ is the magnetic permeability. $\Re_{steel}$ is the magnetic reluctance of the steel core and of the steel blade facing the electromagnet while $\Re_{air}$ is the magnetic reluctance of the air gap. $\Re_{steel}$ is much less than $\Re_{air}$ since $\mu_{steel} >> \mu_{air}$. The circuit can then be simplified with the circuit of Figure 3 c) where $\Re_{steel}$ is neglected.

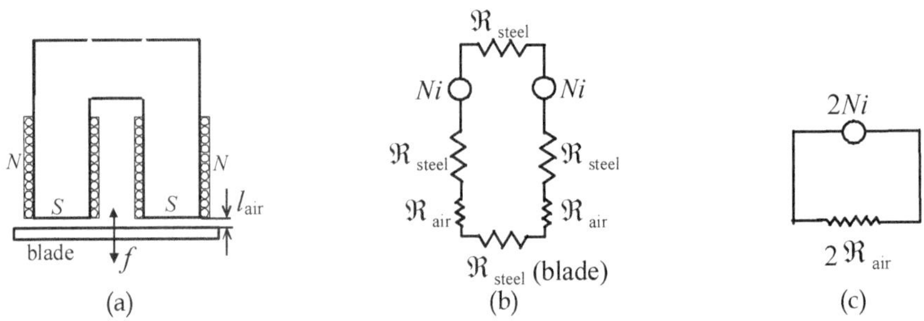

Fig. 3. a) Sketch of the electromagnet facing the steel blade; b) Equivalent magnetic circuit; c) Simplified equivalent magnetic circuit.

The following relationships can be written for the circuit of Figure 3 c):

$$2Ni = 2\Re_{air}\Phi \rightarrow 2Ni = 2\frac{l_{air}}{\mu_{air}S'}\Phi$$
$$2Ni = 2\frac{l_{air}}{\mu_{air}}B \rightarrow B = \frac{Ni \cdot \mu_{air}}{l_{air}} \quad (2)$$

where Φ is the magnetic flux, l_{air} is the air gap, S' is the area of each coil of the EM facing the blade ($S'=S/\sin(\alpha)$), B is the magnetic induction. The attractive force f exerted from the EM on the blade is the sum of two forces f_{coil} exerted by the two branches and is related to the magnetic induction B by the following relationship (principle of conservation of energy):

$$f = 2f_{coil} = 2\left(\frac{B^2 \cdot S'}{2 \cdot \mu_{air}}\right) \quad (3)$$

Substituting Eq.(2) in Eq.(3), the following relationship between the force and the current holds:

$$f = \frac{(Ni)^2 \cdot \mu_{air} \cdot S'}{l_{air}^2} \quad (4)$$

When the electromagnet is fed by an alternating current $i(t) = I\sin(\omega t + \varphi)$, with ω electrical angular frequency $\omega=2\pi f_{el}$ (being f_{el} the electrical frequency in Hz) and φ a given phase shift, the force in Eq.(4) becomes:

$$f = \frac{N^2 \cdot \mu_{air} \cdot S'}{l_{air}^2} I^2 sen^2(\omega t + \varphi) = \frac{N^2 \cdot \mu_{air} \cdot S'}{l_{air}^2}\frac{I^2}{2}\left(1 - \cos(2\omega t + 2\varphi)\right) = f_s + f_a \quad (5)$$

where f_s and f_a are the static and the alternate component of the attractive force. The alternate component of the attractive force becomes:

$$f_a = \frac{(N \cdot I)^2 \cdot \mu_{air} \cdot S'}{2l_{air}^2}\cos(2\omega t + 2\varphi) =$$
$$= F_A\cos(2f_{el} \cdot 2\pi t + 2\varphi) = F_A\cos(f_m \cdot 2\pi t + 2\varphi) \quad (6)$$

The force f_a is the force that excites dynamically the blade at the mechanical frequency $f_m = 2 f_{el}$. The blades are then excited with a "mechanical" frequency f_m that is two times the "electrical" frequency f_{el}.

The force amplitude F_A is determined by the product NI. The current amplitude should be limited to a maximum value of 10A in order to avoid risk for the insulation due to the generation of heat.

In the next paragraph the relationship between the exciting force amplitude F_A and the value of the voltage amplitude V generated by the amplifier is given, since it is of more practical use to know the value of voltage that must be generated by the amplifier in order to excite each blade at a desired force.

2.3 The electrical circuit model

In order to characterize the electrical circuit of the EM the amplitude value of the AC current I is measured for different values of amplitude and frequency of the AC applied voltage V. Figure 4 shows the value of the measured current amplitude I versus the input voltage amplitude V for different frequency values. Note that the trends are linear and for a given voltage value the current (and therefore the amplitude of the force F_A, Eq.(6)) flowing in the EM decreases as the frequency increases. The admittance quantity $Y = I/V$ is diagrammed in Figure 5 (experimental values are presented with black dots). Y is a useful quantity which characterizes the EM in order to know what is the current (and then the actual force) that can be expected given a value of voltage V (representing the generated signal) at a given frequency. This information is of primary importance if controlled force, step-sine tests must be performed. In this case in fact, the voltage must vary during the frequency sweep in order to guarantee the same output current amplitude and therefore the same force amplitude.

Assuming a simple RL model of the electrical circuit the value Y can be calculated as:

$$Y = \frac{1}{\sqrt{R^2 + \omega^2 L^2}} \tag{7}$$

where R is the resistance, L the inductance and ω the angular frequency. The R value is measured at the winding ends (R=0.3 Ω). The L value (5.5 mH, standard deviation 0.7 mH) is estimated from Eq.(7) as a mean value of all the inductances calculated from each measured Y values. The comparison of the calculated trend of Y (solid line) with the "measured" Y in Figure 5 shows a good agreement. The simple RL model can be therefore used to calculate the value of Y also outside the measured points. Substituting Y which includes the information about the electrical frequency, $I=Y(\omega)V$, in Eq.(6):

$$\begin{aligned} f_a &= \frac{(NV)^2 \cdot \mu_{\text{air}} \cdot S'}{2 \cdot \left(R^2 + \omega^2 L^2\right) l_{\text{air}}^2} \cos(2\omega t + 2\varphi) = \\ &= F_A \cos(2\omega t + 2\varphi) \end{aligned} \tag{8}$$

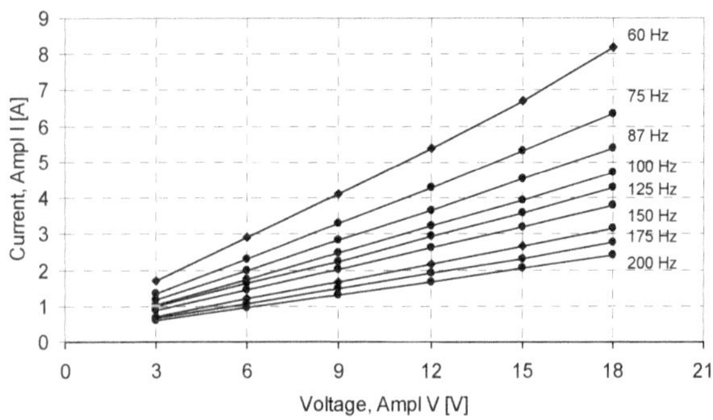

Fig. 4. Measured output current amplitude vs. input voltage amplitude for different input voltage frequency.

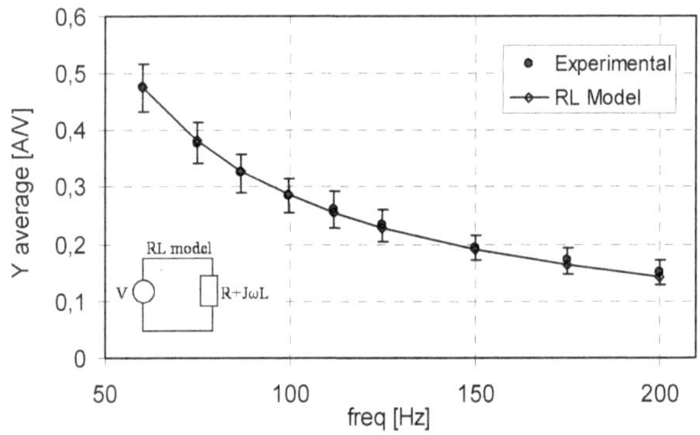

Fig. 5. Comparison of admittance values Y derived from experiments and from *RL* model.

Equation (8) shows the relationship between the voltage amplitude V, the force amplitude F_A and the input electrical frequency (ω):

$$V^2 = \frac{2 \cdot F_a \cdot \left(R^2 + \omega^2 L^2\right) \cdot l_{\text{air}}^2}{\mu_{\text{air}} N^2 \cdot S'} \tag{9}$$

Since $R^2 << \omega^2 L^2$ (R=0.3 Ω, L = 5.5 mH, Berruti et al., 2010, 2011) the term R^2 can be neglected. Considering also that $\omega = 2\pi f_{el}$, Eq. (9) can be rewritten in the two forms:

$$\begin{aligned} V &= \left(\sqrt{\frac{4 \cdot \pi^2 \cdot F_A \cdot L^2 \cdot l_{\text{air}}^2}{\mu_{\text{air}} N^2 \cdot S'}}\right) \cdot f_{el} \\ V &= \left(\sqrt{\frac{4 \cdot \pi^2 \cdot f_{el}^{\ 2} \cdot L^2 \cdot l_{\text{air}}^2}{\mu_{\text{air}} N^2 \cdot S'}}\right) \cdot \sqrt{F_A} \end{aligned} \tag{10}$$

The two relationships in Eq. (10) put in evidence that for a given value of the excitation force amplitude F_A, the corresponding value of the voltage V feeding the EM changes proportionally with f_{el}. In the same way, for a given frequency value f_{el}, the voltage amplitude value V changes proportionally with $\sqrt{F_A}$. These two linear trends of Equation (10) (V vs. f_{el} and V vs. $\sqrt{F_A}$) will be confirmed by the experimental tests as shown in the following sections. The formulas presented in this paragraph are useful to understand the dependence of the output parameter of the exciter (force amplitude F_A varying harmonically with a given frequency f_m) from the input parameter (the voltage amplitude V varying harmonically with a given frequency f_{el}) and to decide the dimension of the EM according to the maximum force that is generated, but are not sufficient to determine the exact numerical values since some parameters like the inductance L, the air gap l_{air} or the active cross section in front of the air gap S' are not known with precision. For this reason a direct calibration of the EMs is required as described in the following paragraph.

2.4 Calibration of the EMs

The EMs used in the multiple excitation system are nominally identical, however, in order to check the EMs manufacture quality, each EM unit is calibrated on a calibration bench shown in Figure 6 a). A force transducer (1), carrying a ferromagnetic anchor (2), is connected to an inertial mass (3). The anchor has the same local shape and size of the disk blade that is tested. The part of the inertial mass close to the anchor is made of aluminum (4) in order to not interfere with the magnetic flux. The EM faces the anchor with a given gap (2 mm). A control software developed in Labview is here used to adjust the harmonic input voltage of the EM amplifier in order to obtain the desired force amplitude (with a tolerance of ±1%) measured by the force transducer connected to the anchor.

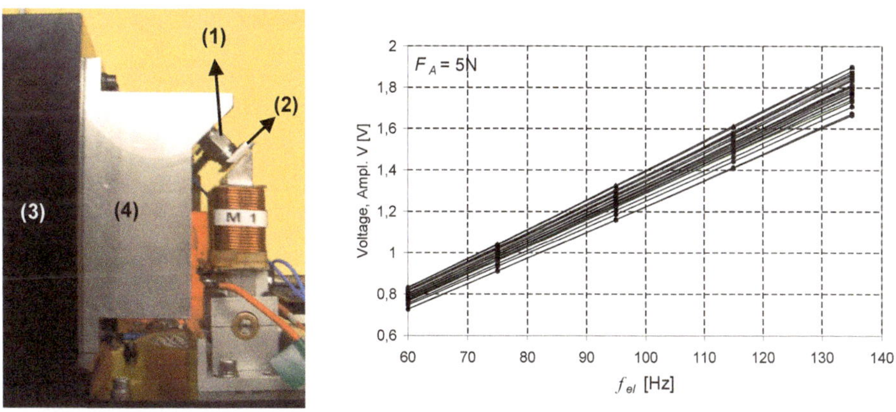

(a) (b)

Fig. 6. a) The EM on the calibration bench; b) The 24 EMs calibration curves.

In Figure 6 b) the calibration curves defined as the input voltage amplitude (V) vs. the input electrical frequency (f_{el}) are shown for the 24 electromagnets that are used later in the application by keeping the force amplitude F_A = 5N constant on the anchor. The repeatability error on the measured voltage values obtained by repositioning the same EM on the rig is 2% maximum. Despite the good calibration repeatability and the nominal equal shape, the calibration curves of the 24 EMs are different from each other. The highest voltage difference for the same electrical frequency is 13% with respect to the mean value . The difference of the calibration curves is associated to the EM manufacture process. The linear trend of the calibration curves in Figure 6 b) confirms the validity of the relationship of Eq.(10) where the amplitude voltage V is proportional to the electrical frequency f_{el} for a given force amplitude F_A.

2.5 The control of the excitation force amplitude

The control of the excitation force amplitude is obtained by means of a device, called Force Measuring ElectroMagnet (FMEM) which allows the measurement of the force exerted by one EM (Firrone & Berruti, 2011). The device is shown in Figure 7 a). The FMEM includes an EM nominally equal to the other EMs which is mounted on a particular support carrying a piezoelectric force transducer (1). The transducer measures the horizontal component (F_M) of the force (F_A) produced by the EM and applied to the blade. The reaction of the vertical component of F_A as well as the mass of the EM is supported by the vertical leaf spring at the base of the FMEM. By knowing the relationship occurring between F_A and F_M at different frequencies, it is possible to control the force F_A applied to the blade by controlling the force F_M measured by the force transducer.

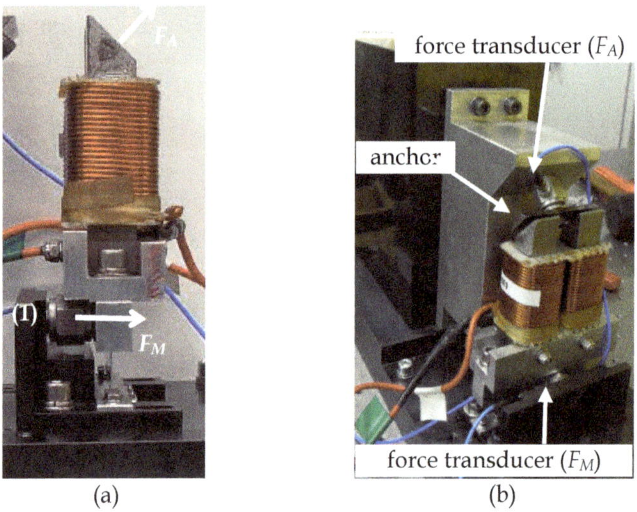

Fig. 7. a) The force measuring electromagnet (FMEM); b) the FMEM on the calibration bench.

The FMEM needs a special calibration process in order to establish the relationship between the force F_A and the force F_M measured by the force transducer (1) at different frequencies. The FMEM is therefore calibrated on the same calibration bench used for the other EMs as shown in the picture of Figure 7 b).

The FMEM faces the anchor with a given gap (2 mm). Once the harmonic force amplitude F_A is controlled by the software to a given value, the force amplitude F_M is measured by the FMEM transducer and recorded. The plot F_M/F_A vs. f_{el} is the calibration curve shown in Figure 8.

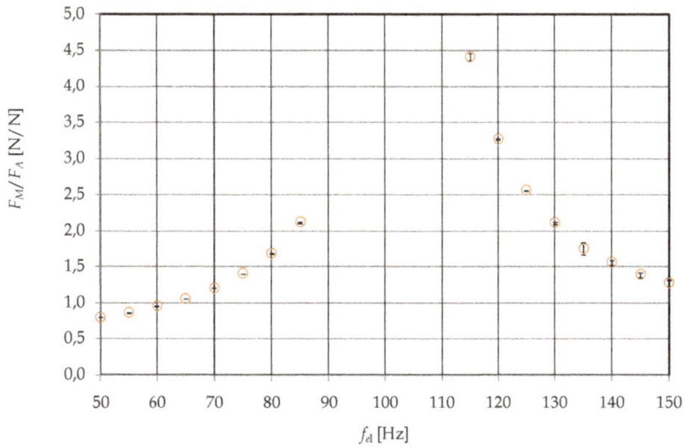

Fig. 8. The FMEM calibration curve.

The curve of Figure 8 shows that the FMEM has a natural frequency within the range of interest. A hammer test confirmed the presence of a natural frequency of the FMEM at 206 Hz (which corresponds to f_{el} = 103 Hz). In order to avoid the resonance condition of the FMEM that can lead to a poor measurement repeatability, the FMEM will be employed only in the ranges of frequency outside the resonance (f_{el} = 50→75 Hz and f_{el} = 140→150 Hz).

2.5.1 Reliability of the FMEM on the calibration bench

In order to estimate the reliability of the force amplitude value imposed and controlled by means of the FMEM, different calibration tests are performed on the calibration bench of Figure 7 b). The tests are listed below.

1. Check of the force control. Calibration curves are measured for different nominal values of the force on the anchor (F_{Anom} =1, 2, 3 , 4, 5, 6 N) with the same air gap of 2 mm. As a result, for each F_{Anom} a calibration curve like that of Figure 8 is obtained. An "inverse calibration" is then performed on the same bench in order to simulate the same control force procedure that will be used on site under the bladed disk. In this case the FMEM transducer that measures F_M is controlled by the software. The set-point controlled by the software at different frequencies is $F_M = (F_M/F_A) \cdot F_{A,nom}$ where $F_{A,nom}$ is the nominal force value and (F_M/F_A) is the calibration curve. When the FMEM transducer reaches the target value F_M, the actual force $F_{A,act}$ exerted on the anchor is measured. The difference between the measured value $F_{A,act}$ and the nominal value $F_{A,nom}$ proved to be always within 1% of $F_{A,nom}$ for all the tested frequencies confirming the quality of the force control procedure.
2. Check of the FMEM repeatability. $F_{A,nom}$ is set to 5 N, and the air gap to 2 mm. Different calibration curves F_M/F_A are obtained after repositioning the FMEM on the calibration bench. The calibration curve F_M/F_A plotted in Figure 8 is the average result $(F_M/F_A)_{AVG}$. The "inverse calibration" is then performed by controlling the force value F_M measured by the transducer on the FMEM support. The set-point controlled by the software at

different frequencies is calculated as $F_M=(F_M/F_A)_{AVG}\cdot F_{A,nom}$. When the FMEM transducer reaches the target value F_M, the actual force $F_{A,act}$ exerted on the anchor is measured. The error of $F_{A,act}$ with respect to the nominal force value of 5N is 3% in the worst case.

It can then be concluded that after this calibration procedure it is possible to obtain the desired value of force amplitude F_A on the anchor with a precision inside 3% by controlling the force F_M on the FMEM support.

3. The test rig

The excitation system is designed to study the complex dynamics of bladed disks for turbomachinery applications. In detail, the aim is to collect a database of measurements on real or dummy bladed disks in order to highlight the effect of mistuning and friction damping. The experimental results will be compared with results from numerical simulation tools. The multiple excitation system described above is set up on the test rig *Octopus* designed to test bladed disks carrying underplatform dampers (UPDs) between the blade platforms (Berruti, 2010). The test rig shown in Figure 9 a) can be divided into two parts, i.e., the central support carrying the real or dummy disk and an outer ring supporting the arm structures used to couple the bladed disk with the UPDs. The required number of arms depends on the number of blades of the specific disk. In this case the test rig is prepared to test a dummy integral bladed disk (*blisk*) with 24 blades, therefore a number of N_b=24 arms must be provided. Nevertheless, the outer ring can be used in part or can be extended depending on the specific disk and its number of blades. Figure 9 a) shows the test rig with the dummy blisk (1) already mounted on it.

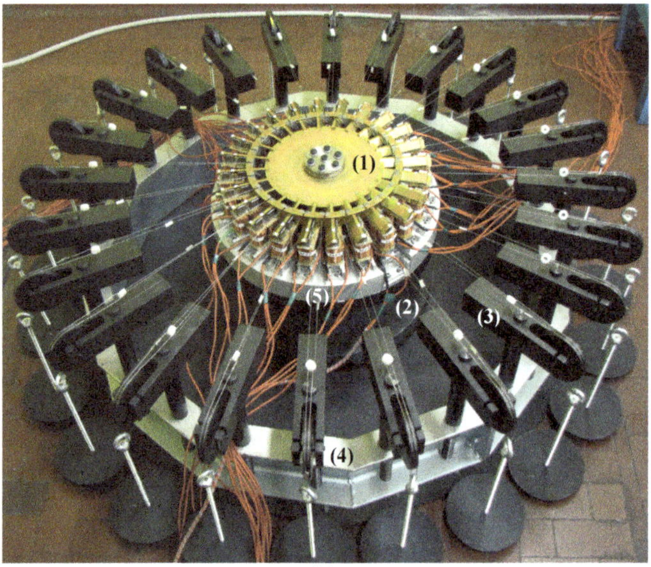

(a)

(b)

Fig. 9. a) The test rig Octopus; b) detail of the UPD.

The blisk is constrained on the central fixture (2), which is basically a big inertial mass (about 400kg) with two coaxial cylinders mounted on a circular plate laid on a rubber sheet. The arm structures (3) support one pulley each and are equally spaced around the fixture. Two wires attached to each UPD (Figure 9 b)) pass over the arm supporting the pulley and are connected to a loading plate which carries dead weights simulating the centrifugal force on the UPD. The pulley has low friction ball bearing and is mounted on a pin tightened on the circular outer ring (4) (about 1.5m diameter) centered with respect to the disk fixture. Each arm carrying the pulley can rotate around the pin in order to align with precision the wires along the radial direction of the centrifugal force acting on the UPDs. The outer ring is fixed to the floor and is not directly connected to the central fixture in order to minimize the transmission of vibrations from the fixture to the ring. The dummy blisk is centered on the fixture by means of a hub integral with the blisk which is tightened with seven screws using a counter torque wrench in order to save the cyclic symmetry of the structure as much as possible. The rig was designed to hold 30kg (294N) for each damper which is a realistic centrifugal force on a real UPD having a mass of about 5g.

3.1 The dummy disk

The dummy disk was manufactured in order to fulfil two requirements: i) to test the electromagnetic excitation system and the control software of the rotating force and ii) to study the dynamic behaviour of a cyclic structure with underplatform dampers (UPDs) under a specific engine order *EO* excitation in order to validate the numerical model of the blisk with friction contacts between blade platforms and UPDs. The FE model of the blisk is shown in Figure 10 a). The mode shapes of a bladed disk are characterized by a number of *Nodal Diameter* (*ND*), i.e., lines passing for the center of the bladed disk where the modal displacement is null. As an example, Figures 10 b)-d) show modes at ND=0,2,4. In particular, only one mode is associated to ND=0 (the so called umbrella mode) and to $ND=N_b/2$ while two modes share the same *ND* when $0<ND<N_b/2$. In this case the natural frequencies of the two modes are equal and the two modes are orthogonal (their scalar product is null). This information can be resumed in a plot where the natural frequencies are plotted with respect to *ND* (Figure 11). Lines connecting the natural frequencies define the modal families of the blisk where modes share a type of motion of the sector. In this case, for instance, the first family (lower frequencies) is characterized by the first bending mode (1F) of the blade, while the second family (higher frequencies) is characterized by the second bending mode (2F, edgewise) of the blade. Label 'free' refers to the modal analysis of the blisk without damper while 'stick' refers to the modal analysis of the blisk coupled to the cylindrical underplatform damper. The 'stick' condition is simulated by constraining the UPDs to the blade platforms through hinge constraints along the line contacts between the cylindrical surface of the UPD and the flat surface of each platform. A cylindrical cross section of the damper is chosen (Figure 11 b)) in order to guarantee the contact between UPDs and blade platforms the simplest as possible. The design aimed at obtaining the first bending mode well isolate from the second bending mode in order to obtain a simple kinematics of the blade platforms in contact with the UPDs. Moreover, a 45° stagger angle is chosen for the blades since both 1F and 2F modal families must have components along the out-of-plane direction since the motion is detected by a Laser Doppler Vibrometer (LDV) which measures the orthogonal component of the velocity with respect to the plane of the blisk.

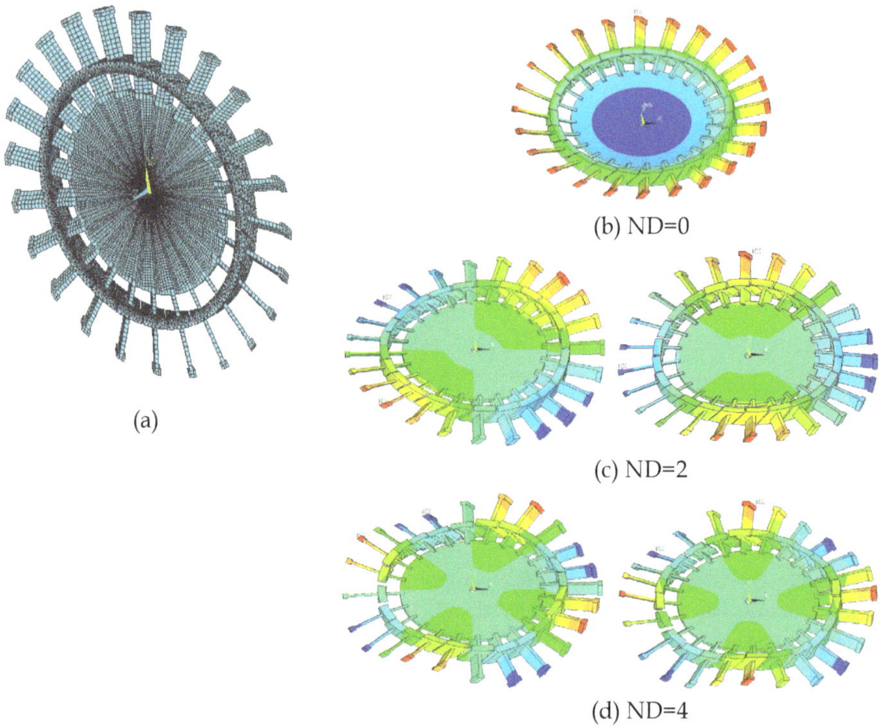

Fig. 10. a) FE model of the blisk; b) umbrella mode (*ND*=0); c) *ND*=2 orthogonal modes; d) *ND*=4 orthogonal modes.

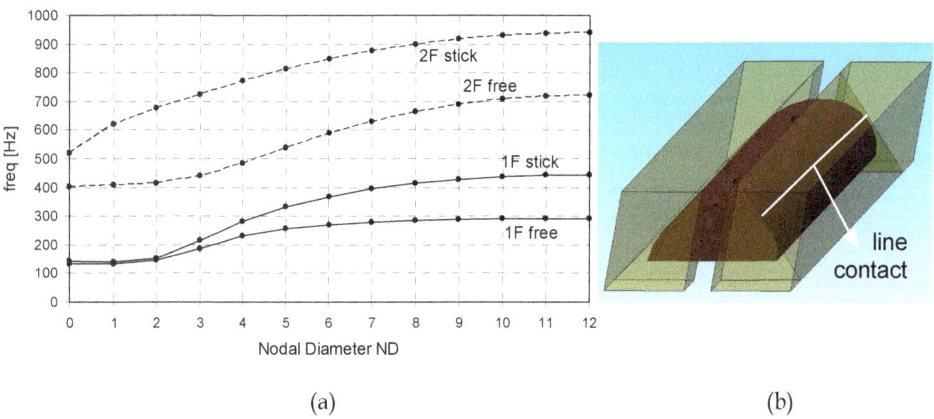

Fig. 11. a) First two bending families, without UPDs (free condition) and with UPDs constrained between the platforms (stick condition); b) line contacts between UPD and platforms.

It can be also noticed that in fully stick condition the highest frequency of the 1F family is under the threshold of 500 Hz. This requirement is determined by the EM currently used; it is possible to see from Figure 5, in fact, that the value of admittance drops for values of electrical frequency higher than f_{el}=200Hz (i.e. 400Hz mechanical frequency). This means that a too high voltages V may be required to excite the blisk for frequencies higher than f_m=400Hz if the required force amplitude F_A is of the order of magnitude of 10N. For this reason the study will be restricted to the first modal family (1F).

4. The excitation system under the disk

In the final configuration, each EM is positioned under the corresponding blade with an air gap of 2mm. However, before placing all the EMs, different tests are performed by mounting only the FMEM under the disk on a housing purposely machined in the aluminum base (Figure 12). In this case the FMEM is facing the blade instead of the ferromagnetic anchor of the calibration bench.

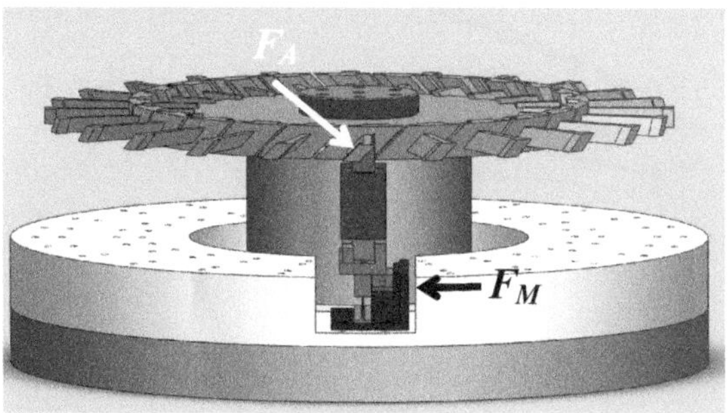

Fig. 12. The FMEM mounted under the disk.

Different tests are performed with the software control in order to obtain the desired $F_{A,nom}$ on the blade. The control is performed by the measurement of F_M (Figure 12). The software adjusts the voltage value until the desired value of force amplitude $F_M = (F_M/F_A)_{AVG} \cdot F_{A,nom}$ is reached by the transducer. This force control is repeated for different $F_{A,nom}$ and for different frequencies f_{el}. In each case the feeding input voltage amplitude V is recorded. In Figure 13 the voltage amplitude V is plotted versus the input electrical frequency f_{el} for different $F_{A,nom}$ values.

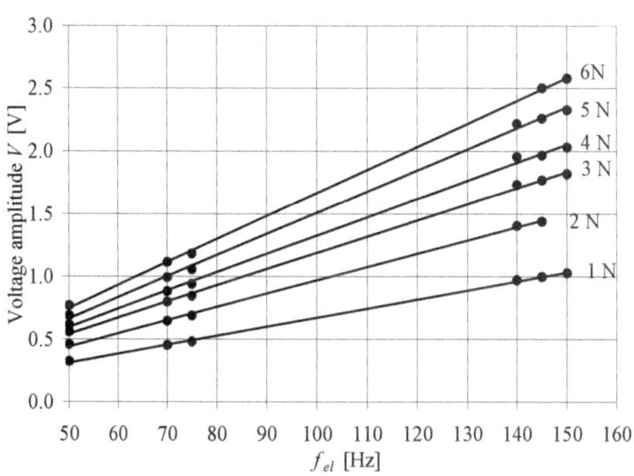

Fig. 13. Feeding voltage amplitude V vs. electrical frequency f_{el} for different excitation force amplitude $F_{A,\,nom}$.

It can be seen that the trend V vs. f_{el} is still linear for a given value of the force amplitude exciting the blade (compare with Figure 6 b) where the same measurement was performed on the calibration bench at F_A=5N). These tests confirm again the first theoretical relationship of Eq.(10). Moreover, the plots V vs. $\sqrt{F_{Anom}}$ (see Figure 14) are also linear for different electrical frequency values f_{el} confirming the second theoretical relationship of Eq.(10).

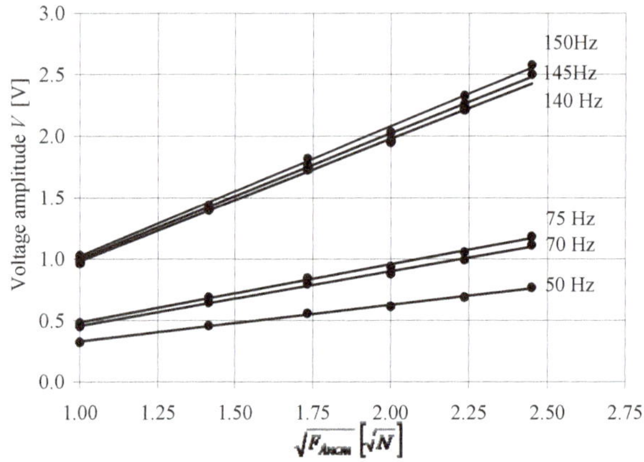

Fig. 14. Feeding voltage amplitude V vs. the square root of the excitation force amplitude F_A for different electrical frequencies f_{el}.

The linear trends V vs. f_{el} and V vs. $\sqrt{F_{Anom}}$ allow to calculate by interpolation the needed input voltage also for force amplitudes or for frequency values different from those employed during the calibration of the FMEM. It is interesting to compare the trends V vs. f_{el} obtained on the calibration bench (dashed lines) and under the disk (solid curves) for two different values of $F_{A,nom}$ (1 N and 5 N) as shown in Figure 15.

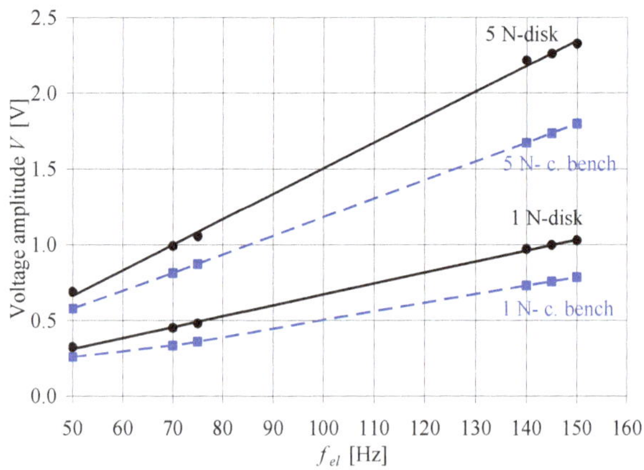

Fig. 15. Comparison of the feeding voltage amplitude V vs. electrical frequency f_{el} when the FMEM is on the calibration bench (dashed line) and under the blisk (solid lines).

It can be noted that higher input voltages are requested by the EM under the disk with respect to the case of the calibration bench in order to obtain the same value of $F_{A,nom}$. The difference is higher for higher frequencies. The difference of the input voltages is due probably to the higher magnetic field loss in the blade connected to the disk compared to the case of the single magnetic anchor on the calibration bench. This voltage difference between the blisk and the calibration bench highlights the importance of having a device like the FMEM which measures the force directly under the disk. The bench calibration of each EM is not enough since the calibration curve does not correspond to the desired $F_{A,nom}$ under the disk.

4.1 Verification of the blisk linearity

The main objective of the tests performed on the dummy blisk is the measurement of the amount of nonlinear friction damping introduced by the UPD. For this reason it must be verified that another source of nonlinear damping is not introduced by the test rig itself and in particular by the contact interface between the blisk and the central hub. The verification of the blisk linearity requires the measurement of the vibration response of the blades. The measurement is performed by means of the scanning LDV which detects the blade velocity along the disk axial direction. One measurement point is chosen for each blade in the same position.

The FMEM is positioned under the bladed disk (as shown in Figure 12) and a harmonic force is applied at different amplitude F_A gradually increasing in order to verify that the blade response measured by the LDV increases proportionally.

The FMEM is fed with an alternate voltage at an electrical frequency f_{el}, the mechanical excitation frequency value f_m (that is two times f_{el}) is chosen as far as possible from both the disk and FMEM natural frequencies in order to avoid calibration complications due to disk or FMEM dynamics. The disk and the FMEM Frequency Response Functions (FRF) measured by impact testing are shown in Figure 16.

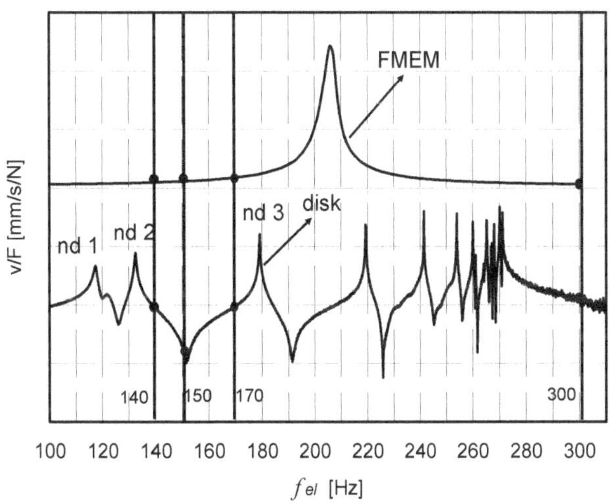

Fig. 16. FRF for the FMEM and for the bladed disk (y-axis is not to scale).

The verification of the disk linearity is performed at two different mechanical excitation frequencies (f_m=150Hz and 300Hz that is, f_{el}=75Hz and 150Hz) which are far from both the disk and FMEM natural frequencies. Once the force control loop has reached the target value F_A at $f_m=2f_{el}$, a trigger signal activates the scanning LDV to measure the velocity amplitude of the 24 blades. The blade velocity amplitudes vs. the blade number (*velocity profile*) can then be plotted.

In Figure 17 and Figure 18 the "velocity profiles" obtained for different excitation force $F_{A,nom}$ amplitude values are plotted in the case of f_{el}=75Hz and in the case of f_{el}=150Hz. It is possible to see that the velocity profile (see red dots for clarity) increases linearly with the force amplitude F_A. The comparison of the ratios of the different force amplitudes applied to the bladed disk and the corresponding average ratios of the velocity values marked with red

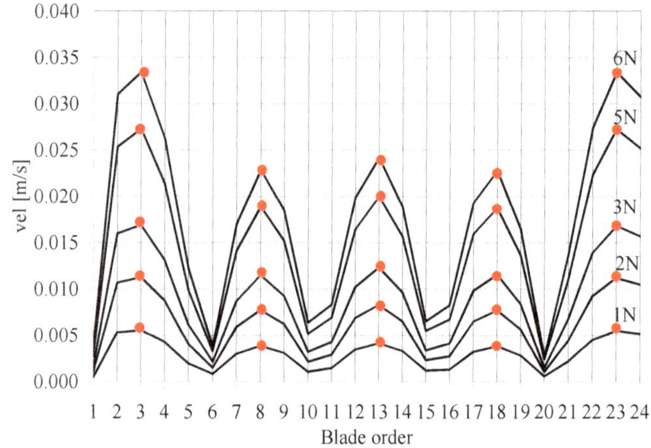

Fig. 17. Blade absolute velocity values for different excitation force amplitude, electrical frequency f_{el}=75 Hz.

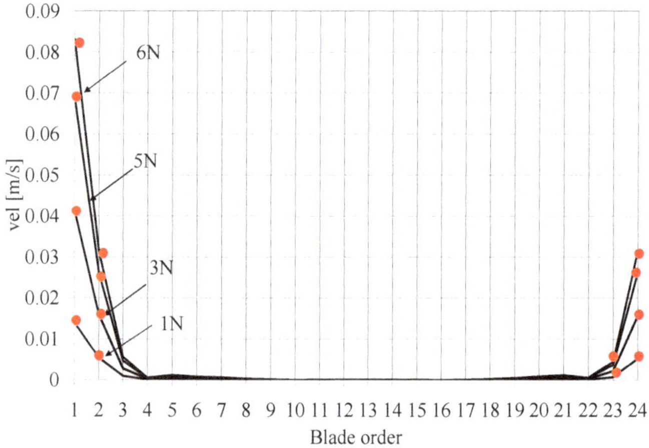

Fig. 18. Blade absolute velocity values for different excitation force amplitude, electrical frequency f_{el}=150 Hz.

dots shows a difference within 2% in the most cases, i.e., within the force value repeatability allowed by the FMEM. As a consequence, the dynamic response of the bladed disk without UPDs can then be considered linear.

4.2 The method for tuning the EMs

In order to obtain the same amplitude of the excitation force for each EMs a tuning process of the magnetic forces is necessary in order to adjust the input voltage for each exciter. The tuning involves an iterative process exciting directly one blade each time and measuring the indirectly excited blade responses. Two system configurations are needed for this calibration:

- configuration 1: only the FMEM is needed under blade 1;
- configuration 2: the 24 EM units are positioned under the bladed disk (after removal of the FMEM).

In the configuration 1 the FMEM is fed with an alternate voltage at the electrical frequency f_{el}. The control software adjusts the voltage amplitude in order to reach the target force value $F_{A,nom}$. As in the case of the blisk linearity verification the scanning LDV measures the velocity amplitude of the 24 blades as a "reference velocity profile" (blade velocity amplitudes vs. the blade number) which is recorded together with the voltage amplitude V and the electrical frequency f_{el}. The mechanical excitation frequencies f_m chosen for the tuning process (140, 150, 170, 300Hz) are highlighted in Figure 16. The chosen mechanical excitation frequency values are far from the disk and from the FMEM natural frequencies. In order to excite the blisk with these excitation frequencies in configuration 1 the FMEM must be supplied with an input voltage with half the value of the electrical frequency (f_{el} = 70, 75, 85, 150Hz). The velocity profiles are plotted in Figure 19 and Figure 20 for the four values of f_{el} feeding the FMEM.

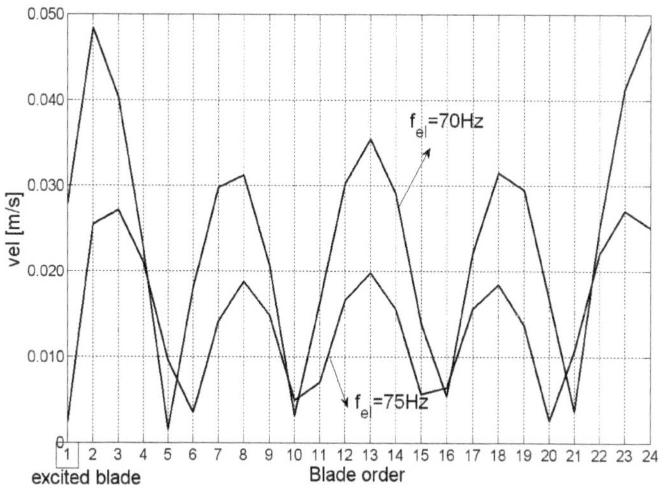

Fig. 19. Velocity amplitude profiles when the FMEM is switched on, force controlled to a nominal force amplitude $F_{A,nom}$=5N, electrical frequency 70Hz and 75Hz.

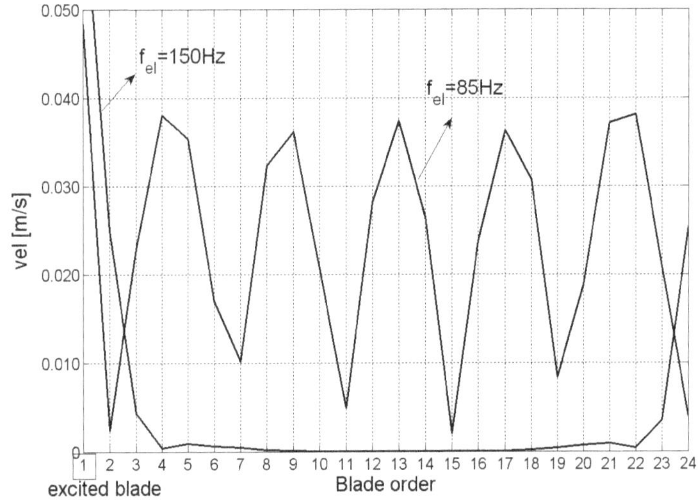

Fig. 20. Velocity amplitude profiles when the FMEM is switched on, force controlled to a nominal force amplitude $F_{A,nom}$=5N, electrical frequency 85Hz and 150Hz.

The velocity profiles of Figure 19 and Figure 20 are kept as reference profiles for the tuning process since they are obtained with a known force amplitude of 5N (measured on site) on the excited blade. Repeatability tests performed after repositioning the FMEM under the disk proved that the repeatability error of the profiles of figure 19 and 20 is lower than 3%.

The system is then set up according to the configuration 2: the FMEM is removed and the 24 EMs are positioned under the blisk. Only one EM per time is switched on. Figure 21 shows the 24 velocity amplitude profiles at the excitation electrical frequency f_{el}=75Hz when the calibration curves of Figure 6 b) are used to set the initial voltage amplitude V for the 24 EMs.

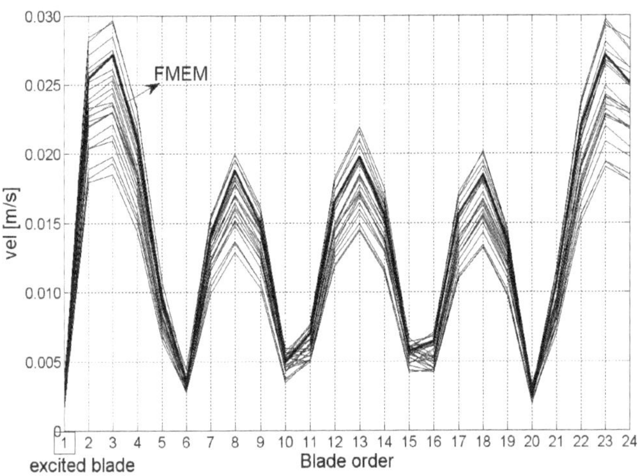

Fig. 21. The 24 velocity profiles, each of them measured with one EM switched on, at f_{el}=75Hz and nominal force amplitude $F_{A,nom}$ =5N, before tuning.

Each velocity profile is measured when only one EM per time excites the structure. The velocity amplitude of the directly excited blade is plotted always in the first position (*Blade Order 1*). On the same figure the velocity reference profile (bold line) obtained with the FMEM at f_{el}=75Hz in the previous step (configuration 1) is plotted. The curves of Figure 21 point out two phenomena:

- there are relevant differences among the velocity profiles when the working EM exciter changes (the standard deviation of the force values compared to the mean value of the force is about 16%),
- the velocity profile obtained by FMEM is inside the profile dispersion of the other 24 EMs.

Considering the dynamics of the blisk linear as proved before and neglecting the presence of mistuning phenomena since the excitation frequency is far from resonance conditions, the differences among the profiles in Figure 22 must be referred to different excitation force amplitudes. These differences are not related to the manufacture of the EM units since they are taken into account in the calibration curves of Figure 6b). Other causes producing different velocity profiles are probably due to i) different positions of the EMs under the disk blades that cause different air gaps, ii) different shape and material properties of the blade compared to the anchor used for calibration. These differences among the EMs highlight the importance of having a FMEM measuring the real force amplitude on site (i.e. under the disk). The 24 EMs can than be tuned to the reference FMEM. The tuning process aims at adjusting the voltage on the 24 EMs until they produce the same velocity profiles of the blisk as the one measured with the FMEM. A method is here proposed starting from the velocity profiles shown in Figure 21. The voltages are adjusted in order to obtain 24 velocity profiles overlapping the FMEM reference profile as shown in Figure 22.

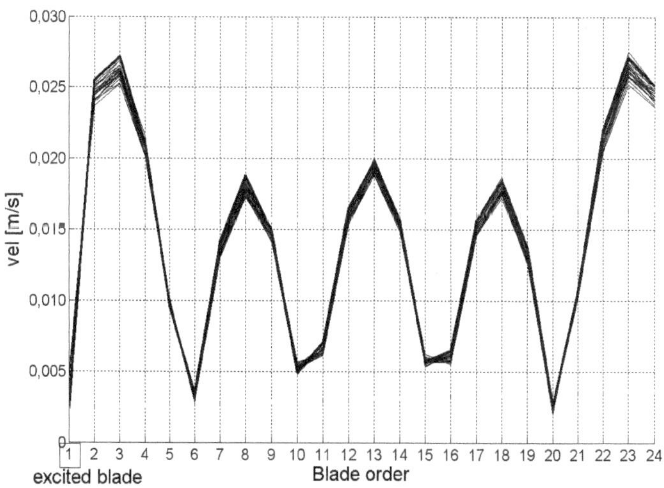

Fig. 22. The 24 velocity profiles, each of them measured with one EM switched on, at f_{el}=75Hz and nominal force amplitude $F_{A,nom}$=5N, after tuning.

In detail the method works as follows. Let $\dot{X}_{be}$ and $\dot{X}_{bFMEM}$ be the measured velocity amplitudes of the blade corresponding to the blade order b when respectively the EM e and the FMEM is switched on. Considering the system linear and assuming the FMEM as a reference, the ratio $\dot{X}_{be} / \dot{X}_{bFMEM}$ is proportional to the excitation force ratio $F_{A,e} / F_{A,FMEM}$ where $F_{A,e}$ is the actual excitation force amplitude on the blade facing the EM e and $F_{A,FMEM}$ is the nominal value of the force (in this case 5N).

For each response of Figure 21 (corresponding to the e-th EM working) the correction factor k_e is calculated as:

$$k_e = mean\left(\frac{\dot{X}_{be}}{\dot{X}_{bFMEM}}\right), \quad b = 1,\ldots,N_b \tag{11}$$

In order to get the same force amplitude for each EM unit the force is corrected as $F_{A,e}{}^{*}=F_{A,e}/k_e$ by adjusting the input voltage as $V_e^* = V_e / \sqrt{k_e}$ since the voltage is proportional to the square root of the force (Eq.10). The velocity profiles plotted in Figure 22 are measured after two iterations of the tuning process at f_{el}=75Hz. The dispersion of the force amplitude corresponding to the different velocity profiles with respect to the nominal value of 5N are less than 2%. It can be concluded that after this tuning process each of the 24 EMs produces on the excited blade the same force amplitude of 5N with a dispersion of 2%. The same calibration procedure is repeated for the other electrical frequency values (f_{el} =70, 85, 150Hz) in order to characterize the linear trend of the voltage as a function of frequency when the same force $F_M=(F_M/F_A)_{AVG} \cdot F_{A,nom}$ is required, as it was done in the calibration bench (Figure 13) with the FMEM. The voltage values at frequencies not directly tuned can be found by linear interpolation since the experimental evidence verifies the linear trends of Eq.(10).

It must be noted that the tuning method here proposed is very efficient since it avoids the calibration of one EM at a time on a separate bench which is time consuming. In fact, the tuning process does not require a particular starting voltage values of the EMs, in particular it is not necessary to start with voltage values coming from the EMs calibration (Figure 6b).

5. Example of test results

The test campaign is performed for the modes at ND=2, 4 without dampers (free, linear configuration) and with UPDs (nonlinear configuration). Each mode is excited with a travelling wave with an engine order EO=ND. First, two forced responses at two different excitation frequencies are shown for the blisk without damper (free system), second the FRFs are shown in terms of mobility, for the two ND at 7 different values of force amplitude F_{Anom} =0.1, 0.2, 0.3, 0.6, 1, 2.5, 5N when UPDs are mounted. Each FRF is the envelope of the maximum value of the FRF of the 24 blades at each frequency. The test results are selected since they show the capability of the rig to detect both the effect of the friction damping on the blisk dynamics and the presence of mistuning.

5.1 Forced response of the blisk without UPDs under the travelling force

A rotating force is generated in order to excite the ND=2 rotating mode shape. For this reason the voltage signal is generated for each EM in order to obtain a force pattern on the EMs equal to:

$$f_{a,n} = F_{Anom} \cos\left(f_m \cdot 2\pi t + (n-1)\frac{2\pi}{N_b} EO\right) \tag{12}$$

$$n = 1,\ldots,N_b \qquad f_m = 2 f_{el}$$

where *EO* is chosen equal to 2 in order to excite the ND=2 rotating mode shape. The force amplitude F_{Anom} is in this case very low since the system is linear (F_{Anom}=0.2N). The response is measured by measuring one point per blade by means of the scanning LDV in order to measure the rotating mode shape along the hoop direction of the blisk. The voltage signal of the first EM is used as a trigger in order to sinchronize the 24 signals measured by the LDV in 24 different time instants. The blisk is excited at f_m=134Hz which is far from the natural frequency f_{1F}=131Hz of the ND=2 rotating mode shape, first family. Figures 23 a)-d) show the

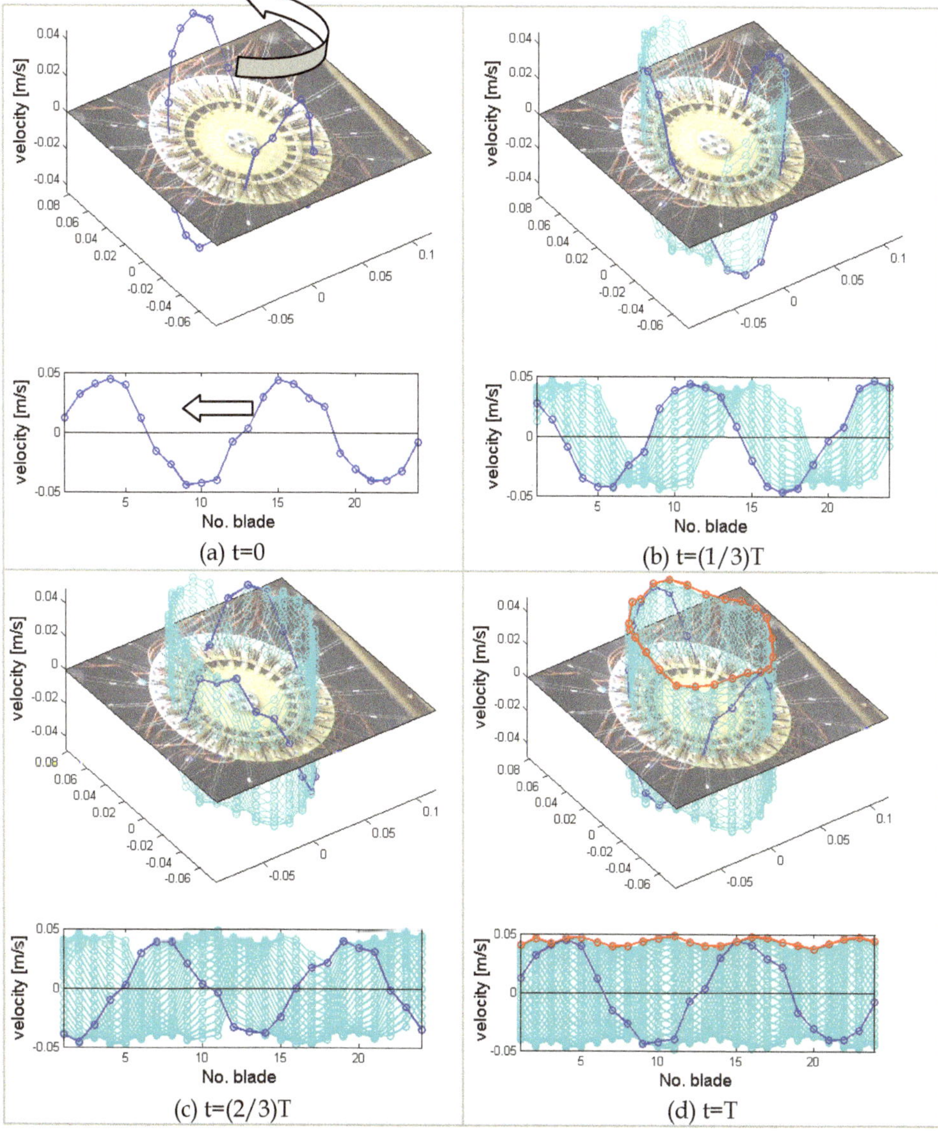

Fig. 23. Rotating Deformed Operative Shape, ND=2, EO=2, far from resonance.

Operative Deformed Shape (ODS) of the blisk for four equally spaced time istants within a complete rotation of the travelling force $T=EO(1/f_m)$. It is possible to see how the two orthogonal mode shapes as seen in the simulation of Figure 10 c) combines to generate a rotating mode shape at ND=2 following the rotating force. Figure 23 d) shows all the steps after one rotation is complete, the red line groups the maximum amplitude that each blade reaches in order to prove that the difference in amplitude is acceptable. If the same test is performed at a frequency which is closer to the natural frequency of the 1F mode, ND=2 (Figure 24 a)-d)), it is possible to see that the blades vibrate with different amplitudes proving the existence of inherent mistuning due to the asymmetries produced by the manufacture or the constraint. However, the mistuning can be considered 'small' since it only perturbs the ND=2 shape of the rotating mode which can be still identified and does not localize the energy of the system to a small number of blades.

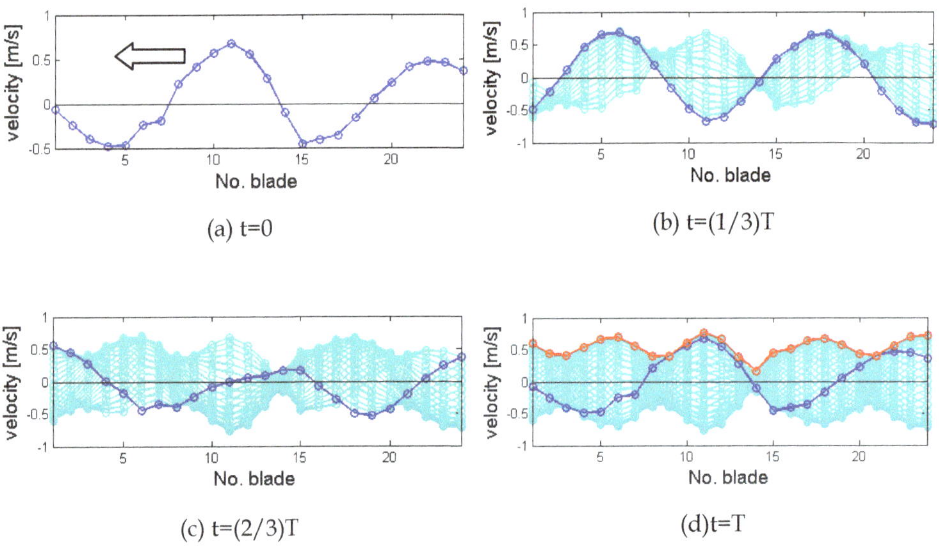

Fig. 24. Rotating Deformed Operative Shape, ND=2, EO=2, resonance condition.

5.2 The forced response with friction damping

The UPDs are positioned in their cavities and each of them is loaded with an equivalent centrifugal load of about 50N (5kg of dead weights) (Firrone et al., 2011). In Figure 25 the free (linear) and the nonlinear FRFs at ND=2 are plotted as the envelope of the maximum value of the FRF of each single blades at each frequency. The FRFs show that the amount of damping depends on the force amplitude value F_{Anom}. When F_{Anom} is low (F_{Anom}=0.1N) the relative displacement at the contacts is not large enough to provide a large amount of damping. At the same time the UPDs act more as an additional constraint whose main effect is to stiffen the structure as demonstrated by the increase of the frequency value of the peaks with respect to the free response. As the excitation force amplitude increases (F_{Anom}=0.2N, 0.3N), more

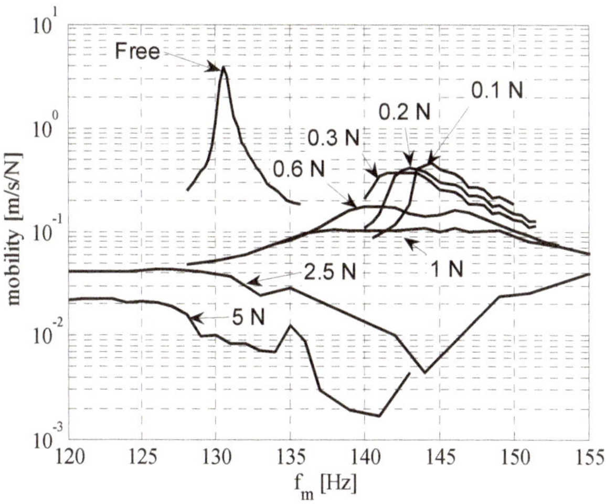

Fig. 25. FRF of the bladed disk (free and with UPDs) for different amplitude values F_A of the excitation force. CF =50N, EO=ND=2.

damping is provided to the bladed disk, a general reduction of the vibration amplitudes of the blades is observed and the distance of the peak frequency from the corresponding *'free'* value decreases since large relative displacements at the contacts globally determine a less stiff constraint for the blades. As the excitation becomes larger and larger (F_A=1N, 2.5N, 5N) the smooth peak is almost flattened while the frequency peak tends toward values lower than that of the free response since the UPDs act as an additional mass of the blisk. These results can be summed up as shown in Figure 26 in terms of the optimization curve where the frequency and the amplitude values of the response peaks are plotted vs. the variable parameter CF/F_{Anom}. It is possible to see that a minimum response of the blisk amplitude mobility exists corresponding to an optimum combination of damper mass (i.e., CF) and amount of excitation (i.e., F_{Anom}).

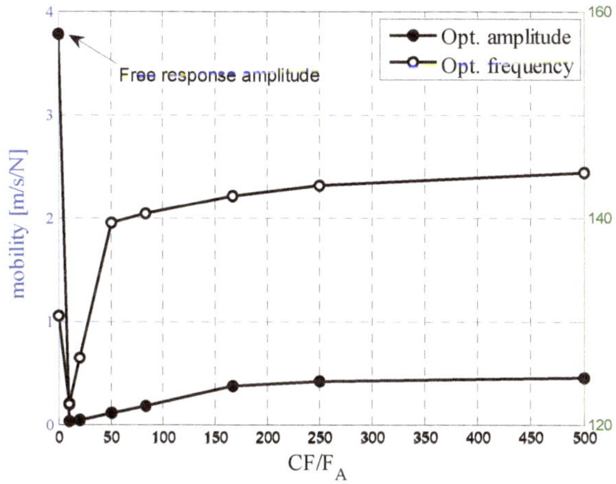

Fig. 26. Optimization curves at ND=2.

5.3 The forced response with mistuning introduced by the friction dampers

In order to show the effect of mistuning introduced by the UPD, the rotating mode shape ND=4 is excited with a travelling force at EO=4. In Figure 27 two peaks are clearly visible for low values of F_{Anom} (F_{Anom}= 0.1, 0.2, 0.3N) for f_m=316Hz and 345Hz. The presence of two peaks instead of one is an index of presence of mistuning since the lack of cyclic symmetry causes the split of the natural frequency of the 'twin' modes at ND=4. This can be verified by plotting the Operative Deformed Shape (ODS) of each blade when the maximum absolute value of the velocity during the blisk vibration is reached. The two stationary mode shapes vibrate at resonant conditions for two different frequency values and the rotating response of the blisk is no more obtained.

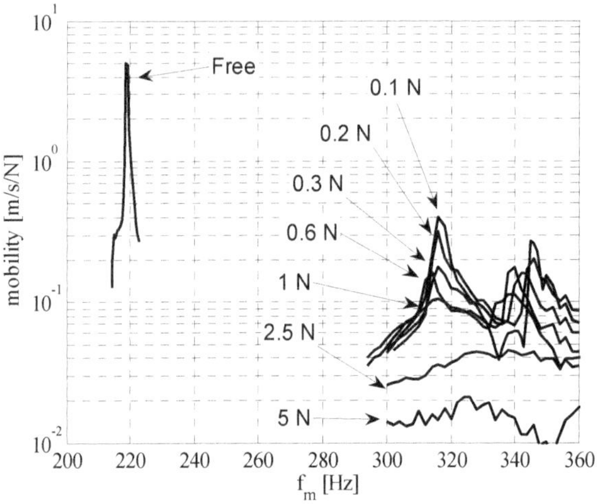

Fig. 27. FRF of the bladed disk (free and with UPDs) for different excitation force amplitude values. CF =50N. EO=4.

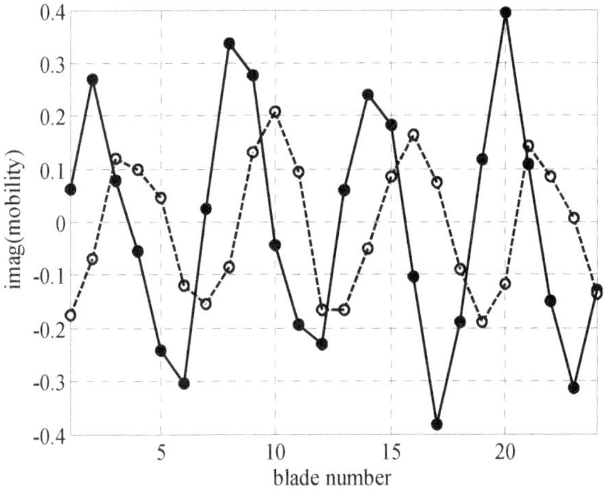

Fig. 28. Maximum amplitude of the ODS (mobility) for the different blades at 316Hz (solid line) and 345Hz (dotted line), F_{Anom} =0.1N, EO=4.

In particular, the ODS for the two peak frequencies are plotted for the case F_{Anom}= 0.1N in Figure 28. It can be noted that each maximum amplitude of the ODS represents a mode shape ND=4 (4 maxima are present). Moreover the two mode shapes of Figure 28 are orthogonal. This consideration puts in evidence that the two peaks of Figure 27 are the mistuned 'twin' modes at ND=4.

Figure 27 suggests another important consideration. By increasing F_{Anom} more damping is provided in the system and the two mistuned peaks become flat and disappear confirming that the right amount of damping can mitigate the mistuning phenomenon. The test rig proved then to be able to put in evidence the interaction of friction damping and mistuning.

6. Conclusions

The design, calibration procedure and testing of a complete (hardware and software) non-contact excitation system is presented. The system aims at performing accurate measurements of the dynamic behaviour of bladed disks in presence of mistuning and nonlinearity due to friction contacts. The system yields accurate results and it is characterized by three key features.

First, the purposely designed EMs generate force amplitudes that can be considered high for a noncontact excitation system. Each EM can reach a force amplitude up to 10 N (up to 300Hz excitation frequency of the system) and 5 N (up to 600 Hz). The several features affecting this force value are highlighted.

Second, thanks to a novel calibration method performed on site under the disk to iteratively calibrate the force, differences of force amplitudes among the blades are highly reduced (2% variation among force amplitudes). The method is efficient since it avoids the calibration of each EM (one by one) separately which is time consuming and in any case it proved to be inaccurate when the EMs are mounted under the blisk.

Third the value of the force amplitude exciting the blades is known with good accuracy (less than 5% error with respect to the nominal value). This feature is essential for stepped sine, force controlled tests to investigate the nonlinearity due to friction contacts. A device carrying one of the EMs and instrumented with a force transducer is designed, constructed and calibrated for this purpose. The device called FMEM (Force Measurement ElectroMagnet) is able to measure the force directly on site under the disk and is used as a reference for the other EMs during calibration.

The travelling excitation system is applied to the test rig Octopus where an integral bladed disk carries underplatform dampers (UPDs) which introduce nonlinearities due to friction.

The test campaign provided an example of results which proved that the system is capable of an engine order type excitation at different controlled force amplitudes. The developed noncontact travelling excitation system together with a noncontact laser Doppler vibrometer measurement of the response highlight 1) the effectiveness of the UPDs in reducing the blades vibration amplitude at different EOs, 2) the interaction of the damping provided by the UPDs and the introduction of mistuning.

7. Acknowledgment

The work described in this paper has been developed within the PRIN and CORALE projects (National Interest Research Projects).

8. References

Avalos, J. & Mignolet, M. P. (2008) On damping entire bladed disks through dampers on only a few blades, GT-2008-51446, *Proceedings of the of ASME Turbo Expo*, June 9-13, Berlin, Germany.

Beirov, B., Kühhorn, A., Schrape, S. (2008). A discrete model to consider the influence of the air flow on blade vibrations of an intergral blisk compressor rotor, *Proceedings of the of ASME Turbo Expo*, June 9-13, Berlin, Germany.

Beirov, B., Kuhhorn, A., Nipkau, J. (2009). On the influence of strain gauge instrumentation on blade vibrations of integral blisk compressor rotors applying a discrete model, GT2009-59207, *Proceedings of the of ASME Turbo*, June 8-12, Orlando, Florida.

Berruti, T. (2010). A test rig for the investigation of the dynamic response of a bladed disk with underplatform dampers, *Mechanics Research Communications*, vol 37, doi:10.1016/j.mechrescom.2010.07.008, pp 581-583.

Berruti, T., Firrone, C.M., Gola, M.M. (2010). A test rig for non-contact travelling wave excitation of a bladed disk with underplatform dampers, GT2010-22879, *Proceedings of the of ASME Turbo*, Glasgow, UK., pp.1-9.

Berruti, T., Firrone, C.M., Gola, M.M. (2011). A test rig for non-contact travelling wave excitation of a bladed disk with underplatform dampers. *Journal of Engineering for Gas Turbines and Power*, Vol 133, pp. 032502-1-8, ISSN 0742-4795, DOI: 10.1115/1.4002100.

Castanier, M. P. & Pierre, C. (2006). Modeling and analysis of mistuned bladed disk vibration : Status and emerging directions. *Journal of Propulsion and Power*, Vol. 22, Issue 2, pp.384-396, DOI: 10.2514/1.16345. ISSN 0748-4658.

Firrone, C.M., Berruti, T. (2011). An electromagnetic system for the non-contact excitation of bladed disks, *Experimental Mechanics*, DOI: 10.1007/s11340-011-9504-1.

Firrone, C.M, Berruti, T., Gola, M.M. (2011). On force control of an engine order type excitation applied to a bladed disk with underplatform dampers, IDETC2011-47255, *Proceedings of the International Design Engineering Technical Conferences (IDETC)*, Washington, pp. 1-10.

Chang, J. Y. & Wickert, J. A. (2002). Measurement and analysis of modulated doublet mode response in mock bladed disks", *Journal of Sound and Vibration*, Vol.250, Issue 3, pp-379-400, doi:10.1006/jsvi.2001.3942.

Gilbert, C., Kharyton, V., Thouverez, F., Jean, P. (2010). On forced response of a rotating integrally bladed disk: predictions and experiments, GT2010-23610, *Proceedings of the of ASME Turbo Expo*, June 14-18, Glasgow, UK.

Gotting, F., Sextro, W., Panning, L., Popp, K. (2004). Systematic mistuning of bladed disk assemblies with friction contacts, GT2004-53310, *Proceedings of the of ASME Turbo Expo*, June 14-17, Vienna, Austria.

Heath, S. & Imregun, M. (1998). A survey of Blade Tip-Timing Measurement Techniques for turbomachinery Vibration", *Journal of Engineering for Gas Turbines and Power*, Vol.120, Issue 4, pp.784-791, doi:10.1115/1.2818468.

Heinz, C., Schatz, M., Casey, M. V., Stuer, H. (2010). Experimental and analytical investigations of a low pressure model turbine during forced response excitation, GT2010-22146, *Proceedings of the of ASME Turbo*, June 14-18, Glasgow, UK.

Jones, K.W. & Cross, C.J. (2003). Travelling Wave Excitation System for Bladed Disks", *Journal of Propulsion and Power*, Vol. 19, pp.135-141, ISSN 0748-4658.

Judge, J., Ceccio, S. L., Pierre, C. (2003). Traveling-wave excitation and optical measurement techniques for non-contact investigation of bladed disk dynamics, *The Shock and vibration digest*, Vol. 35, Issue 3, pp.183-910, ISSN 1741-3184.

Kenyon, J.A. & Griffin, J.H. (2003). Experimental Demonstration of Maximum Mistuned Bladed Disk Forced Response. *Journal of Turbomachinery*, Vol. 125, Issue 4, pp. 673-681. doi:10.1115/1.1624847.

Kruse, M. J. & Pierre, C. (1997). An Experimental Investigation of Vibration Localization in Bladed Disks, Part I: Free Response", *Proceedings of the 42nd ASME Gas Turbine & Aeroengine Congress, User's Symposium & Exposition*, Orlando, Florida.

Kruse, M. J. & Pierre, C. (1997). An Experimental Investigation of Vibration Localization in Bladed Disks, Part II: Forced Response", *Proceedings of the 42nd ASME Gas Turbine & Aeroengine Congress, User's Symposium & Exposition*, Orlando, Florida.

Petrov, E., Hennings, H., Di Mare, L., Elliott, R. (2010). Forced response of mistuned bladed discs in gas flow: a comparative study of predictions and full-scale experimental results. *Journal of Engineering for Gas Turbines and Power*, Vol. 132, Issue 5, 052504, doi:10.1115/1.3205031. ISSN: 0742-4795.

Pierre, C., Judge, J., Ceccio, S. L., Castanier, M. P. (2002). Experimental Investigation of the Effects of Random and Intentional Mistuning on the Vibration of Bladed Disks, *Proceedings of the 7th National High Cycle Fatigue Conference*, Palm Beach, Florida.

Prchlik, L., Misek, T., Kubin, Z., Duchek, K.. (2009). The measurement of dynamic vibration modes and frequencies of a large LP bladed disc, GT2009-60002, *Proceedings of the of ASME Turbo Expo*, June 8-12, Orlando, Florida, USA.

Rice, T., Bell, D., Singh, G. (2007). Identification of the stability margin between safe operation and the onset of blade flutter, GT2007-27462, *Proceedings of the of ASME Turbo Expo*, May 14-17, Montreal, Canada.

Sever, I., A. (2004). Experimental validation of turbomachinery blade vibration predictions, *PhD Thesis*, Department of Mechanical Engineering, London.

Strehlau, U. & Kuhhorn, A. (2010). Experimental and numerical investigations of HPC blisks with a focus on travelling waves, GT2010-22463, *Proceedings of the of ASME Turbo*, June 14-18, Glasgow, UK.

Szwedowicz, J., Gilbert, C., Sommer, T. P., Kellerer, R. (2006). Numerical and experimental damping assessment of a thinwalled friction damper in the rotating set-up with high pressure turbine blades", *Proceedings of the of ASME Turbo Expo*, May 8-11, Barcelona, Spain.

Szwedowicz, J., Secall-Wimmel, T., Dünck-Kerst,, P., Sonnenschein, A., Regnery, D., Westfahl, M. (2007). Scaling concept for axial turbine stages with loosely assembled friction bolts: the linear dynamic assessment, part I, GT2007-27502, *Proceedings of the of ASME Turbo Expo*, May 14-17, Montreal, Canada.

Thomas, D. L. (1979). Dynamics of rotationally periodic structures. *International Journal for numerical methods in engineering*, Vol.14, pp. 81-102.

7

Shape Measurement by Phase-Stepping Method Using Multi-Line LEDs

Yoshiharu Morimoto[1], Akihiro Masaya[1],
Motoharu Fujigaki[2] and Daisuke Asai[3]
[1]*Moire Institute Inc.,*
[2]*Wakayama University,*
[3]*Hikari,*
Japan

1. Introduction

Three-dimensional shape measurement is used in various fields such as robotics and inspection of industry production. It is requested to measure shape with high speed and high accuracy. It is also requested to make the system compact and low-cost. It is, however, difficult to satisfy these requests simultaneously for 3D shape measurement. In order to measure shape in real-time, Kato et al. proposed a method using phase-shifting electronic moiré pattern (Kato et al., 1997). The authors developed integrated phase-shifting method for real-time shape measurement by a grating projection method using phase-shifting (Morimoto et al., 1999, Fujigaki & Morimoto, 2003). However, they showed only the phase distribution of the contour lines.

In almost all conventional shape measurement methods, the optical systems are modelled and the parameters of the model are obtained with a calibration process using calculation of the geometrical parameters of optical devices such as positions of lens centres of a camera and a projector. However, the model cannot contain all of the information of the optical system such as lens distortion, intensity error of a projected grating, brightness linearity of a projector and a camera, etc. The less information causes measurement errors. Furthermore, it is time-consuming to calculate the spatial coordinates using the parameters.

In order to obtain a 3D-shape with grating projection method, the authors previously proposed whole-space tabulation method (WSTM) (Fujigaki & Morimoto, 2008a, Fujigaki et al., 2008b, 2009, Morimoto et al., 2009). The relationship between the coordinates and the phase of the grating recorded at each pixel of a camera is obtained as calibration tables in a three-dimensional space by experiment beforehand. Therefore the analysis is very fast because of looking at the calibration tables from the phase information at the pixel without any complex calculation. It provides fine resolution even when the phase distribution of the grating is not linear.

The tabulation method is extended to a case of inaccurate phase-shifting of a grating by a projector using five-line light emitted diode (LED) light sources (Morimoto et al., 2010a, 2010b). In this paper, a grating projector with nine-line LED light sources is newly developed. The phase-shifting usually uses a grating projector with a high resolution stage or a liquid crystal display (LCD) projector. It is expensive and limits the speed of phase-shifting. The light-power efficiency of an LED is very high. The size of the light source is very small. It is easy and very fast to control the power and the switch on/off timing. The nine-line LED light sources are set in front of a grating. The switch for each LED line of the light sources is 'switch on'. The phase of the grating shadow is shifted by changing the switches for LEDs into 'on/off' in synchronization with the phase-shifting and the recording with the camera sequentially. Using the LED light sources for a grating projector, it is possible to make phase-shifting without any moving devices. The projector is low cost and high-speed. The authors call the method 'phase-stepping method using multi-line LEDs.' Even when the positions of the LED light sources are not so accurate, the error is almost cancelled by using the calibration tables obtained by the WSTM with the same experimental setup. This method will satisfy for shape measurement system high-speed, high accuracy, compact and low-cost.

In this paper, the theory and the system of shape measurement using the WSTM and the phase analysis methods using multi-line LED light-sources are explained. Furthermore, some experimental results of shape measurements using the system are shown.

2. Phase-shifting method in grating projection method

2.1 Grating projection method

Figure 1 shows a schematic system for grating projection method. A grating with a cosinusoidal brightness distribution is projected by a projector. The grating projected on an object is deformed according to the shape of the object. The deformed grating is recorded by

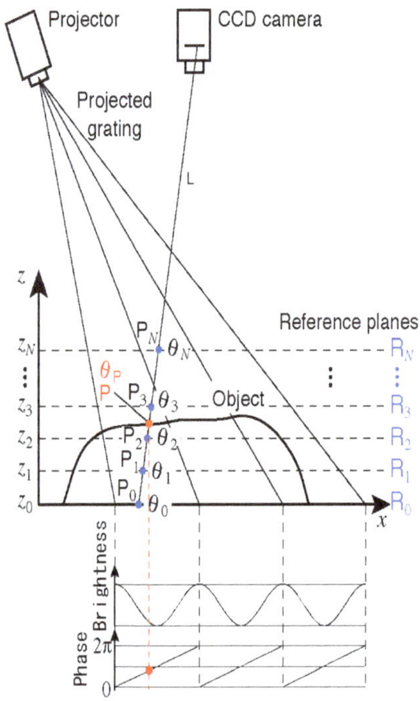

Fig. 1. Schematic view of grating projection system

a camera and analyzed the deformed grating to obtain the shape. As analysis method of the deformed grating, phase shifting method which is the most popular accurate method, is used as mentioned next section.

2.2 Phase-shifting method

In shape measurement using grating projection method, phase analysis of the projected grating provides accurate results. In order to analyze the phase, there are many methods (Takeda & Mutoh, 1983, Asundi & Zhou., 1999, Sitnik & Kujawinska, 2002, etc.). Phase-shifting method is the most popular method. Let us explain the phase-shifting method. Figure 2(a) shows phase-shifted grating patterns projected onto an object. Figure 2(b) shows the brightness (intensity) distributions along the horizontal center line of Fig. 2(a). The intensity distribution is almost cosinusoidal.

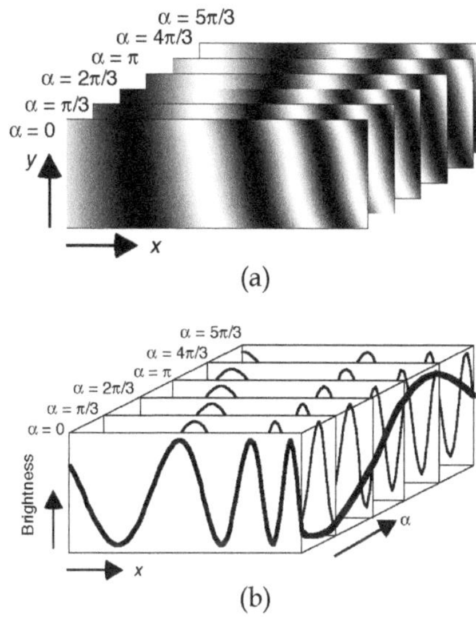

Fig. 2. Phase analysis of projected grating using phase-shifting grating images; (a) Phase-shifted grating pattern projected onto object; (b) Brightness distribution along horizontal center line of Fig. (a)

When a grating with a cosinusoidal brightness distribution is projected or displayed on a reference plane or an object, and the phase of the grating is shifted N times for one cycle, the k-th phase-shifted grating images can be expressed as follows:

$$I_k(x,y) = I_a(x,y)\cos[\theta(x,y) + k\frac{2\pi}{N}] + I_b(x,y) \quad (k = 0,1,\ldots,N-1) \tag{1}$$

where $I_b(x, y)$ represents the background brightness in the image, which is insensitive to the change in phase. $I_a(x, y)$ represents the amplitude of the grating brightness and $\theta(x, y)$ is the initial phase value. The phase distribution of the grating pattern can be obtained as follows (Srinivasan et al., 1984, 1985).

$$\tan\theta(x,y) = -\frac{\sum_{k=0}^{N-1} I_k(x,y)\sin(k\frac{2\pi}{N})}{\sum_{k=0}^{N-1} I_k(x,y)\cos(k\frac{2\pi}{N})} \tag{2}$$

Usually N is selected as 3 or 4. If the number N of the phase-shifted moiré patterns is larger, the phase analysis becomes more accurate. In this study, N =4 and 6 are selected.

3. Whole-Space Tabulation Method (WSTM) (Fujigaki et al., 2008a, 2008b, 2009, Morimoto et al., 2009)

In grating projection method, the coordinates of a point on an object is usually calculated using the geometric positions of the lenses of a camera and a projector. If the distortion of lenses is considered, the calculation is complex and time-consuming. Therefore several methods using a reference plane or reference planes were proposed for accurate shape measurement (Zhou & Su, 1994, Asundi & Wensen, 1999, Su et al., 2004, Ha et al., 2004, Yen et al., 2006). For real-time shape measurement, it is required to calculate spatial coordinates from the phase values of the projected grating in short time. The authors previously proposed the calibration methods which used two or multiple reference planes (Fujigaki & Morimoto, 1996, Shinke et al., 2001, Ri et al., 2005). The calibration method was extended to a very large number of planes. All the relationships between the phase of the projected grating and the spatial coordinates can be obtained as calibration tables for each pixel of the camera. The method is called 'whole-space tabulation method' (WSTM). Let us explain the method.

Figure 1 also shows a schematic view of a shape measurement system by grating projection method and explanation of the principle of the calibration method using multiple reference planes. An LCD panel as a reference plane is set on a linear stage. The LCD panel is covered with a scattered film. The scattered film functions as a screen when a grating pattern is projected from the projector onto the reference plane. It is used to make calibration tables on the relationship between the phase θ and the z coordinate. The scattered film also functions as a backward screen when a grating patterns are displayed on the LCD panel to determine the x and y coordinates.

The LCD reference plane is set vertically to the z-direction. It is translated in the z-direction step by step. A camera and a projector are arranged and fixed in front of the reference plane. The grating is projected on the reference plane at first, and it is also projected on an object to be measured later. By translating the reference plane along the z-axis, a pixel of the camera records the intensities as images at the points P_0, P_1, P_2 .... and P_N, on the reference planes R_0, R_1, R_2.... and R_N, respectively. And also, by recording the phase-shifted grating images, the phase distribution along the visual line L of a pixel of the camera is obtained.

In order to obtain the x and y coordinates on the reference plane, the phase-shifted gratings displayed on the LCD panel are taken by the camera. From these phase-shifted images, the calibration tables are formed to obtain the x, y and z coordinates from the phase θ at each pixel as shown in Figs. 3 and 4.

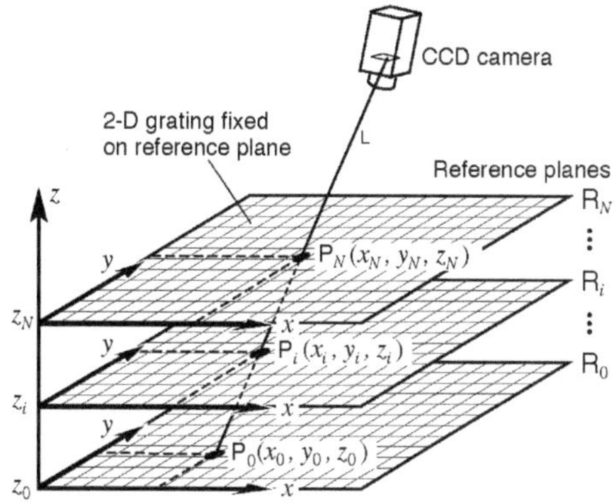

Fig. 3. Schematic view of calibration tables to obtain x, y and z coordinates from phase in whole space tabulation method

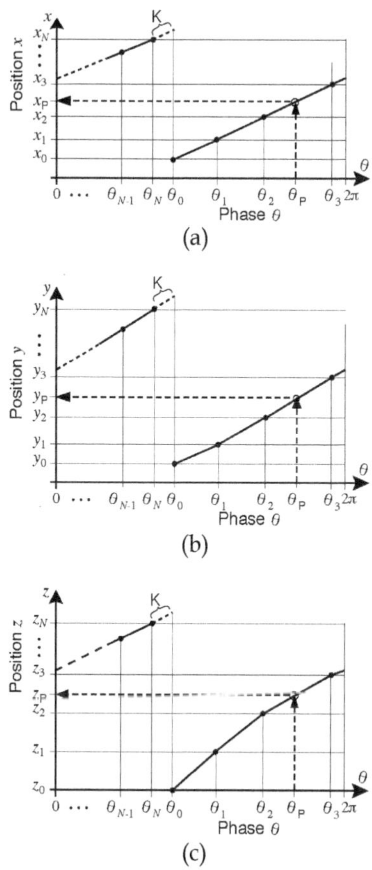

Fig. 4. Calibration tables to obtain x, y and z coordinates from phase at each pixel point of camera; (a) Table of phase θ and x; (b) Table of phase θ and y; (c) Table of phase θ and z

In the measurement procedure, an object is placed between the reference planes R_0 and R_N. Phase-shifted gratings are projected onto the object and the phase distributions of the grating are analyzed from the phase-shifted grating images. The 3-D coordinates at each pixel are obtained from the phase value in high-speed by referring the calibration tables for each pixel.

This method is called the WSTM as mentioned already. It excludes the effect of lens distortion and intensity error of the projected grating in measurement results theoretically. Tabulation makes short-time measurement possible because the 3D coordinates are obtained by looking at the calibration tables from the phase at each pixel point of the camera and it does not require any time-consuming complex calculation.

4. Phase-stepping method by changing light sources

4.1 Geometrical relationship among phase, phase-stepping amount and intensity of projected grating

Let us explain the principle of the phase-stepping method using multi-line LED light sources. Figure 5 shows that a grating is projected from the light sources. This figure shows in the case of four line LED light sources. If each line LED light source is switched on and off one by one sequentially, the shadow of the grating is moved.

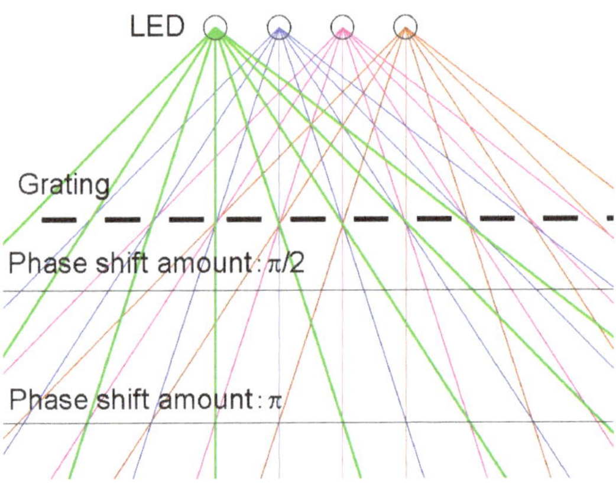

Fig. 5. Shadow of grating is moved when each line LED is switched on and off one by one sequentially. Phase stepping amount is function of vertical distance from LED light source in phase-stepping method using multi-line LED light sources

The amount of phase shifts changes with the z-directional distance from the light source to each point in Fig. 5. The positions with the amounts of phase shift, $\pi/2$ and π, are shown as horizontal straight lines, respectively.

The equations of the geometric relationship among light sources, a grating and an object are derived using Fig. 6. In this figure, five line LED light sources are shown as an example. Now, the subscript number n is given to each line LED light source L, and the n-th line light source is set to L_n (n=0, 1, 2, 3, 4). The light sources are mutually parallel and the five lines have regular intervals with the pitch l. The plane including the five LED line light sources is

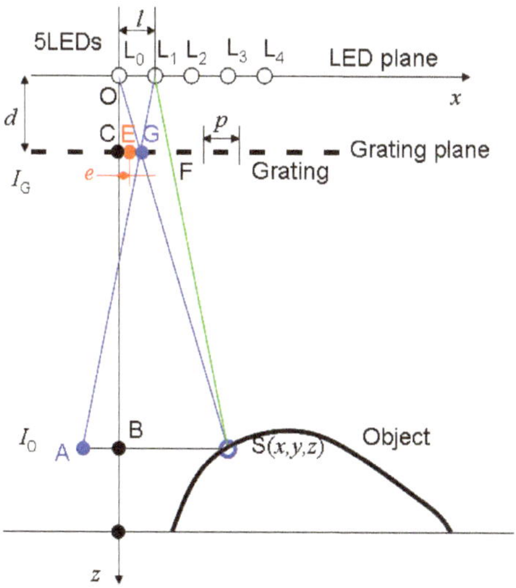

Fig. 6. Schematic of grating projection system and object in shape measurement system using five line LED light sources

called an LED plane where z=0 with the field (x, y). The position of the first LED line L_0 is the origin O of the x, y, and z coordinates. In the LED plane, the x direction is normal to the LED lines, and the y-direction is the parallel to the LED lines. The z-direction is normal to the LED plane. The brightness distributions of the five line LEDs are assumed as uniform and equal in the x- and y-directions in the observation region at fixed z.

A grating plane is parallel to the LED plane, and the distance between the LED plane and the grating plane is d. The grating lines with pitch p drawn on the grating plane are parallel to the LED lines. The intersection of the grating plane and the z-axis is set to C and the distance between C and the origin E, i.e., the center E of the first grating line is set to e.

The transmissivity intensity distribution of the grating at z=d is denoted by the following equation.

$$I_g = a_g \cos\{\Phi\} + b_g = a_g \cos\{\frac{2\pi}{p}(x_g - e)\} + b_g \tag{3}$$

where Φ is the phase, a_g is the amplitude, b_g is the intensity of background, x_g is the x-directional coordinate at the grating plane, that is, x_g is the x-directional distance from the point C.

When the light source L_0 is switched on and the shadow of the grating is projected at the position S(x, y, z), the intensity at the position S(x, y, z) is expressed by the following equation.

$$I_0 = a_g \frac{d^2}{z^2} \cos\{\frac{2\pi}{p}(\frac{d}{z}x - e)\} + b_g \frac{d^2}{z^2} \tag{4}$$

where it is considered that the intensity at a point is inversely proportional to the square of the distance z. As shown in Fig. 6, the point G of the grating is projected at the point S on the object. Only the n-th line LED light source L_n is switched on one by one, and it is considered that the shadow of the grating is projected on the object.

The intensity distribution of the shadow projected through the grating from the light source L_n has the following distribution shown at z.

$$I_n = a_g \frac{d^2}{z^2} \cos[\frac{2\pi}{p}\{(\frac{d}{z}x - e) + nl(1 - \frac{d}{z})\}] + b_g \frac{d^2}{z^2} = a\cos\{\Phi + n\Psi\} + b \quad (5)$$

where

$$a = a_g \frac{d^2}{z^2} \quad (6)$$

$$b = b_g \frac{d^2}{z^2} \quad (7)$$

$$\Phi = \frac{2\pi}{p}(\frac{d}{z}x - e) \quad (8)$$

$$\Psi = \frac{2\pi l}{p}(1 - \frac{d}{z}) \quad (9)$$

4.2 Relationship among phase, phase-shift amount and height

When the shift amount Ψ is obtained, the z-coordinate is determined using Eq. (9) as follows.

$$z = \frac{2\pi l d}{2\pi l - \Psi p} \quad (10)$$

When the phase Φ is obtained, the z-coordinate is determined using Eq. (8) as follows.

$$z = \frac{2\pi dx}{p\Phi + 2\pi e} \quad (11)$$

From Eq. (10) or (11), the z-coordinate is obtained.

4.3 Phase analysis by phase-stepping method by changing light sources

As shown in the case of $\Psi=\pi/2$ and $\Psi=\pi$ in Fig. 5, there are the positions where the phase shift amounts are $2\pi/N$ (N is an integer). At the positions, the phase analysis becomes possible using Eq. (2). At the other positions, there are some errors in the phase value obtained using Eq. (2). However, if the relationship between the calculated phase θ and z is a single-valued function at some region, the phase determination is possible in the region by combining the WSTM. The height is exactly obtained by looking up the tables of the WSTM even if the phase shifting method has a little error.

5. Experiment

5.1 Experimental setup

In this study, some experiments are performed to confirm the performance of the phase-stepping method using line LED light sources. Figure 7 shows the schematic optical system of the experiment. Figure 8 shows the photograph of the LED board with 9 line LED light sources we developed.

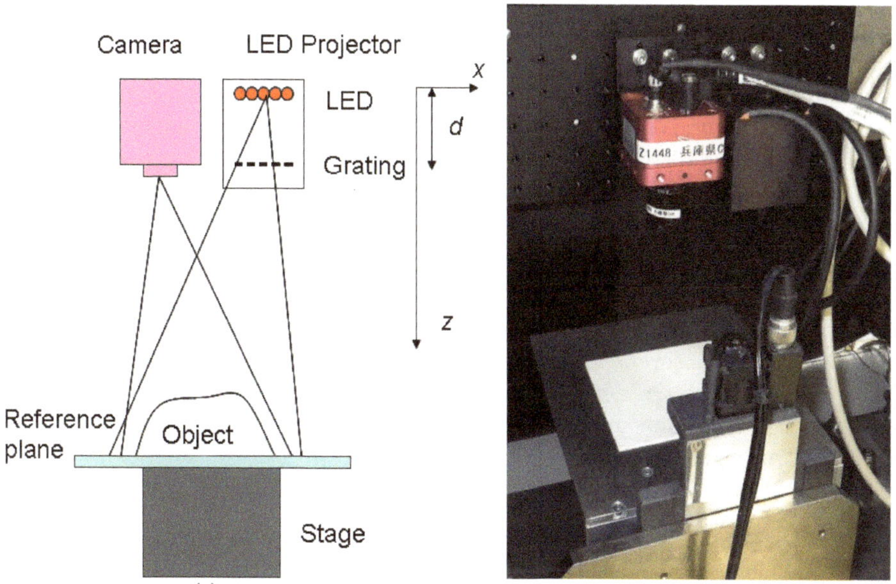

(a) (b)

Fig. 7. Optical system used in experiment; (a) Schematic view; (b) Photograph

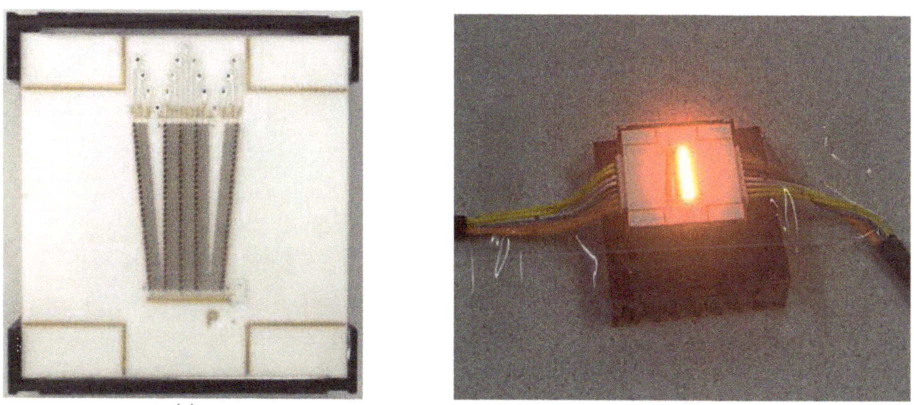

(a) (b)

Fig. 8. Photographs of the developed line LED light sources; (a) Board with 9 LED lines and slanting 6 lines; (b) Photo when switching on a line

The LED has nine lines perpendicular to an x-axis, and slanting six lines. Here, the perpendicular lines are used. The pitch l of the LED lines is 0.5 mm. One line has 30 LED tips of a 0.35-mm square. A Ronchi grating is used as a grating. The pitch p of the grating is 0.5 mm. The interval d between the LED plane and the grating plane is 30 mm. The experiments are performed where the base level of z is 180 mm.

5.2 Experimental results

The experimental results were shown in Figs. 9 to 14.

Figure 9 shows the intensity distributions in the x direction in the center of the y direction of the grating projected on the reference plane when the lighting line LED light sources were changed. In this figure, only five distributions are shown. .Although there is some random noise, the intensity distributions are almost cosinusoidal and the phase shift is also performed at equal intervals. Although this random noise is decreased by time averaging, any averaging is not used in this study to confirm the performance.

At first, only the first four lines of LED were used for the phase-stepping method and the phases are calculated using N= 4 in Eq. (2). The phase distribution along the x-axis on the reference plane at z=180mm is shown in Fig. 10(a). Although the phase distribution is repeated from $-\pi$ to π several times, the relationship between z and phase at a point of the reference plane for a 2π phase interval is shown in Fig. 10(b). The relationship is considerably curved although the relationship is almost straight line in theoretical treatment. However, by combining the WSTM, the errors will be cancelled by using the same optical system when the tables are used. Then the measurement is performed during a 2π phase interval without unwrapping of phase.

In the same manner, the phase distribution in the case of N=6 using the first six lines of LED is shown in Fig. 11. In both of the distributions in the x-direction and the z-direction, the relationships are almost linear. Thus, the linearity is different according to the value N.

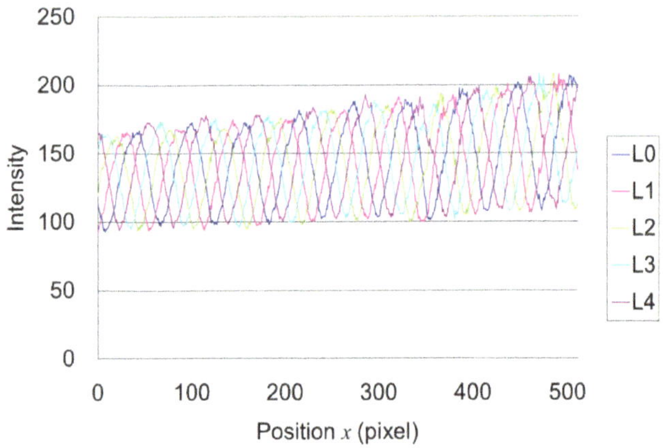

Fig. 9. Intensity distributions along x-axis of projected grating on reference plane when changing light sources

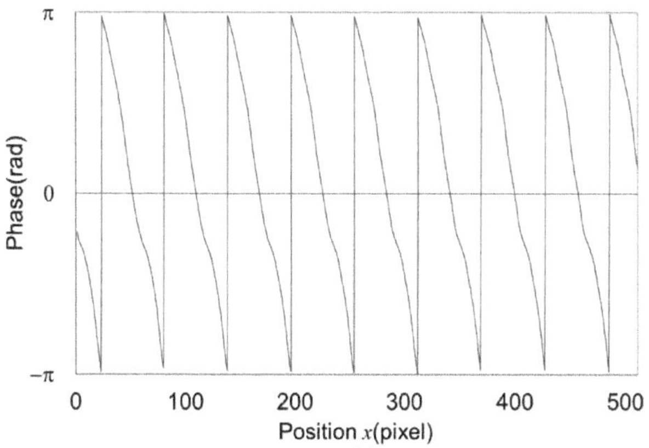

(a)

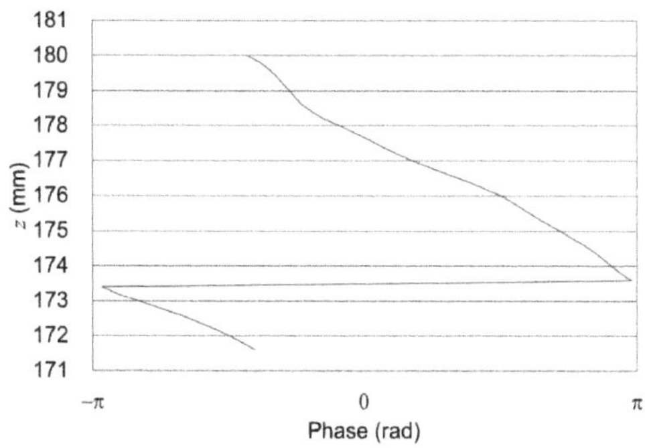

(b)

Fig. 10. Phase distributions along x-direction and z-direction when phase calculation is performed using N=4; (a) x-directional phase distribution; (b) z-directional phase distribution

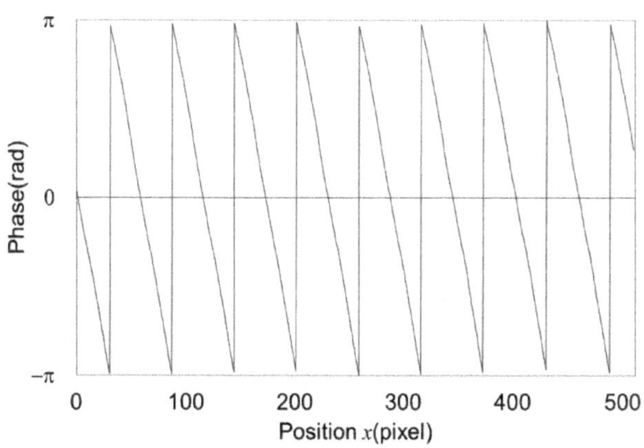

(a)

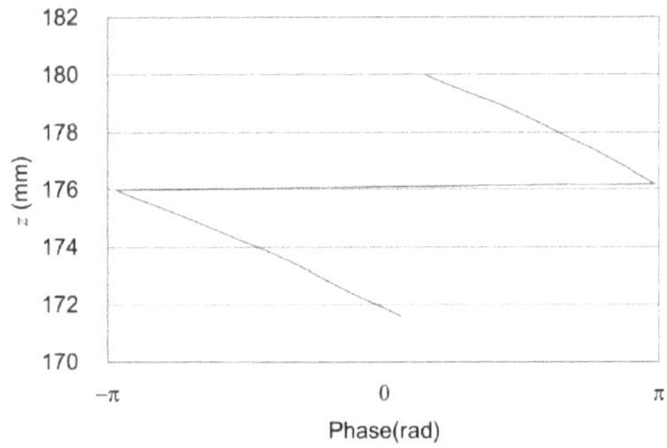

(b)

Fig. 11. Phase distributions along x-direction and z-direction when phase calculation is performed using N=6; (a) x-directional phase distribution; (b) z-directional phase distribution

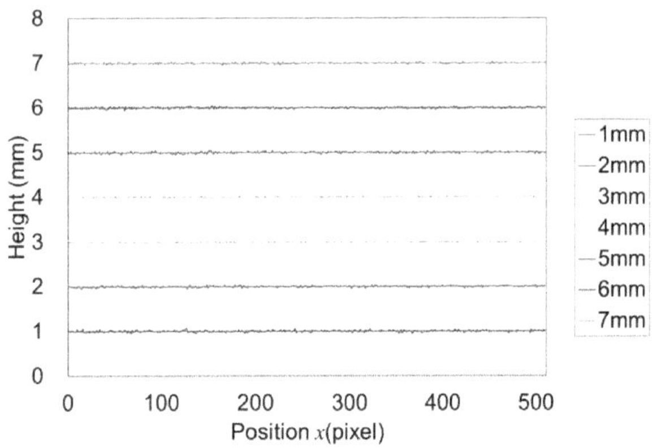

Fig. 12. Height distributions when the height is changed from reference plane at z=180mm (in case of N=4)

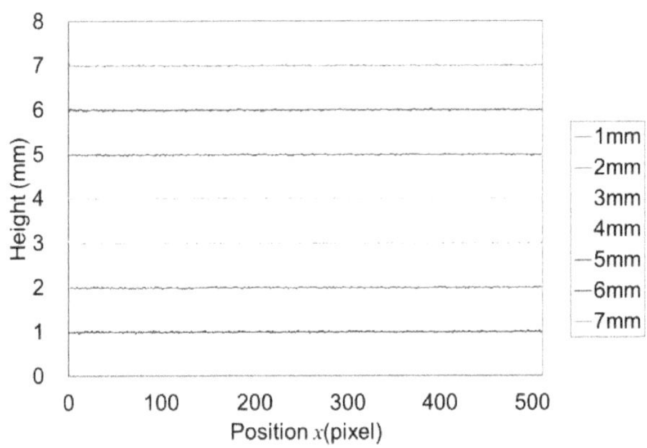

Fig. 13. Height distributions when the height is changed from reference plane at z=180mm (in case of N=6)

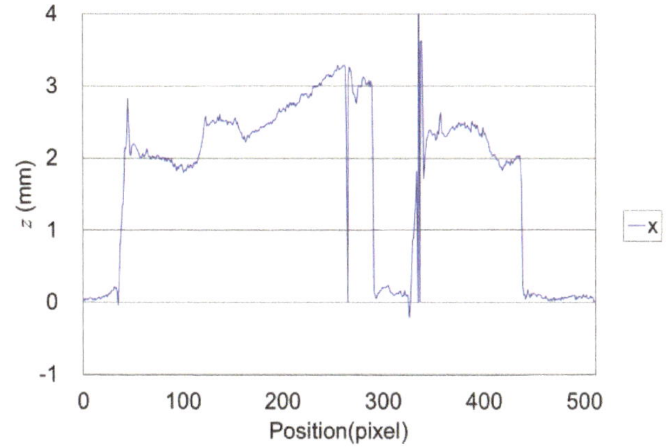

(a)

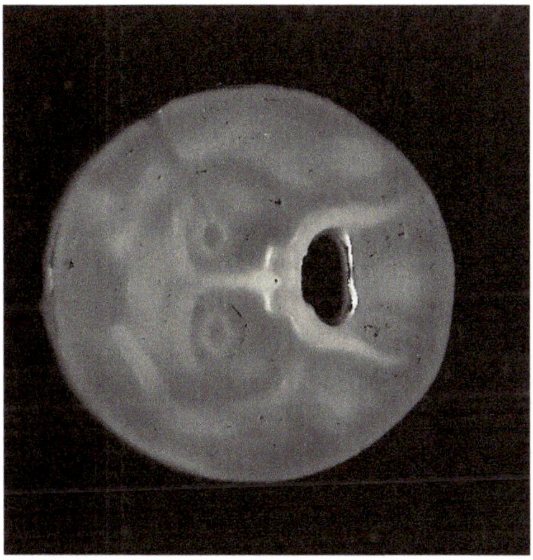

(b)

Fig. 14. Measurement result of mold of face; (a) Height distribution along center line; (b) Height distribution

Next, height measurement was performed. As the specimen, the reference plane was used. The results are shown in Figs. 12 and 13 and Tables 1 and 2. The average errors are less than 5 μm, and the standard deviations are 11~19 μm. Although the standard deviation in the case of N=6 is better than the one in the case of N=4, both have good accuracy.

As an example of shape measurement, the result of shape measurement of a face model cast is shown in Fig. 14. Although the measurement value shows unusual values in the edge part in small areas, almost all over the object is measured by the proposed method.

Height	1mm	2mm	3mm	4mm	5mm	6mm	7mm
Average	1.003	2.002	3.002	4.002	5.001	6.001	7.000
Error	0.003	0.002	0.002	0.002	0.001	0.001	0.000
Standard deviation	0.019	0.018	0.017	0.017	0.017	0.015	0.017

Table 1. Results of height measurement (in case of N=4)

Height	1mm	2mm	3mm	4mm	5mm	6mm	7mm
Average	1.005	2.003	3.004	4.002	5.001	6.002	7.000
Error	0.005	0.003	0.004	0.002	0.001	0.002	0.000
Standard deviation	0.013	0.013	0.012	0.012	0.012	0.012	0.011

Table 2. Results of height measurement (in case of N=6)

6. Conclusions

A new shape measurement method called phase-stepping method using multi-line LED light sources is proposed. A special LED array with nine line LEDs was developed. By using whole-space tabulation method (WSTM), a system using the multi-line LED light sources was developed. It could shift the phase of the projected grating easily. The principle and a system of the shape measurement using the phase stepping method using LEDs were shown. It was applied to three-dimensional shape measurement, and height has been measured with sufficient accuracy less than 20 μm. It is possible to make a high speed, high precision, low cost, small size and wide-dynamic range system. It also excludes lens distortion and intensity error of the projected grating in measurement results theoretically.

7. Acknowledgments

This study was supported by Hyogo COE Program Promotion Project financially. We appreciate the support.

8. References

Asundi, K. & Zhou, W. (1999a). Mapping Algorithm for 360-deg Profilometry with Time Delayed Integration Imaging, *Optical Engineering, 38,* pp. 339-344.

Asundi, A. & Wensen, Z. (1999b). Unified Calibration Technique and its Applications in Optical Triangular Profilometry, *Applied Optics, 38-16,* pp. 3556-3561.

Fujigaki M. & Morimoto, Y. (1996). Shape Measurement by Grid Projection Method without Influence of Aberration of Lenses, *Post Conference Proc. of the VIII International Congress on Experimental Mechanics SEM*, pp. 167-172.

Fujigaki M. & Morimoto, Y. (2003). Real-time Shape Measurement by Integrated Phase-Shifting Method Using Ronchi Grating, *Trans. Jpn. Soc. Mech. Eng., (in Japanese), 69-685, C,* pp. 2374-2380.

Fujigaki M. & Morimoto, Y. (2008a). Shape Measurement with Grating Projection Using Whole-Space Tabulation Method, *Journal of JSEM (in Japanese), 8-4,* pp. 92-98.

Fujigaki, M., Takagishi, A., Matui T. & Morimoto, Y. (2008b). Development of Real-Time Shape Measurement System Using Whole-Space Tabulation Method, *SPIE International Symposium, Proc. SPIE 7066,* 706606.

Fujigaki M., Masaya, A., Murakami, R. & Morimoto, Y. (2009). Accuracy Improvement of Shape Measurement Using Whole-Space Tabulation Method, *Proc. of ICEM2009 held in Singapore.*

Ha, T., Takaya, Y. & Miyoshi, T. (2004). High-precision On-machine 3D Shape Measurement Using Hypersurface Calibration Method, *Proc. SPIE, 5603,* pp. 40-50.

Kato, J., Yamaguchi, I., T. Nakamura & Kuwashima, S. (1997). Video-rate Fringe Analyzer Based on Phase-shifting Electronic Moire Patterns, *Applied Optics,* Vol. 36, No. 32, pp. 8403-8412.

Morimoto, Y., Fujigaki M. & Toda, H. (1999). Real-Time Shape Measurement by Integrated Phase-Shifting Method, *Proc. of SPIE, Vol. 3744,* pp. 118-125.

Morimoto, Y., Fujigaki M. & Masaya, A. (2009). Shape Measurement by Grating Projection and Whole-space Tabulation Method, *Proc. of ISOT2009 held in Istanbul*

Morimoto, Y., M. Fujigaki, M. Masaya, A. & Amino, Y. (2010a). Shape Measurement by Whole-space Tabulation Method Using Phase-shifting LED Projector, *Proceedings of International Conference on Advanced Phase Measurement Methods in Optics and Imaging, Monte Verita, Locarno, Switzerland, 17 to 21 May.*

Morimoto, Y., Fujigaki, M. Masaya, A. & Oura , Y. (2010b). Shape Measurement by Light-stepping Method Using LEDs, *ISOT 2010 International Symposium on Optomechatronic Technologies, 25-27 October 2010, Toronto, Canada*

Ri, S., Fujigaki M. & Morimoto, Y. (2005). Phase Reliability Evaluation in Phase-shifting Method Using Fourier Transform for Shape Measurement, *Optical Engineering, 44-8,* 083601-1-8.

Shinke, A., Morimoto, Y. & Fujigaki, M. (2001). Development of Accurate Shape Measurement System Using Multiple Reference Planes, *Proceedings of APCFS & ATEM '01,* pp. 811-814.

Sitnik, R. & Kujawinska, M. (2002). Digital Fringe Projection System for Large-volume 360-deg Shape Measurement, *Optical Engineering, 41-2,* pp. 443-449.

Srinivasan, V., Liu, H. C. & Halioua, M. (1984). Automated phase-measuring profilometry of 3-D diffuse objects, *Appl. .Opt. 23,* pp. 3105-3108.

Srinivasan, V., Liu, H. C. & Halioua, M. (1985). Automated phase-measuring profilometry: a phase mapping approach,, *Appl. Opt. 24,* pp. 185-188.

Su, X., Song, W, Cao, Y. & Xiang, I. (2004). Phase-height Mapping and Coordinate Calibration Simultaneously in Phase-Measuring Profilometry, *Optical Engineering, 43,* pp. 708-712.

Takeda, M. & Mutoh, K. (1983). Fourier Transform Profilometry for the Automatic Measurement of 3-D Object Shapes, *Applied Optics, 22-24,* pp. 3977-3982.

Yen, H. N., Tsai D. M. & Yang, J. Y. (2006). Full-field 3-D measurement of solder pastes using LCD-based Phase Shifting Techniques, IEEE Transactions on Electronics Packaging Manufacturing, 29-1, pp. 50-57.

Zhou, W-S. & Su, X.-Y. (1994). A Direct Mapping Algorithm for Phase-measuring Profilometry," Journal of Modern Optics, 41-1, pp. 89-94.

8

Electro-Luminescence Based Pressure-Sensitive Paint System and Its Application to Flow Field Measurement

Yoshimi Iijima and Hirotaka Sakaue
Japan Aerospace Exploration Agency, Japan

1. Introduction

Pressure-sensitive paint (PSP) relates a static or oxygen pressure in a testing fluid to a luminescent signal. It uses a photophysical process of oxygen quenching [Lakowicz]. A PSP measurement system requires an illumination source to excite the PSP that can be a xenon lamp, an LED array, and a laser. These are a point illumination that creates a distribution of the illumination on a PSP coated surface. This distribution results in a pressure-independent luminescence from the PSP surface. By rationing with a reference image, the pressure-independent luminescence is cancelled [Liu and Sullivan]. The reference image is obtained under a constant pressure. It is only related to the pressure-independent luminescence mainly due to the distribution of the illumination and a movement and/or distortion in a testing article. The rationing method is valid when there is no movement and/or distortion in the article. If there is such a case, a misalignment of the images may occur that causes a substantial error to a PSP measurement. In addition to the illumination problem, a PSP in general has a temperature dependency [Liu and Sullivan]. This also creates an error to a PSP measurement. For example, a platinum-porphyrin based PSP changes about 1% of the luminescent signal by 1 °C change in temperature. This is equivalent to the 1 kPa change in the luminescence signal.

Electro-luminescence (EL) gives a surface illumination instead of a point illumination. This may reduce the misalignment error. An inorganic EL has a potential to spray on a testing article. This will open our application to provide a testing article with an illumination layer mounted. Previous studies reported that an EL has a temperature dependency [Airaghi *et al*, Schulze *et al*]. The illumination output of the EL increased with an increase of the temperature. The temperature dependency is opposite to that of a conventional PSP. The combined system of the PSP and EL may reduce the overall temperature dependency of the PSP system.

In this chapter, we introduce an EL-based PSP system. A spectral characterization as well as a spatial uniformity of an EL is described. The temperature and pressure calibrations of the system are included. The resultant system is demonstrated to obtain a pressure distribution created by an impinging jet.

2. Background

2.1 Inorganic electro-luminescence

An electro-luminescence (EL) uses a luminescent particle, which is electrically excited to give an illumination output [Destriau]. An EL can be categorized as an inorganic and organic EL based on a material of the particle and an illumination mechanism. An inorganic EL, which is commercially available, uses a phosphor as a luminescent particle. It can be a powder or a thin-film. A powder-type EL can be applicable to spray on a testing article. Fig. 1 shows a schematic of a powder-type inorganic EL. It consists of multi-layers in addition to a phosphor layer.

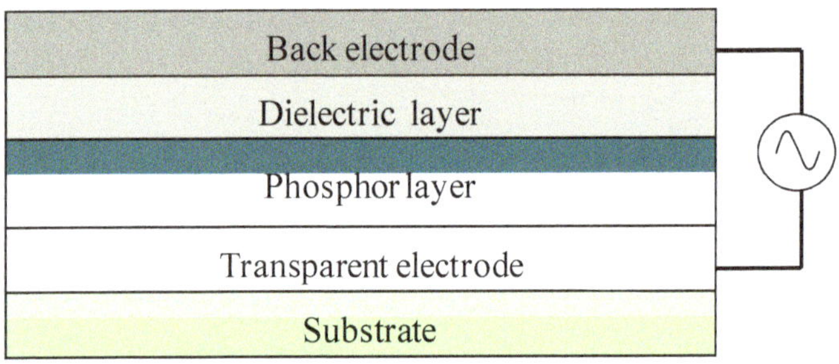

Fig. 1. Multi-layer construction of an electro-luminesence (EL)

2.2 Pressure-Sensitive Paint (PSP)

Based on the oxygen quenching, the luminescent signal, I, can be related to a static pressure by using the Stern-Volmer equation [Liu and Sullivan].

$$\frac{I_{ref}}{I} = A_P + B_P P \quad (1)$$

Where A_P and B_P are calibration coefficients. As a pressure calibration, B_P denotes the pressure sensitivity, σ. A high σ gives more sensitive to the pressure.

$$\sigma = \left.\frac{d\left(I_{ref}/I\right)}{dp}\right|_{p=p_{ref}} = B_P \ (\%/\text{kPa}) \quad (2)$$

A temperature characterization of a PSP can be described as the second order polynomial in Eq. (3):

$$\frac{I}{I_{ref}} = c_{T0} + c_{T1}T + c_{T2}T^2 \quad (3)$$

Here, c_{T0}, c_{T1}, and c_{T2} are calibration constants at atmospheric conditions. The luminescent signal is denoted as I, and T denotes the temperature. The subscript *ref* denotes the reference conditions. The reference conditions were at 100 kPa and 23 °C. Throughout the temperature and pressure calibrations as well as the application to the flow measurement, we used these reference conditions. We defined the temperature dependency, δ, which is a slope of the temperature calibration at the reference conditions.

$$\delta = \left.\frac{d\left(I/I_{ref}\right)}{dT}\right|_{T=T_{ref}} = c_{T1} + 2c_{T2}T_{ref} \quad (\%/^{\circ}\mathrm{C}) \tag{4}$$

If the absolute value of δ is large, it tells us that the change in I over a given temperature change is also large. This is unfavorable condition as a pressure sensor of a PSP. On the contrary, zero δ means the PSP is temperature independent, which is a favorable condition.

3. Development of Electro-Luminescence based Pressure-Sensitive Paint system

3.1 Characterization setup

Fig. 2 shows the characterization setup for an EL, PSP, and EL-PSP system. We can control the temperature and pressure inside a test chamber. An EL, PSP, and EL-PSP was placed in the chamber under controlled temperature and pressure. The EL was excited by a sinusoidal voltage (Yokogawa, FG300) that creates an AC input. The input was amplified (NF Electronic Instrument, High Speed Power Amplifier 4055) to give the EL illumination. We set the illumination frequency at 1 kHz throughout this chapter. A spectrometer (Hamamatsu Photonics, Photonic Multichannel Analyzer) was used to obtain the luminescent output related to the wavelength. When obtaining the PSP output, we placed a high-pass filter of 600 nm in front of the spectrometer to exclude the illumination. The characterization setup also used a 16-bit CCD camera (Hamamatsu Photonics) to acquire the PSP intensity. Depending on the pressure dye used for a PSP, we placed a band-pass filter in front of the camera to acquire the PSP image (section 3.3). As a comparison, a xenon lamp with a band-pass filter of 400 ± 50 nm was used as a conventional illumination source.

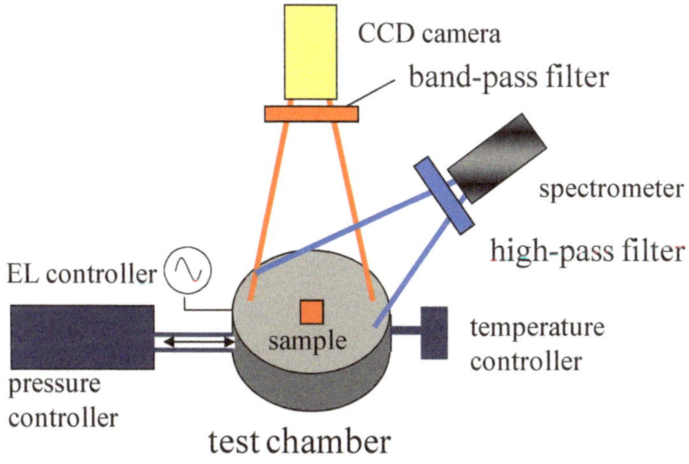

Fig. 2. Schematic of the characterization setup

3.2 EL characterization

Fig. 3 shows the temperature spectra of the inorganic EL used (Nippon Membrane). A peak exists at 500 nm with a broad spectrum from 400 to 650 nm. The output was normalized at the peak illumination at the reference temperature. As we can see, the illumination output increases with an increase of the temperature.

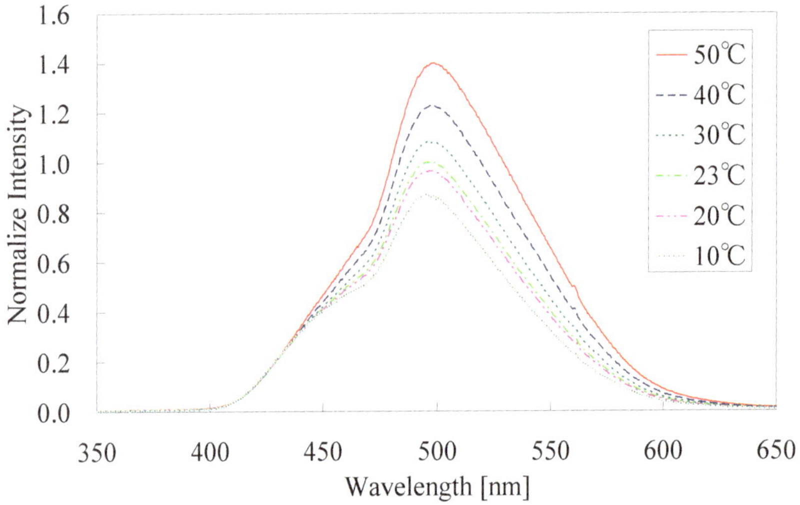

Fig. 3. The temperature spectra of the inorganic EL used

We integrated the spectrum from 400 to 600 nm to determine the illumination intensity. Fig. 4 (a) and (b) show a temperature and pressure calibration. The temperature calibration was obtained under atmospheric conditions. For the temperature calibration, Eq. (3) was fitted to the calibration data. For the pressure calibration, Eq. (1) was fitted to the calibration data. The values of δ and σ were 1.1 %/°C and 0 %/kPa, derived from Eqs. (4) and (2), respectively. As seen in the illumination spectra (Fig. 3), the intensity increased with an increase of the temperature. As described in section 3.3, the temperature dependency of the EL was opposite to that of the PSP; the EL increases its intensity with temperature but the PSP decreases its intensity. On the other hand, the intensity was independent of the pressure.

Fig. 5 (a) shows the illumination uniformity of the EL. As a comparison, an illumination image was obtained from the xenon lamp (Fig. 5 (b)). The images were acquired by the CCD camera with the fixed distance from the illuminated surface. The same area of 30-mm in square was shown. This includes non-illuminated area, which is shown as dark area in Fig. 5. The EL has the size of 25-mm square. For the xenon lamp, it illuminated the EL surface, while the EL was switched off. An illumination tip of a xenon lamp was adjusted to place at the edge of the EL square. The dark area was used as the minimum illumination to normalize the illumination output. The mean output at the center of the illuminated area was used as the maximum illumination for the normalization. Here, d denotes the length of the image area to extract the trend in the illumination in the horizontal axis. Because a xenon lamp is a point illumination source, it showed a non-uniform illumination over the square. On the other hand, the EL showed a uniform illumination over the square. To compare the illumination uniformity, cross sectional distributions given in lines 1 through 4 were shown below each illumination image. The difference from the area averaged value of illumination was shown. As we can see, the EL showed fairly identical in the distributions, while the xenon source showed variations in the illumination.

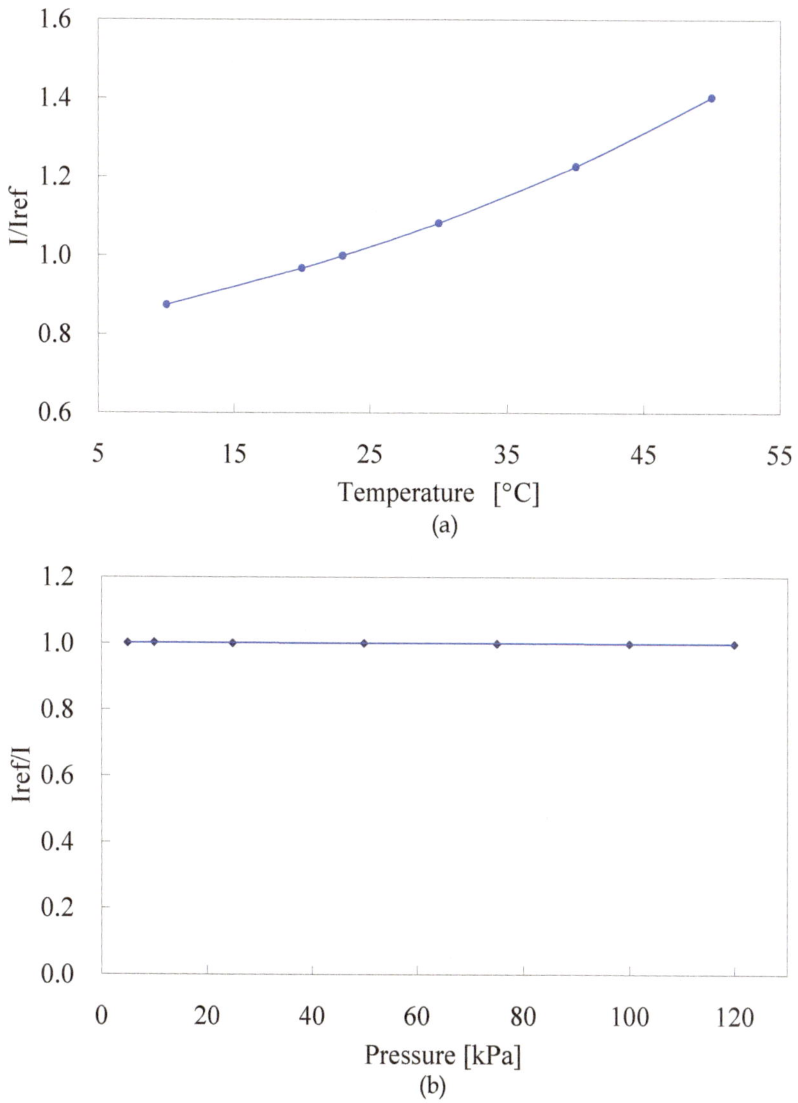

Fig. 4. (a). The temperature calibration of the EL used; (b). The pressure calibration of the EL used

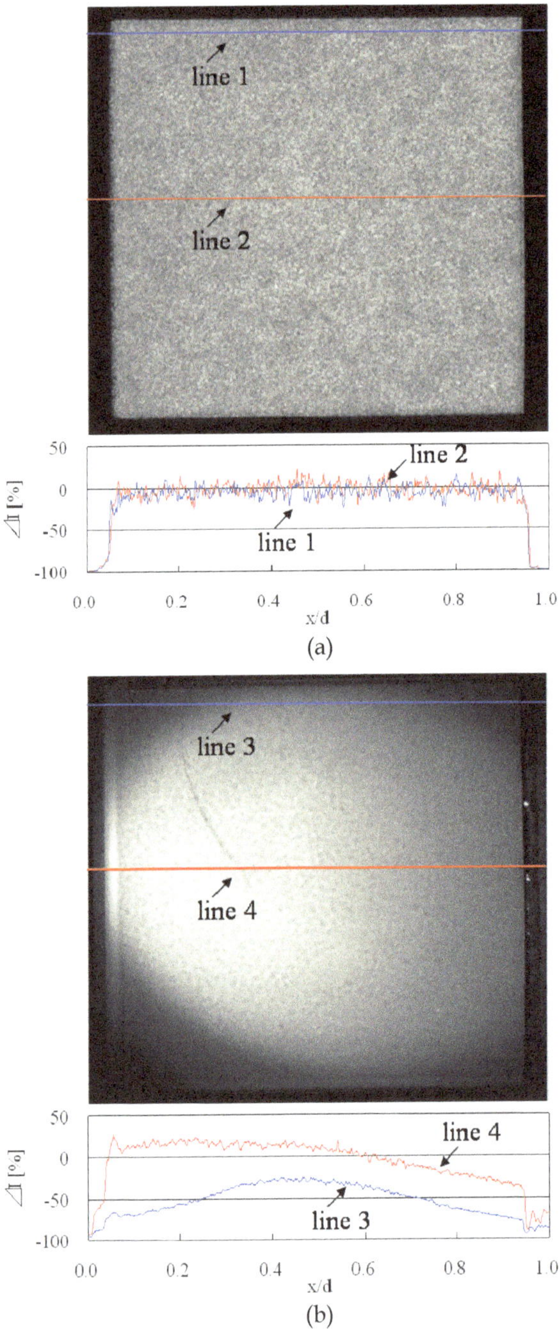

Fig. 5. (a). Illumination uniformity of the EL; (b). Illumination uniformity of a xenon lamp. ΔI denotes the difference in the illumination outputs

3.3 EL-PSP characterization

To tailor the EL as an illumination source for a PSP system, we need to exclude the overlaid spectra between the EL and a PSP emission. We used platinum porphyrin (PtTFPP) and bathophen ruthenium (Ru(dpp)) as a pressure dye in a PSP component, which is commonly used for a conventional PSP system. We inserted a band-pass filter (Fujifilm, BPB50) to exclude the spectral overlay. It is a sheet filter made of a tri acetyl cellulose with its thickness of 90 μm. Fig. 6 shows the transmittance of the band-pass filter. This filter cuts off from 580 to 680 nm, which corresponds to the emission of the PSPs for the EL-PSP system.

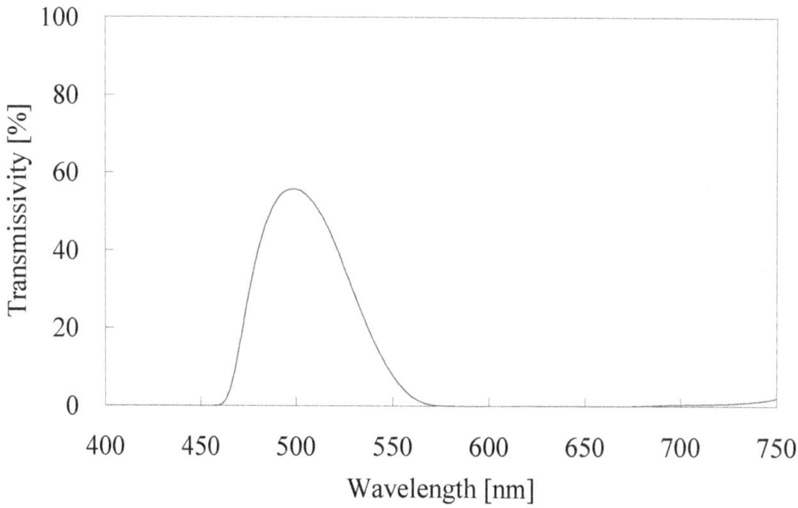

Fig. 6. Transmittance (%) of a sheet band-pass filter to prevent a spectral overlay between the EL illumination of PSP emission.

The developed EL-PSP system consists of the EL, the band-pass filter, and a PtTFPP based PSP. This PSP uses poly-IBM-co-TFEM as a polymer. Similar to PtTFPP, Ru(dpp) based EL-PSP system consist of the EL, the same band-pass filter, and a Ru(dpp) based PSP. This PSP uses RTV118 as a polymer. Overall thickness of the EL-PSP layer was 0.9 mm. In the present case, an EL-PSP layer was provided. It can be applied on a simple curvature, such as a cylinder. If a spraying process would be developed for an inorganic EL, it would be possible to spray the EL-PSP on a complex geometry. Fig. 7 showed the spectral outputs of PtTFPP based EL-PSP system and Ru(dpp) based EL-PSP system, respectively. The former emitted at 650 nm peak, while the latter emitted a broad region with a peak at 600 nm. Note that, due to the sheet filter, the EL illumination did not exist over 580 nm. Because the PSP emission was on top of this filter, the emission was not influenced. Therefore, the PSP emission of Ru(dpp) showed its emission below 580 nm.

Fig. 8 shows the temperature calibrations of the EL-PSP systems. We used a band-pass filter of 650 ± 20 nm in front of the CCD camera to determine the luminescent intensity for PtTFPP based EL-PSP systems, while a band-pass filter of 600 - 750 nm for Ru(dpp) based EL-PSP system. As a comparison, the results of a conventional PSP system were shown. Here, a conventional PSP system uses the same luminophore and polymer as the EL-PSP

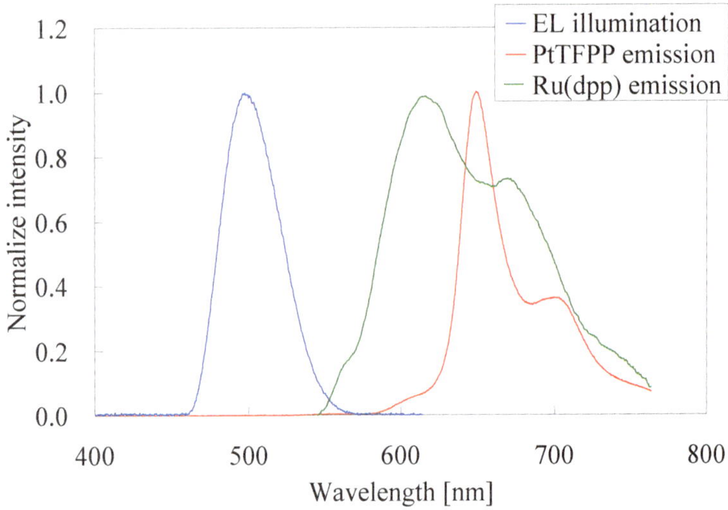

Fig. 7. Luminescent spectra of PtTFPP based EL-PSP system and Ru(dpp) based EL-PSP system

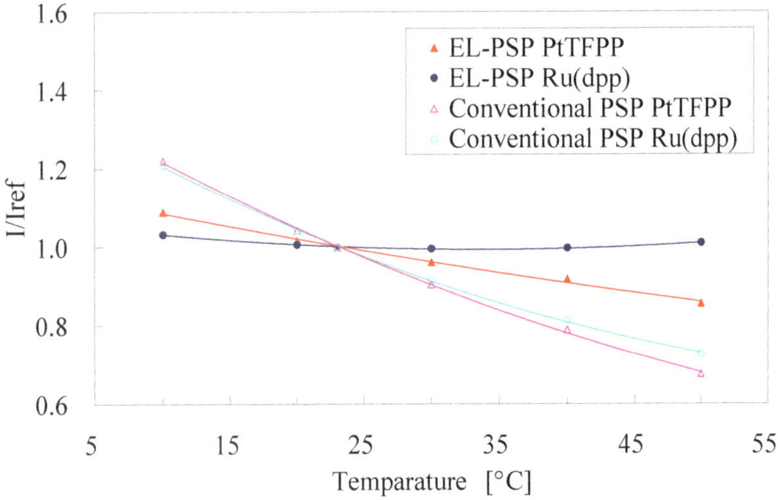

Fig. 8. The temperature calibration of the EL-PSP systems compared to that of a conventional PSP system

systems but uses a xenon lamp as an illumination. PtTFPP based EL-PSP system showed δ of -0.6 %/°C, while the corresponding conventional PSP showed -1.3 %/°C, respectively. This EL-PSP system could reduce δ by 54%. On the other hand, Ru(dpp) based EL-PSP system showed δ of -0.1 %/°C, while the corresponding conventional PSP showed -1.2 %/°C, respectively. This EL-PSP system could reduce δ by 92%. The temperature calibrations tell that the combination of an EL and PSP greatly influences the temperature dependency.

Fig. 9 shows the pressure calibrations of the EL-PSP systems. Similar to the temperature calibration, the pressure calibrations of conventional PSPs were shown for comparisons. As we can see from the calibration results, similar trends can be seen for all results. Based on Eq. (2), σ of the EL-PSP systems and conventional PSPs were 0.8 %/kPa.

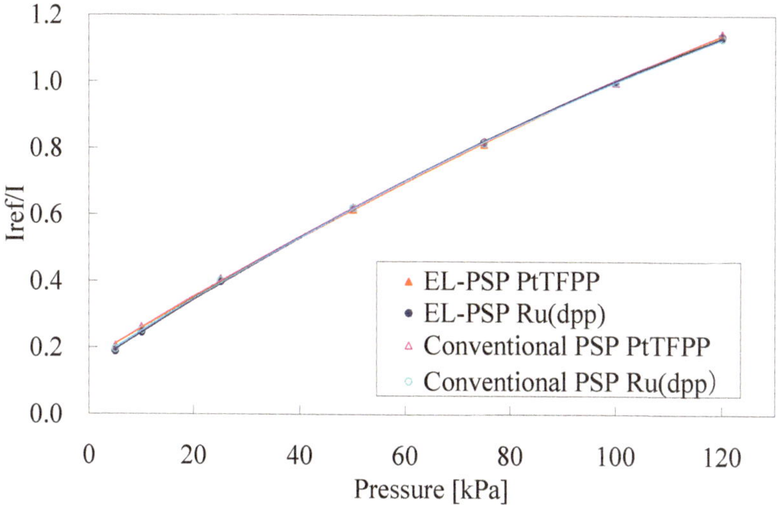

Fig. 9. The pressure calibration of the EL-PSP systems compared to that of a conventional PSP system

4. Demonstration

4.1 Demonstration setup

We applied the developed system to a sonic jet impingement for comparison. A schematic of the experimental setup is shown in Fig. 10. A sonic jet from a 2-mm orifice was impinged on the EL-PSP surface. The same camera and optical filter in calibrations (section 3.1) were used to acquire the luminescent image from the system. The camera was placed above the system as shown in Fig. 10. To discuss a temperature dependency of the measurement, we acquired two reference images: one at the ambient temperature of 23 °C, and the other at a higher temperature of 30 °C. These reference images were acquired when there was no flow that provided the constant pressure at 100 kPa over the EL-PSP surface. As a comparison, a conventional PSP system was used to acquire the same flow. In this system, a xenon lamp through 400 ± 50 nm instead of the EL illumination was used as an excitation. The EL was switched off during this measurement.

4.2 Global pressure measurement and discussion

Fig. 11 (a) and (b) show pressure maps obtained from PtTFPP based and Ru(dpp) based EL-PSP systems. *A priori* pressure calibration obtained in Fig. 9 was used to convert the luminescent image to the pressure map. The ambient reference was used as a reference image. It showed a clear diamond shock pattern created by the sonic jet impingement. We can notice small scratch-like spots in the pressure maps. These were created during the preparation process of the EL-PSP layer, which can be improved to layer the EL, the sheet filter, and PSP.

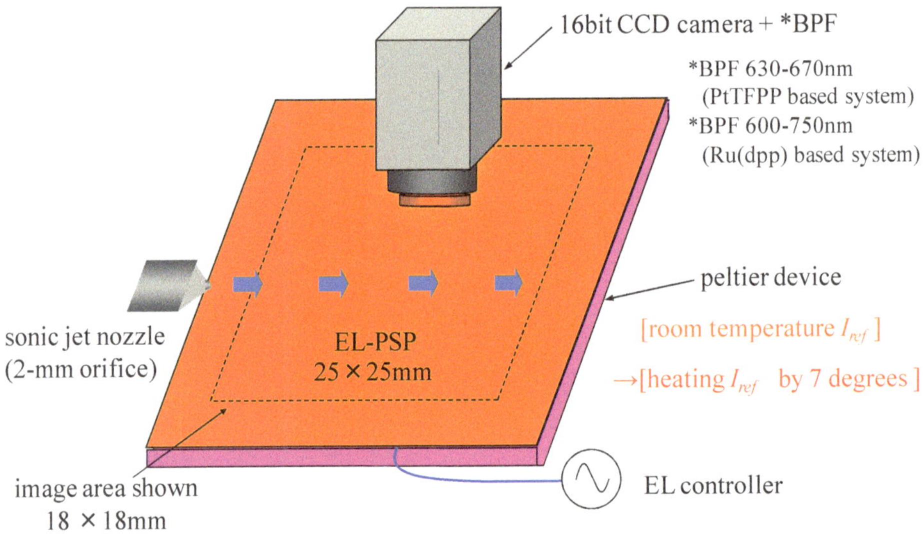

Fig. 10. Schematic description of the demonstration setup

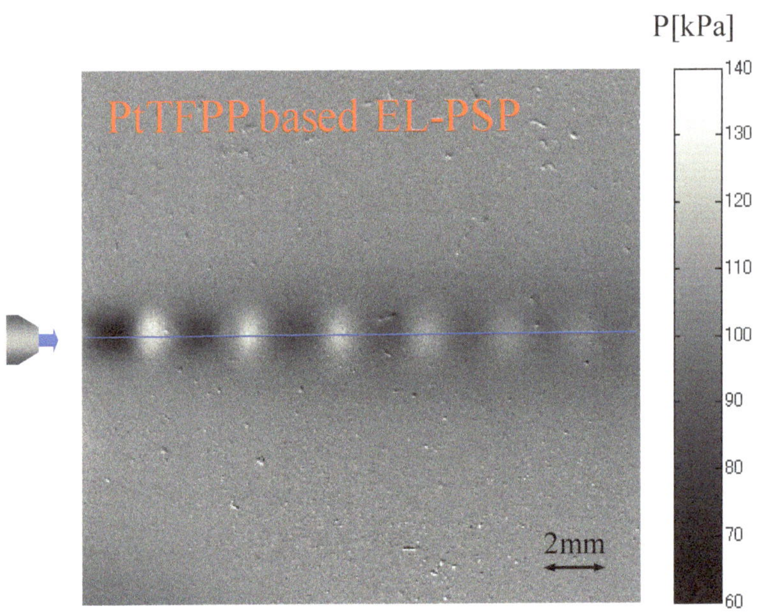

(a)

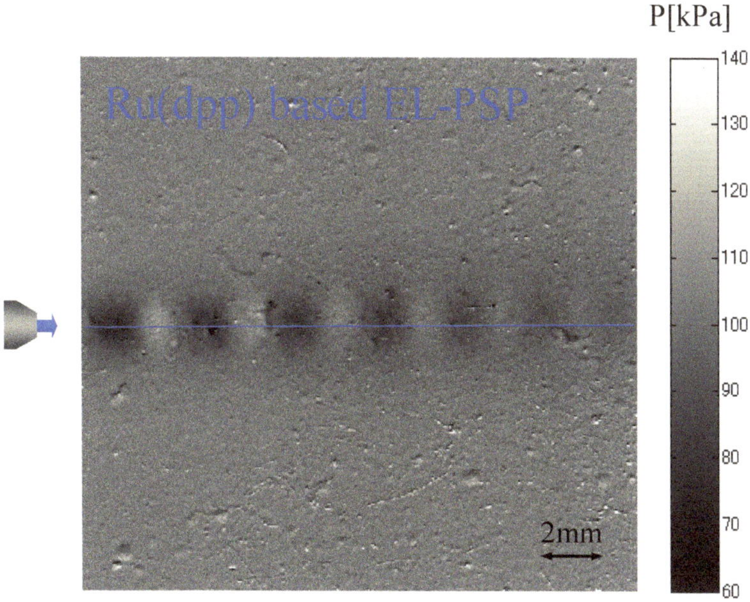

(b)

Fig. 11. Pressure maps created by a sonic jet impingement obtained from (a) PtTFPP based EL-PSP system and (b) Ru(dpp) based EL-PSP system

Fig. 12 (a) through (d) show cross-sectional pressure distributions along the centerline of the jet impinged surface. The location of the centerline was shown as a solid line in Fig. 11. Here $L = 0$ denotes the left edge of the solid line in Fig. 11. Fig. 12 (a) showed the distribution from PtTFPP based EL-PSP system, and Fig. 12 (b) showed the results from the conventional PSP system, respectively. Similarly, Fig. 12 (c) and (d) are a comparison between Ru(dpp) based EL-PSP system and Ru(dpp) based conventional PSP. The scratch-like spots discussed in Fig. 11 was obvious for the pressure distributions from Ru(dpp) based EL-PSP system. For each figure, the pressure distributions obtained from the pressure map results under the ambient and the high temperature reference were shown. As increasing the temperature, the luminescent intensity was reduced (Fig. 8). The high temperature reference, therefore, gave lower luminescent ratio in the left hand side of Eq. (1). This results in an underestimated pressure. We can see that both systems showed the underestimation of the pressure distributions. For the EL-PSP system, the pressure distribution showed smaller change in the pressure measurement compared to that of the conventional PSP system. The difference in the pressure measurement was an average of 3 kPa for PtTFPP based EL-PSP system, while the corresponding conventional PSP was an average of 12 kPa. There was a 75% improvement in the temperature dependency for the pressure measurement obtained from PtTFPP based EL-PSP system. Because Ru(dpp) based EL-PSP system showed very small temperature dependency (-0.1 %/°C, section 3.3), pressure distributions obtained from the different references were almost identical. The difference in the pressure measurement was an average of 1 kPa. On the other hand, Ru(dpp) based conventional PSP showed the difference of an average of 15 kPa. For Ru(dpp) based EL-PSP system, there was a 93% improvement in the temperature dependency.

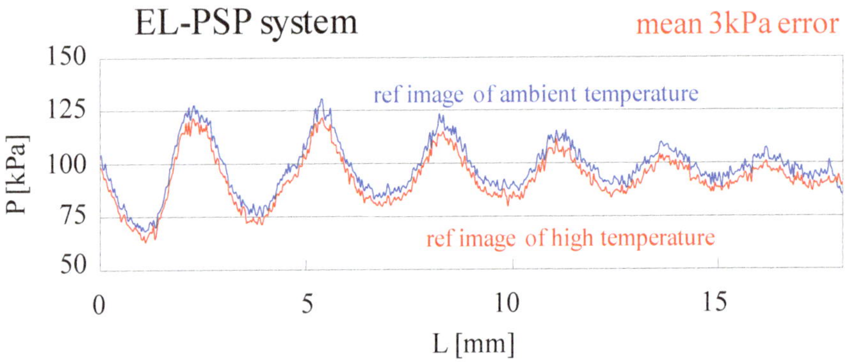
EL-PSP system
mean 3kPa error
150
125
100
75
50
P [kPa]
ref image of ambient temperature
ref image of high temperature
0
5
10
15
L [mm]

(a)

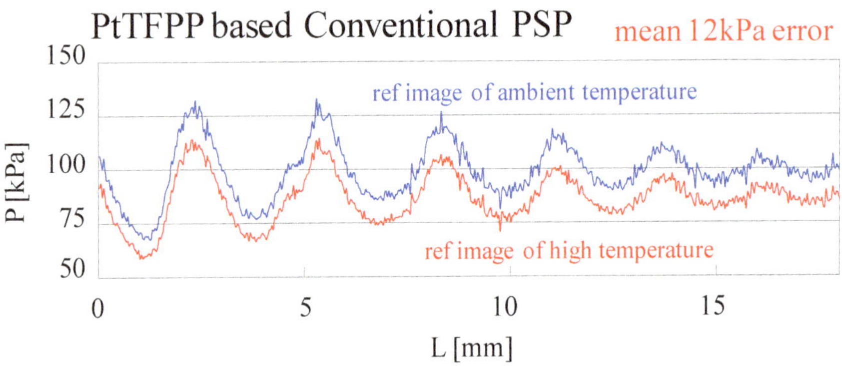
PtTFPP based Conventional PSP
mean 12kPa error
150
125
100
75
50
P [kPa]
ref image of ambient temperature
ref image of high temperature
0
5
10
15
L [mm]

(b)

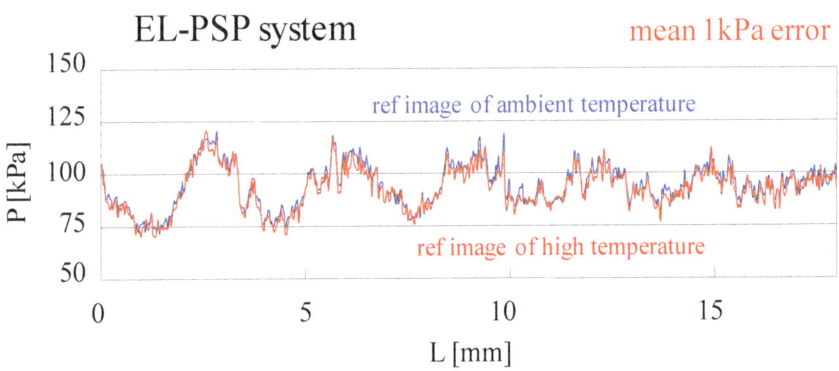
EL-PSP system
mean 1kPa error
150
125
100
75
50
P [kPa]
ref image of ambient temperature
ref image of high temperature
0
5
10
15
L [mm]

(c)

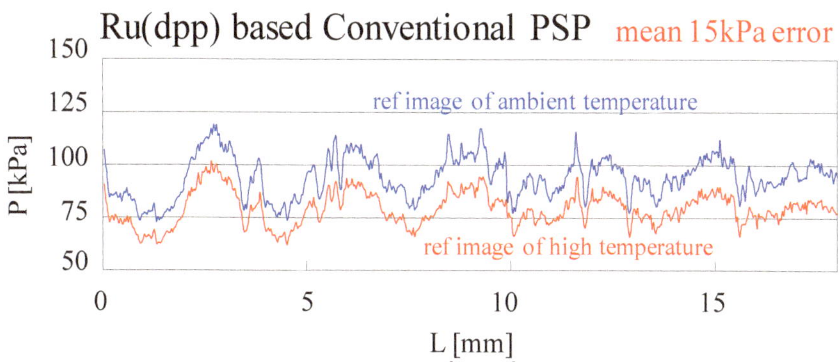

(d)

Fig. 12. Cross-sectional pressures obtained from (a) PtTFPP based EL-PSP system, (b) PtTFPP based conventional PSP system, (c) Ru(dpp) based EL-PSP system, and (d) Ru(dpp) based conventional PSP system. Pressures under ambient temperature and a high temperature were shown for each figure

The reference image was shifted about 2 mm towards the downstream to provide a misalignment of the image. This changed the distribution in the pressure independent luminescence between the reference and jet impinged images. The shifted reference was aligned by an image processing as commonly used for a PSP measurement. Fig. 13 (a) and (b) show the pressure maps obtained from PtTFPP based EL-PSP system and conventional PSP system with the image alignment process, respectively. Because the 2-mm width of the upstream area was not acquired it was shown as a shaded area. The EL-PSP system can provide a pressure distribution as seen in Fig. 11 (a). This tells us that a uniform illumination of the EL can reduce the pressure-independent luminescence and extract the pressure distribution. On the other hand, a point illumination of the conventional PSP system showed a distribution which is not caused by the pressure (Fig. 11 (a)). The processed image could not remove the pressure independent luminescence caused by a non-uniform illumination. This was because the distribution in the pressure independent luminescence obtained by the point illumination was different from the reference and jet impinged images. By using the EL as a surface illumination, the EL-PSP system greatly reduced the misalignment error.

5. Conclusion

We introduced a pressure-sensitive paint (PSP) measurement system based on an electro-luminescence (EL) as a surface illumination. This consisted of an inorganic EL as the illumination, a band-pass filter, and a PSP. The band-pass filter, which passed below 580 nm, was used to separate an overlay of the EL illumination and the PSP emission. In this champter, two types of PSPs were used to construct the EL-PSP system. One is platinum porphyrin (PtTFPP) based PSP, which gave the emission at 650-nm peak. The other is bathophen ruthenium (Ru(dpp) based PSP, which gave a braod emission with a peak at 600 nm. The EL showed an opposite temperature dependency to that of the PSPs; the illumination intensity of the EL increased with increase temperature with the temperature dependency of 1.1 %/°C. The EL gave a uniform illumination compared to that of a point illumination source such as a xenon lamp.

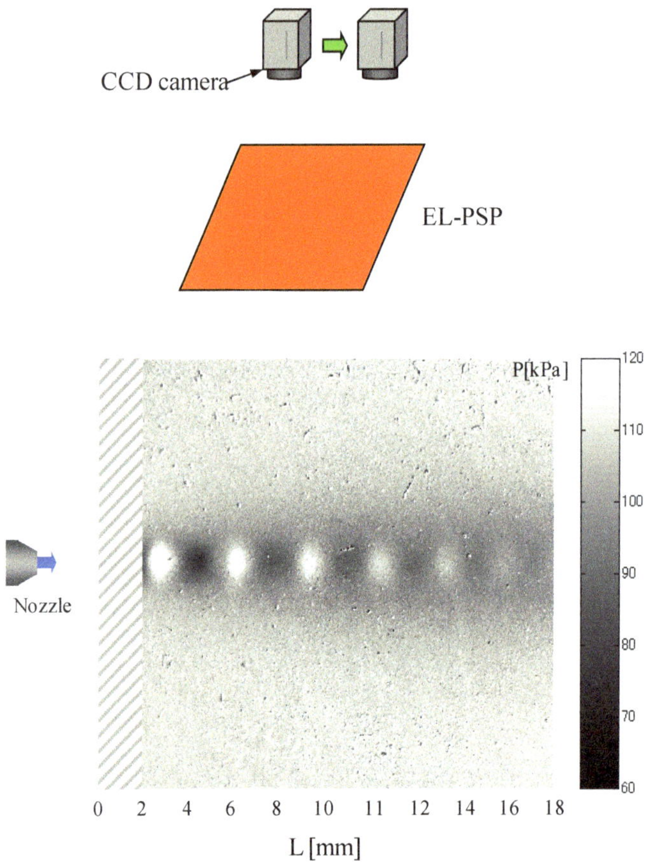
CCD camera
EL-PSP
P[kPa]
120
110
100
90
80
70
60
Nozzle
0 2 4 6 8 10 11 12 14 16 18
L [mm]

(a)

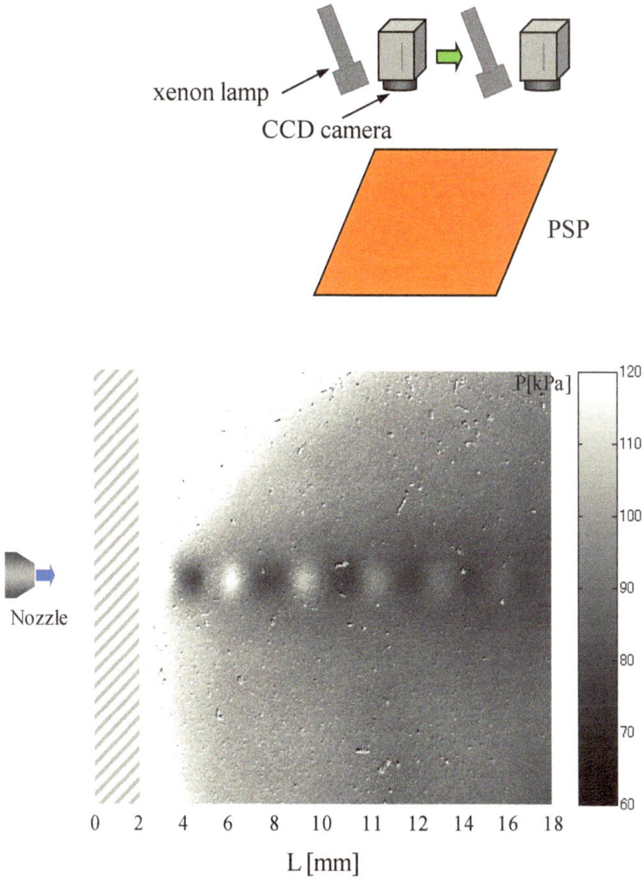

(b)

Fig. 13. Pressure maps processed from a misaligned reference image obtained from (a) PtTFPP based EL-PSP system and (b) PtTFPP based conventional PSP system

A combination of EL and PSP greatly influenced the temperature dependency, while a minimal influence could be seen for the pressure sensitivity. Under atmospheric conditions, PtTFPP based EL-PSP system reduced the temperature dependency by 54% compared to that of a conventional PSP system. For Ru(dpp) based EL-PSP system, the temperature dependency was greatly reduced by 92% compared to that of a conventional PSP system. Both EL-PSP systems showed the pressure sensitivity of 0.8 %/kPa, which was the same for conventional PSP systems.

An application of the EL-PSP systems to a sonic jet impingement showed that the systems demonstrated the reduction of the temperature dependency compared to that of the conventional PSP system. The temperature dependency was reduced by 75% in a pressure measurement for PtTFPP based EL-PSP system, while by a reduction of 93% was obtained for Ru(dpp) based EL-PSP system. Because of a uniform illumination, the EL-PSP system could reduce an image misalignment error compared to that of the conventional PSP system.

6. References

B. Schulze, C. Klein, ICIASF05, Sendai, Japan, August 29-September 1, 274 - 282, (2005)

G. Destriau, *J. Chem. Phys.*, 33, 587, 1936

J. R. Lakowicz, Principles of Fluorescence Spectroscopy, Kluwer Academic/Plenum Publishers, New York, USA, 1999, Chapter 1

Liu, T. & Sullivan, J. P. (2004). *Pressure and Temperature Sensitive Paints*, Springer, ISBN 3-540-22241-3, Heidelberg, Germany (Chapter 1 and Chapter 3)

S. Airaghi, T. Rösgen, M. Guille, Proceedings of 11th International Symposium on Flow Visualization, University of Notre Dame, Notre Dame, Indiana, USA, paper id number 198, Aug. 9 - 12, (2004)

9

Calibration of Measuring Systems Based on Maximum Dynamic Error

Krzysztof Tomczyk
Cracow University of Technology, Faculty of Electrical and Computer Engineering, Poland

1. Introduction

This chapter presents methods and algorithms for determining the maximum errors which can be generated by measuring systems in reference to their standards. The application of such errors in the process of calibration of measuring systems intended for measurement of dynamic signals is discussed in detail. The problem of maximum errors lies in the fact that it is impossible to analyze the full range of all possible dynamic input signals. Therefore we find out the one that represents all signals. It is the signal generating errors of maximum value.

The existence and availability of signals maximizing both the integral-square error and the absolute value of error are discussed, and relevant solutions are presented. Moreover, the constraints imposed on the input signal are analyzed. These constraints refer to the magnitude as well as maximum rate of signal change. The latter constraint is applied in order to match the dynamic properties of the signal to the dynamic properties of the system under test.

In the relevant literature it was proved that the signals maximizing the errors in question always achieve one of the constraints imposed on it. If only the signal amplitude is constrained, the maximizing signal is always of the 'bang-bang' type, while in the case of two constraints relating to the magnitude and the rate of change such a signal can be only of triangular or trapezoid shape with slopes resulting from the values of constraints.

For the integral-square error, no analytic solution referring to the shape of the maximizing signal has been found so far. This is why the solution of this problem presented in the chapter is based on the application of the genetic algorithm method. This method guarantees that the results are obtained in minimized calculation time which depends only on the assumed number of population, the value of the stop condition and number of assumed switches.

In the case of the absolute error, the analytical formulae and algorithm which give precise results and can be realized in a very short calculation time are considered. The signal maximizing this criterion is presented in the form of figures showing the successive steps of proceeding as well as the error shape corresponding to this signal.

Fig. 1 presents a block diagram of our calibration process.

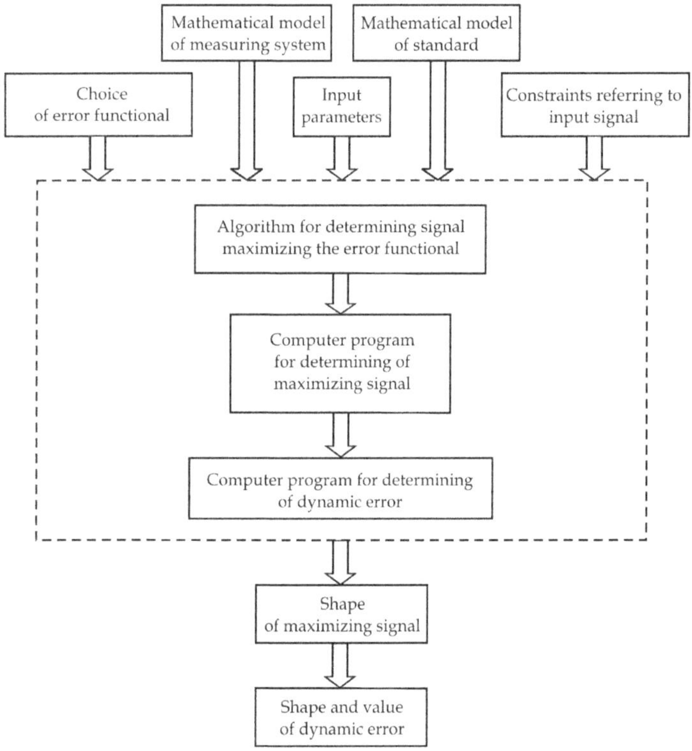

Fig. 1. Block diagram of calibration process of measuring systems based on maximum dynamic error.

In this chapter respective procedures and essential mathematical operations, which allow carrying out the calibration process according to the block diagram presented above, are discussed in detail. In part two the input-output relations of measuring systems are discussed, part three is devoted to the synthesis of the mathematical model of standard and, in part four error functionals are presented and analyzed. The problem of existence and attainability of maximizing signals is discussed in part five, while parts six and seven present the procedures for determining signals maximizing the integral-square error and the absolute value of error.

The optimization of the *minimax* type on the basis of maximizing signals is presented in part eight. Such optimization is performed in two steps. In the first step the input signal maximizing the assumed error functional is determined, while in the second step the model parameters optimization based on the obtained signal is performed.

The identification problem of the mathematical model of measuring system is omitted in this chapter as it does not concern the theory of calibration. It was assumed that this model exist and is given in the form of transfer function.

The practical application of the presented theory and algorithms is illustrated on the example of low-pass measuring systems.

2. Input-output relations of measuring systems

Let output signal $y(t)$ of calibrated system be described by integral convolution

$$\begin{gathered} y(t)=\int_0^t x(\tau)h(t-\tau)d\tau,\ t\in[0,T] \\ h(t)=h_m(t)-h_s(t) \end{gathered} \tag{1}$$

where: $x(t)$ - input signal, $h_m(t),\ h_s(t)$ - impulse responses of system and its standard.

The convolution (1) can be presented in a discrete form as a sum

$$\begin{gathered} y(n)=\Delta\sum_{k=0}^{n} x(k)h(n-k),\ \ n=0,1,...,N-1 \\ \Delta -\ \text{sample interval of } T \end{gathered} \tag{2}$$

or in matrix form

$$\begin{bmatrix} y_0 \\ y_1 \\ y_2 \\ \dots \\ y_{N-1} \end{bmatrix} = \Delta \begin{bmatrix} h_0 & 0 & 0 & \dots & 0 \\ h_1 & h_0 & 0 & \dots & 0 \\ h_2 & h_1 & h_0 & \dots & 0 \\ \dots & \dots & \dots & \dots & \dots \\ h_{N-1} & h_{N-2} & h_{N-3} & \dots & h_0 \end{bmatrix} \begin{bmatrix} x_0 \\ x_1 \\ x_2 \\ \dots \\ x_{N-1} \end{bmatrix} \tag{3}$$

In many cases it is convenient to present the output $y(t)$ by means of the following formula

$$y(n)=\frac{1}{n+1}\sum_{m=0}^{n}\left[\sum_{k=0}^{n} x(k)e^{-i\frac{2\pi}{n+1}mk}\sum_{k=0}^{n} h(k)e^{-i\frac{2\pi}{n+1}mk}\right]e^{i\frac{2\pi}{n+1}mn}\Delta \tag{4}$$

or by the Fourier transform

$$\begin{gathered} Y(i\omega)=X(i\omega)H(i\omega) \\ Y(i\omega)=F\{y(n)\},\ X(i\omega)=F\{x(n)\}\ ,H(i\omega)=F\{h(n)\} \end{gathered} \tag{5}$$

where F presents the DFT or FFT (Tomczyk, 2011).

Multiplication of the Fourier transformations in (5) gives

$$\begin{gathered} \operatorname{Re} Y(i\omega)=\operatorname{Re} X(i\omega)\operatorname{Re} H(i\omega)-\operatorname{Im} X(i\omega)\operatorname{Im} H(i\omega) \\ \operatorname{Im} Y(i\omega)=\operatorname{Im} X(i\omega)\operatorname{Re} H(i\omega)+\operatorname{Re} X(i\omega)\operatorname{Im} H(i\omega) \end{gathered} \tag{6}$$

From Eq. (6) we get

$$y(n)=IF\left[\operatorname{Re} Y(i\omega)+j\operatorname{Im} Y(i\omega)\right]\Delta \tag{7}$$

where IF denotes inverse DFT.

In state space we have

$$\begin{gathered} y(n+1) = \mathbf{\Phi}\, y(n) + \mathbf{\Psi}\, x(n) \\ y(\cdot) - \text{ state vector}, y(\cdot) \in \Re^{p}, p - \text{ number of state variables} \\ x(\cdot) - \text{ input (control) vector}, x(\cdot) \in \Re^{p}, q - \text{ number of inputs} \end{gathered} \tag{8}$$

Presenting (8) in the matrix form, we have

$$\begin{bmatrix} y(n+1) \\ \vdots \\ y_p(n+1) \end{bmatrix} = \begin{bmatrix} \varphi_{1,1} & \cdots & \varphi_{1,p} \\ \vdots & & \vdots \\ \varphi_{p,1} & \cdots & \varphi_{p,p} \end{bmatrix} \begin{bmatrix} y(n) \\ \vdots \\ y_p(n) \end{bmatrix} + \begin{bmatrix} \psi_1 \\ \vdots \\ \psi_p \end{bmatrix} x(n) \tag{9}$$

where

$$\begin{gathered} \mathbf{\Phi}_{(pxp)} = e^{\mathbf{A}\Delta} \\ \mathbf{A}_{(pxp)} - \text{state matrix} \end{gathered} \tag{10}$$

and

$$\begin{gathered} \mathbf{\Psi}_{(pxq)} = \int_0^{\Delta} e^{\mathbf{A}t} dt\, \mathbf{B} = \mathbf{A}^{-1}(e^{\mathbf{A}\Delta} - \mathbf{I})\mathbf{B} \\ \mathbf{B}_{(pxq)} - \text{ input (control) matrix}, \mathbf{I}_{(pxp)} - \text{unit matrix} \end{gathered} \tag{11}$$

3. Synthesis of mathematical model of standard

The calibration procedures for measuring systems intended for measurement of static quantities are commonly known and have been used for a long time in measuring practice. They cover the hierarchy of standards and calibrations circuits.

For measuring systems intended for measurement of dynamic quantities neither legal regulations concerning hierarchy of their accuracy nor specific calibrating procedures have been worked out so far. It results from the fact that for such systems input signals are time dependent, often undetermined, and whose shapes cannot be predicted in advance. The second reason is that the problem of standards for such systems has not been solved till now, because various measuring systems fulfill different objective functions (Tomczyk&Sieja, 2006). For such a situation we expect that models of standard will satisfy mathematical notation of these functions. In the dynamic measurement selection of standard parameters can be realized by means of optimization methods, which assure the conditions of non-distortion transformation, or by means of ideal filters of a band-pass which corresponds to the range of the work of the calibrated system. For low-pass systems this standard is given by

$$H_s(i\omega) = \begin{cases} c \cdot e^{-st_0} & for\ 0 \le \omega \le \omega_m \\ 0 & for\ \omega \ge \omega_m \end{cases} \tag{12}$$

Impulse response of (12) equals

$$h_s(t) = \frac{c}{\pi}\omega_m Sa\left[\omega_m\left(t - t_0\right)\right] \tag{13}$$

where: c - amplification coefficient, t_0 - filter delay, ω_m - cut-off frequency.

4. Error functionals

Dynamic errors applied in our calibration are defined in different function spaces by means of chosen functionals. If the values of these functionals are determined by means of maximizing signals, it means that they will be valid for any dynamic signals which can appear at the input of the calibrated system. In this way all the possible input signals are included in this special, maximizing one.

The process of maximum error determination requires a special input signal to be used, which warrants that the error determined with it will always be higher or at least equal to the value generated by any other signal. The solution of this problem needs to prove that there exists a signal maximizing the chosen error functional at all, and if so, to elaborate a suitable procedure for its determination. The maximum values of error can constitute the base to work out the hierarchy of accuracy for instruments for dynamic measurement.

The integral-square error

$$I(x) = \int_0^T y^2(t)\,dt \tag{14}$$

and absolute value of error

$$D(x) = |y(t)| \tag{15}$$

will be discussed as an example.

The choice of the functionals (14) and (15) results from the fact that they are the most common in many different engineering domains.

5. Existence and attainability of maximizing signals

In (Layer, 2002, Layer&Tomczyk, 2010) it was proved that signals maximizing $I(x)$ and $D(x)$ functionals in $[0,T]$ always achieve one of the constraints imposed on them.

The constraints in question concern the magnitude a and rate of change ϑ. For systems described by means of linear differential equations, the constraint of magnitude $x(t)$ results from the measuring range of this system. Imposition on the signal of only this one constraint gives the solutions of 'bang-bang' type. As a result unexpected great values of errors are generated. It is affected by the dynamics of 'bang-bang' signals which have infinitely high derivative values in the instants of switching. Outside of these instants, they have constant values. Because of this, the 'bang-bang' signals are not matched to the dynamics of physically existing systems, since they can only transmit signals of a limited value of the rate of change. In order to match the dynamic behaviour of the input signals an additional constraint resulting from the dynamic properties of the calibrated system should be imposed. Proper matching is obtained by restricting the maximum rate of change ϑ of the signal.

5.1 Constraints of signal

In the case of two simultaneous constraints a and ϑ, the maximizing signal has a triangular shape with the slopes inclination $\pm\alpha$,

$$\vartheta = tg\alpha \tag{16}$$

or a trapezoid shape with the slopes inclination $\pm\alpha$ and the height a.

We can analyze here two different premises referring to the rate of change ϑ value. The first one refers to the time domain

$$\vartheta \leq \max_{t\in[0,T]} \left|\dot{x}(t)\right| \leq \max_{t\in[0,\infty]} \left|h(t)\right| \tag{17}$$

where $h(t)$ denotes the impulse responses of the system.

In the frequency domain constraint ϑ results from the maximum frequency of the band-pass of system ω_m. In this case we have

$$\vartheta \leq \max_{t\in[0,T]} \left|(a\sin\omega_m t)'\right| = a\,\omega_m. \tag{18}$$

where A denotes amplitude.

6. Procedure for determining signals maximizing the integral-square error

For the integral-square error (14) good results in the determining of signals maximizing this functional are achieved by the application of genetic algorithm. Fig. 2 presents a flowchart of this algorithm.

If $I_i(x) > I_h(x)$, the switch $s1$ takes the position 1 and the value of $I_i(x)$ is assigned to $I_h(x)$. Then $I_h(x)$ is stored in the buffer memory. Parallel to this operation, the vector of switching signal $x_i(t)$ and the output $y_i(t)$ are saved in the memory.

For $i = N$ the switch $s2$ changes the position from 0 to 1 and the values $x_i(t)$, $y_i(t)$ and $I_h(x)$ are assigned to $x_0(t)$, $y(t)$ and $I_{\max}(x)$, respectively.

The algorithm presented in Fig.3 includes three main operations: reproduction, crossing and mutation.

In the first step, the initial population $(c_{11}, c_{12}, ..., c_{1n})$ including an even number of chromosomes is selected at random. Each chromosome consists of detectors $(d_{11}, d_{12}, ..., d_{1m}), (d_{21}, d_{22}, ..., d_{2m}), ..., (d_{n1}, d_{n2}, ..., d_{nm})$, which correspond to the switching times of x_i signal. For each chromosome the error (14) is determined. Then on the basis of the obtained results, adaptation coefficient I_N is calculated as

$$I_N = I_{c_1} + I_{c_2} + .. + I_{c_n} \tag{19}$$

This coefficient presents the total error generated by all chromosomes (Tomczyk, 2006).

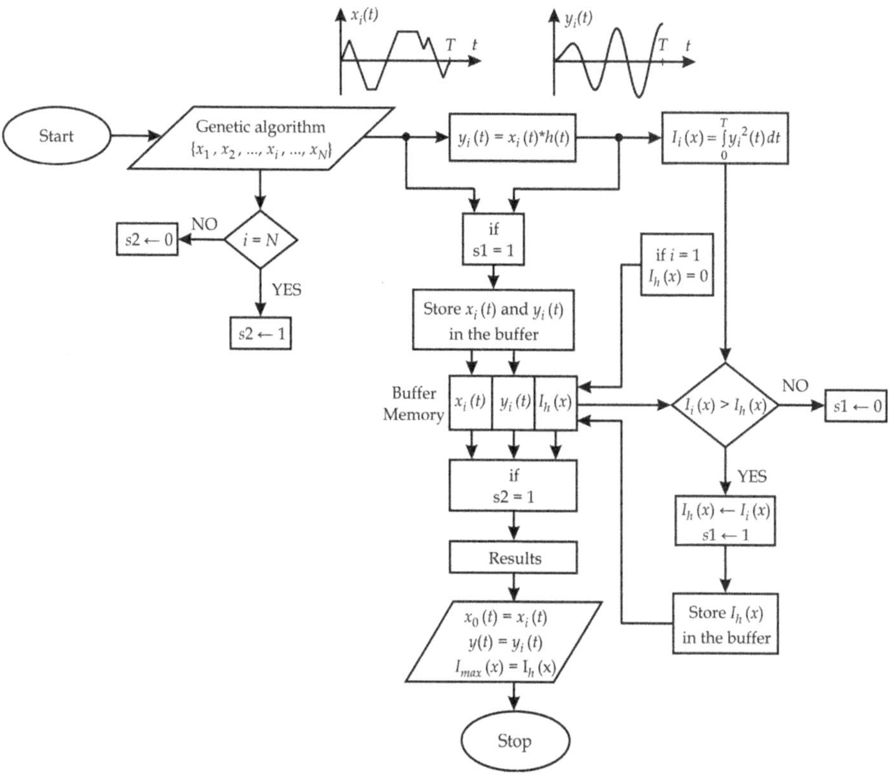

Fig. 2. Flowchart of calibration procedure

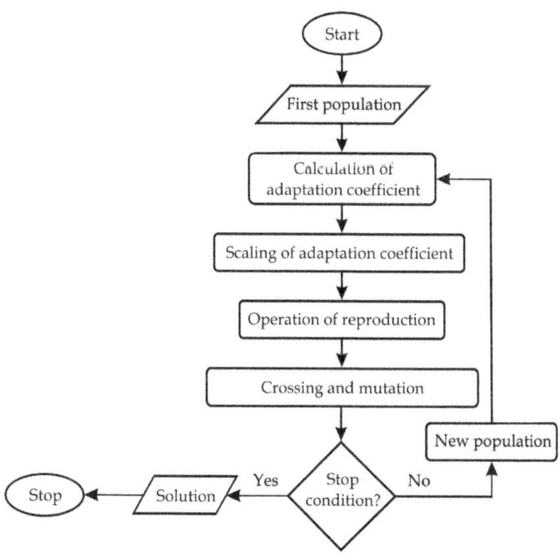

Fig. 3. Flowchart of genetic algorithm

In order to form the next populations, it is necessary to determine the percentage share $\tilde{I}_c$ of each chromosome in I_N

$$\begin{aligned} \tilde{I}_{c_1} &= \frac{I_{c_1}}{I_N}100\% \\ &\vdots \\ \tilde{I}_{c_n} &= \frac{I_{c_n}}{I_N}100\% \end{aligned} \tag{20}$$

Relations (19) and (20) allow for estimation of the usefulness of each chromosome in the population. The higher $\tilde{I}_c$ value of the given chromosome gives more probability to include it in the next population.

When the difference between the obtained values of adaptation coefficients is too small, it is necessary to carry out the scaling operation of the adaptation coefficient.

The next step is the operation of reproduction. According to the probability calculated by (20), the chromosomes from the initial population are selected at random. Depending on the value of $\tilde{I}_c$ coefficient, chromosomes have a bigger or smaller chance to be included in the next population. There are several ways of selection chromosomes for the new population. The most popular way is represented by the roulette wheel method. The results of random selection are rewritten to the new descendant population $(c_{21}, c_{22}, ..., c_{2n})$.

The next step is the crossing process. The chromosomes from the new population are joined in pairs in a random way and for the given crossing probability P_c are crossed or not.

The crossing is carried out according to the following formulae (Tomczyk, 2006):

The first crossing of paired d_{i1} and d_{j1} detectors gives

$$\begin{aligned} \tilde{d}_{i1} &= (1-\alpha)d_{i1} + d_{j1} \\ \tilde{d}_{j1} &= \alpha\, d_{i1} + (1-\alpha)d_{j1} \end{aligned} \tag{21}$$

where: $\tilde{d}_{i1}$ is a descendant detector of i chromosome, and $\tilde{d}_{j1}$ is a descendant detector of j chromosome.

The coefficient α in (21) is selected according to

$$\begin{aligned} \alpha_{i\min} &= \frac{-d_{i1}}{d_{j1}-d_{i1}}, \quad \alpha_{i\max} = \frac{d_{i2}-d_{i1}}{d_{j1}-d_{i1}} \\ \alpha_{j\min} &= \frac{-d_{j1}}{d_{i1}-d_{j1}}, \quad \alpha_{j\max} = \frac{d_{j2}-d_{j1}}{d_{i1}-d_{j1}} \end{aligned} \tag{22}$$

where: $\alpha_{i\min}$ and $\alpha_{i\max}$ present the minimum and maximum limit of α changeability for the detector from i chromosome, while $\alpha_{j\min}$ and $\alpha_{j\max}$ present minimum and maximum limit of α changeability for the detector from the j chromosome.

The changeability range of α is contained in the range between 0 and the third value calculated by means of (22) minus this value multiplied by Δ (sample interval of T).

Then the α value is selected at random from the above range, and is substituted into (21).

The second crossing of detectors d_{im} from i chromosome and d_{jm} from the j chromosome, where $k = 2,3,..., m$, gives

$$\begin{aligned} \tilde{d}_{ik} &= (1-\alpha) d_{ik} + d_{jk} \\ \tilde{d}_{jk} &= \alpha\, d_{ik} + (1-\alpha) d_{jk} \end{aligned} \tag{23}$$

where $\tilde{d}_{ik}$ is the k detector of i descendant chromosome and $\tilde{d}_{jk}$ is the k detector of j descendant chromosome.

Now, the coefficient α is selected according to

$$\begin{aligned} \alpha_{i\min} &= \frac{d_{ik-1} - d_{ik}}{d_{jk} - d_{ik}}, \quad \alpha_{i\max} = \frac{d_{ik+1} - d_{ik}}{d_{jk} - d_{ik}} \\ \alpha_{j\min} &= \frac{d_{jk-1} - d_{jk}}{d_{ik} - d_{jk}}, \quad \alpha_{j\max} = \frac{d_{jk+1} - d_{jk}}{d_{ik} - d_{jk}} \end{aligned} \tag{24}$$

For the i chromosome, the mutation operation is described by

$$\begin{aligned} \tilde{d}_{ik} &= (\tilde{d}_{i(k+1)} - \tilde{d}_{i(k-1)})\, \alpha + \tilde{d}_{i(k-1)} \\ &\alpha \in [0, 1], \; k = 1,2,...,m \end{aligned} \tag{25}$$

The time calculation of genetic algorithm can be reduced significantly if a stop condition is applied. It stops the calculation, if the value of $I_h(x)$ stored in memory does not change during a given number of iterations. (Tomczyk, 2006).

7. Algorithm for determining signals maximizing the absolute error

7.1 Signals constraint on magnitude

In the case of $D(x)$ error (15) the maximum $|y(t)|$ occurs for $t = T$

$$\max|y(t)| = y(T) = \int_0^T x(\tau) h(T-\tau)\, d\tau \tag{26}$$

if

$$x(\tau) = x_0(\tau) = \text{a}\; sign[h(T-\tau)] \tag{27}$$

where a is the magnitude of $x(\tau)$.

Replacing τ by t in (26), we obtain

$$|y(t)| = y(T) = \int_0^T x(t) h(T-t)\, dt \tag{28}$$

and $x_0(t)$ maximizing (28) has now the form

$$x_0(t) = \text{a}\ sign[h(T-t)] \tag{29}$$

Substituting (29) into (28) gives

$$\max|y(t)| = y(T) = \text{a}\int_0^T |h(T-t)|\,dt = \text{a}\int_0^T |h(t)|\,dt \tag{30}$$

which can be computed in the very simple way.

7.2 Signals constraint on magnitude and rate of change

Let

$$x(t) = \int_0^t \phi(\tau)d\tau \tag{31}$$

then

$$y(T) = \int_0^T h(T-\tau)\int_0^t \phi(\tau)\,d\tau\,dt \tag{32}$$

Constraints related to $x(t)$ and $\upsilon(\tau)$ are as follows

$$|\int_0^t \phi(\tau)d\tau| = |x(t)| \le \text{a} \tag{33}$$

and

$$|\phi(t)| = |\dot{x}(t)| \le \vartheta \tag{34}$$

Changing the integration order in (32), we have

$$y(T) = \int_0^T \phi(\tau)\int_\tau^T h(T-t)\,dt\,d\tau \tag{35}$$

and after replacing τ for t, we get finally

$$y(T) = \int_0^T \phi(t)\int_t^T h(T-\tau)\,d\tau\,dt \tag{36}$$

A formula (36) proves that $\phi(t)$ maximizing $y(T)$ has the maximum magnitude $\phi(t) = \pm\vartheta$ by virtue of the formula (34) if

$$\phi(t) = sign\int_t^T h(T-\tau)\,d\tau \tag{37}$$

and $\phi(t) = 0$ in such subintervals, for which (33) between the switching moments has the form

$$|\int_0^t \phi(\tau)d\tau| > \text{a} \tag{38}$$

Using the equations (32)-(38), we can determine signal $x_0(t)$ in the following cases

First case

When $|\int_0^t \upsilon_0(\tau)d\tau \neq \mathbf{a}|$ for ε varying in the intervals $[0,+\vartheta]$ and $[0,-\vartheta]$, (Fig. 4a,b), where $\upsilon_0(t) = +\vartheta$ for $\int_t^T h(T-\tau)d\tau > +\varepsilon$ and $\upsilon_0(t) = -\vartheta$ for $\int_t^T h(T-\tau)d\tau < -\varepsilon$, than the signal $x_0(t)$ is determined in three steps, according to Eqs. (39)-(47)..

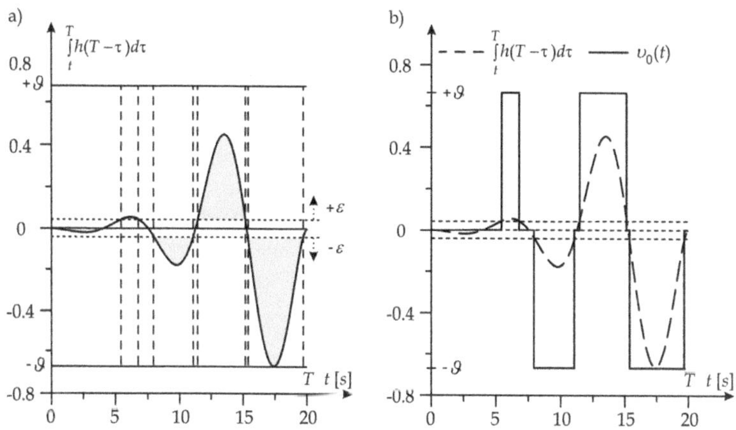

Fig. 4. Exemplary functions $\int_t^T h(T-\tau)d\tau$ and $\upsilon_0(t)$

During the first step, the 'bang-bang' functions $\upsilon_1(t)$ of the magnitude $\pm\vartheta$ are determined with switching moments resulting from (37) - Fig.5a

$$\begin{aligned} \upsilon_1(t) &= +\vartheta \quad \text{if } \phi(t) > 0 \\ \upsilon_1(t) &= -\vartheta \quad \text{if } \phi(t) < 0 \end{aligned} \tag{39}$$

In the second step, we obtain the function $\upsilon_2(t)$ by integrating $\upsilon_1(t)$ - Fig. 5b.

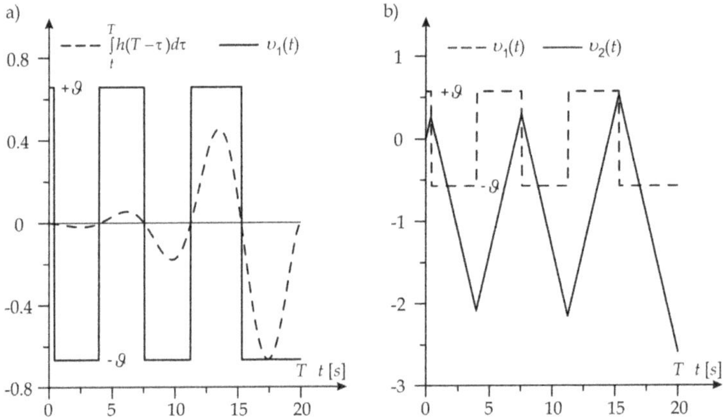

Fig. 5. Exemplary functions $\int_t^T h(T-\tau)d\tau,\ \upsilon_1(t)$ and $\upsilon_2(t)=\int_0^t \upsilon_1(\tau)d\tau$

Function $\upsilon_2(t)$ in particular switching intervals $t_1, t_2, ..., t_m$ of $\upsilon_1(t)$ is given by the following relations

for $t \le t_1,\ m=1$

$$\upsilon_2(t) = \vartheta \cdot t \tag{40}$$

for $t_1 < t \le t_2,\ m=2$

$$\upsilon_2(t) = \vartheta\, t_1 - \vartheta\,(t-t_1) \tag{41}$$

for $t_i < t \le t_{i+1},\ i=2,3,...,m,\ t_{m+1}=T,\ m$ - number of switchings

$$\upsilon_2(t) = \vartheta\, t_1 + \vartheta \sum_{j=2}^{i} (-1)^{j-1}\,(t_j - t_{j-1}) + (-1)^i\, \vartheta\,(t-t_i) \tag{42}$$

In the last step, we determine the function $\upsilon_3(t)$ on the basis of $\upsilon_2(t)$:

$$\begin{aligned} \upsilon_3(t) &= \pm\vartheta \quad \text{if} \quad |\upsilon_2(t)| \le \text{a} \\ \upsilon_3(t) &= 0 \quad \text{if} \quad |\upsilon_2(t)| > \text{a} \end{aligned} \tag{43}$$

Finally, through integration of $\upsilon_3(t)$, we obtain the signal $x(t)=x_0(t)$ and this is the aim. This operation is shown in Fig. 6a.

During the intervals in which $\upsilon_3(t)=\pm\vartheta$, the signal shape is triangular, with the slope $\pm\vartheta$. In the intervals in which $\upsilon_3(t)=0$, the signal is a constant and has the magnitude $\pm a$.

For m switching moments of $\upsilon_3(t)$ the value of error is described by the following equations:

for $m=1$

$$y(T)=\frac{o_1}{t_1}\int_0^{t_1} h(T-\tau)\tau\, d\tau + \frac{o_T-o_1}{T-t_1}\int_{t_1}^{T} h(T-\tau)\,(\tau-t_1)\, d\tau + o_1\int_{t_1}^{T} h(T-\tau)\, d\tau \tag{44}$$

for m ≥ 2

$$y(T)=\frac{o_1}{t_1}\int_0^{t_1} h(T-\tau)\tau\, d\tau + \sum_{i=2}^{m}\left[\frac{o_i-o_{i-1}}{t_i-t_{i-1}}\int_{t_{i-1}}^{t_i} h(T-\tau)(\tau-t_{i-1})\, d\tau + o_{i-1}\int_{t_{i-1}}^{t_i} h(T-\tau)\, d\tau\right]+$$
$$+\frac{o_T-o_m}{T-t_m}\int_{t_m}^{T} h(T-\tau)\,(\tau-t_m)\, d\tau + o_m\int_{t_m}^{T} h(T-\tau)\, d\tau \tag{45}$$

where $o_m = x_0(t_m)$, $o_T = x_0(T)$.

Fig. 6b presents the signal $x_0(t)$ and the error $y(t)$ corresponding to it

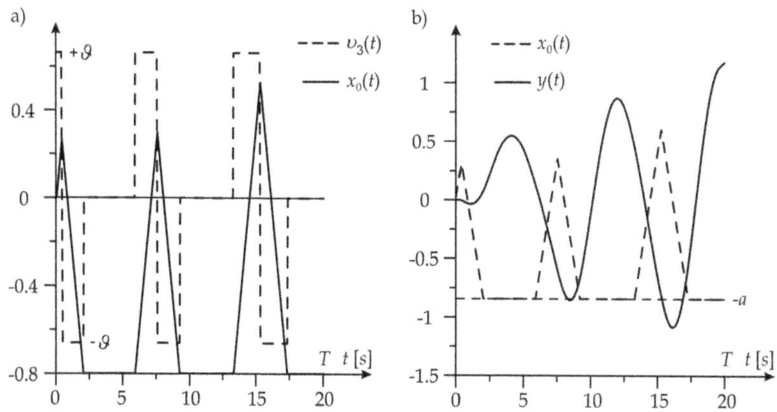

Fig. 6. Function $\upsilon_3(t)$, signal $x_0(t)=\int_0^t \upsilon_3(\tau)d\tau$ and error $y(t)$

The equations (44-45) in the discrete form are defined by:
for $n = 1$

$$y(N)=\frac{o_1}{t_1}\Delta\sum_{k=0}^{n_1} h(N-k)k + \frac{o_N-o_1}{N-n_1}\Delta\sum_{k=n_1}^{n_1} h(N-k)(k-n_1) + o_1\Delta\sum_{k=n_1}^{N} h(N-k) \tag{46}$$

for $n \geq 2$

$$y(N)=\frac{o_1}{n_1}\Delta\sum_{k=0}^{n_1} h(N-k)\, k + \sum_{i=2}^{m}\left[\frac{o_i-o_{i-1}}{n_i-n_{i-1}}\Delta\sum_{k=n_{i-1}}^{n_1} h(N-k)(k-n_{i-1}) + o_{i-1}\Delta\sum_{k=n_{i-1}}^{n_1} h(N-k)\right]$$
$$+\frac{o_N-o_m}{N-n_m}\Delta\sum_{k=n_m}^{N} h(N-k)(k-n_m) + o_m\Delta\sum_{k=n_m}^{N} h(N-k) \tag{47}$$

Second case

If $\vartheta \cdot T \le \mathrm{a}$ then the signal $x_0(t)$ is given directly by

$$x_0(t) = \vartheta \int_0^t sign[h(T-\tau)]d\tau \tag{48}$$

and the error equals

$$y(t) = \int_0^t x_0(\tau)h(t-\tau)d\tau \tag{49}$$

Equation (49) in the discrete form is defined by

$$y(N) = \sum_{k=0}^{N} x(k)h(n-k)\Delta, \text{ for } n=0,1,...,N-1 \tag{50}$$

Fig. 7 presents the signal $x_0(t)$ and error $y(t)$ obtained by (48)-(49)

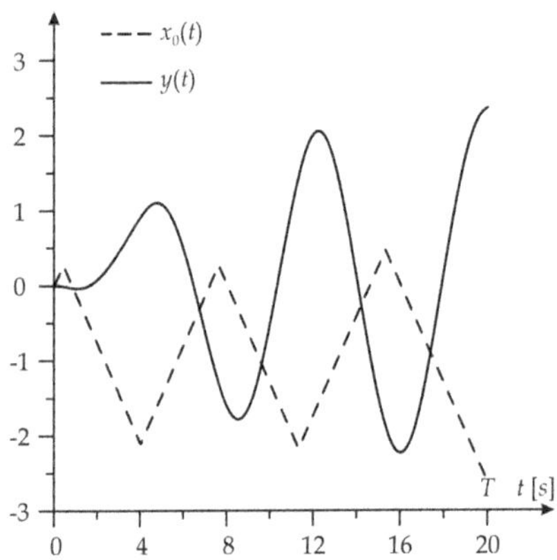

Fig. 7. Signal $x_0(t)$ and error $y(t)$

Third case

If $\vartheta \cdot T > \mathrm{a}$ then the signal $x_0(t)$ is determined indirectly by means of the functions $\upsilon_4(t)$ - $\upsilon_7(t)$

$$\upsilon_4(t) = \frac{\vartheta}{2\mathrm{a}} \int_0^t h(T-\tau)(t-\tau)d\tau \tag{51}$$

$$\upsilon_5(t) = \mathrm{a} \cdot sign[\upsilon_4(T-t)] \tag{52}$$

$$\begin{aligned} &\upsilon_6(t) = \upsilon_5(t) && \text{if } t \le \frac{2\mathrm{a}}{\vartheta} \\ &\upsilon_6(t) = \upsilon_5(t) - \upsilon_5\left(t - \frac{2\mathrm{a}}{\vartheta}\right) && \text{if } \frac{2\mathrm{a}}{\vartheta} < t \le T \end{aligned} \tag{53}$$

$$\begin{aligned} &\upsilon_7(t) = \upsilon_6(t)\frac{\vartheta}{\mathrm{a}} && \text{if } t \le \frac{2\mathrm{a}}{\vartheta} \\ &\upsilon_7(t) = 0 && \text{if } \frac{a}{\vartheta} < t \le \frac{2\mathrm{a}}{\vartheta} \\ &\upsilon_7(t) = \upsilon_6(t)\frac{\vartheta}{2\mathrm{a}} && \text{if } \frac{2\mathrm{a}}{\vartheta} < t \le T \end{aligned} \tag{54}$$

Equation (51) in the discrete form is given by

$$\upsilon_4(n) = \frac{\vartheta}{2\mathrm{a}}\Delta\sum_{k=0}^{n} h(n-k)(n-k), \;\; n = 0,1,\ldots,N-1 \tag{55}$$

The functions $\upsilon_4(t)$ and $\upsilon_5(t)$ are shown in Fig. 8a, while $\upsilon_5(t)$ and $\upsilon_6(t)$ in Fig. 8b.

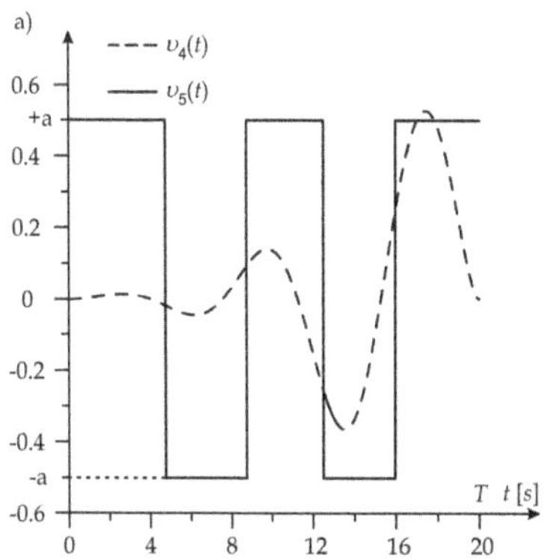

Fig. 8(a). Functions $\upsilon_4(t)$ and $\upsilon_5(t)$.

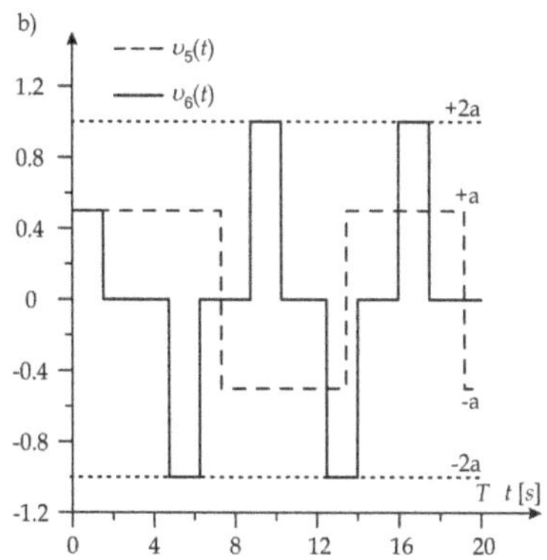

Fig. 8(b). Functions $\upsilon_5(t)$ and $\upsilon_6(t)$

Fig. 9a presents the function $\upsilon_7(t)$ and the signal $x_0(t) = \int_0^t \upsilon_7(\tau)d\tau$, while the signal $x_0(t)$ and the error $y(t)$ corresponding to it are shown in Fig. 9b (Layer&Tomczyk, 2009; Layer&Tomczyk, 2010).

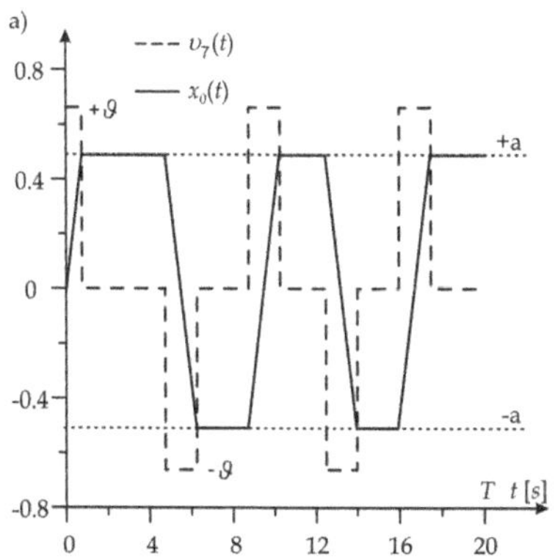

Fig. 9(a). Function $\upsilon_7(t)$ and signal $x_0(t)$.

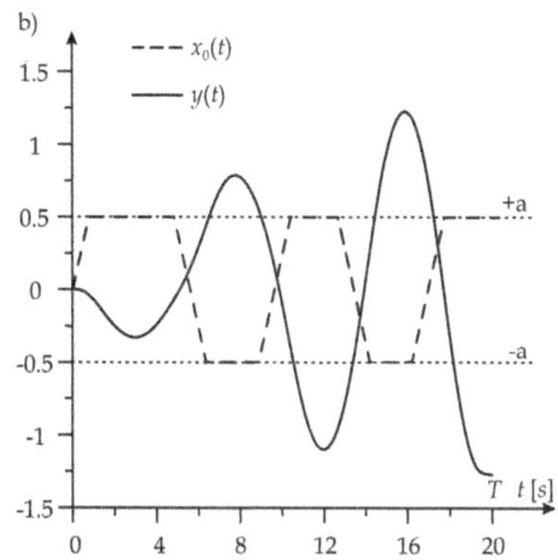

Fig. 9(b). Signal $x_0(t)$ and error $y(t)$

8. Maximum errors in optimization of models of measuring systems

In the technical domain the application of higher order models usually gives a possibility to better map the dynamic properties of real systems. On the other hand, the analysis of such models is most often difficult and time-consuming. Hence a tendency has developed to replace a higher order model by a simplified one. The class and order of such model are adopted in an arbitrary way. Its parameters can be determined by means of methods that minimize the mapping error, considering all possible input signals. Therefore it is suggested to solve the problem by using one signal which maximizes the error. In this way, the mapping error being determined is credible for any input. The application of the *minimax* method by means of the Levenberg-Marquardt algorithm (Tomczyk, 2009) gives good results.

This method includes two main numerical computation stages. At the first stage the $x_0(t)$ signal maximizing the error (Fig.10) is determined.

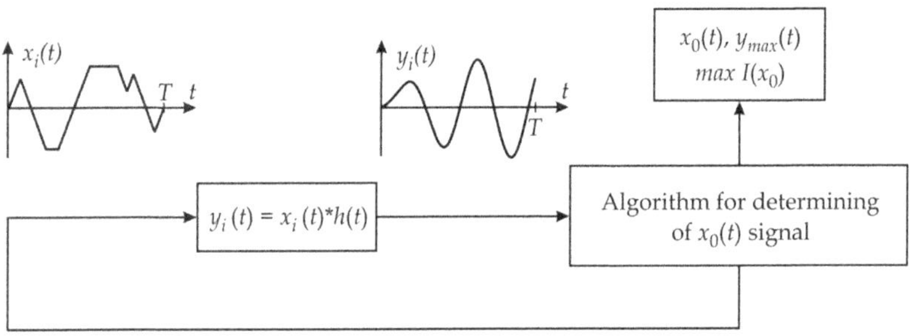

Fig. 10. Block diagram for determining $x_0(t)$ signal

At the second stage the class and order of the simplified model are adopted. For this model by means of $x_0(t)$ signal the parameters minimizing the error are determined.

The Levenberg-Marquardt algorithm is an iterative method, in which the vector of unknown parameters is determined in $k+1$ step by the following formula

$$\Xi_{k+1} = \Xi_k^{\ T} - \left[\mathbf{J}^T(\Xi_k,t)\mathbf{J}(\Xi_k,t) + \mu_k I \right]^{-1} \mathbf{J}^T(\Xi_k,t)y(\Xi_k,t) \tag{56}$$

with the error

$$I = \int_0^T y^2(\Xi_k,t)dt \tag{57}$$

where

$$y(\Xi_k,t) = \int_0^t h(t-\tau)x(\tau)d\tau \tag{58}$$

and

$$\mathbf{J}(\Xi_k,t)=\begin{bmatrix} \frac{\partial y(\Xi_k,t_1)}{\partial \xi_1} & \frac{\partial y(\Xi_k,t_1)}{\partial \xi_2} & \cdots & \frac{\partial y(\Xi_k,t_1)}{\partial \xi_m} \\ \frac{\partial y(\Xi_k,t_2)}{\partial \xi_1} & \frac{\partial y(\Xi_k,t_2)}{\partial \xi_2} & \cdots & \frac{\partial y(\Xi_k,t_2)}{\partial \xi_m} \\ \vdots & \vdots & \vdots & \vdots \\ \frac{\partial y(\Xi_k,t_n)}{\partial \xi_1} & \frac{\partial y(\Xi_k,t_n)}{\partial \xi_2} & \cdots & \frac{\partial y(\Xi_k,t_n)}{\partial \xi_m} \end{bmatrix} \tag{59}$$

The notations in (56)-(59) are as follows:

- $k=1,2,...,p,\ p$ - number of iteration loops
- $\mathbf{J}_{(n x m)}(\Xi_k,t)$ - Jacobian matrix
- $\mathbf{I}_{(m x m)}$ - unit matrix
- μ_k - scalar, its value changes during iteration
- $\Xi=[\xi_1,\xi_2,...,\xi_m]$ - model parameters searched for.

The Levenberg-Marquardt algorithm is used for computation in the following stages:

Stage 1, for k=1

1. Assume the initial values of the parameters of vector Ξ_k
2. Assume the initial value of the coefficient μ_k (e.g. $\mu_k=0.1$)
3. Solve the matrix equation (59) and calculate (58)
4. Calculate the value of error (57)
5. Determine the parameters of vector Ξ_{k+1}, following (56).

Stage 2 and further steps, for $k=2,3,...,p$

1. Update the values of the parameters of vector Ξ_k
2. Solve the matrix equations (59), calculate (58) and (56)
3. Calculate the value of error (57)

Compare the values of error (57) for the step k and the step $k-1$. If the result is $y(\Xi_k,t)\geq y(\Xi_{k-1},t)$, multiply μ_k by the specified value $\lambda\in\Re$ (e.g. $\lambda=10$) and return to the step 2 of stage 2. If the result is $y(\Xi_k,t)<y(\Xi_{k-1},t)$ divide μ_k by the value λ and return to the step 1 of stage 2.

Fig. 11. presents a block diagram of *minimax* optimization by means of the Levenberg-Marquardt algorithm.

The iteration process can be finished if in the consecutive stage the decrease in the value of error (57) is very small and insignificant. Then we fix $\mu_k=0$ and determine the final result for the parameters of vector Ξ. If the coefficient μ_k is high, it means that the solution is not satisfactory, the values of the parameters of vector are not optimum and the value of error (57) is not at minimum level. At this point we can assumed

$$\mathbf{J}^T(\Xi_k,t)\mathbf{J}(\Xi_k,t)<<\mu_k\mathbf{I} \tag{60}$$

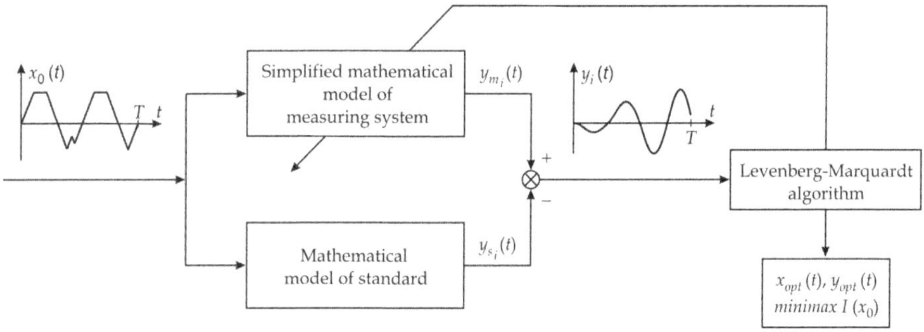

Fig. 11. Block diagram of *minimax* optimization by means of the Levenberg-Marquardt algorithm

In such a case the formula (60) leads to the steepest descent method

$$\Xi_{k+1} = \Xi_k^{\ T} - \frac{1}{\mu_k}\mathbf{J}^T(\Xi_k,t)y(\Xi_k,t) \tag{61}$$

If the value of coefficient μ_k is small, it means that the values of the parameters of vector Ξ are close to the optimum solution. Then

$$\mathbf{J}^T(\Xi_k,t)\mathbf{J}(\Xi_k,t) >> \mu_k\mathbf{I} \tag{62}$$

and the Levenberg-Marquardt algorithm is reduced to the Gauss-Newton method

$$\Xi_{k+1} = \Xi_k^{\ T} - \left[\mathbf{J}^T(\Xi_k,t)\mathbf{J}(\Xi_k,t)\right]^{-1}\mathbf{J}^T(\Xi_k,t)y(\Xi_k,t) \tag{63}$$

The selection of coefficient values μ_k and λ depends on the adopted programs and selected software.

9. Example of application

As an example let us consider the fourth order Butterworth, Tchebyhev and Bessel analog filters and integral-square error (14). These filters have the following transfer functions:

$$H_{Bt4}(s) = \frac{c}{\left[\left(\frac{s}{\omega_m}\right)^2 + 1.8478\frac{s}{\omega_m} + 1\right]\left[\left(\frac{s}{\omega_m}\right)^2 + 0.7654\frac{s}{\omega_m} + 1\right]} \tag{64}$$

$$H_{T4}(s) = \frac{c}{\left[5.5339\left(\frac{s}{\omega_m}\right)^2 + 2.1853\frac{s}{\omega_m} + 1\right]\left[1.2039\left(\frac{s}{\omega_m}\right)^2 + 0.1964\frac{s}{\omega_m} + 1\right]} \tag{65}$$

$$H_{Be4}(s) = \frac{c}{\left[0.4889\left(\frac{s}{\omega_m}\right)^2 + 1.3397\frac{s}{\omega_m} + 1\right]\left[0.3890\left(\frac{s}{\omega_m}\right)^2 + 0.7743\frac{s}{\omega_m} + 1\right]} \tag{66}$$

As a reference let us adopt the mathematical model of ideal low-pass filter given by (12) and (13) for $c=1$ and $\omega_m = 100\ {}^{rad}\!/_{s}$.

In order to determine the signals maximizing the error (14) let us adopt:

- magnitude $a = 0.2V$,
- rate of change , as a maximum absolute value of $\vartheta = \int_0^t h(\tau)d\tau$, where $h(t)$ is the difference between the impulse responses of filter and its standard: $\vartheta_{Bt4} = 1.11\frac{V}{s}$, $\vartheta_{T4} = 1.36\frac{V}{s}$ and $\vartheta_{Be4} = 1.01\frac{V}{s}$ for Butterworth, Tchebyhev and Bessel filters respectively,
- $T = 0 - 0.5s.$,
- discretization step $\Delta = 0.001s$.

The obtained switching times of signals errors and their values for Butterworth, Tchebychev and Bessel filters are presented in Tables 1-3.

$T[s]$	Butterworth filter	
	$a=0.2V,\ \vartheta = 1.11\frac{V}{s}$	
	$x_0(t)$	$I(x_0)$
0	0	0
0.05	ϑ_+ [0.0, 0.045s.], ϑ_- [0.045, 0.05s.]	$8.67 \cdot 10^{-7}$
0.1	ϑ_+ [0.0, 0.095s.], ϑ_- [0.095, 0.1s.]	$4.57 \cdot 10^{-5}$
0.15	ϑ_+ [0.0, 0.13s.], ϑ_- [0.13, 0.15s.]	$2.68 \cdot 10^{-4}$
0.2	ϑ_+ [0.0, 0.185s.], $+a$ [0.185, 0.2s.]	$7.55 \cdot 10^{-4}$
0.25	ϑ_+ [0.0, 0.2s.], $+a$ [0.2, 0.25s.]	$19 \cdot 10^{-4}$
0.3	ϑ_+ [0.0, 0.185s.], $+a$ [0.185, 0.275s.], ϑ_- [0.275, 0.3s.]	$25 \cdot 10^{-4}$
0.35	ϑ_+ [0.0, 0.05s.], ϑ_- [0.05, 0.285s.], $-a$ [0.285, 0.35s.]	$39 \cdot 10^{-4}$
0.4	ϑ_+ [0.0, 0.08s.], ϑ_- [0.08, 0.345s.], $-a$ [0.345, 0.4s.]	$57 \cdot 10^{-4}$
0.45	ϑ_+ [0.0, 0.11s.], ϑ_- [0.11, 0.405s.], $-a$ [0.405, 0.45s.]	$79 \cdot 10^{-4}$
0.5	ϑ_+ [0.0, 0.135s.], ϑ_- [0.135, 0.455s.], $-a$ [0.455, 0.5s.]	$113 \cdot 10^{-4}$

Table 1. Time switchings and values of integral-square error for Butterworth filter

T[s]	Tchebychev filter $a=0.2,\ \vartheta=1.36\frac{V}{s}$ $x_0(t)$	$I(x_0)$
0	0	0
0.05	ϑ_+ [0.0, 0.045s.], ϑ_- [0.045, 0.05s.]	$4.52\cdot10^{-6}$
0.1	ϑ_+ [0.0, 0.095s.], ϑ_- [0.095, 0.1s.]	$4.61\cdot10^{-5}$
0.15	ϑ_+ [0.0, 0.12s.], ϑ_- [0.12, 0.15s.]	$4.01\cdot10^{-4}$
0.2	ϑ_+ [0.0, 0.15s.], $+a$ [0.15, 0.2s.]	$12\cdot10^{-4}$
0.25	ϑ_+ [0.0, 0.15s.], $+a$ [0.15, 0.25s.]	$25\cdot10^{-4}$
0.3	ϑ_+ [0.0, 0.045s.], $+a$ [0.045, 0.24s.], ϑ_- [0.24, 0.3s.]	$36\cdot10^{-4}$
0.35	ϑ_+ [0.0, 0.07s.], ϑ_- [0.07, 0.29s.], $-a$ [0.29, 0.35s.]	$59\cdot10^{-4}$
0.4	ϑ_+ [0.0, 0.1s.], ϑ_- [0.1, 0.35s.], $-a$ [0.35, 0.4s.]	$87\cdot10^{-4}$
0.45	ϑ_+ [0.0, 0.125s.], ϑ_- [0.125, 0.4s.], $-a$ [0.4, 0.45s.]	$122\cdot10^{-4}$
0.5	ϑ_+ [0.0, 0.15s.], ϑ_- [0.15, 0.45s.], $-a$ [0.45, 0.5s.]	$173\cdot10^{-4}$

Table 2. Time switchings and values of integral-square error for Tchebychev filter

T[s]	Bessel filter $a=0.2,\ \vartheta=1.01\frac{V}{s}$ $x_0(t)$	$I(x_0)$
0	0	0
0.05	ϑ_+ [0.0, 0.045s.], ϑ_- [0.045, 0.05s.]	$8.55\cdot10^{-8}$
0.1	ϑ_+ [0.0, 0.095s.], ϑ_- [0.095, 0.1s.]	$5.66\cdot10^{-5}$
0.15	ϑ_+ [0.0, 0.12s.], ϑ_- [0.12, 0.15s.]	$2.74\cdot10^{-4}$
0.2	ϑ_+ [0.0, 0.15s.], $+a$ [0.15, 0.2s.]	$7.28\cdot10^{-4}$
0.25	ϑ_+ [0.0, 0.15s.], $+a$ [0.15, 0.25s.]	$18\cdot10^{-4}$
0.3	ϑ_+ [0.0, 0.045s.], $+a$ [0.045, 0.24s.], ϑ_- [0.24, 0.3s.]	$25\cdot10^{-4}$
0.35	ϑ_+ [0.0, 0.07s.], ϑ_- [0.07, 0.29s.], $-a$ [0.29, 0.35s.]	$37\cdot10^{-4}$
0.4	ϑ_+ [0.0, 0.1s.], ϑ_- [0.1, 0.35s.], $-a$ [0.35, 0.4s.]	$51\cdot10^{-4}$
0.45	ϑ_+ [0.0, 0.125s.], ϑ_- [0.125, 0.4s.], $-a$ [0.4, 0.45s.]	$70\cdot10^{-4}$
0.5	ϑ_+ [0.0, 0.15s.], ϑ_- [0.15, 0.45s.], $-a$ [0.45, 0.5s.]	$99\cdot10^{-4}$

Table 3. Time switchings and values of integral-square error for Bessel filter

Fig. 12 presents signals $x_0(t)$ and errors $y(t)$ as well as Fig. 13 illustrates the integral-square error as function of time duration $T = 0-0.5$ s. for $\Delta T = 0.05$s.

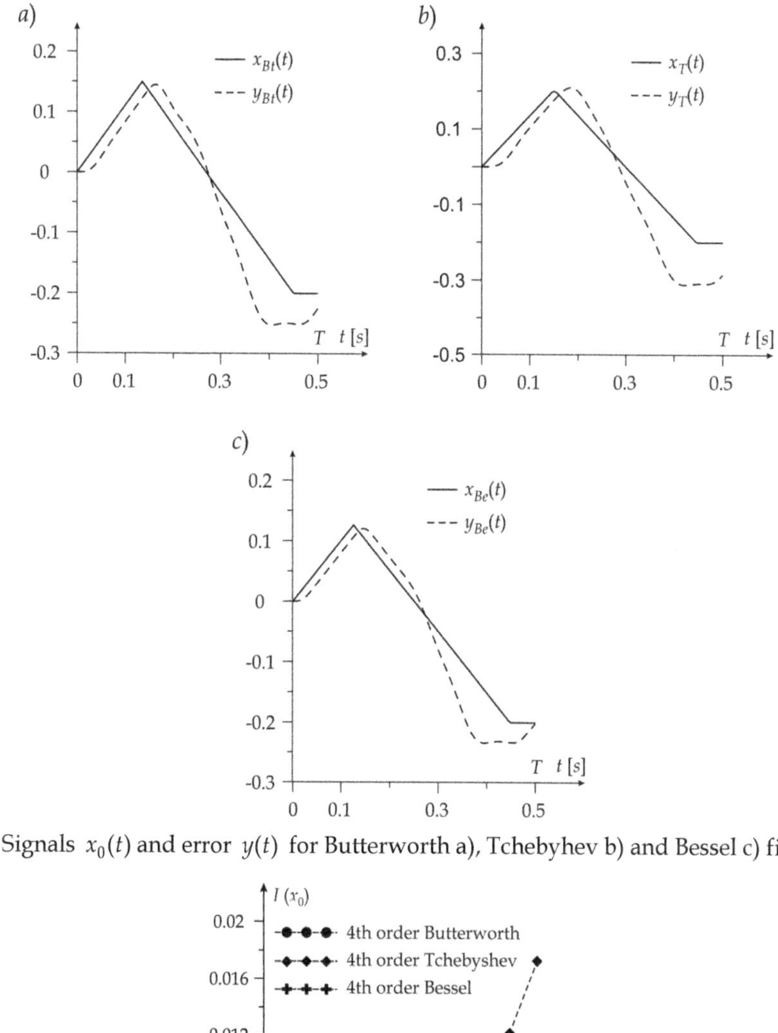

Fig. 12. Signals $x_0(t)$ and error $y(t)$ for Butterworth a), Tchebyhev b) and Bessel c) filters.

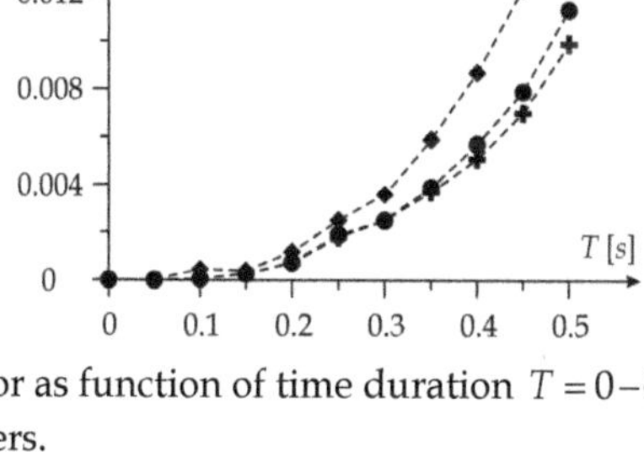

Fig. 13. Integral-square error as function of time duration $T = 0-0.5$ s. for Butterworth, Tchebychev and Bessel filters.

From Fig. 13 it follows that for $T < 0.2s.$ signal $x_0(t)$ generates similar values of error for all filters. In the interval $0.2s. < T < 0.35\ s.$ the highest value of error generates the Tchebyshev filter, however for the Butterworth and Bessel filters the values of error are similar. For $T > 0.35s.$ the higher error generates the Tchebyshev filter, whereas the Bessel filter generates the lowest value.

10. Conclusion

The calibration of systems based on the theory of maximum dynamic errors, makes it possible to create accuracy classes for such dynamic systems and the hierarchy of dynamic accuracy resulting from them. These hierarchies are valid regardless of the shape of the measured dynamic signal that can appear at the input of the investigated system.

The calibration methods presented in this chapter can be widely applied in many different domains, eg. in electrical metrology (transducers, strain gauge amplifiers, filters, etc), geophysics (accelerometer and vibrometer systems), medicine (electroencephalograph and electrocardiograph systems), meteorology (autocomparators, autobridges), etc.

11. References

Layer, E. (2002). *Modelling of Simplified Dynamical Systems,* Springer-Verlag, ISBN 3-540-43762-2, Berlin Heidelberg

Layer, E.; Tomczyk, K. (2009). *Determination of non-standard input signal maximizing the absolute error,* Metrology and Measurement Systems. Vol. XVII, no. 2, pp. 199-208, ISSN 0860-8229

Layer, E.; Tomczyk, K. (2010). *Measurements, Modelling and Simulation of Dynamic Systems,* Springer-Verlag, ISBN 978-3-642-04587-5, Berlin Heidelberg

Tomczyk, K. (2006). *Application of genetic algorithm to measurement system calibration intended for dynamic measurement,* Metrology and Measurement Systems. Vol. XIII, no. 1, pp. 93-103, ISSN 0860-8229

Tomczyk, K. (2009). Levenberg-Marquardt Algorithm for Optimization of Mathematical Models according to Minimax Objective Function of Measurement Systems, Metrology and Measurement Systems. Vol. XVI, no. 4, pp. 599-606, ISSN 0860-8229

Tomczyk, K. (2011). *Procedure for correction of the ECG signal error introduced by skin-electrode interface,* Metrology and Measurement Systems. Vol. XVIII, no. 3, pp. 461-470, ISSN 0860-8229

Tomczyk, K.; Sieja, M. (2006). *Acceleration transducers calibration based on maximum dynamic* error, Czasopismo Techniczne 3-E, Cracow University of Technology, pp. 37-49, ISSN 0011-4561

10

Computer-Based Measurement System for Complex Investigation of Shape Memory Alloy Actuators Behavior

Marek Kciuk and Grzegorz Kłapyta
Silesian University of Technology, Faculty of Electrical Engineering, Department of Mechatronics, Poland

1. Introduction

1.1 Basic information

Shape Memory Alloys belong to group of so called "SMART" materials, which became more and more popular last years. The smartness of that kind of materials is defined as possibility of its particular properties stimulation using physical or chemical way. Shape Memory Alloy (SMA) is able to change its size (shape) when temperature is changing. Shape Memory Effect (SME) is macroscopic effect of martensitic transition which overlap inside the material. There are two characteristic states of SMA: low temperature - martensitic - state and high temperature - austenitic - state. Material can be easily plastically deformed by relatively small mechanical force in martensitic state. In opposite, material is strong and very hard to deform in austenitic state - it is only elastic deformation viable. During martensitic transition internal crystal order changes, which causes change of material's properties. Material returning to its natural shape during reverse martensitic transition (from austenitic to martensitic state) can put pressure or perform mechanical work. Complex behavior of SMA materials can be illustrated by thermo-mechanical characteristic (Fig.1).

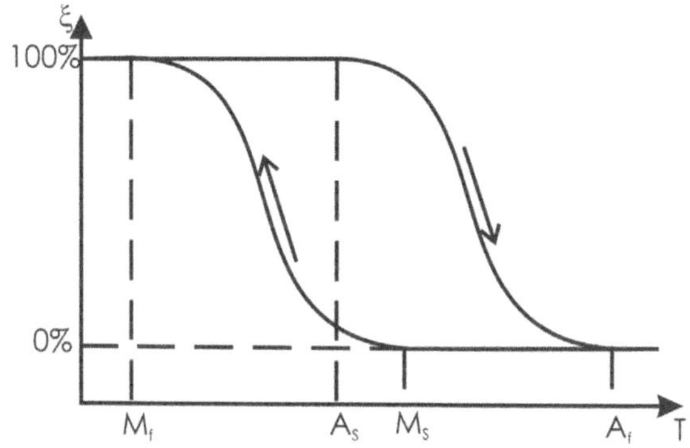

Fig. 1. Thermo-mechanical characteristic of Shape Memory Alloy [1].

Martensitic transition is not isothermal transition. The characteristic is highly nonlinear and exhibit permanent hysteresis loop. There are four important and specific temperature points: M_S and M_f are direct martensitic transition start and finish temperature, while A_S and A_f are reverse martensitic transition start and finish temperature. The highest temperature (A_f) means that all the material is already in austenitic phase; the lowest temperature (M_f) means that all the material is in martensitic phase. ξ is measure of percentage content of molecules in martensitic phase in the volume of the material. For all the temperatures between M_f and A_f the content martensitic fraction (ξ) of material is undefined.

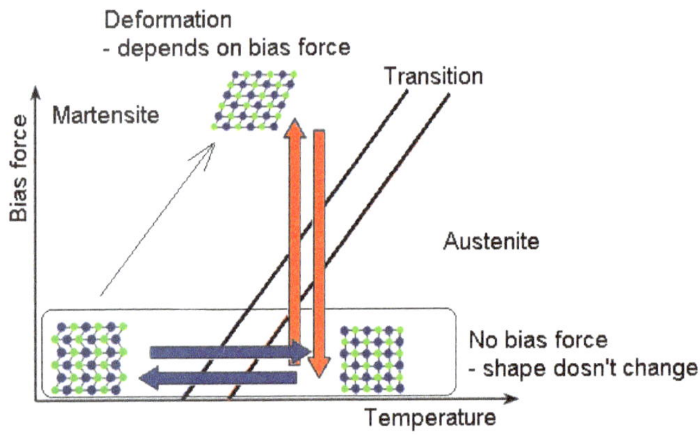

Fig. 2. Shape of SMA material via bias force.

The bias force is important for shape and for martensitic transition parameters. If there is no bias force the macroscopic shape in martensitic state is the same as in austenitic state. If there is a bias force martensitic shape is deformed and is different than austenitic shape. It is presented in Fig. 2.

Important temperatures of martensitic transitions strongly depend on mechanical bias force. The higher bias force is, the higher important temperatures are (there is linear relationship – see Fig.3). Characteristic temperatures under no bias force are marked with "0" subscript.

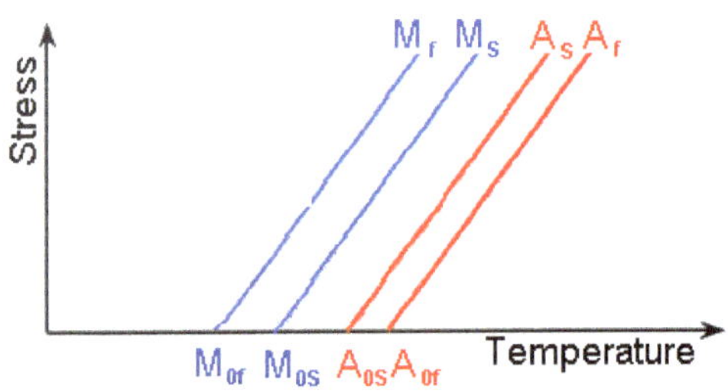

Fig. 3. Characteristic temperatures of martensitic transition via bias force.

Therefore there are two possible ways of transition - one stimulated by temperature, second - stimulated by bias force (respectively blue and red arrows in Fig.2). Temperature stimulated transformation enables to achieve three types of shape: austenitic shape (natural shape, ordered structure of material), martensitic twinned shape (natural shape, unordered structure of material), martensitic deformed shape (deformed shape, unordered structure of material. Bias force stimulated transformation is isothermal and it starts with low bias force and temperature above A0f. Increased mechanical force causes that characteristic temperatures shifts. It is so called superelasticity and enables to achieve only two types of shape: austenitic shape and deformed martensitic shape. This phenomena is presented in Fig.4.

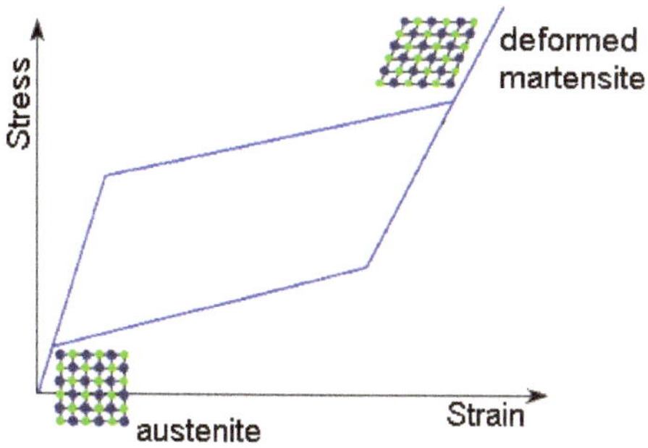

Fig. 4. Superelasticity effect of SMA material.

1.2 Modeling of Shape Memory Alloys

SMAs are already well known and frequently used materials. Different types of mathematical models describe SME behavior. Most popular is modeling based on thermodynamical laws, using different types of free energy equations i.e. Helmholtz's, Gibbs', or Landau'a - Devonshier's [3,6]. There are also mezoscopic models (describing groups of molecules), micromechanical models (describing molecules behavior) and macroscopic models, describing whole material. Ikuta's model describes the state of material by two elements: elastic element and suspension element -the factors are changing with phase changes. There are several types of Ikuta's models (i.e. two layers model, multilayer model, running layers model etc.) [4,5].

Thermodynamical models are the most popular but they are very difficult to implement. They need a lot of parameters and it is hard to use them in control systems.

1.3 Applications

There are two main application groups: SMA works as a sensor or as an actuator. There is also third group of applications, which combine both previously mentioned together.

SMAs are most frequently used as temperature sensors in household, industry devices or air conditioning. Natural hysteresis loop is useful advantage for these kinds of sensors. Threshold temperature mainly depends on material. It is possible to adjust this temperature in some range by mechanical bias force changing.

SMA are also used as active (controlled) or passive (not controlled) actuators. Passive actuators exploit the phenomenon of superelasticity. They are used in civil engineering to protect buildings bridges etc. (e.g. as seismic protection). Passive SMA actuators can suppress vibrations. See Fig.5.

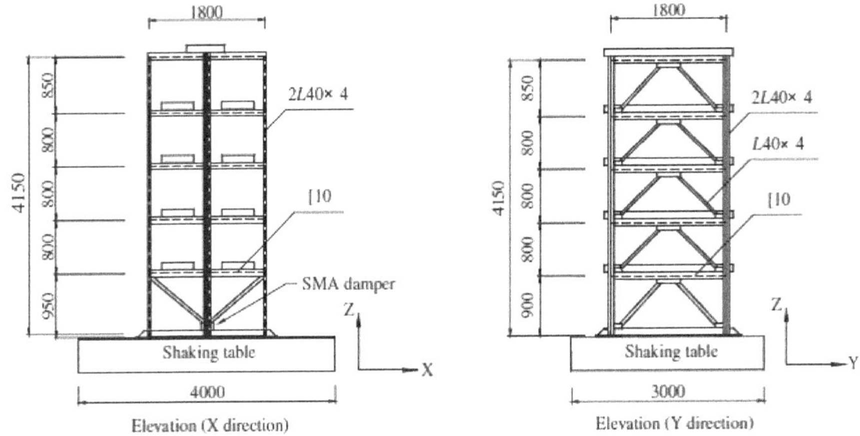

Fig. 5. Comparison of stress distribution of protected and not protected building [15].

Active SMA actuators are popular in mechatronic applications as unconventional drives. They are light-weight actuators with high power to weight ratio. They are able to replace sophisticated electro-mechanical devices in simple tasks. They generate linear displacement in direct way. Sparkles work ensures the safety of work in flammable and wet areas. It is also possible to build light-weight robots (or robotic tools) actuated by SMA. Unconventional robotic hand driven by SMAs is described in [7]. Two types of unconventional mobile robots (snake robot and net robot) are presented in [8, 9]. It is possible to build micro robots made in MEMS technology, actuated by SMAs [10].

2. Measuring stand

Measuring stand for obtaining the complex SMA characteristics was designed and built in Department of Mechatronics at Silesian University of Technology. The stand allows for measurements of static as well as dynamic characteristics of SMA wires. Multiparameter simultaneous measurement enables comparison of different characteristics and allows for assessment of impact of particular parameters on the studied phenomena. Main assumptions for measuring stand project were:

- resistive heating of wire with current constraint,
- possibility of mechanical load changing,
- temperature measurement using infrared camera,
- optical measurement of wire displacement,
- simultaneous measurement of current, voltage, wire displacement and temperature,
- maximum wire length - 900mm,
- minimal mechanical load - 75g,
- maximum wire displacement - 80 mm,
- automatic measurement process with data acquisition system.

Measurement system presented in this paper allows to investigate complex electro-thermo-mechanical characteristics. It is possible to measure electro-mechanical characteristics (length of actuator vs. wire current $L = f(I)$, wire resistance vs. length of actuator $R = f(L)$, etc.), thermo-mechanical characteristic (wire length vs. temperature of actuator $L = f(T)$), electro-thermal characteristic (temperature of actuator vs. supplying current $T = f(I)$), etc. The idea of measurement system is described in Fig.6.

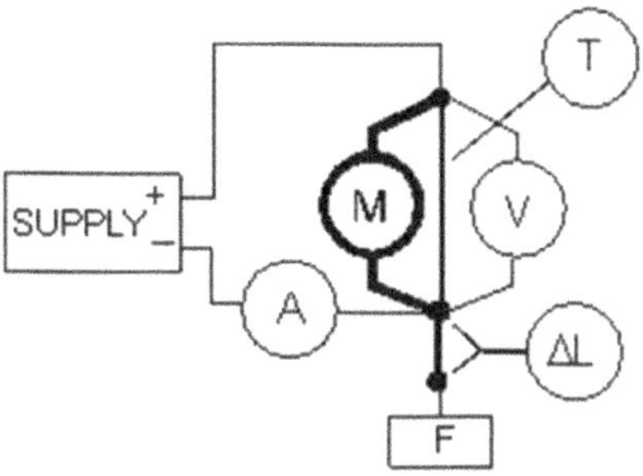

Fig. 6. Idea of measurement system.

There are six main modules: mechanical construction (M) with bias force (F), supply module, and measurement modules: current module (A), voltage module (V), temperature module (T), length changing measurement module (ΔL). Measurement stand is controlled via PC computer in LabVIEW environment. Detailed construction of each module is described below.

Mechanical construction: Design of a stand (Fig.7) allows to place SMA wire (1) between stiff crosspiece (2) and moving piston (3). Gravitational mechanical load can be increased by using additional masses and allows for constant load during measurement process. Base (4), frame (5), and the crosspiece (2) are made of aluminum profiles. Infrared camera (6) is mounted to the base at constant level but it can be shifted horizontally to allow picture sharpness regulation. Electrical circuit is electrically separated from aluminum base by PCV distances. To improve the quality of infrared measurement part of the stand was covered with black cardboard sleeve. In window of Fig.7a. you can see also displacement sensor placed in the bottom part of stand.

Heat supply is done in electrical way, hence SMA wire is a good conductor. According to the Joule-Lenz Law the wire temperature rise is nearly proportional to the square of flowing current. To set the constant temperature in SMA wire the supply system should allow for current stabilization. The power supply PSH-3620A (36V, 20A) was chosen to feed the system. It allows for current as well as for voltage stabilization. Stabilization is switched automatically according to Ohm's law. Measurement stand is closed inside black cartoon shield which guarantees steady air conditions and is helpful for temperature measurements using infrared camera.

Measurement of electrical parameters is done by three precise multimeters Rigol DM3052. It allows for simultaneous measurement of current and voltage of SMA actuator and voltage signal of displacement sensor. Data acquisition and control signals are realized by PC computer. All elements of the measuring system are connected and synchronized by GPIB network (IEEE 488.2 standard). Control program for static measurements was written in G language (LabVIEW environment). For dynamic measurements data recording is performed using digital oscilloscope Tektronix MSO 2024 equipped with four separated channels and digital filter.

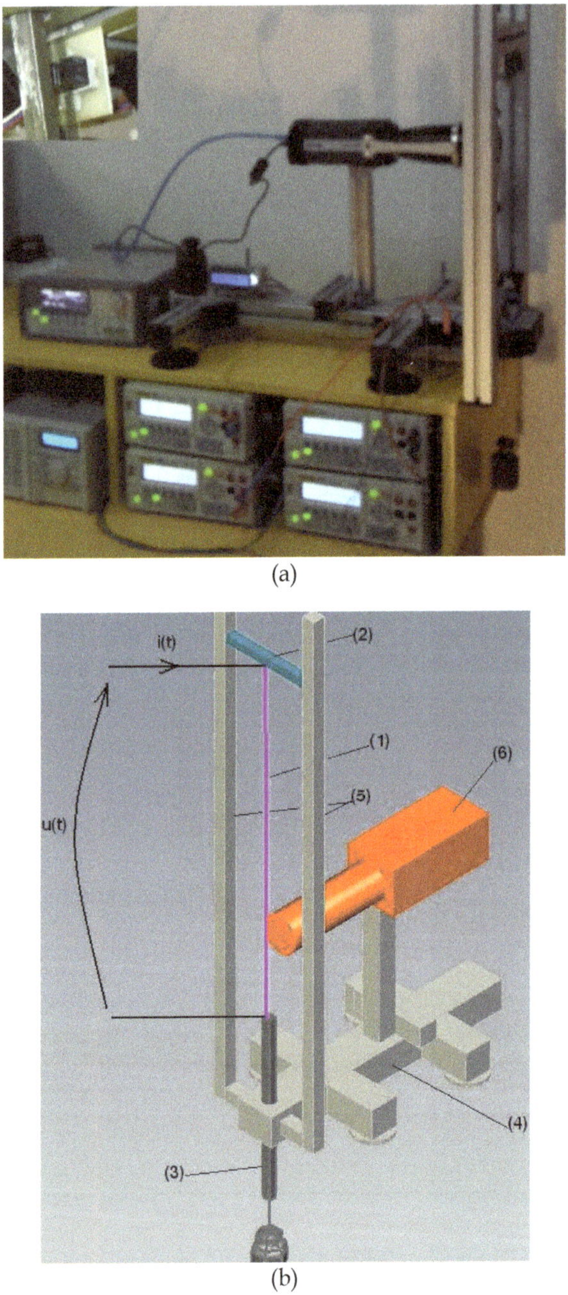

Fig. 7. Measuring stand: (a)real view; (b)3D model

Temperature measurement is realized using infrared camera Flir A325. It has resolution of 320x240 pixels and can record pictures up to 60Hz. For measuring temperature of very thin wires (50-500μm) special macroscopic lens (closeup 1x) were used. Its surface resolution is 25μm. Camera signal is send to PC computer by Gigabit Ethernet connection (with maximum speed of 1GB/s). IR camera software finds in the image point of the highest temperature and acquires it as the actual temperature of the wire. Additionally, ambient temperature is measured all the time by digital thermometer chip DS18B20, however its result is only displayed but not recorded. Fig.8 presents infrared camera view for SMA wire with 308μm diameter.

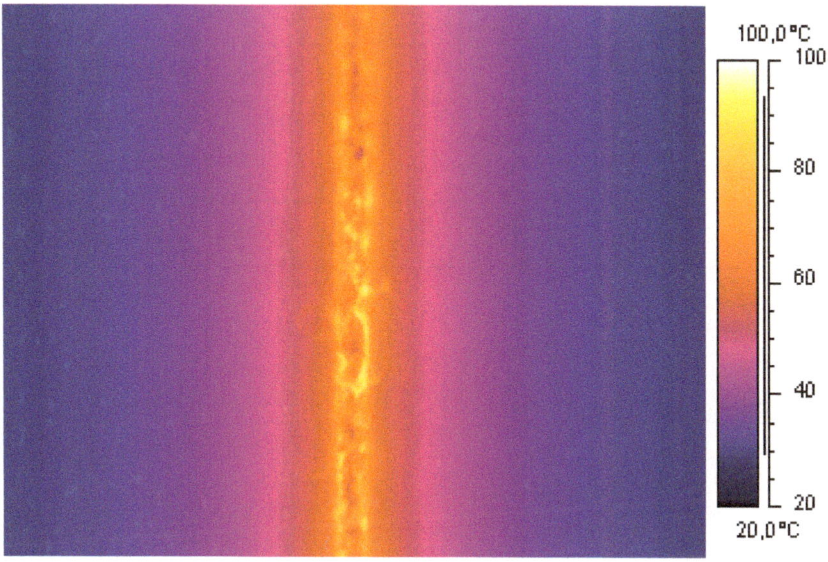

Fig. 8. Infrared camera view for SMA wire with 308μm diameter.

Displacement measurement module is working in two steps. First "displacement-voltage" converter converts displacement into voltage signal, next voltage signal is measured by multimeter. Module consists of optical sensor (converting piston movement into series of pulses), pulses counter, PWM signal generator (PWM signal fulfillment is proportional to number of pulses), lowpass filter and buffer for conditioning signal. Amplitude of converter's external voltage is measured by digital multimeter Rigol DM3052. Movement detection is realized by integrated quadrature optical sensor H9720 and measuring tape with 85 μm resolution scale. Quadrature optical sensor H9720 detects direction of movement based on two geometrically shifted signals (A and B). Pulse counting and PWM generating is realized by 8-bit microcontroller AVR ATmega88. Microcontroller count pulses from one input while second input is used only for direction detection. Pulse counting is realized by detection of (both) signal ramps using external interrupt INT. When ramp is detected both inputs are compared. If their signals have the same value - counter is decreased, if not - it is increased. Block diagram presenting displacement-voltage converter is shown in Fig.9b while idea of pulse counting (with direction detection) is shown in Fig.10.

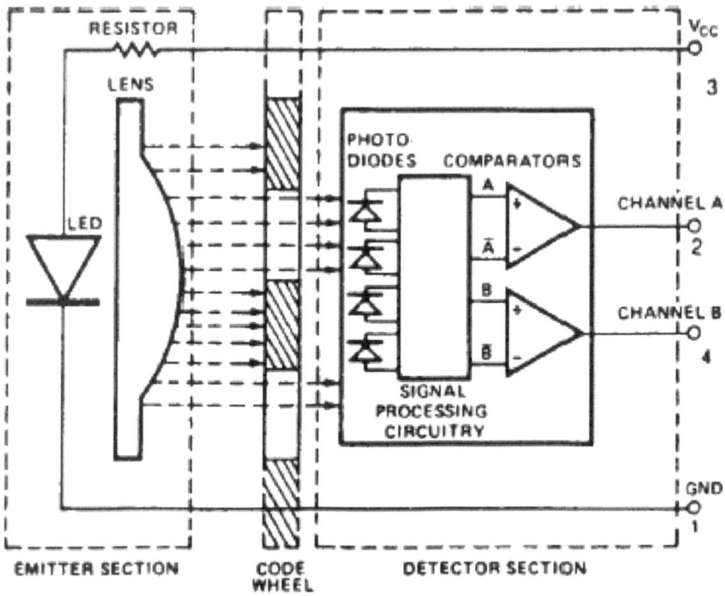

Fig. 9(a). Block diagram of Q9720 sensor. [11]

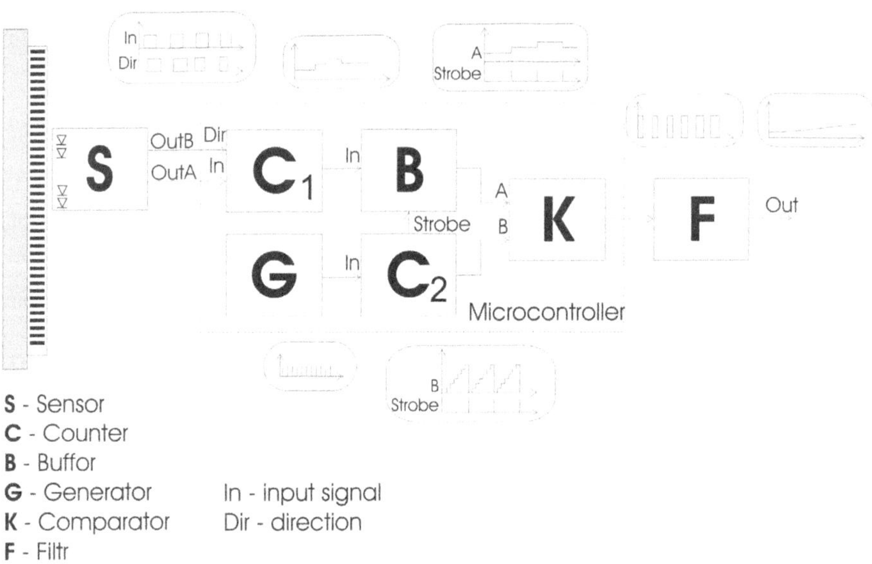

Fig. 9(b). Block diagram of displacement-voltage converter.

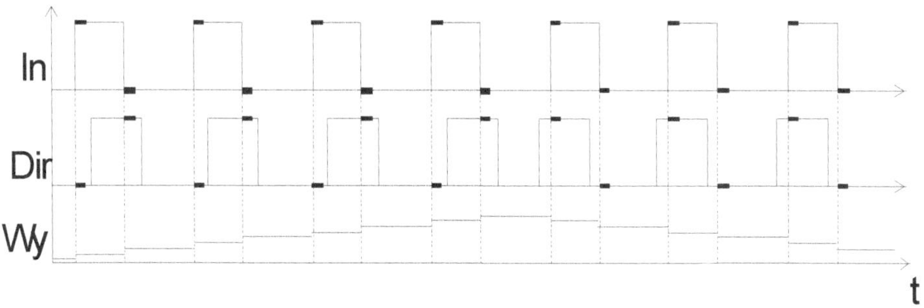

Fig. 10. The idea of pulse counting.

Sensor (S) generate series of pulses (LS) which are counted by counter (C1) and compared in comparator (K) with number of generator (G) pulses (LG) counted by counter (C2). Counter (C2) works in one direction only - the value is increased to overrun the counter. Time needed to overflow this counter determines period of generated PWM signal. PWM signal frequency is 3,5 kHz. Unless LG < LS in comparator its output is set to "high" in other case it is "low". Sensor and external microcontroller interrupt are working asynchronously. Output of pulse counter (LS) is buffered to avoid errors caused by unpredictable changes of counted value. Changing of number of pulses in buffer is possible only in moment of overflow of counter L2 (Strobe is the signal controlling this operation). This solution is hardware-implemented in timers built-in to AVR microcontrollers. The role of lowpass RC filter connected to output of PWM generator is to convert digital PWM signal into analog (average) signal. Cutoff frequency of the filter is CF=384Hz. Output buffer (operational amplifier with k=1) is conditioning signal and assure high impedance of filter load.

Control software was written in G language (LabVIEW environment). It enables to control wide variety of measurement instruments and protocols. The most popular available protocols are RS-232, USB, FireWire, LAN, GPIB, etc.). All measurement instruments, instead of IR camera which use Giga Ethernet, communicate each other by GPIB. One communication protocol enables synchronization between instruments.

Measurement scheme is divided into two loops. First loop increase heating current from minimum to maximum value with specified step; second loop decreases heating current from maximum to minimum value with the same step (See Fig.11):

$$\begin{cases} i(j) = I_{MIN} + j \cdot \Delta I & \text{- current increasing} \\ i(j) = I_{MIN} + (N - j) \cdot \Delta I & \text{- current decreasing} \end{cases} \tag{1}$$

where:
$i(j)$ - step current,
j - number of step $j \leq N$,
N - number of steps,
ΔI - current step difference.

Values: N, j, ΔI are set by the user. There are ten measurement points per each step. Time between consecutive steps is set as step delay. This two direction measurement process allows for obtaining characteristics of both - direct and reverse - transitions.

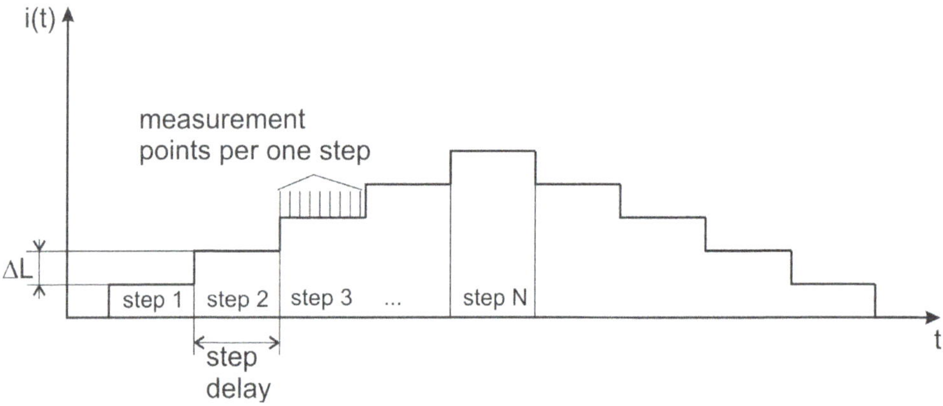

Fig. 11. The idea of measurement process.

Measurement process is triggered simultaneously using "send to group" command. Registered values of each multimeter are then acquired in serial way. Measurement algorithm of parallel triggering and serial acquisition is presented in Fig.12. Controlling algorithm is shown in Fig.13.

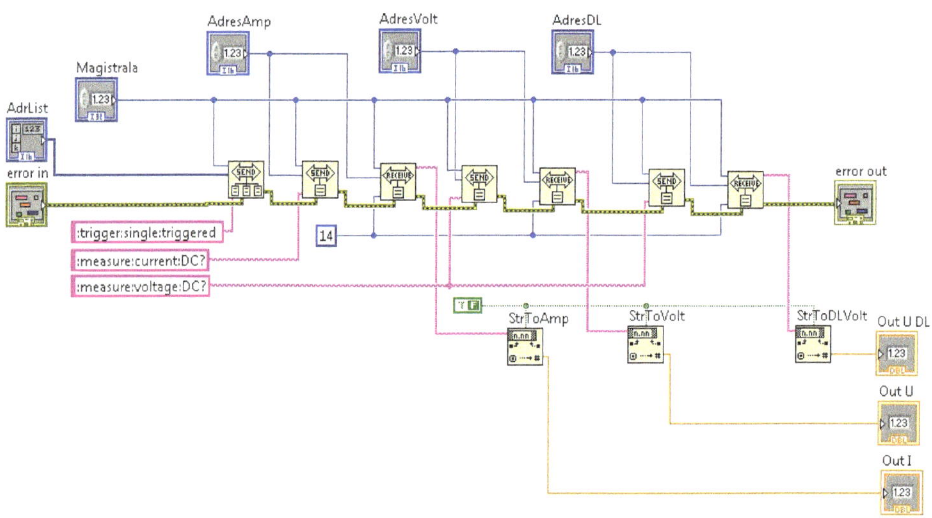

Fig. 12. Data acquisition subprogram with parallel triggering.

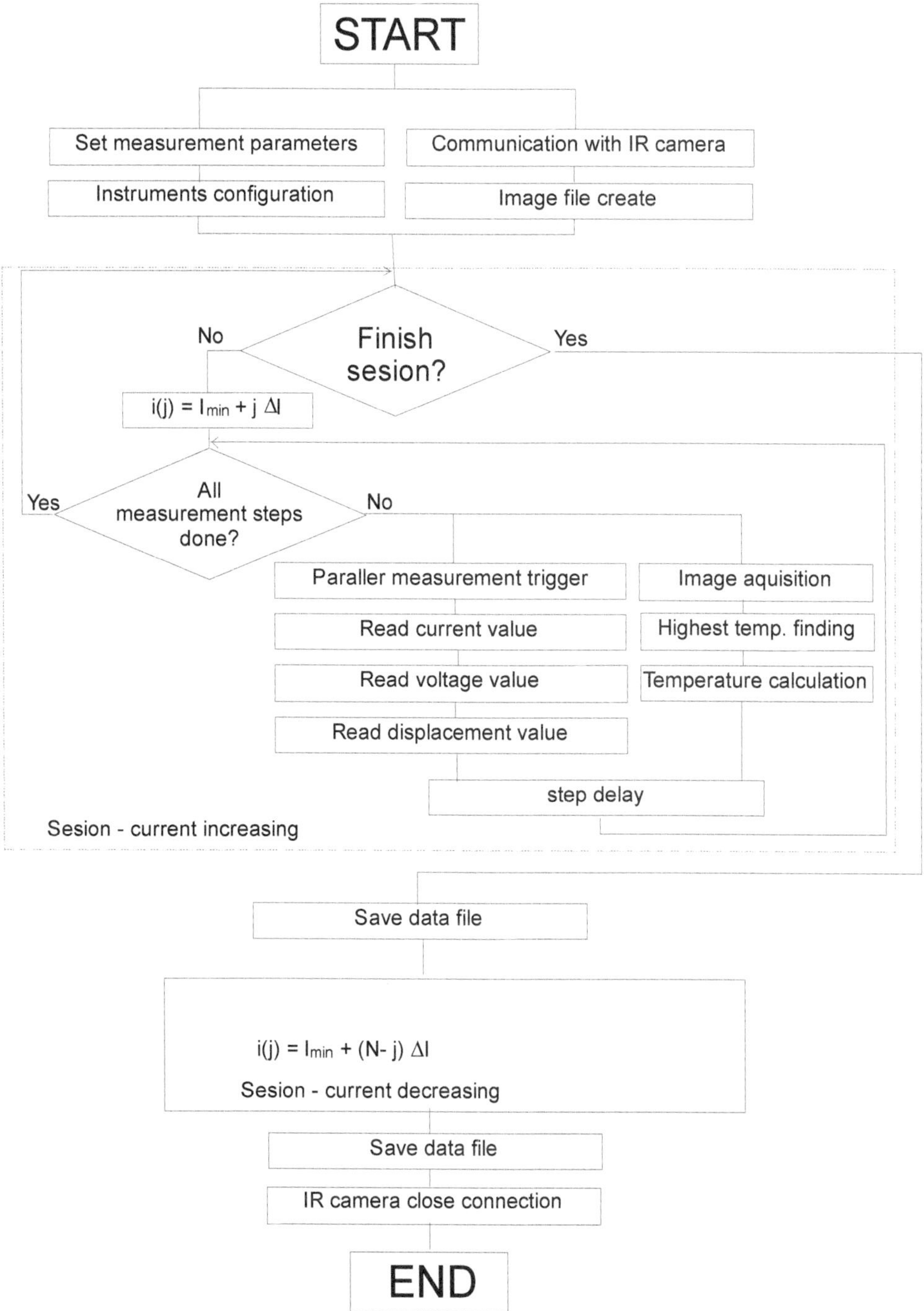

Fig. 13. Measurement algorithm.

3. Measurements accuracy

Supply current and measurements taken in the stand are executed with a specified accuracy. In the case of supply module, accuracy is related to heating current stability. Analysis of measurement accuracy of the measuring system, built with manufactured components (power supply, multimeters), is associated with the analysis of accuracy of each instrument. Accuracy of all instruments are presented in Tab. 1.

Instrument	Measurement range	Accuracy
Power supply	20A	≤ 0,2% + 10mA
	Resolution	10mA
Multimeter	200mV DCV	0,003 + 0,003
	2V DCV	0,002 + 0,0006
	20V DCV	0,002 + 0,0004
	200mA DCI	0,02 + 0,002
	1A DCI	0,02 + 0,016

Table 1. Accuracy of measuring instruments.

Heating-current stability is checked by the multimeter. Amendment for current stability is done according to following equation:

$$I = I_{set} + 5mA \pm 4mA \tag{2}$$

where:

I_{set} - current set by control unit of power supply.

Current measurement accuracy is 1mA.

Accuracy of SMA displacement consist of two parts. The first part is connected with "displacement-voltage" converter. The second one is connected with multimeter accuracy. Accuracy of "displacement-voltage" converter consist of accuracy of measurement tape - 85μm and accuracy of PWM generator. 8-bits PWM generator has 256 steps. One step is equal to 6,3mV. Accuracy of multimeter (for this measurement range) is 1mV, so it is much higher. Resulting accuracy of displacement measurement is ±1 step of measuring tape, it is equal to ±0,17mm.

4. Results and analyses

Designed measurement system allows for wide research program. All measurement results presented below concern one type of actuator. It is F2000, its nominal load is 2kg.

Some example results showing interesting characteristics and obtained relationships are shown below. All characteristics were measured at constant conditions of heat dissipation. Ambient temperature was 25±2 Celsius degrees. Cartoon shield avoided external air movement around the tested element. One static measurements cycle took approximately five hours (each step took nearby three minutes - depending on wire's diameter). Each of dynamic measurements was done in approximately ninety seconds. Measurements is finished after reaching steady state, which is defined as no change in displacement over a period of time longer than ten seconds.

4.1 Steady-state characteristics

Example result of steady-state electro-mechanical characteristic is presented in Fig.14.

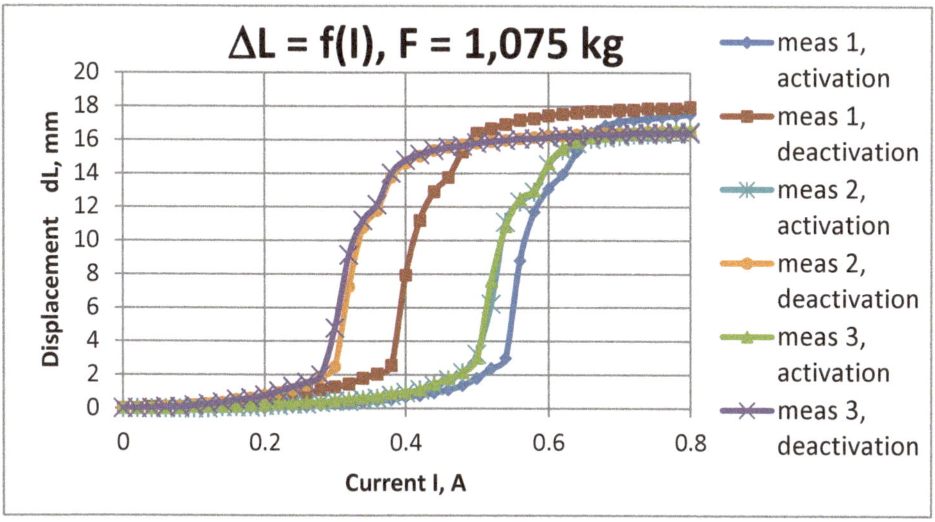

Fig. 14. Steady-state electro-mechanical characteristic.

There are three measurements called "meas 1", "meas 2", "meas 3", increasing current procedure is called "activation", decreasing current procedure is called "deactivation". Results of second and third measurement process are almost identical. Repeatability of measurement results has been analyzed. Difference between first and other results is connected with hysteresis phenomenon and history of actuator. Based on the test measurements the complete measurement procedure was determined. Full measurement procedure is presented below:

- Set the SMA actuator in the measurement stand,
- Load SMA actuator with proper mechanical bias,
- Calibration of actuator,
- Perform series of measurements.

Calibration is the process of rapid activation and deactivation of mechanically loaded actuator performed three times. It causes history cleaning before proper measurement process. Displacement measuring module is reset after calibration process. Example results obtained using developed measurement procedure are shown in Fig.15.

During measurements the question raised - what if to reverse temperature change before the transition is complete. So incomplete hysteresis loop measurements were done as well. Exemplary results are presented in Fig.16. There are results for different maximum heating currents changing in range of 0,5A to 0,8A. Full hysteresis loop is achieved for current value equal to 0,8A - it is the reference measurement. Transition starts when heating current value is about 0,4A. Each measurement was done three times. Time of martensitic transformation is independent of the rate of change of the factors causing it [6]. It means that you cannot accelerate the transition by increasing temperature (current). So process of activation performs identically each time, but halting the supply of heat energy needed to activate the actuator, stops, and even change the direction of transition.

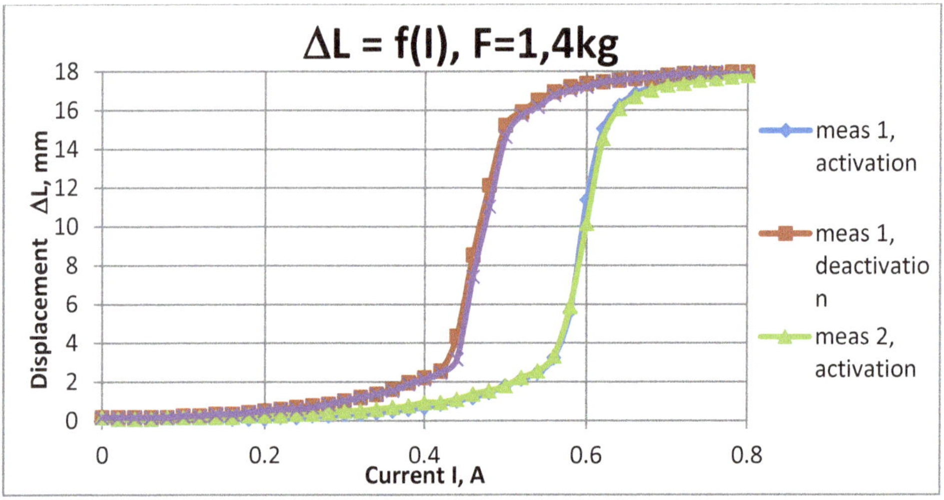

Fig. 15. Steady-state electro-mechanical characteristics acquired using developed measurement procedure (with actuator calibration).

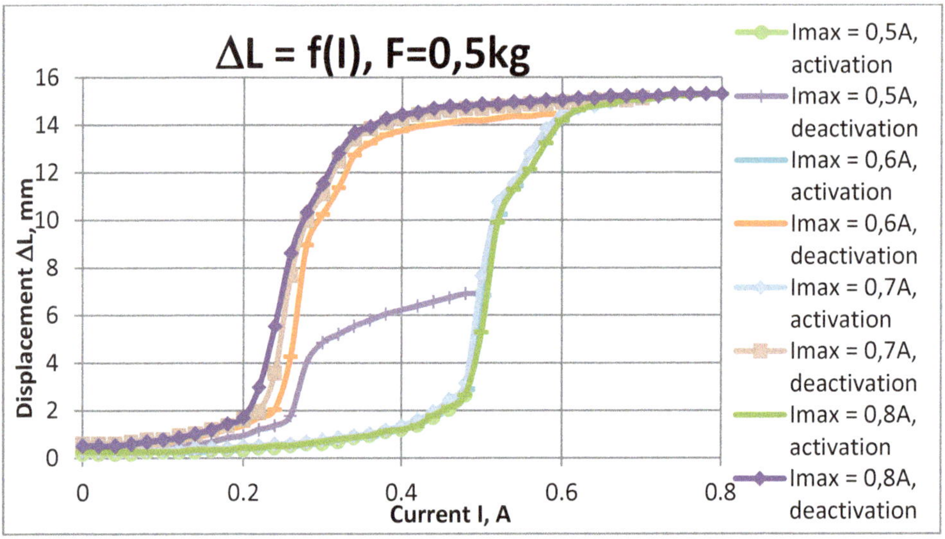

Fig. 16. Incomplete hysteresis loop measurements.

The next important characteristic of SMA actuator is load characteristic. Exemplary load characteristic is presented in Fig.17. According to the theory, increasing the mechanical load causes a linear increase of characteristic temperatures of martensitic transition while maintaining a constant width of the hysteresis loop.

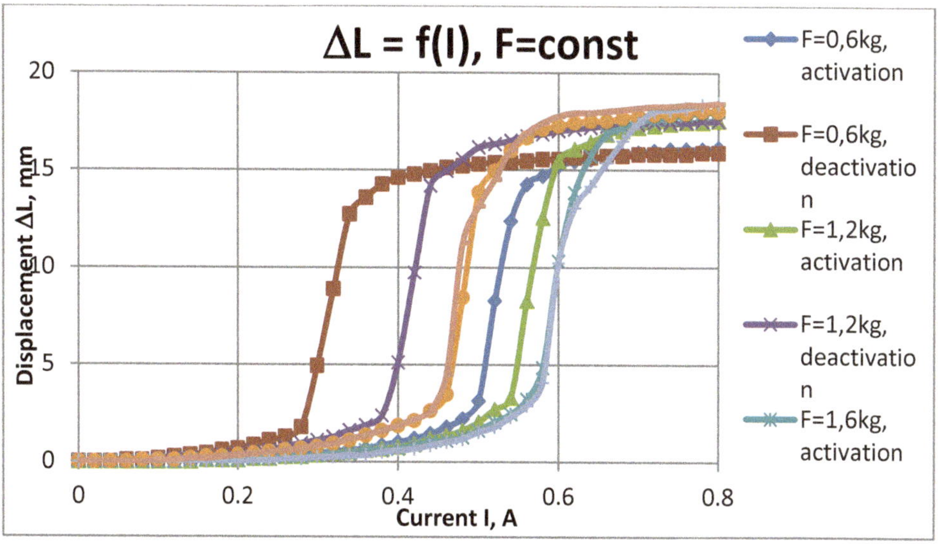

Fig. 17. Load influence on the process of martensitic transition.

It can be easily observed that higher mechanical load causes higher displacement. It is natural phenomenon connected with plastic deformation in martensitic state. Width of hysteresis loop changes according to mechanical load. For example width of hysteresis loop for 0,6kg is 0,2A (0,52A-0,32A), while it is 0,12A (0,6A-0,48A) for bias force equal to 1,6kg and 2,0 kg. It causes not linear change of important temperatures of martensitic transition.

Next interesting characteristic is electro-thermal characteristic showing relation between current and wire temperature (for constant ambient temperature). It is shown in Fig.18.

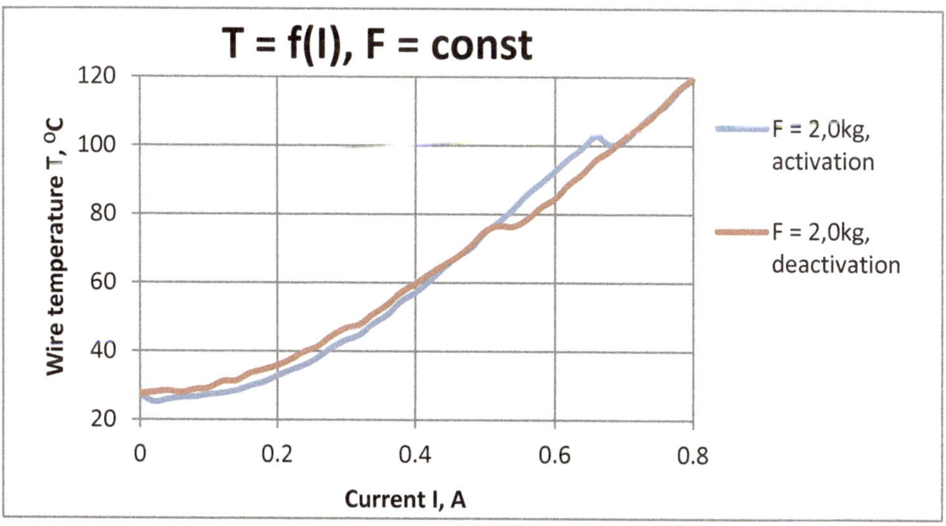

Fig. 18. Temperature vs. current characteristic, for constant ambient temperature.

Temperature characteristics of the martensitic state is a quadratic function of current. In the case of austenite, the situation is similar. A small number of measurement points in austenite state prevents the determination of the equation function. The most interesting range is the martensitic transition phase. There are different cases of temperature behavior in this state. Temperature is constant or even decreases. This is caused by consumption of thermal energy by transition process. Similar results were mentioned in [13].

Very useful characteristic is resistance as a function of heating current. In some applications the characteristics of resistance as a function of the length of the actuator are used to estimate the actual length of the actuator [12]. There are two types of resistance characteristics. First of them - resistance in function of heating current - is presented in Fig.19.

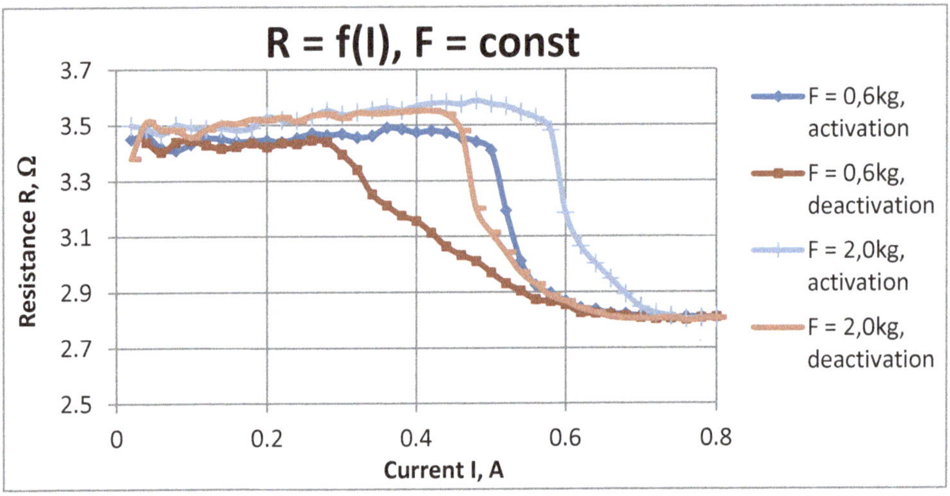

Fig. 19. Resistance vs. Current.

It is easy to notice the difference of characteristics for small and big mechanical loads (close to nominal bias force). The general rule is better noticeable for the deactivation process (direct martensitic transition). Characteristics shift toward higher currents with increasing mechanical load. This is a trend similar to the phenomenon of characteristic temperatures rising with increasing mechanical load.

Another phenomenon that can be observed is linear change of resistance with extending the wire in activation process (Fig.20). It is independent of mechanical load while for deactivation the function is linear only for mechanical load close to the nominal load. This phenomenon allows for quite accurate displacement approximation using measurements of resistance.

Simultaneous multiparameter measurement enables to plot different characteristics and a comprehensive assessment of the impact of occurring phenomena. Fig.21. shows the characteristics of displacement and temperature as a function of heating current.

The most energy consuming part of the martensitic transition is finish of activation and start of deactivation. It can happen (as can be seen in above picture) that the wire temperature decreases in spite of constant delivering of energy.

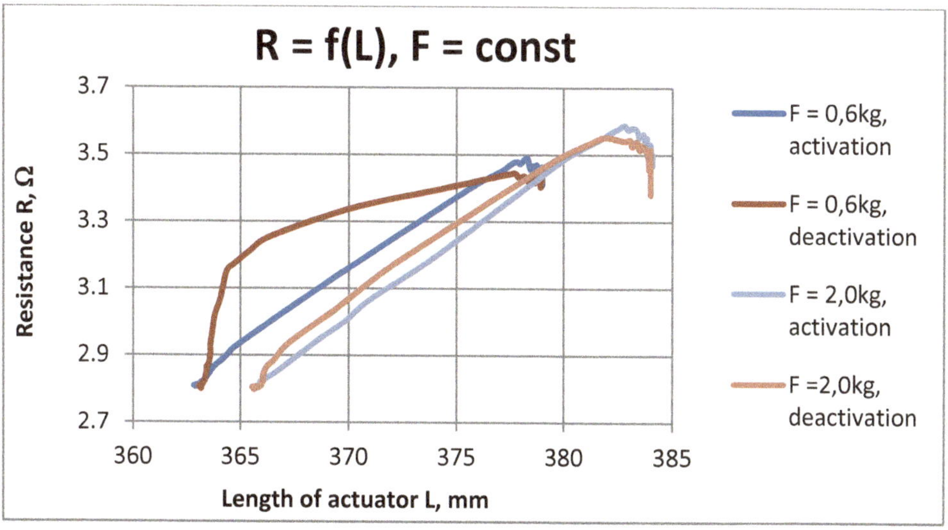

Fig. 20. Resistance vs. length of actuator.

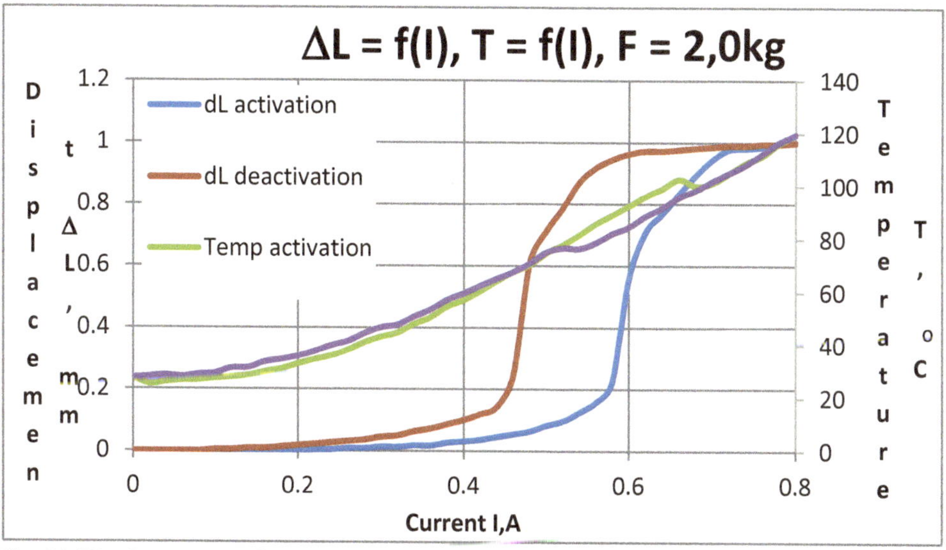

Fig. 21. Displacement and temperature characteristic vs. heating current.

4.2 Dynamic characteristics

Dynamic characteristics was acquired for two cases - switch on and switch off the heating current. In the first case dynamics of activating process is measured, in the second - dynamics of deactivating. Different mechanical bias forces and different current values have been considered. Example results are given in Fig. 22 - 25.

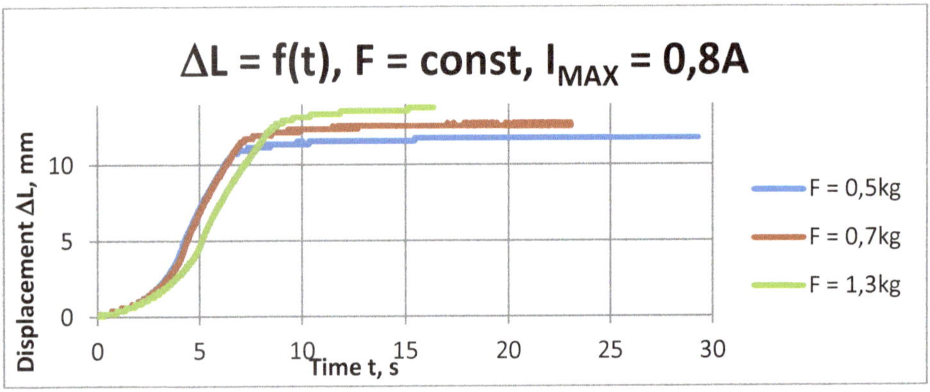

Fig. 22. Activation process for different mechanical bias forces.

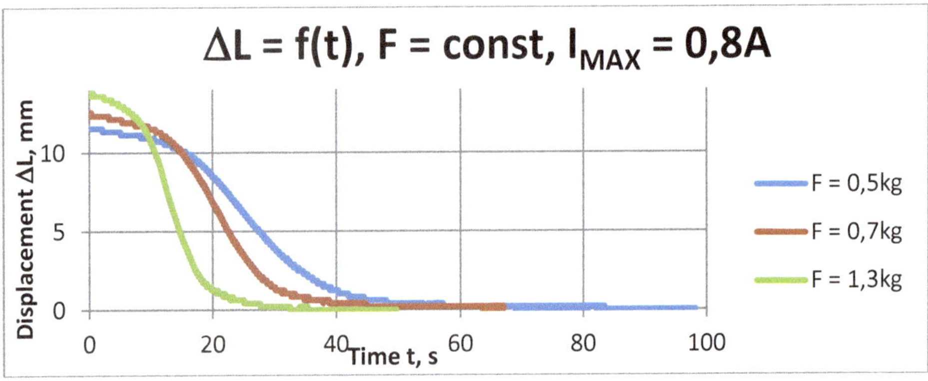

Fig. 23. Deactivation process for different mechanical bias forces.

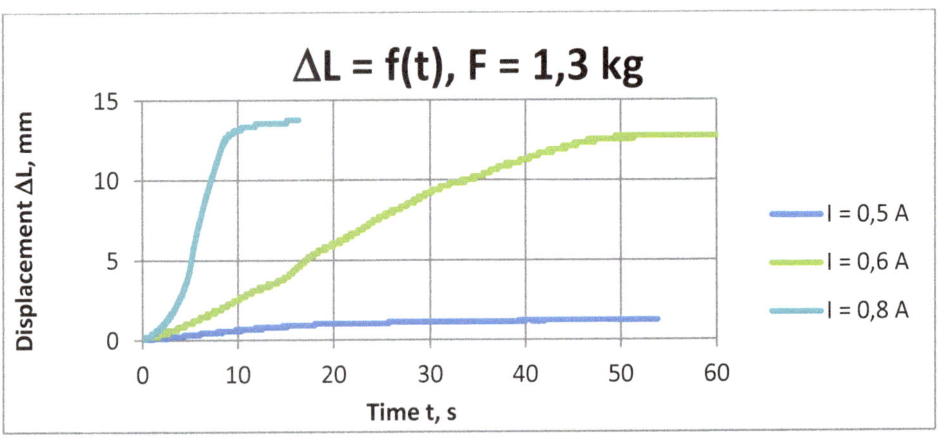

Fig. 24. Activation process for different heating currents.

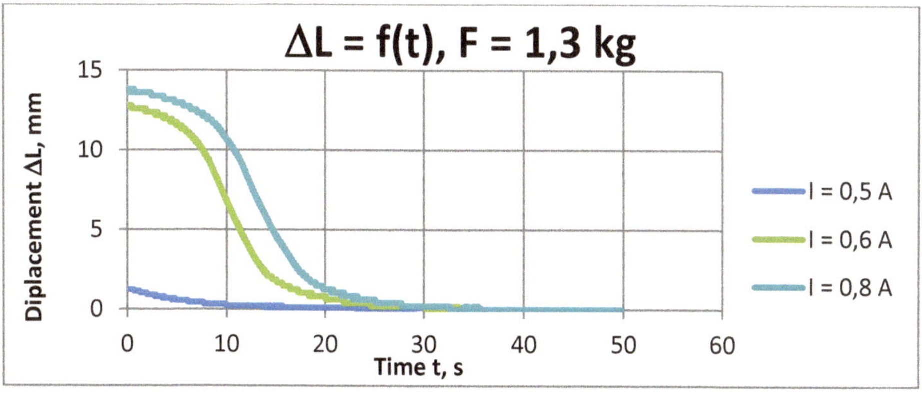

Fig. 25. Deactivation process for different heating currents.

Deactivation characteristics in Fig.23. are the result of unconstrained SMA actuator cooling. It is easy to notice the difference between time of activation and deactivation of the actuator. The higher force is, the faster actuator recovers. There is small difference in activation time, so mechanical load does not significantly affect the activation time but influence final displacement value.

Heating current value significantly influences the speed of the activation process. Current value 0,8A causes full activation, lower values does not. There is not big difference in final displacement between heating current values 0,6A and 0,8A, but time is important factor. The slope of the deactivation curve does not depend on the starting point but is strongly connected with conditions of heat dissipation.

5. Conclusions

Designed measurement stand can be very helpful in research of Shape Memory Effect. It allows for simultaneous multiparameter measurement which enables to draw various characteristics of SMA wire actuators. Presented characteristics prove the utility of designed measurement systems. It is possible to compare different characteristics of the SMA actuator for one forced parameter (e.g. comparison of the displacement and wire temperature as a function of heating current (Fig.21)). Using some dependences that occur among certain physical quantities it is also possible to calculate some additional parameters and draw additional characteristics (e.g. resistance as function of current or actuator length (Fig.18 and Fig.19)).

Design of the stand allows for performance of reach research program. Measurement process is performed in semiautomatic way. User has to set measurement parameters described in (1) and start the process. All current changes and data acquisition is done automatically which increase measurement accuracy. Automatic operation save a lot of time and eliminates the distortions arising from the presence of people close to measurement system (e.g. air movement around heated wire). Moreover process parameters can be changed remotely (using Internet) which means that it is possible to obtain whole family of characteristics without physical access to the stand. The use of virtual measuring instruments and computer control makes it easy to develop the measuring stand. Some planned developments of the measuring stand would give:

- the ability to automatically adjust mechanical load,
- the opportunity to ask about specific mechanical load parameters (time or displacement),
- possibility to adjust air movement,
- automatic measurement of environmental conditions (temperature, pressure, humidity),
- possibility of automatic adjusting environmental conditions.

The stand was co-financed with governmental founds (project number N N510 353036).

6. References

[1] Kciuk M.: *Computer based measurement system for shape memory alloy.* IX International Workshop for Candidates for Doctor's Degree OWD'2007, Wisła 20-23.10.2007, str. 177-180,

[2] Koji Ikuta, Masahiro Tsukumoto, Shigeo Hirose: *Shape memory alloy servo actuator system with electric resistance feedback and application for active endoscope.* CH2555-1/88/0000/0427$01.00 0 1988 IEEE

[3] Krzysztof Biereg: *Porównanie wybranych równań konstytutywnych stopów z pamięcią kształtu.* Modelowanie inżynierskie, Gliwice 2006.

[4] Koji Ikuta, Masahiro Tsukamoto, Shigeo Hirose: *Mathematical model and experimental verification of Shape Memory Alloy for designing micro actuator.* IEEE 1991

[5] Koji Ikuta, Hidenori Shimizu: *Two dimensional mathematical model of Shape Memory Alloy and intelligent SMA-CAD.* IEEE 1993.

[6] Ziółkowski A.: *Pseudoelastyczność stopów z pamięcią kształtu badania doświadczalne i opis teoretyczny.* Praca habilitacyjna, IPPT PAN, Warszawa 2006.

[7] Kyu-Jin Cho, Harry H. Assada: *Multi-axis SMA actuator array for driving anthropomorphic robot hand.* International Conference on Robotics & Automation, Spain, April, 2005.

[8] Liu C.Y., Liao W. H.: *A snake robot using Shape Memory Alloys.* International Conference on Robotics and Biomimetics Shenyang, China, August 22-26, 2004, Proceedings of the 2004 IEEE.

[9] Masaru Fujii, Hiroshi Yokoi: *Modeling and movement control of mobile SMA-Net.* International Symposium on Computational Inteligence in Robotics and Automation, Kobe, Japan, July, 16-20, 2003, Proceedings 2003 IEEE.

[10] Ikuta K.: *Micro/miniature Shape Memory Alloy actuator.* CH2876-1/90/0000/2156$01.00 0 1990 IEEE.

[11] Technical data for sensor H9720 *http://www.avagotech.com/docs/AV02-0511EN*

[12] Zurbitu J., Kustov S., Zabaleta A., Cesari E., Aurrekoetxea J.: *Thermo-mechanical behavior of NiTi at impact. Shape Memory Alloys.* ISBN 978-953-307-106-0, publishing date: October 2010

[13] Kciuk M., Kłapyta G.: *Koncepcja stanowiska pomiarowego do wyznaczania charakterystyk elektro-termo-mechanicznych stopów z pamięcią kształtu (SMA).* Materiały XIII Sympozjum „Podstawowe Problemy Energoelektroniki, Elektromechaniki i Mechatroniki" PPEEm 2009, Wisła 14-17.12.2009, str. 242-247.

[14] Hui Li, Chen-Xi Mao Jin-Ping Ou: *Experimental and theoretical study on two types of Shape Memory Alloy devices.* Earthquake Engineering And Structural Dynamics 19 September 2007 in Wiley InterScience (www.intersciecnce.wiley.com).

11

Determining Exact Point Correspondences in 3D Measurement Systems Using Fringe Projection – Concepts, Algorithms and Accuracy Determination

Christian Bräuer-Burchardt, Max Möller, Christoph Munkelt, Matthias Heinze, Peter Kühmstedt and Gunther Notni
Fraunhofer IOF Jena, Jena, Germany

1. Introduction

Fringe projection systems for contactless surface measurements are increasingly used in applications of industrial inspection, quality control, rapid prototyping, archaeology, and in medical applications. Such systems provide increasing accuracy together with shorter measurement time and increasing data volume. The demands of increasing data volume, accuracy, and measurement speed can be satisfied using a more powerful calculation technique and better algorithms.

Calculation of 3D surface points by fringe projection systems is typically based on triangulation known from photogrammetry. An extensive review gives e.g. Luhmann (Luhmann et al., 2006). This involves the task of finding homologous points describing points in different images depicting the same original measuring point. In fringe projection systems the finding of homologous points typically uses phase information (Creath, 1986) of the points. The phase value characterizes the origin of a projected ray in the projector image plane. The projector is geometrically considered as an inverse camera (Schreiber & Notni, 2000).

When an optical contactless 3D measurement system has to be designed or selected among different alternatives some different basics should be clear. It must be known which objects should be measured, how fast the measurement should run, and how accurate the measurement has to be. If these aspects are plain, the design or selection may start. As the calculation of the resulting 3D points on the object's surface is based on triangulation, the procedure of finding corresponding points is crucial. This procedure has mainly two aspects: uniqueness and precision of the localized position of the corresponding points. In this work the various methods to obtain unique point correspondences are not described in detail, but rather the second aspect (precision) should be dealt with. Uniqueness of point correspondences is obtained by phase unwrapping (Sansoni et al., 1999; Reich et al., 2000). This can be realized e.g. by the use of multiple spatial frequencies (Li et al., 2005), temporal phase unwrapping methods (Zhang et al., 1999), or use of Gray code sequences (Sansoni et al., 1999). Due to its unambiguousness, the usage of Gray code leads to robust results.

An extensive survey over coded structured light techniques to solve the correspondence problem which is the basis for 3D surface reconstruction is given by Battle (Battle et al., 1998).

Zhang and Yau suggest a real-time coordinate measurement (Zhang & Yau, 2006) where phase unwrapping is realized by determination and tracking of a marker. An interesting method for phase unwrapping using at least two cameras is presented by Ishiyama (Ishiyama et al., 2007a). There the number of possible correspondences is drastically reduced by back-propagation of the correspondence candidates into the image of the second camera. Ishiyama gives another suggestion (Ishiyama et al., 2007b) for 3D measurement using the invariance of cross-ratio of perspective projection. Young (Young et al., 2007) suggests the use of the limitation of the measuring volume in order to reduce the search area for corresponding points on the epipolar line to segments achieving a reduction of the projected binary code by careful placement of additional cameras (or additional measuring positions). Li (Li et al., 2009) uses this approach in combination with the multi-frequency technique in order to realize real-time 3D measurements.

We presented an algorithm (Bräuer-Burchardt, 2011a) describing how one can obtain unique point correspondences using geometric constraints. This algorithm works when certain conditions as restricted measurement volume depth hold.

A number of works were published from our working group concerning accuracy determination of phase values and phase value based measurements (Notni & Notni, 2003; Bräuer-Burchardt et al., 2010). These works show the importance of theoretical error estimation as well as the performance of experimental error analysis of fringe projection based 3D surface measurement systems.

Different arrangements of camera(s) and projector(s) in 3D measuring systems using fringe projection have been recently proposed (Maas, 1992; Ishiyama et al., 2007a). Arising methods are based on triangulation (Luhmann et al., 2006) between two or bundle adjustment between more than two viewpoints (sensor positions), or between two camera positions within one sensor position. The most common triangulation procedure to obtain 3D measurement data in fringe projection systems is between two cameras or one camera and one projector (Schreiber & Notni, 2000; Reich, 2000).

We consider the case of having one or two cameras and one projector in our sensor head. We distinguish three kinds of establishing correspondences using one sensor position. The first one is the simple observation of a phase value (Creath, 1986; Schreiber & Notni, 2000) at a certain camera coordinate. The corresponding point in the projector image is directly yielded by the observed phase value. This method is called CP mode (one camera and one projector are used for triangulation). Another method is to search for the observed phase value in the second camera image denoted by CC (two cameras are used for triangulation). The third one called VR mode (virtual raster) uses a virtual projection grid and finds the phase values of the grid points in the two camera images. This method corresponds to the technique described by Reich (Reich et al., 2000).

In this work these different methods are compared regarding measurement accuracy, handling, and sensitivity against image disturbances. The measuring principles are briefly presented and advantages and disadvantages are discussed.

A model describing the random error of the resulting 3D measurement data was developed and confirmed by the results of simulation experiments and real data measurements obtained by two different measuring devices. The results are analysed and discussed and an outlook to future work is given.

2. Situation and measuring principles

Recently several fringe projection systems for 3D surface determination for different measurement objects were developed at our institute (Kühmstedt et al., 2005; Munkelt et al., 2005; Kühmstedt et al., 2007; Kühmstedt et al. 2007b; Bräuer-Burchardt et al., 2011b). Some of them are shown in figs.1 and 2. They are based on the projection and observation of one or two 90° rotated fringe sequences consisting of sinusoidal fringe patterns (between 3 and 16 images) and optionally of a Gray code sequence (usually 5 to 7 images). From these sequences phase values are determined using a 3-, 4-, 6-, 8-, or 16-phase algorithm (Kühmstedt et al., 2007). Phase values are used together with the extrinsic and intrinsic parameters of the optical components to determine the 3D point coordinates of the reconstructed object surface.

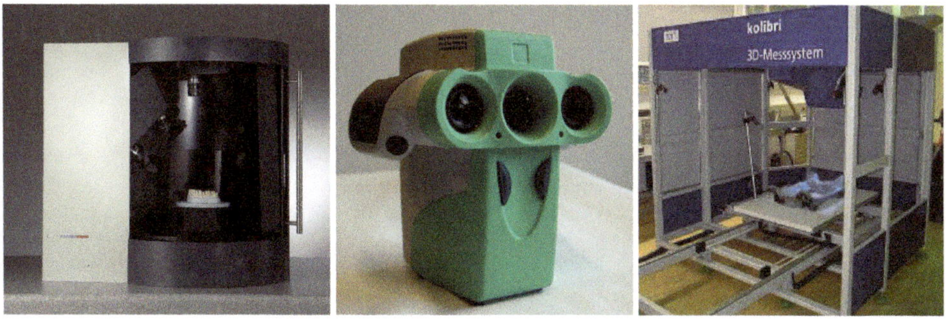

Fig. 1. Several fringe projection based measurement systems developed at Fraunhofer IOF: kolibri flex mini, kolibri CORDLESS, kolibri step (from left to right)

The use of epipolar constraint in order to find point correspondences is a typical approach in phototogrammetry (Luhmann et al., 2006). It reduces the task to a correspondence problem between two one-dimensional vectors.

Let us consider using a sensor consisting of two cameras C_1 and C_2 and one projector P in a fix geometric arrangement (see fig. 2 right).

In the following the basic principles of fringe projection based 3D surface measurement should be briefly explained.

2.1 Phasogrammetry

Phasogrammetry is the connection of the mathematical principles of photogrammetry and fringe projection. The classical approach of fringe projection is described e.g. by Schreiber and Notni (Schreiber & Notni, 2000), which has been extended depending on several applications (Reich et al., 2000; Chen & Brown, 2000). The principle should be briefly explained as follows. A fringe projection unit projects one or two perpendicular, well defined fringe sequences onto the object, which is observed by one or more cameras. These sequences may consist of a binary code sequence as the Gray code (Sansoni et al., 1999; Thesing, 2000) and a sequence of up to 16 sinusoidal fringe patterns. The so called rough phase value (Schreiber & Notni, 2000), and in combination with the Gray code the unique unwrapped phase value (Sansoni et al., 1999) is obtained using the sequence of sinusoidal patterns. Unique phase values are used to realize point correspondences in order to obtain measurement values by triangulation (Luhmann et al., 2006).

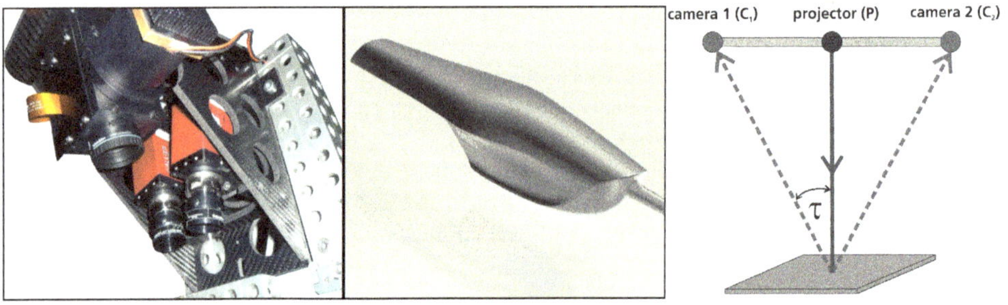

Fig. 2. High speed sensor HS (left), intraoral sensor DirectScan (middle), and typical sensor arrangement (right)

The rough phase value is the phase position within one period and the unique phase value is obtained by phase unwrapping leading to a monotone series of phase values (if error free) in a certain direction.

Phase values are produced in order to identify projector image coordinates or to produce virtual landmarks (Kühmstedt et al., 2007). These markers may be used both for calibration of the system and for the calculation of the 3D measurement data.

One sequence of fringe images is processed resulting in phase images $\Phi_{i,x}$ for each measuring position and each camera C_i. After rotation of the fringe pattern by 90°, the sequence may be projected and observed again resulting in a second phase image $\Phi_{i,y}$ for each camera. The phase values $\phi_{i,x}$ and $\phi_{i,y}$ (see fig. 3) correspond to image coordinates in the projector plane. The resulting 3D points are obtained by triangulation between the coordinates of the camera and the projector, or between corresponding points of two cameras. This can be regarded as standard procedure in photogrammetry (Luhmann et al., 2006).

2.2 Stereo vision and epipolar geometry

Using active stereo vision, images of the object are captured from two different perspectives. Pairs of image coordinates resulting from the same object point (the homologous points) have to be identified. The object can be reconstructed by triangulation using these points. In the case of active stereo vision a single intensity pattern or a sequence of patterns is projected onto the measurement object. There are several techniques to identify the homologous points in both cameras as e.g. described by Luhmann (Luhmann et al., 2006).

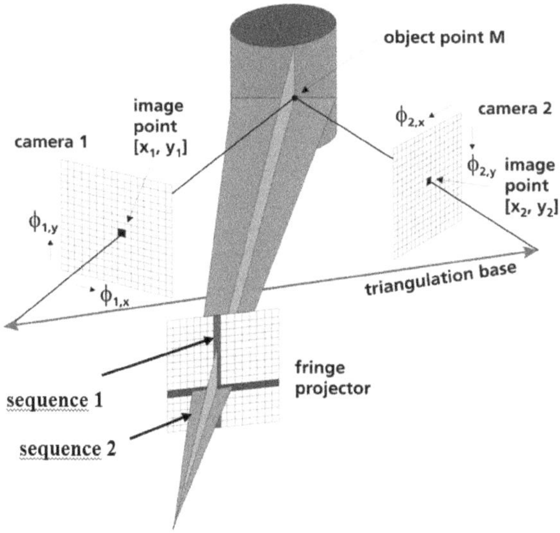

Fig. 3. Principle of phasogrammetry

Epipolar geometry is a well-known principle which is often used in computer vision when stereo systems are present. It is characterized by an arrangement of two cameras observing almost the same object scene. The projection centres O_1 and O_2 of the two cameras define together with an object point M a plane E in the 3D space (see fig. 4). The images of E are corresponding epipolar lines concerning M. When the image point m_1 of M is selected in the image I_1 of camera C_1, the corresponding point m_2 in the image I_2 of camera C_2 must lie on the corresponding epipolar line. This restricts the search area in the task of finding corresponding points. In the following we assume a system consisting of two cameras C_1 and C_2 and one projector P in a fixed arrangement.

Epipolar constraint usage is advantageous compared to using second fringe projection direction because only 50% of the recorded images are necessary. However, the calibration should be accurate and stable because correspondence finding is performed only on a line. Systematic errors of the sensor geometry lead to systematic errors of the position determination of the corresponding point to be found (Bräuer-Burchardt et al., 2011c).

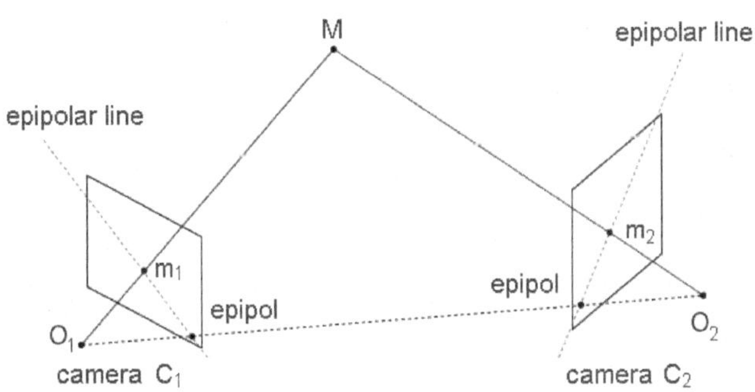

Fig. 4. Epipolar geometry

2.3 Camera calibration

Camera calibration describes the process of the determination of the intrinsic and extrinsic parameters (including lens distortion parameters) of an optical system. It has been extensively described in the literature, e.g. in (Brown, 1971; Tsai, 1986; Chen and Brown, 2000; Luhmann et al., 2006). Different principles have been applied in order to conduct camera calibration. The choice of the method depends on the kind of the optical system, the exterior conditions, and the desired measurement quality. In case of the calibration of photogrammetric stereo camera pairs, the intrinsic parameters (principal length, principal point, and distortion description) of both cameras should be determined as well as the relative orientation between the cameras.

The position of the camera in the 3D coordinate system is described by the position of the projection centre O = (X, Y, Z) and a rotation matrix R obtained from the three orientation angles ω, ϕ, and κ. Considering stereo camera systems, only the relative orientation between the two cameras (Luhmann et al., 2006) is considered, because the absolute position of the stereo sensor is usually out of interest.

Lens distortion may be considerable and should be corrected by a distortion correction operator D. It can be described by distortion functions or by a field of selected distortion vectors (distortion matrix). The determination of D is performed within the calibration procedure (Tsai, 1986) or separately (Bräuer-Burchardt, 2004).

3. Different concepts of accurate point correspondence determination

Correspondence determination in a stereo system means first the identification and, second, the localization of the point coordinates in the two considered 2D images mapping the same original 3D point.

The first task can be solved using a suitable method which depends on the actual conditions of the measurement. As phase values are concerned as the basis of the point correspondence finding we distinguish between methods producing unique phase values (e.g. using Gray code sequences or multi frequency projection techniques) and periodically repeating rough phase values.

In this work we assume to use a technique leading to unique phase values either in one or two directions. How to obtain unique phase values can be read elsewhere e.g. in (Battle et al., 1998; Zhang et al., 1999; Sansoni et al., 1999; Li et al., 2005; Bräuer-Burchardt et al., 2011a). Assume in the following that the uniqueness problem is solved.

In this section different concepts of correct point localization should be considered. These concepts distinguish e.g. concerning the hardware effort (one or two cameras), the costs, and the error sensitivity.

3.1 The CP method

The CP method is the simplest method to localize point correspondences and realize 3D point calculation.

Finding point correspondences is performed as follows. The pixels of camera C_1 are the coordinates x_1, y_1 of a point $p_1 = (x_1, y_1)$ and the observed phase values (ξ and η) at p_1

determine the coordinates of the corresponding point in the projector image plane. Together with the intrinsic camera parameters of C_1 and P and the relative orientation between C_1 and P, the 3D point is calculated by triangulation (see fig. 3).

For this method only one camera is used and no algorithmic effort for finding corresponding points is necessary. Hence this is also the simplest method concerning the computational effort. Additionally, because only two major hardware sensor components (one camera and one projector) are necessary, such sensors become lightweight and cheap. These aspects are advantageous whereas the dispensation with the second camera may be unfavourable. Figure 5 (left) shows a sketch of the CP arrangement.

The random error of the 3D point measurement is determined by the uncertainty of the phase determination and its scaling depends on the triangulation angle (see section 5).

3.1.1 CP algorithm

Precondition: Projection and observation (by one camera C_1) of a fringe image series consisting of two 90° rotated sequences subsequently producing two phase images Φ_x and Φ_y (*)

Input: two phase images Φ_x and Φ_y (*) for the observation camera C_1

Consider all pixels of the image I_1 of camera C_1. Let p_i=(x_i, y_i) the coordinates of point p_i.

- Read the phase values ξ_i and η_i at p_i: q_i =(ξ_i, η_i)
- Perform triangulation between p_i and q_i to obtain 3D point P_i

Output: point cloud {P_i}

(*) Remark: The CP algorithm works also using only one projected fringe direction. Then, triangulation is performed between a ray and a plane in the 3D space.

3.2 The CPE method

The CPE method is the simple extension of the CP mode according to epipolar geometry. The second projected fringe direction is omitted. Instead of it the geometric information of the epipolar lines is used. The second coordinate of the observed phase value results from the position of the epipolar line in the projector image plane as illustrated by fig. 5 (right).

The advantages of the CPE method are the same as those of CP together with a reduced image sequence (half the number of images) meaning shorter projection and observation time and subsequently leading to shorter measurement time. Additionally, calculation time is reduced, too. The disadvantage is a reduced robustness of the coordinate determination, mainly determined by the epipolar lines. This can be, however, prevented by additional control measurements as e.g. suggested by the authors (Bräuer-Burchardt et al., 2011c).

3.2.1 CPE algorithm

Precondition: Projection of a fringe image series in one direction subsequently producing one phase image Φ_x and observation (by one camera C_1)

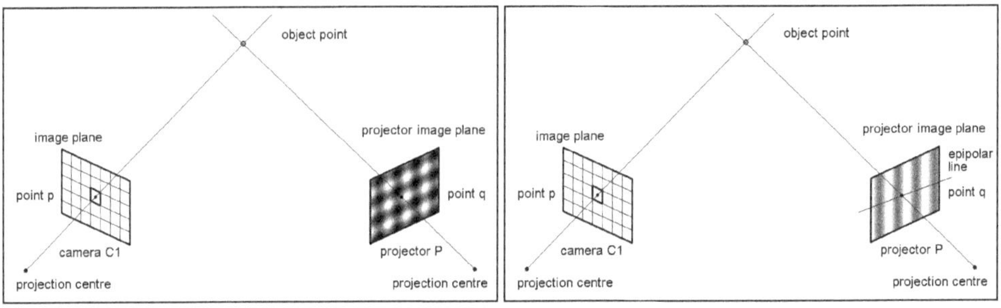

Fig. 5. Projector and camera arrangement for the CP (left) and the CPE (right) algorithms

Input: one phase image Φ_x for the observation camera C_1

Consider all pixels of the image I_1 of camera C_1. Let $p_i=(x_i, y_i)$ the coordinates of point p_i.

- Calculate the epipolar line g_i in the projector image plane corresponding to p_i
- Read the phase value ξ_i at p_i:
- Find the position $q_i =(\xi_i, \eta_i)$ of phase value ξ_i on g_i
- Perform triangulation between p_i and q_i to obtain 3D point P_i

Output: point cloud $\{P_i\}$

3.3 The CC method

The CC method (see Munkelt et al., 2005; Kühmstedt et al., 2007) uses corresponding points in the images of the two cameras C_1 and C_2 (see fig.1). For finding the point correspondence camera C_1 is disclaimed as primary camera. The camera pixels of C_1 are the coordinates of the point p_1 = (x_1, y_1). The phase values ϕ = (ξ, η) at p_1 are searched for in the image of camera C_2. This search may use bilinear or bicubic interpolation and determines the exact position p_2 = (x_2, y_2) of (ξ, η) with subpixel accuracy in the image of camera C_2 (see fig. 6). Together with the intrinsic camera parameters of C_1 and C_2 and the relative orientation between C_1 and C_2 the 3D point is calculated by triangulation.

The same procedure can be performed using C_2 as primary camera and determining the position of the corresponding point in the C_1 image with subpixel accuracy.

The main advantage of the CC technique over the CP method is the possibility to correct errors influenced by disturbed illumination or characteristic line errors. If the illumination is disturbed in both images in the same manner the error is similar and the correspondence will be found correctly despite the disturbance. This effect was reported by Munkelt (Munkelt et al., 2005).

The main disadvantage is the additional camera leading to higher weight and bigger volume of the sensor, higher costs, and an increased calculation effort.

3.3.1 CC algorithm

Precondition: Projection and observation (by two cameras C_1 and C_2) of a fringe image series consisting of two 90° rotated sequences subsequently producing two phase images Φ_x and Φ_y for each camera

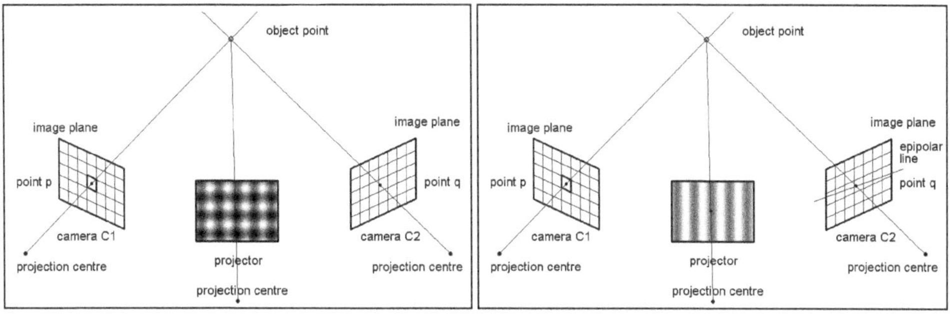

Fig. 6. Sensor arrangements for the CC (left) and the CCE (right) algorithms

Input: four phase images Φ_{1x}, Φ_{1y}, Φ_{2x} and Φ_{2y} for both observation cameras C_1 and C_2

Consider all pixels of the image I_1 of camera C_1. Let $p_i=(x_i, y_i)$ the coordinates of point p_i.

- Read the phase values ξ_i and η_i at p_i: $\phi_i = (\xi_i, \eta_i)$, i=1,2
- Find coordinates $q_i=(u_i, v_i)$ in the image of C_2 with phase values $\phi_l = (\xi_i, \eta_i)$
- Perform triangulation between p_i and q_i to obtain 3D point P_i

Output: point cloud $\{P_i\}$

3.4 The CCE method

The CPE method is the extension of the CP mode according to epipolar geometry. Instead of the second projected fringe direction the geometric information of the epipolar lines in the image I_2 of camera C_2 concerning the considered image point coordinates p=(x, y) are used.

The advantage of the method is the connection of the advantages of the CC method together with the epipolar constraint use. The disadvantages are analogous.

3.4.1 CCE algorithm

Precondition: Projection of a fringe image series in one direction subsequently producing one phase image Φ_{1x} and Φ_{2x} for each of the two cameras C_1 and C_2

Input: phase images Φ_{1x} and Φ_{2x} for the observation cameras C_1 and C_2

Consider all pixels of the image I_1 of camera C_1. Let $p_i=(x_i, y_i)$ the coordinates of point p_i.

- Calculate the epipolar line g_i in image I_2 of camera C_2 corresponding to p_i
- Read the phase value ξ_i at p_i:
- Find the position $q_i = (\xi_i, \eta_i)$ of phase value ξ_i on g_i
- Perform triangulation between p_i and q_i to obtain 3D point P_i

Output: point cloud $\{P_i\}$

3.5 The VR method

The VR method (VR means virtual raster) is derived from the CC method. Instead of taking the measured phase values at the pixels of one of the two cameras, a virtual phase raster is defined according to the desired 3D point density in the resulting point cloud. The position of every phase raster point is determined in the image of the C_1 camera as well as in the image of the C_2 camera (see fig.2, right). If both positions $p_1 = (x_1, y_1)$ and $p_2 = (x_2, y_2)$ are detectable, the triangulation can be performed and the resulting 3D point can be calculated.

The advantage of the VR method is the possibility to choose an arbitrary spatial resolution of the resulting 3D measurement data. However, it should be noticed that neighbouring resulting points may be not independent. Here a particular careful error analysis based e.g. on the work of Notni (Notni & Notni, 2003) should be performed.

3.5.1 VR algorithm

Precondition: Projection and observation (by two cameras C_1 and C_2) of a fringe image series consisting of two orthogonal sequences subsequently producing two phase images Φ_x and Φ_y for both cameras C_1 and C_2

Input: four phase images Φ_{1x}, Φ_{1y}, Φ_{2x} and Φ_{2y} for both observation cameras C_1 and C_2

Consider all phase values $\phi_i = (\xi_i, \eta_i)$ of the selected raster.

- Find coordinates $p_i=(x_i, y_i)$ with the phase values $\phi_i = (\xi_i, \eta_i)$ in C_1
- Find coordinates $q_i=(u_i, v_i)$ with the phase values $\phi_i = (\xi_i, \eta_i)$ in C_2
- Perform triangulation between p_i and q_i to obtain 3D point P_i

Output: point cloud $\{P_i\}$

3.6 The VRE method

The VRE method is obtained by extension of the VR method by epipolar constraint. It can be achieved by fixing the raster of the searched phase values. However, the raster of the selected epipolar lines must be fixed, too. Usually the raster step width must be different because one direction is driven by the phase and the other by a metric default value.

Hence it is doubtful whether the usage of the VRE method would be meaningful. Too few experiments have been performed yet. A really convincing contention could only be given, if further analysis and experiments would have been successfully performed.

3.6.1 VRE algorithm

Precondition: Projection of a fringe image series in one direction subsequently producing one phase image Φ_x for each camera C_1 and C_2

Input: two phase images Φ_{1x} and Φ_{2x} for the observation cameras C_1 and C_2

Predefine a bundle of corresponding epipolar lines $g_{i,1}$ and $g_{i,2}$ in both camera images

Consider all phase values ξ_i of the selected raster

- Find coordinates $p_i=(x_i, y_i)$ with the phase value ξ_i on $g_{i,1}$
- Find coordinates $q_i=(u_i, v_i)$ with the phase value ξ_i on $g_{i,2}$
- Perform triangulation between p_i and q_i to obtain 3D point P_i

Output: point cloud $\{P_i\}$

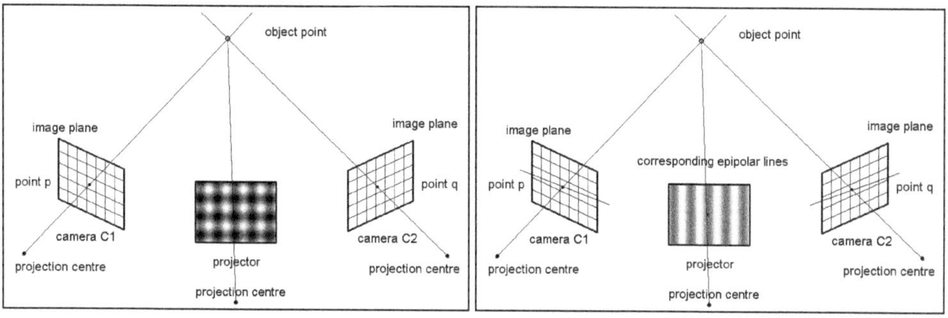

Fig. 7. Sensor arrangements for the VR (left) and the VRE (right) algorithms

4. Random and systematic errors at point correspondence localization

In order to evaluate and to compare all the different correspondence finding methods, an estimation of the error of the reconstructed 3D points was performed. The 3D error of a reconstructed point depends on a number of influences including phase measurement error, characteristic line error, calibration error, distortion, and others, having a systematic and a random component. All error sources that are not depending on the correspondence algorithm are assumed to have the same influence on the measurement accuracy of all considered methods.

Assuming that the determined calibration is fixed and the error of the calibration occurs as a systematic error, the phase noise has the main influence on the random error of the resulting 3D point. All other relevant error sources are expected to be systematic and are not considered here. A detailed analysis of the phase error is described by Notni (Notni & Notni, 2003).

4.1 Error models

As suggested by Rivera-Rios (Rivera-Rios et al. 2000) we assume a Gaussian independent random error $\Delta p=(\Delta x, \Delta y)$ of the observed point coordinates in the camera images. This leads to a certain random error $\Delta P=(\Delta X, \Delta Y, \Delta Z)$ of the measured 3D coordinate of the reconstructed point, additionally depending on the triangulation angle τ. This error is not symmetric in X, Y, and Z. The smaller the triangulation angle the higher is the longitudinal error compared to the lateral one. The amount of the error in the different coordinates depends on the orientation of the cameras in the world coordinate system W, too.

Assume the orientation of the world coordinate system with Z-axis along the boresight of the sensor obtained by a transform of W yielding W'. Then the standard deviations in X-, Y- and Z-direction of the measured point cloud characterize the lateral and longitudinal aspects of the random error of the measurement.

However, in the case of our correspondence methods CP and CC the point clouds characterizing the error of one measured point are deformed to a flat (situated nearly in a plane) point cloud. This is because one coordinate (the phase value or one camera pixel) is fixed. In the case of VR the shape of the error point cloud is the intersection of two cones (see fig. 9).

The extensions to CPE, CCE, and VRE by epipolar constraint reduce the shape of the error clouds again: to a line segment in the CPE and CCE case and to a flat ellipse in the VRE case (see figs. 8 and 9). Initially, this means a reduction of the random error by use of epipolar constraint. However, a systematic error may additionally occur in the case of a calibration error. This error, however, may be reduced by some additional effort as shown by the authors (Bräuer-Burchardt et al., 2011c).

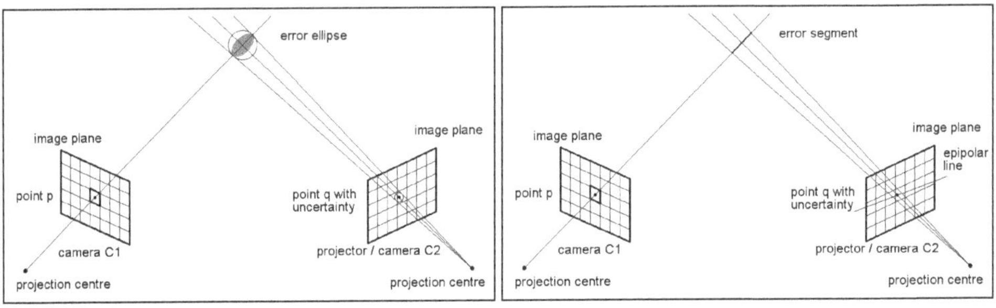

Fig. 8. Error model for CP/CC (left) and CPE/CCE (right)

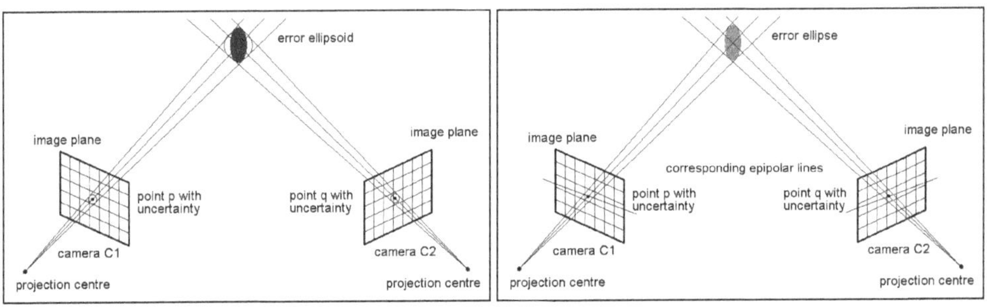

Fig. 9. Error model for the VR (left) and the VRE (right) algorithms

In order to estimate the random error of a single point measurement the following assumptions are made. Assume that the calibration is fix and error free, and a possible distortion error is completely corrected. The error Δp of the camera coordinate of a certain point representing a given phase value should be normally distributed with expectation value $E(\Delta p)=0$ and a certain variance $\sigma_c^2(\Delta p)$ which should be equal in X- and Y- direction. The amount of this error depends on the phase noise (depending on the measuring conditions, number of images in the sequence (Notni & Notni, 2003), and the parameters of the current filter operators (e.g. Gaussian).

Let σ_c be the standard variation of the random error of the pixel coordinate after all filtering operations, m the magnification of the mapping and M the inverse magnification M=1/m.

Let us consider first the case CP/CC and the transformed world coordinate system W′. Because of the fixed coordinate in the primary camera, the random error of the measured 3D point depends only on the random error in one camera image and the triangulation angle τ. We consider the standard deviations $\sigma_x(\Delta p)$, $\sigma_y(\Delta p)$, and $\sigma_z(\Delta p)$ of the calculated 3D point coordinates. We obtain

$$\sigma_x \approx 0; \quad \sigma_y \approx \frac{1}{2}\sigma_c \cdot M \cdot fac; \quad \sigma_z \approx \frac{1}{\sin(\tau)}\sigma_c \cdot M \cdot fac \tag{1}$$

where *fac* is set to *fac* = 1 for both the CP and the CC mode. Rather it should be *fac* = 1 for CP and $fac = \sqrt{2}$ for the CC method because of the double uncertainty of the phase value. However, a number of experiments showed that the random error of CC is equal to that of CP mode using the same triangulation angle. We assume that usually some error sources have the same influence to the phase error in both images of cameras and hence this causes the reduction of $fac = \sqrt{2}$ to *fac* = 1 also in the CC mode.

Considering CPE and CCE we obtain

$$\sigma_x \approx 0; \quad \sigma_y \approx 0; \quad \sigma_z \approx \frac{1}{\sin(\tau)}\sigma_c \cdot M \cdot fac \tag{2}$$

for both the CPE and the CCE mode.

Considering VR and assuming an independent random error Δp with E(Δp)=0 and $\sigma_c^2(\Delta p)$ of the same amount in the two camera images, we obtain

$$.\sigma_x = \frac{1}{\sqrt{2}}\sigma_c \cdot M; \quad \sigma_y = \frac{1}{\sqrt{2}}\sigma_c \cdot M; \quad \sigma_z = \frac{\sqrt{2}}{\sin(\tau)}\sigma_c \cdot M \tag{3}$$

and for VRE we get

$$.\sigma_x \approx 0; \quad \sigma_y \approx 0; \quad \sigma_z = \frac{\sqrt{2}}{\sin(\tau)}\sigma_c \cdot M \tag{4}$$

As we can see considering formulas (1) to (4) CPE and CCE provide the best estimation of the random 3D error because the error only depends on the uncertainty of the phase determination along the epipolar line. However, an additional systematic error can occur due to the uncertainty of the epipolar line position. It may be erroneous due to calibration errors caused by thermic or mechanic influences as shocks or vibrations. However, the stability of the calibration can be checked, and hence this error can be estimated and minimized as it was shown by the authors (Bräuer-Burchardt et al., 2011c). CP/CC and CPE/CCE errors are mainly influenced by the triangulation angle τ.

5. Experiments and results

5.1 Simulations

In order to estimate the random error of a 3D point measurement and to confirm the theoretic assumptions according to formulas (1) and (2), the following simulation experiment was performed. For a given geometric situation of the sensor (extrinsic and intrinsic parameters of the camera(s) and the projector, triangulation angle between the two relevant image normal vectors) a meaningful value of the random error Δp of the camera coordinates was chosen. It was chosen a normally distributed random value Δp = (Δx, Δy) with expectation value (0, 0) for the camera coordinate uncertainty. The variance was chosen according to realistic conditions by experiment. The error ΔE of the calculated 3D point coordinate is given as (ΔX, ΔY, ΔZ) error vector in the virtual world coordinate system W′. The alignment of W′ is according to the orientation of the boresight of the contributing cameras (C and P or C_1 and C_2, resp.).

Error estimation will be achieved by analysis of the covariance matrix of the coordinate values of the point clouds. These point clouds are those of the measured or simulated points in the world coordinate system W according to the calibration data. The variances of the error vector components are obtained as the eigenvalues ev_1, ev_2, and ev_3 of the covariance matrix. Hence the random error of the 3D point calculation can be described by the standard deviations in the virtual world coordinate system W′ corresponding to the square roots of the eigenvalues:

$$ls_x = \sqrt{ev_1}\,; \quad ls_y = \sqrt{ev_2}\,; \quad ls_z = \sqrt{ev_3} \tag{5}$$

The simulation of a measurement with random error was performed 400 times. The resulting 3D point cloud was analysed using covariance matrix leading to the 3D point error characteristic (σ_x, σ_y, and σ_z). For the VR case the image coordinate of the pixels of camera C_1 were disturbed by a 2D error value with the same statistical characteristics (normal distribution, zero expectation value, same variance as for C_2), too.

The standard deviations of the three coordinates in the transformed world coordinate system W′ according to (1) and (2) and the simulation results using 400 measurements are given in table 1 showing that it holds approximately $\sigma_x = ls_x$, $\sigma_y = ls_y$, and $\sigma_z = ls_z$.

exp. status	quantity \ mode	CP	CPE	CC	CCE	VR	VRE
	τ [°]	11,2	11,2	24,0	24,0	24,0	24,0
model	σ_x [µm]	0,0	0,0	0,0	0,0	1,6	1,6
	σ_y [µm]	1,2	0,0	1,2	0,0	1,6	0,0
	σ_z [µm]	11,9	11,9	5,7	5,7	8,0	8,0
simulation	ls_x [µm]	0,0	0,0	0,0	0,0	1,7	1,6
	ls_y [µm]	1,2	0,0	1,3	0,0	1,7	0,0
	ls_z [µm]	13,0	12,8	6,0	5,9	8,0	8,1
experiment	ls_x [µm]	0,003	0,003	0,003	0,003	1,0	1,0
	ls_y [µm]	1,1	0,003	1,0	0,003	1,2	0,005
	ls_z [µm]	11,0	11,0	6,0	6,0	9,4	9,4

Table 1. Model data, simulation data, and experimental data of measurements using different methods

5.2 Experiments on real measurements

In order to evaluate the different correspondence finding methods several experiments determining the accuracy of the measurements were performed. For the measurements the "kolibri flex mini" system was used. This system (see fig.1) is a table top system with a sensor head consisting of two cameras (1.4 Mpixels) and one projector. The working distance is about 300 mm and the measuring volume is 90 mm (diameter) x 25 mm (height).

The measuring accuracy given by the standard deviation of the surface points on a plane is about 5 µm.

The first experiment was a measurement of a single point in repeated measurements in order to confirm the simulation experiment described in the previous section. The number of measurements was 400. The resulting *ls*-values of all measurements are listed in table 1 and confirm the theoretic assumptions.

The next experiment was performed in order to confirm the assumption about the dependence of the random error on the triangulation angle. Different triangulation angles were realized by a modification of the "kolibri flex mini" system. Each measurement with one fix triangulation angle and using the VR mode was repeated at least 100 times. A range between 5 and 40° was realized for the triangulation angle τ.

The results of the variation of the triangulation angle are shown by a plot of the accuracy values in fig. 10.

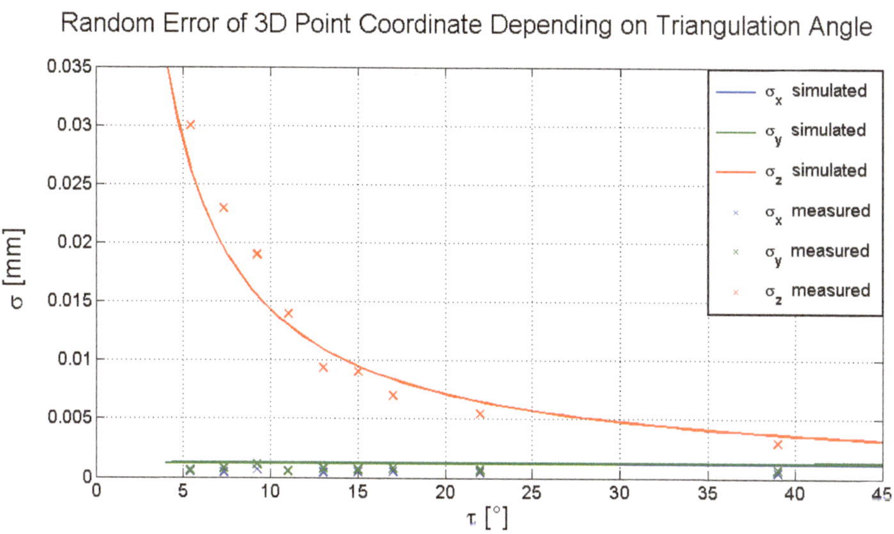

Fig. 10. Plot of accuracy of VR mode depending on triangulation angle τ

6. Discussion

The results of the experiments and the simulations mainly confirm the theoretic analysis of the measuring accuracy. Especially in the CP/CPE and CC/CCE modes the correlation between the results is very high. Using the VR/VRE modes, too few experiments have been performed in order to make significant statements. Here, additional effects may influence the random error and lead to a difference to our model. This will be, however, analysed in our future work.

The completeness of a measurement is directly dependent on the triangulation angle. The smaller the triangulation angle the bigger is the number of measured 3D points because shadowing and occlusion effects are reduced. Otherwise, if triangulation angle becomes too small, σ_z error increases. However, a difference of the completeness between a CP/CPE and a CC/CCE measurement with $\tau_{CC} = 2\tau_{CP}$ also depends heavily on the shape of the measuring object.

In general, CP/CPE mode and CC/CCE mode with the same triangulation angle $\tau_{CC} = \tau_{CP}$ provide the same results in the accuracy. However, having disturbances in the phase generation which similarly occur in both camera images (e.g. by indirect illumination, irregular reflexes, phase generation errors, characteristic line errors, or not corrected distortion of the projector optics) the error using the CC/CCE mode may be considerable smaller than using CP/CPE mode as described by Munkelt (Munkelt et al., 2005).

The results are helpful for the decision of the method to be used taking into account the typical measuring objects, the demands on the measurement, and the costs already in the phase of system design.

A combination of the methods may improve a measurement using only one method concerning accuracy and/or completeness in certain cases depending on the measuring object as suggested by the authors (Bräuer-Burchardt et al., 2010). However, this requires additional algorithmic effort in the generation of software tools which has not been realized yet for our systems. It also extends the calculation time and will not be used in fast speed applications. However, first tests of the combination of the CP and VR methods have been successfully performed in order to obtain data of regions which were hidden or in the shade using only CC mode.

Together with the results described by Kühmstedt (Kühmstedt et al., 2009), a comprehensive prediction of the expected error of 3D measurement systems using fringe projection can be made depending on the properties of the measuring device, the exterior conditions and the measuring objects.

As already mentioned in section 3 the main advantage of the CP/CPE method is the minimal effort in equipment and calculation in contrast to the usage of the CC/CCE mode. There, however, illumination caused errors may be reduced by the principle (Munkelt et al., 2005).

The extension by use of the epipolar constraint is always meaningful because it saves half of the recording time and reduces the random error in direction perpendicular to the epipolar lines. However, it should be realized that the calibration is correct and stable which can be checked by additional measurements.

In order to design a new sensor the choice of the selected point localization principle depends on several aspects like requested measurement accuracy, field of application, measurement conditions (e.g. illumination, sensor mounting, measuring objects), costs, sensor weight, and so on. Hence, this work cannot give the optimal way to perfect measurement data but it may help to understand the influences on the measurement accuracy.

7. Summary and outlook

In this work some different correspondence localization methods which are usually applied in 3D measurement systems using fringe projection were compared concerning the measuring accuracy. A theoretic model for the random error of the measured 3D point coordinates was established. A simulation tool was developed and applied. The results obtained by these estimations were confirmed by real measurement data.

Future work is addressed on several aspects. First, the accuracy determination of the VR/VRE modes should be improved and the error model should be potentially extended. Here a number of new experiments must be performed.

Second, a combination of some of the methods should be merged into one measuring system in order to obtain more complete results in the case of difficult measuring object geometries. Here, new algorithms which realize an automatic 3D data fusion, a smooth passage between regions of different measuring uncertainties, and an optimization of the connected data must be developed.

Third, the VR/VRE methods can be extended in order to obtain equidistant distributed 3D points as result of a measurement. This involves the realization of an adaptive virtual phase raster for the point corresponding algorithms. Here, information about the shape of the measuring object is necessary that may be obtained in a preliminary rough measurement.

8. References

Battle J., Mouaddib, E. & Salvi, J. (1998). Recent progress in coded structured light as a technique to solve the correspondence problem: a survey. PR, Vol. 31, No 7, 963-982

Bräuer-Burchardt, C. (2004). A Simple New Method for Precise Lens Distortion Correction of Low Cost Camera Systems. Pattern Recognition (Proc 26th DAGM-Symposium), Springer, 570-77

Bräuer-Burchardt, C., Möller, M., Munkelt, C., Kühmstedt, P. & Notni, G. (2010). Comparison and evaluation of correspondence finding methods in 3D measurement systems using fringe projection. Proc. SPIE Vol. 7830, 783019-1 - 783019-9

Bräuer-Burchardt, C., Munkelt, C., Heinze, M., Kühmstedt, P. & Notni, G. (2011a). Using Geometric Constraints to Solve the Point Correspondence Problem in Fringe Projection Based 3D Measuring Systems. Proc. ICIAP 2011, Part II, Springer LNCS, 265-74

Bräuer-Burchardt, C., Breitbarth, A., Kühmstedt, P., Schmidt, I., Heinze, M. & Notni, G. (2011b). Fringe Projection Based High Speed 3D Sensor for Real-Time Measurements. Proc. SPIE, Vol. 8082, 808212-1 - 808212-8

Bräuer-Burchardt, C., Kühmstedt, P. & Notni, G. (2011c). Error compensation by sensor re-calibration in fringe projection based optical 3D stereo scanners. Proc. ICIAP 2011, Part II, Springer LNCS, 363 -73

Brown, D.C. (1971). Close-range camera calibration. Photogram. Eng. 37(8), 855-66

Chen, F. & Brown, G.M. (2000). Overview of three-dimensional shape measurement using optical methods. Opt. Eng. 39, 10-22

Creath, K. (1986). Comparison of phase-measurement algorithms. Surface characterization and testing. Proc. SPIE 680, 19–28

Ishiyama, R., Sakamotom, S., Tajima, J., Okatani, T. & Deguchi, K. (2007a). Absolute phase measurements using geometric constraints between multiple cameras and projectors. Applied Optics, Vol. 46, Issue 17, 3528-3538

Ishiyama, R., Okatani, T. & Deguchi, K. (2007b). Precise 3-d measurement using uncalibrated pattern projection. Proc. IEEE Int. Conf. on Image Proc., vol. 1, 225-228

Kühmstedt, P., Heinze, M., Himmelreich, M., Bräuer-Burchardt, C. & Notni, G. (2005). Optical 3D sensor for large objects in industrial application. Proc. SPIE 5856, 118-127

Kühmstedt, P., Munkelt, C., Heinze, M., Himmelreich, M., Bräuer-Burchardt, C. & Notni, G. (2007). 3D shape measurement with phase correlation based fringe projection. Proc. SPIE vol. 6616, 66160 B-1-B-9

Kühmstedt, P., Bräuer-Burchardt, C., Munkelt, C., Heinze, M., Schmidt, I., Hintersehr, J., Notni, G. (2007b). Intraoral 3D scanner. Proc. SPIE vol. 6762, 67620 E-1-E-9

Kühmstedt, P., Bräuer-Burchardt, C. & Notni, G. (2009). Measurement Accuracy of Fringe Projection Depending on Surface Normal Direction. Proc. SPIE vol. 7432, 743203,1-9

Li, E.B., Peng, X., Xi, J., Chicaro, J.F., Yao, J.Q. & Zhang, D.W. (2005). Multi-frequency and multiple phase-shift sinusoidal fringe projection for 3D profilometry. Optics Express. 13, 1561-1569

Li, Z., Shi, Y. & Wang, C. (2009). Real-time complex object 3D measurement. Proc. ICCMS, 191-194

Luhmann, T., Robson, S., Kyle, S. & Harley, I. (2006). Close range photogrammetry. Wiley Whittles Publishing

Maas, H.-G. (1992). Robust automatic surface reconstruction with structured light. IAPRS, Vol. XXIX, Part B5, 709-713

Munkelt, C., Kühmstedt, P., Heinze, M., Süße, H. & Notni, G. (2005). How to detect object-caused illumination effects in 3D fringe projection. Proc. SPIE vol. 5856, 632-639

Notni, G. H. & Notni, G. (2003). Digital fringe projection in 3D shape measurement - an error analysis. Proc SPIE 5144, 372-80

Reich, C., Ritter, R. & Thesing, J. (2000). 3-D shape measurement of complex objects by combining photogrammetry and fringe projection. Opt. Eng. 39, 224–231

Rivera-Rios, A.H. & Marefat, M. (2005). Stereo Camera Pose Determination with Error Reduction and Tolerance Satisfaction for Dimensional Measurements. Proc. IEEE Int Conf on Robotics and Automation, 423-428

Sansoni, G., Carocci, M. & Rodella, R. (1999). Three-dimensional vision based on a combination of Gray-code and phase-shift light projection: Analysis and compensation of the systematic errors. Applied Optics, Vol. 38, Issue 31, pp. 6565-6573

Schreiber, W. & Notni, G. (2000). Theory and arrangements of self-calibrating whole-body three-dimensional measurement systems using fringe projection techniques. Opt. Eng. 39, 159-169

Tsai, R. (1986). An efficient and accurate camera calibration technique for 3-D machine vision. IEEE Proc CCVPR, 364-74

Young, M., Beeson, E., Davis, J., Rusinkiewicz, S. & Ramamoorthi, R. (2007) Viewpoint-coded structured light. Proc. CVPR, 1-8

Zhang, H., Lalor, M.J. & Burton, D.R. (1999). Spatiotemporal phase unwrapping for the measurement of discontinuous objects in dynamic fringe-projection phase-shifting profilometry. Applied Optics. 38, 3534-3541

Zhang, S. & Yau, S.T. (2006). High-resolution, real-time 3D absolute coordinate measurement based on a phase-shifting method. Optics Express, Vol. 14, Issue 7, 2644-2649

12

Design Methodology to Construct Information Measuring Systems Built on Piezoresonant Mechanotrons with a Modulated Interelectrode Gap

Alla Taranchuk and Sergey Pidchenko
Khmelnitsky National University, Ukraine

1. Introduction

One of the perspective tendencies of development of information measuring systems is applying the sensors of physical magnitudes on basis of quartz resonators (QR) of special type, the stimulation of piezoelement in which is accomplished in the vacuum or filled with dielectric gap. The changes of geometry (modulation) of the gap between piezoelement and electrode (electrodes) of quartz resonator leads to the shift of personal resonance frequencies of QR, as well as the oscillatory system in general, the part of which it is. Despite some substantial advantages of presented way of QR control, piezoresonance sensors (PRS) of given type were not widely used, which, to the authors' opinion, is caused by inappropriate study of theoretical and practical material of construction of PRS of given type. The material of present chapter fills in the gap in the study.

Present chapter covers the theoretical issues, projecting and applying in highly informative measuring systems of piezoresonator with modeled under the influence of mechanical force the interelectrode gap and mobile electrode in the form of membrane, which is named piezoresonance mechanotron (PRMT) (Kolpakov et al., 2009). We present mechanic scheme of PRMT, examine its basic constructive and electrical characteristics, describe the regimes of movement of mobile electrode. We propose approximate method of linearization of graduation characteristic characteristics of PRMT while its use in primary measuring transducers nonelectric magnitudes. Mechanic scheme, mathematical model and the results of numeric simulation of a PRMT membrane bending flexure using ANSYS - programme and a thermal PRMT model as well as numeric MATHLAB/ FEMLAB - analyses of temperature - frequency characteristics under the excitation power variation of a PRMT piezoelement and outer temperature change are presented. We present real life piezoresonance mechanotron designs to be used as high precision surplus air pressure transducers and the results of the development research of their precision characteristics. Examples of PRMT applications as well as methods of secondary transducing of its output signals in real life information measuring systems to define human hemodynamic par meters are presented.

2. Quartz resonator as an element of precision primary measuring transducers of nonelectric magnitudes - Quartz resonator frequency control

Quartz resonator (QR) represents electro - mechanical oscillatory system, conventional equivalent scheme of which is shown in Fig. 1. Dynamic elements L_1 - equivalent inductivity, C_1 - equivalent capacity and R_1 - equivalent resistance of losses are caused by presence of direct and reverse piezoeffects and resonance peculiarities of piezoelement in quartz resonator. As the quality factor of quartz resonators is very high and comprises tens and hundreds thousands for usual resonators and several millions for precision ones, equivalent dynamic branch of quartz resonator has a sense only in narrow band of frequency next to resonance, and in other area of frequency equivalent electric resistance of QR is determined by static capacity C_0 (Cady, 1946).

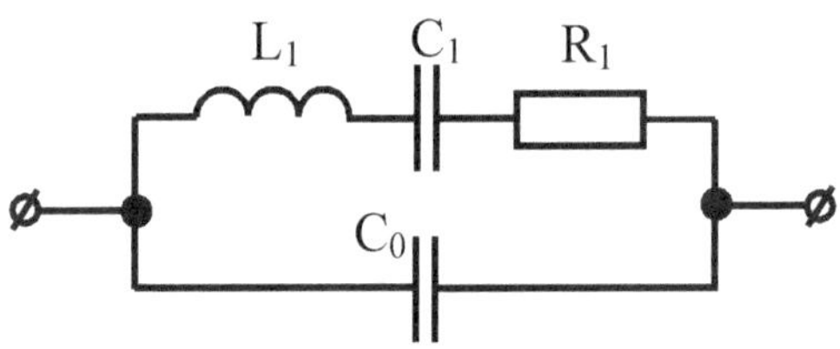

Fig. 1. Equivalent scheme of quartz resonator

The basic parameters of quartz resonator are the frequencies of successive and parallel resonance

$$f_1 = \frac{1}{2\pi\sqrt{L_1 C_1}} \text{ and } f_a = f_1\sqrt{1+m}\,; \tag{1}$$

the quality factor and the width of interresonance interval

$$Q = \frac{2\pi f_1 L_1}{R_1} = \frac{1}{2\pi f_1 C_1} \text{ and } \frac{f_a - f_1}{f_1} \approx 0.5m\,, \tag{2}$$

where $m = C_1/C_0$ - is capacity factor.

At the same time dynamic inductivity L_1 is connected with equivalent mass of QR, and the dynamic capacity C_1 is connected with elastic peculiarities (equivalent compliance).

Control the frequency in using QR as the piezoresonant sensor of physical magnitudes, as it can be seen from the analyses of equivalent resonator circuit, can be realized:

- by the influence on equivalent compliance (C_1) or on equivalent mass (L_1) of oscillatory system or on both mentioned parameters at the same time. In accordance with (1) in this case both resonance frequencies are varied;
- by the varieties of active losses (R_1). Increasing the decrement of damping $\delta = 1/Q$, as is known, decreases the frequency own oscillations in accordance with the correlation $w = \sqrt{w_0^2 - \delta^2}$.

- by managing the meaning of value of inter- electrode capacity C_0 or capacity correlation $m = C_1/C_0$.

The first method, which is based on the modulation of equivalent compliance C_1 or the mass L_1 of oscillatory system, is widely used in piezoresonant sensors (PRS) of pressure, micromoving, microweighing, humidity, gas analysis and others (Malov, 1989; Kolpakov & Pidchenko, 2011).

As for construction of PRS the high quality factor resonators are used, in which the decrement of damping is so small that it does not practically influence the resonance frequencies, the second method of managing on frequency did not find wide application.

The third method is interesting from the point of view of applying in resonators of special type (piezoresonant mechanotron), in which between piezoelement and electrode (electrodes) there is the gap, vacuum-processed or filled with material with dialectic permeability ε . Changes of gap geometry (modulation), as well as varying its electric characteristics, leads to the shift of resonance frequency of oscillatory system as a whole.

The given way, in comparison with the first, in some cases has certain advantages, for example, at measurements of micromovings and physical magnitudes led by it. It is connected with an exception of direct mechanical influences on a sensor piezoelement that leads to change (deterioration) of its characteristics as highly quality and highly stable element of oscillatory system of the quartz crystal oscillator.

3. Design methodology of piezorezonant mechanotrons built on quartz resonators with a modulated interelectrode gap

Introduced PRMT is related to measuring converters with control of its resonance frequency by changing magnitude of gap between launch electrodes of piezoelectric element (PE).

Change of parameters of oscillatory system under action physical magnitude (moving and reducible magnitude) is used in PRMT. PRMT design map is shown in Fig. 2, (a), where x_0 - initial gap; x_m - magnitude stroke mobile electrode (membrane); $x = x_0 - x_m$ - current value of modulated interelectrode gap (Kolpakov F., Pidchenko S., Taranchuk A., et al, 2009).

Connection of capacitive parasitics with minor gap capacity enabled to reduce influence on the resonance frequency of oscillatory system. Modulated gap application, i.e. measuring capacitor connection inwards construction allows raising capacitive-ratio, and therefore tuning range of oscillatory system.

Piezoresonant mehanotron with modulated interelectrode gap (PRMT with MIG) equivalent electric circuit is shown in Fig. 2, (b). Following descriptions are made: L_1, R_1, C_1 - PE dynamic equivalent parameters on the fundamental mode of the thickness-shear vibrations; C_{PE} - PE static capacity; C_1/C_{PE} - capacitive ratio (for AT-cut crystal $m = (4...7) \cdot 10^{-3}$); $C_0 = C_m + C_{par}$ - capacitive parasitics with mounting capacity.

Rather important is the fact, that parasitic capacitive is connected through small capacitive gap and therefore it has little influence onto resonance frequency oscillatory system.

PRMT by its electric equivalent circuit is similar to any capacitor the operated quartz resonator. However, the use of modulated gap, e.i. connection of measuring compensator into the construction, dives the chance to increase considerably the capacitor relation, and therefore, the band of oscillatory system restructure. For example, on frequency 10 MHz under the gap modulation we can get the frequency deviation (25…30) kHz.

Let's define PRMT characteristics. Circuit equivalent resistance is:

$$Z = \frac{R_1 k_1 + j\omega L_1 k_2 - \frac{j}{\omega} k_4}{j\omega C_0 \left(R_1 \left(1 + C_{PE} k_3\right) + j\omega L_1 \left(1 + C_{PE} k_3\right) + j\left(\frac{C_{PE}}{C_0 C_1} + \frac{k_5}{\omega} \right) \right)}, \tag{3}$$

where $k_1 = 1 - \frac{C_{PE}}{C_g}$; $k_2 = 1 + \frac{C_{PE}}{C_g}$; $k_3 = \frac{1}{C_g} - \frac{1}{C_0}$; $k_4 = \frac{1}{C_1} + \frac{1}{C_g} + \frac{C_{PE}}{C_g}$;

$k_5 = \frac{1}{C_0} - \frac{1}{C_g} - \frac{1}{C_1} - \frac{C_{PE}}{C_g C_1}$ - coefficients.

In conditions of a resonance $I_m(Z) = 0$, we obtain:

$$n_4 \omega^4 + n_2 \omega^2 + n_0 = 0, \tag{4}$$

where $n_4 = L_1 C_0$, $n_2 = \frac{C_{PE} R_1^2 \left(C_0 - C_g\right)}{L_1 \left(C_g + C_1\right)}$, $n_0 = \frac{C_0}{C_1 L_1 C_g}$.

Solving (4) for ω^2 can be written as follows:

$$\omega_{r,a}^2 = \frac{C_{PE} R_1^2 \left(C_0 - C_g\right)}{2 C_0 L_1 \left(C_1 + C_g\right)} \pm \frac{1}{2 L_1 C_0} \sqrt{\left(\frac{C_{PE} R_1^2 \left(C_0 - C_g\right)}{L_1 \left(C_g + C_1\right)} \right)^2 - \frac{4 C_0^2}{C_1 C_g}}. \tag{5}$$

For typical valuations of QCR parameters: $R_1 = 10$ Ohm, $C_1 = 22$ pF, $L_1 = 11{,}5\,\mu H$, $C_0 = (1\ldots10)$ pF , $C_{PE} = (2...3)$ pF, $C_b = (0.1...1)$ pF. On basis of measurement results characteristics of QR control are built. They are shown in Fig. 3.

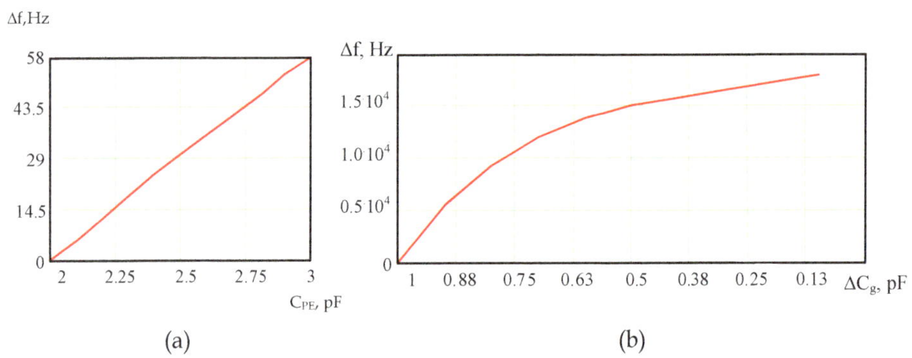

Fig. 2. (a) Dependence of absolute detuning on C_{PE}; (b) Dependence of absolute detuning on ΔC_g

The analysis of received dependence shows, that the use of frequency control with the help of interelectrode gap rather effective and under little changes of volume of clearance allows to get information deviation of frequency of order of tens KHz.

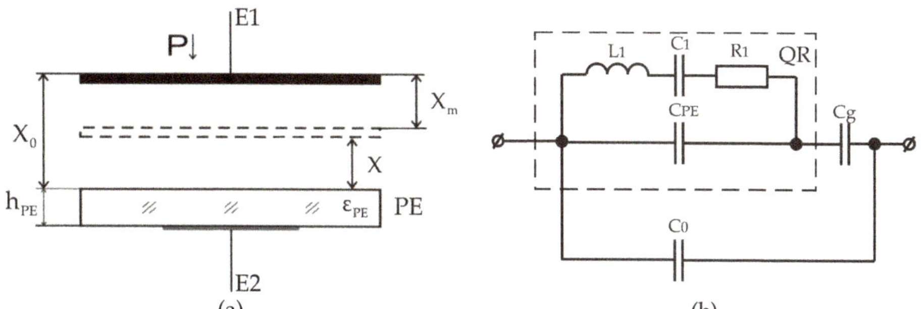

Fig. 3. (a) PRMT with MIG design map; (b) Equivalent electric circuit of PRMT

For real PRMT with piezoresonator AT - cut (diameter of piezoelement $d_{PE} = 18$ mm, electrode diameter is $d_e = 8$ mm, nominal frequency of piezoelement $f_n = 10.0$ MHz, $F_0 = 6500$ Hz) experimentally was received modulation (graduation) characteristic, shown in Fig. 4.

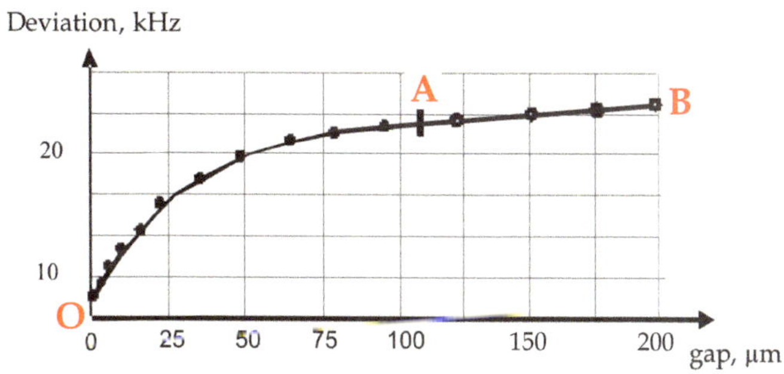

Fig. 4. Graduation characteristic of PRMT with MIG

It is obvious that it can be symbolically divided into two sectors (see Fig. 4). The first one, OA - is the area with maximum deviation and nonlinearity. The second one, AB - is the linear area of the curve, which is characterized by relatively small frequency changes. This allows to talk about two working conditions of PMRT: nonlinear and linear. The work in nonlinear conditions allows to achieve maximum resolution capability, in linear - maximum linearity of modulation characteristic under the demodulation of measuring signal by the linear frequency detector. Using PE in PMRT of pressure it is reasonable to use nonlinear conditions, which provides maximum sensitivity of PMRT, but allows to linearity graduation characteristic.

As opposed to PMRT it is possible to use external capacitive frequency control (ECFC), under which in the capacity of sensor the capacitive sensitive element, playing a role of C_g, is used.

We propose the comparative characteristic of external capacitive frequency control and control of modulation of interelectrode gap.

Equivalent electric circuit of QR with ECFC shown in Fig. 5. Standard piezoresonanator is used and is connected sequentially with tuning inductance L_t and capacitive sensor of pressure $C_t(p)$ (see Fig. 5).

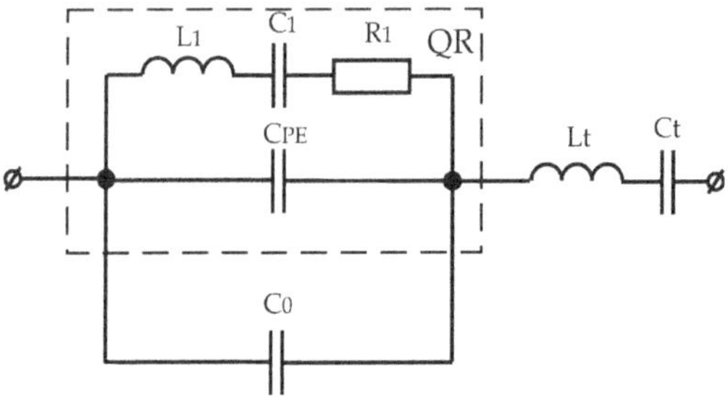

Fig. 5. Equivalent electric circuit of QR with external capacity control

Comparing Fig. 1, (a) and Fig. 5, we can mark, that parasitic capacitance C_0, including also capacitance of assembling, on schemes it is included on a miscellaneous. Under external capacitive control it is connected parallel to static capacitive of PE. As frequency deviation is directly proportional to the capacitor relation

$$m = \frac{C_1}{C_{PE}} = \frac{C_1}{C_{PE} + C_0}, \tag{6}$$

so controlling the oscillatory system with an increase C_0 considerably worsens.

As controlling the modulation of interelectrode gap C_0 (see Fig. 2) is connected through small capacitance of gap and practically does not influence the size of static capacitance, i.e.

$$m' = \frac{C_1}{C_{PE}}. \tag{7}$$

From here a relative benefit on controllability of a variant with gap modulation

$$k_c = \frac{m'}{m} = 1 + \frac{C_0}{C_{PE}}. \tag{8}$$

This fact is confirmed by the calculation of deviation of frequency for two types of control according to formulae (5) (see Fig. 6).

Comparing these two types of control (see Fig. 6) shows multiple benefit of PMRT on the conversion conductance and the preference of its use. At this time in some devices of not high accuracy, for their reduction in price, the ECFC can also be used (Pidchenko, Kolpakov, Akulinichev, 2000).

4. Linearization of a PRMT modulation characteristic based on fractional-linear approximation of graduation characteristic

Having written down (5) in terms of graduation characteristics, we'll receive expression for resonant frequency PRMT:

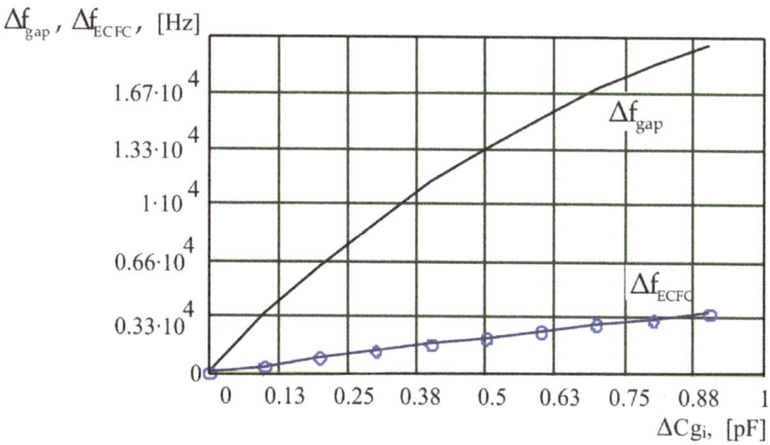

Fig. 6. Frequency mismatches comparison of ECFC and PRMT with MIG

$$f(x) = f_0 \sqrt{1 + \frac{m}{1 + \frac{h_{PE}}{\varepsilon_{PE} x}}} \approx f_0 \left[1 + \frac{0.5m}{1 + \frac{h_{PE}}{\varepsilon_{PE} x}} \right], \tag{9}$$

where f_0 - frequency rating of oscillating system with $x = 0$, m - capacity ratio; h_{PE} - PE dimension which determine frequency (thickness); ε_{PE} - dielectric conductivity of PE material; x - current gap value.

Let us introduce the PE parameter $a = \frac{h_{PE}}{\varepsilon_{PE}} - \frac{N}{f_0 \varepsilon_{PE}}$, where N - is frequency coefficient (for AT cut $N = 1661$ kHz·mm). From (9) we get the relative frequency deviation of measuring transducer (MT):

$$\delta_F(x) = \frac{f(x) - f_0}{f_0} = \frac{0.5mx}{x + a}. \tag{10}$$

Taking into account that $x = x_0 - x_i$ (10) it can be written in such a way

$$\delta_F(x_m) = \frac{0.5m(x_0 - x_m)}{(x_0 + a) - x_m}. \tag{11}$$

Fractional - linear function (9) has such look:

$$\delta_F(x_m) = \frac{a_1(1+a_2x_m) - a_2(a_0+a_1x_m)}{(1+a_2x_m)^2} = \frac{a_0+a_1x_m}{1+a_2x_m}, \quad (12)$$

where a_0, a_1, a_2 are the coefficients of approximation of normalized fractional characteristic:

$$a_0 = \frac{0.5mx_0}{x_0+a};\ a_1 = -\frac{0.5m}{x_0+a};\ a_2 = -\frac{1}{x_0+a}. \quad (13)$$

We can show at PMRT with parameters, which are character to pressure transducer. Let $x_0 = 0.1$ mm, $m = 6.29 \cdot 10^{-3}$, $a = 0.03691$ mm, $x_i \in [0...0.06]$ mm. We substitute initial data into (13), find coefficient of approximation of normalized fractional characteristic: $a_0 = 2.297 \cdot 10^{-3}$, $a_1 = -2.297 \cdot 10^{-2}$ 1/mm, $a_2 = -7.304$ 1/mm. Hence:

$$x_m = x_{m\,max} = 0.06 \text{ mm},\ \delta_F(x_m) = 1.6357 \cdot 10^{-3};$$

$$x_m = {x_{m\,max}}/{2} = 0.03 \text{ mm},\ \delta_F(x_m) = 2.059 \cdot 10^{-3};$$

$$x_m = x_{m\,min} = 0 \text{ mm},\ \delta_F(x_m) = 22.97 \cdot 10^{-3}.$$

Complete reorganization of output frequency of MT under given parameters of oscillatory system is 22.97 kHz, and $\Delta f = 0$ corresponds own frequency of quartz plate and corresponds $f_0 = 10$ MHz.

For PMRT of micromovings the variable $x \in [0, x_0]$ and normalized graduation characteristic corresponds (10). In accordance with this

$$\delta_F(x) = \frac{a_0+a_1x}{1+a_2x}, \quad (14)$$

where $a_0 = 0$, $a_1 = 0.5\frac{m}{a}$, $a_2 = \frac{1}{a}$.

Having differentiated (14), we get the slope S_F of characteristic PMRT:

$$S_F = \frac{a_1 - a_2a_0}{(1+a_2x)^2}. \quad (15)$$

We write also the formula for measuring the equivalent dynamic resistance of quartz oscillatory system R_{ekv} under variation of interelectrode gap:

$$R_{ekv} = R_0\left(1 + {C_0}/{C_{air}}\right)^2 = R_0\left(1 + {x\varepsilon_{PE}}/{h_{PE}}\right)^2 = R_0\left(1 + {x}/{a}\right)^2, \quad (16)$$

where R_0 is a dynamic resistance of under $x = 0$, C_0 is static size of QR, C_{air} the size of air gap.

These parameters are necessary while projecting PMRT, and approximation is essential for the synthesis of linear graduation characteristic (Kolpakov, Pidchenko, Hilchenko, 1999).

5. Accounting for a PRMT electrodes nonparallelity

All given above correlations are received in assumption that electrodes of PMRT are strictly parallel to each other, however it's practically unrealistic. Firstly, it is extremely difficult to provide the parallelism of electrodes while producing PMRT; secondly, the slightest mechanical stress, which appears in the construction under the effect of outer factors can lead to micron defects.

Taking into account everything said before, we take into account nonparallelism, defining the condenser capacity with nonparallel flat electrodes (see Fig. 7) with the radius R of static and $R' > R$ of movable.

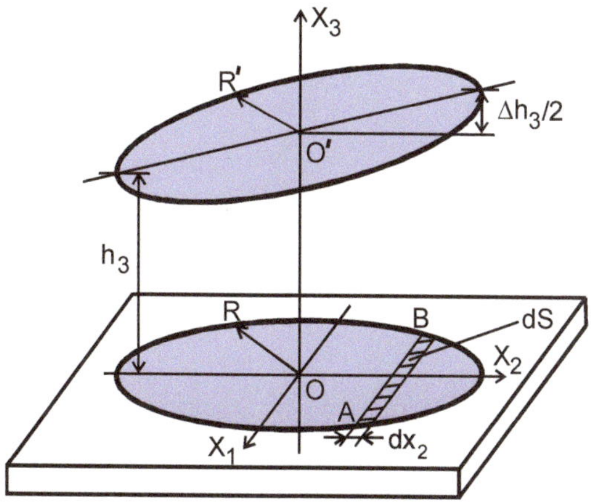

Fig. 7. Record of nonparallelism of PE and flat electrode

The area of the element with the width dx_2 equals

$$dS = AB \cdot dx_2 = 2\sqrt{R^2 - x_2^2} \cdot dx_2 \cdot \tag{17}$$

The distance to appropriate element of upper cover equals

$$x_3 = OO' + \frac{\Delta h_3}{2} \cdot \frac{x_2}{R} = h_3 + \frac{\Delta h_3}{2} + \frac{\Delta h_3}{2R} \cdot x_2 \cdot \tag{18}$$

Elementary capacity between elementary areas

$$dC = \frac{\varepsilon\varepsilon_0 2\sqrt{R^2 - x_2^2} \cdot dx_2}{(h_3 + \Delta h_3 / 2) + x_2 \cdot \dfrac{\Delta h_3}{2R}}, \tag{19}$$

whence full capacity of condenser with nonparallel relatively to the axis x_2 electrodes is

$$C = \int_{-R}^{R} dC = 2\varepsilon\varepsilon_0 \int_{-R}^{R} \frac{\sqrt{R^2 - x_2^2} \cdot dx_2}{(h_3 + \Delta h_3/2) + x_2 \Delta h_3 / 2R} \cdot \tag{20}$$

Implementing the transformation of under the integral formula (20), we move from the form

$$\int_{-R}^{R} dC = S\int_{-R}^{R} \frac{\sqrt{Z}\cdot dx_2}{p+x_2}, \qquad (21)$$

where $Z = a + bx_2^2$; $a = R^2$; $b = -1$; $p = \left(\frac{2}{\delta_h}+1\right)R$; $\delta_h = \frac{\Delta h_3}{h_3}$; $S = \frac{4\varepsilon\varepsilon_0 R}{\Delta h_3}$, to the form

$$\int_{-R}^{R} dC = S\left[b\int_{-R}^{R}\frac{x_2 dx_2}{\sqrt{Z}} + p\int_{-R}^{R}\frac{dx_2}{\sqrt{Z}} + \left(a+bp^2\right)\int_{-R}^{R}\frac{dx_2}{(x_2+p)\sqrt{Z}}\right] \qquad (22)$$

Using the table meanings of integrals in (22) and performing some transformations, we get

$$C = \frac{\varepsilon\varepsilon_0 \pi R^2}{h_3 \Phi_{np}}, \qquad (23)$$

where $\Phi_{np} = 0.5 + 0.25\delta_{h3} + 0.5\sqrt{1+\delta_{h3}}$ is the function of nonparallelism; $\delta_{h3} = {\Delta h_3}/{h_3}$.

In the terms of graduating characteristic of QR with MIG the relation

$$\frac{C_3}{C_0} = \frac{h_{PE}}{\Phi_{np}\varepsilon_{PE}x} = \frac{a}{\Phi_{np}x}, \qquad (24)$$

whence normalized graduating characteristic of PMRT in the conditions of measuring the micromovings

$$\delta_F(x) = \frac{0.5m}{1+\dfrac{a}{\Phi_{np}x}} = \frac{0.5m\Phi_{np}x}{\Phi_{np}x+a} = \frac{0.5m\tilde{x}}{\tilde{x}+a}\ . \qquad (25)$$

The analysis for $\Phi_{np} > 1$ and $\delta_{h3} \in [0,1]$ shows that the display of nonparallelism is equivalent to increasing the gap between mobile and static electrodes in Φ_{npi} times for $\delta_{h3} = \delta_{h3i}$.

In the condition of changing the pressure, the normalized graduating characteristic PMRT under the record of nonparallelism of electrodes has the image

$$\delta_F(x_m) = \frac{\tilde{a}_0 + \tilde{a}_1 x_m}{1+\tilde{a}_2 x_m}, \qquad (26)$$

where $\tilde{a}_0 = \frac{0.5m\Phi_{np}x_0}{\Phi_{np}x_0 + a}$, $\tilde{a}_1 = \frac{-0.5m}{\Phi_{np}x_0 + a}$, $\tilde{a}_2 = \frac{-1}{\Phi_{np}x_0 + a}$.

Hence, the increase of nonparallelism of electrodes (rotation of mobile round the axis X_1) is equivalent to an increase of initial gap x_0.

Breaking the flatness of two surfaces usually occurs by means of moving of one of it at the same time relative to two axes X_1 и X_2. That is why in general case the function of nonparallelism can be presented as following

$$\Phi_{np} = \Phi_{np1} + \Phi_{np2}, \tag{27}$$

Where $\Phi_{np1}(\delta_{h31})$ and $\Phi_{np2}(\delta_{h32})$ are partial functions, which allow for rotation of mobile electrode round the axes X_1 и X_2 accordingly. Independence of these rotations define the type Φ_{np1} and Φ_{np2} according to (24).

Therefore, nonparallelism of electrodes of PRMT can be taken into consideration on a stage of its projecting and does not influence linearity of graduating characteristic (Kolpakov, Pidchenko, Taranchuk, et al., 2009).

6. Special features of piezoresonant mechanotrons built on a resonant membrane

Building the measuring force transducers and adduced to it physical quantities (pressure, motion, etc), which use the effect of tensosensitivity QR, is connected with solving complicated problem of connecting force transducers element with quartz piezoelement. Wide possibilities in solving of these tasks are opened while using non- contact frequency control of QR (quartz resonator) by means of modulating the interelectrode gap (Kolpakov, Akulinichev, 1999).

PRMT is shown in Fig. 8 combines the use of two types of control: tensosensitivity of quartz resonator and its control over the modulation of interelectrode gap. Under the activity of applied pressure in the plane of piezoelement, the mechanical strains σ, and as a result of which the PE changes its properties, that is its resonance frequency changes. This is the demonstration of the first control mechanism.

The second control mechanism consists in the following: the resonant membrane, which is jammed on the contour, is under the effect of spread air pressure P. Under the effect of this pressure there happens the PE bending flexure and, as a result, the extension of interelectrode gap x_g, which in its turn leads to the reduction of interelectrode gap capacity C_g. Under the influence of this pressure there is deflection PM and, as

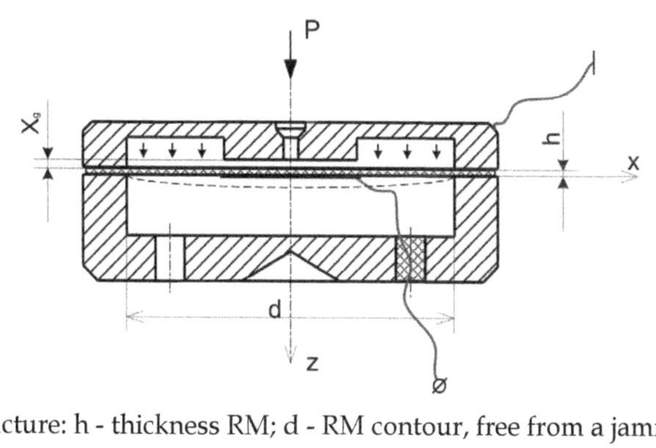

Fig. 8. PRMT structure: h - thickness RM; d - RM contour, free from a jamming

consequence, augmentation of an interelectrode gap that in turn leads to reduction of capacity of an interelectrode gap.

Numerical modeling of tense - strained state RM of round (elliptic) shape, is performed on a basis of the system of engineering analysis ANSYS, which uses the method of finite elements.

Taking into account anisotropic properties of the model the following constants are set: the model of elasticity $E_x = 78$ GPa, $E_y = 85.3$ GPa, $E_z = 92.6$ GPa, coefficient Poisson $\mu = 0.077$ and module of shift $G_{xy} = 42$ GPa.

Taking into account the character of loading (the bending flexure is small in comparison with the plate width) and condition $d/h > 10$ we use the model for thin plate. For such model normal to middle plane until the curve remains normal to this plane after curve as well. That is why the deformation of shift is absent, i.e. $\gamma_{xy} = \gamma_{yz} = 0$.

Taking into account geometry and force symmetry of calculated model, it is enough to look at the sector of membrane, which is limited by main axis, with setting the corresponding limited conditions: absence of moving and rotation in plane; plane of symmetry is xy – $u_z = 0$; plane of symmetry yz – $u_x = 0$; $\gamma_{xy} = \gamma_{yz} = 0$. Such assumptions allow to reduce considerably the dimension of being solved task. Jamming the free edge of plate is modeled by means of fixing the junctions, which are situated in the area of jamming, in the plane xy (see Fig. 9).

The plate is loaded with excess pressure $P = 4 \cdot 10^4$ Pa (300 mmHg), which operating in a plane of a cover.

While solving this task the geometrical nonlinearity - the changes of cylindrical rigidity of the cover in the deformation process is taken into account. This is provided by means of regenerating (recalculation) of matrix of rigidity after each iteration. Solving the system of nonlinearity equations is done with the help of the method of Newton-Rawson with automatic change of iteration step.

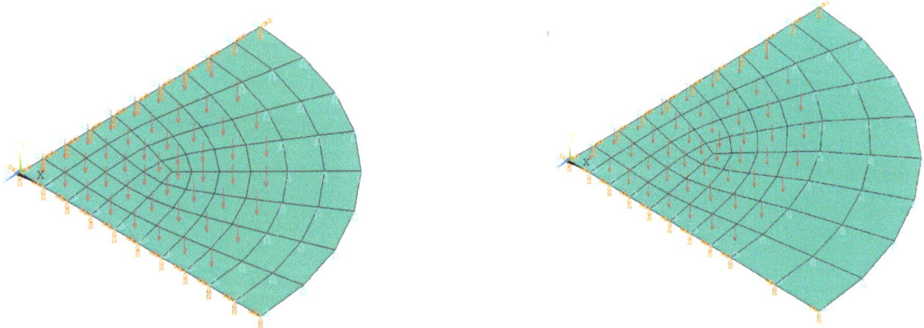

(a) (b)

Fig. 9. Finite - element model of RM with visualization of limit conditions and the enclosed pressure: a - round; b - elliptic

As a result of calculations for RM with round cover of jamming and the parameters $h = 169\ \mu m$ and $d = 11$ mm the values of jamming u_z and normal tension σ_x are received, σ_z (see Fig. 10). The calculation of parameters RM with elliptical cover of jamming was done for $a = 5{,}5$ mm and $b = 4{,}4$ mm (see Fig. 11).

The analyses of received data in the process of numerous modeling shows, that the value of motion RM for round jamming comprises approximately $20\ \mu m$, elliptic - $13\ \mu m$, and the reserve of solidity on mechanical tensions, penetrating into plane of piezoelement, comprises 5.5.

The results of modeling showed high effectiveness of proposed methods on the basis of method of finite elements for performing high - precise calculations of complicated constructive elements of piezoquartz measuring transducers on the stage of their projecting and exploitation, making analytical decision for which is impossible.

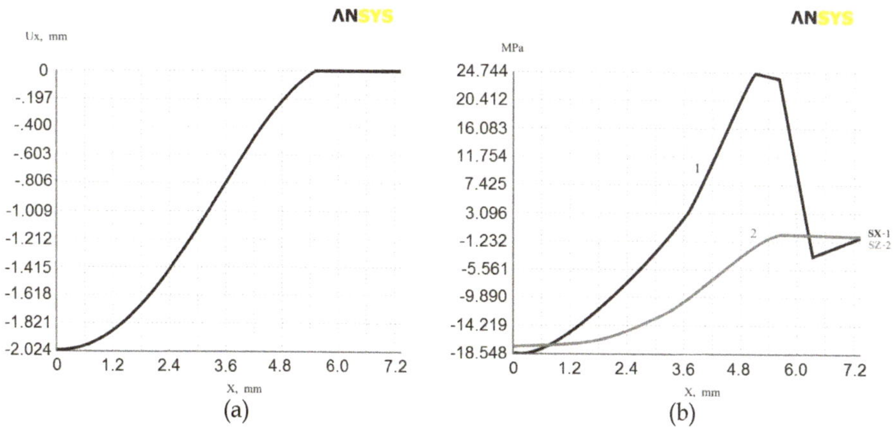

Fig. 10. Calculations of parameters of round RM: (a) bending flexure; (b) tension σ_x (1) and σ_z (2)

7. Analyses of a piezoresonant mechanotron temperature characteristics employing modulated interelectrode gap

In accordance with (12) and (14) temperature sensitivity of QR with MIG is determined by the variations $x_0(T)$, $m(T)$, $\varepsilon_{PE}(T)$, $h_{PE}(T)$. Let us look at factors more detailed, which contribute into the changes of information frequency parameter.

Permittivity of quartz depends very little on orientation of plates, its capacity temperature coefficient (TC) under increasing the temperature to 100°C is little and in general it determines the character of temperature dependence $C_0(T)$ ($TC\varepsilon_{PE} = 5 \cdot 10^{-5}\ /\ {}^{\circ}C \pm 25\%$). Temperature coefficient C_0:

$$TCC_0 = C_{0(20^0)}\left[1 + 5 \cdot 10^{-5}\left(t^{\circ} - 20^{\circ}\right)\right]. \tag{28}$$

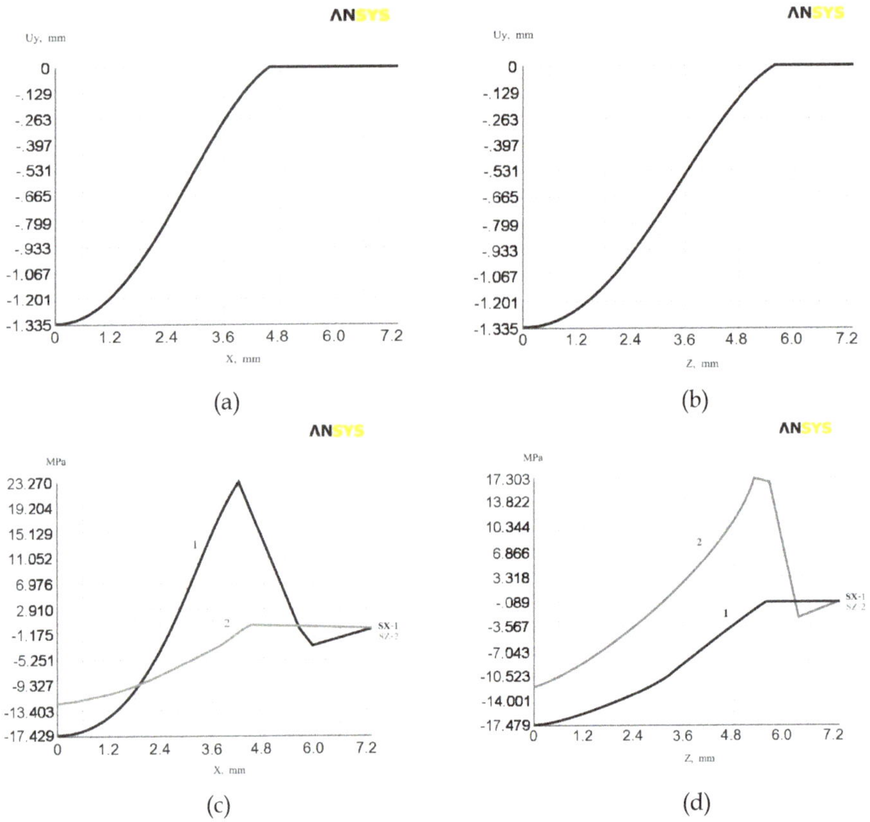

Fig. 11. Calculated parameters of elliptic RM: (a), (b) - bending flexure of membrane on axis a and b of ellipse accordingly; (c), (d) - tension σ_x (1) and σ_z (2) - on axis a and b of the ellipse accordingly

Temperature dependence of dynamic capacity $C_q(T)$ for flat PE AT- cut:

$$TCC_q = C_{q(20^\circ)}\left[1 + 2.5\cdot 10^{-4}\left(t^\circ - 20^\circ\right) + 2\cdot 10^{-7}\left(t^\circ - 20^\circ\right)\right]. \qquad (29)$$

For the wide temperature interval (-60…+100)°C quadratic terms can not be taken into account. Temperature dependence $h_{PE}(T)$ is estimated by the value of temperature coefficient of linear expansion ($TCLE = 10^{-5cm} / ^\circ C$). Dependability $x_0(T)$ is defined by TCLE of metal envelope, glued material, difference of TCLE frame and regulated screw, termal deformations of mobile electrode (membrane). In accordance with (12), taking into account said before,

$$\delta_{1f}(T) = \frac{a_{0(T)} + a_{1(T)}\left(\Delta T\alpha_x + x_0\right)}{1 + a_{2(T)}\left(\Delta T\alpha_x + x_0\right)}, \qquad (30)$$

where $a_{0(T)} = a_0 \dfrac{1+\delta_m+\delta_{x_0}}{\left(1+\delta_{x_0a}\right)}$ $a_{1(T)} = -a_0 \dfrac{1+\delta_m+\delta_{x_0}}{\left(1+\delta_{x_0}\right)\left(1+\delta_{x_0a}\right)}$; $a_{2(T)} = \dfrac{a_{1(T)}}{0.5m\left(1+\delta_m\right)}$.

It is known that detuning of relative frequency of successive resonance leads to a turn of the temperature frequency characteristic (TFC) QR. Dependence of TFC on detuning depends on the change of capacitive relation of resonator in the temperature interval and is calculated

$$\delta_{f_0 mcrp} = 0.5m\alpha_m e_y\left(T-T_0\right), \tag{31}$$

where α_m is the temperature coefficient of capacitive relation (for resonators AT-cut $\alpha_m \approx 3\cdot 10^{-4}$, e_y - is relative detuning as to the frequency of successive resonance of QR with MIG when $x=0$, T- is current value of temperature , T_0 - temperature, relative to which the change TFC is defined. Therefore, the relative change of frequency

$$\delta_{2f}\left(T\right) = \frac{a_{0(T)} + a_{1(T)}x_0}{1+a_{2(T)}x_0}\alpha_m \Delta T. \tag{32}$$

The influence of temperature warp of electrodes and changes x_{0ecv} are taken into account by introducing the correction to the value of initial gap:

$$\delta_{3f}\left(T\right) = \frac{a_{0(T)} + a_{1(T)}\left(x_0+\Delta x\right)}{1+a_{2(T)}\left(x_0+\Delta x\right)}. \tag{33}$$

It is necessary to take into account the changes of frequency, which are connected with temperature instability of oscillator, piezoelement and occurred, because of temperature changes of construction element size, strain load of PE. Therefore

$$\delta_{4f}\left(T\right) = \delta_{fget}\left(T\right) + \delta_{fa}\left(T\right) + K_F\Delta T. \tag{34}$$

Taking into account all said above, we write approximate model of termal- sensitivity of measuring transducers with QR MIG:

$$\delta_f\left(T\right) = \delta_{1f}\left(T\right)+\delta_{2f}\left(T\right)+\delta_{3f}\left(T\right)+\delta_{4f}\left(T\right) = \frac{a_0(T)+a_1(T)(\Delta T\alpha_x + x_0)}{1+a_2(T)(\Delta T\alpha_x + x_0)} + \frac{a_0(T)+a_1(T)x_0}{1+a_2(T)x_0}\alpha_m\Delta T + \\ + \frac{a_{0(T)}+a_{1(T)}\left(x_0+\Delta x_0\right)}{1+a_{2(T)}\left(x_0+\Delta x_0\right)} + \delta_{fget}\left(T\right)+\delta_{fa}\left(T\right)+K_F\Delta T. \tag{35}$$

Having grouped the components, we get

$$\delta_f\left(T\right) = \frac{a_{0(T)}+a_{1(T)}\left(x_0+\alpha_x\Delta T+\Delta x_0\right)}{1+a_{2(T)}\left(x_0+\alpha_x\Delta T+\Delta x_0\right)}\left(1+\alpha_m\Delta T\right)+\alpha_{fget}\left(T\right)+\delta_{fa}\left(T\right)+K_F\Delta T. \tag{36}$$

Dependence of frequency of QR on the temperature is defined as physical properties of crystal element, as well as the size and material of its electrodes, quartz holders, the topology of fastening of a piezoplate. At fast changes of temperature (temperature blow) the main reason of instability of frequency is the presence of temperature gradients in quartz PE and mechanic tension, which appears in the space of piezocrystal, as well as in the place of electrodes placement and fastening of quartz holders under not uniform resonator heating (Kolpakov, Pidchenko, Taranchuk, et al., 2009).

Let's survey geometrical model of node of the quartz holder - piezoresonator. Researched construction (see Fig. 12) usually consists of two parts: quartz piezoplate (1) and quartz holder (2), which is produced of duraluminium DT-16.

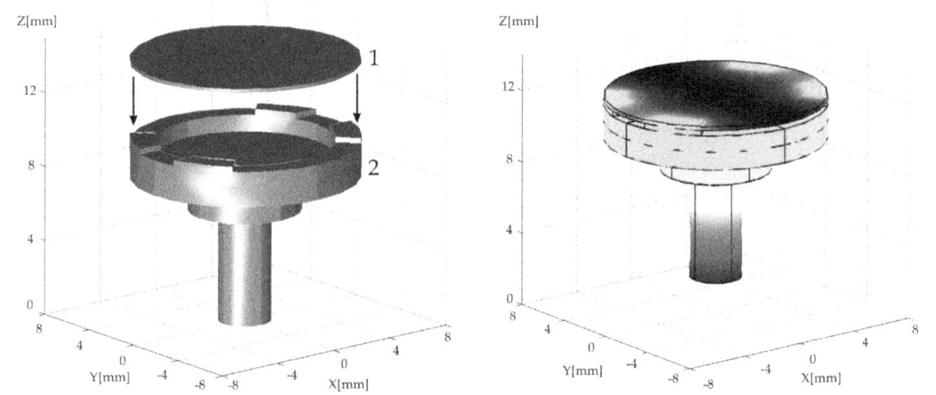

Fig. 12. The construction of a node of quartz holder of piezoresonant sensor with MIG

Solving the equation of term heat conductivity

$$\rho C \frac{\partial T}{\partial t} - \nabla \cdot (k \nabla T) = Q, \tag{37}$$

where ρ is density, C is heat capacity, k is coefficient of heat conductivity, Q is the meaning of heat source, by the method of finite elements in the system of mathematical modeling MATLAB (FEMLAB) under abrupt changes of temperature from 0°C to 50°C (limit conditions of third type) we get the distribution of temperature field of sensor PE in character time moments (see Fig. 13) We use received information for calculating instability of PRMT frequency.

For definition of the temperature induced shifts of frequency piezoelectric resonator we will present the general shift of frequency in a kind

$$\Delta f / f_0 = \left(\frac{\Delta f(T)}{f_0} \right)_{stat} + \left(\frac{\Delta f(T_R(\psi) - T_0)}{f_0} \right)_{dyn}, \tag{38}$$

where $\left(\frac{\Delta f(T)}{f_0} \right)_{stat} = \sum_{i=1}^{3} a_i (T - T_H)^i \Big|_{T=T_{PE_{min}}}$ is static component of instability of PE, which is defined by its static temperature- frequency characters and minimal temperature $T_{PE_{min}}$;

$\left(\frac{\Delta f\left(T_0 - T_R(\psi)\right)}{f_0}\right)_{dyn}$ - is dynamic component of instability, which originates by means of distortion of temperature field of PE.

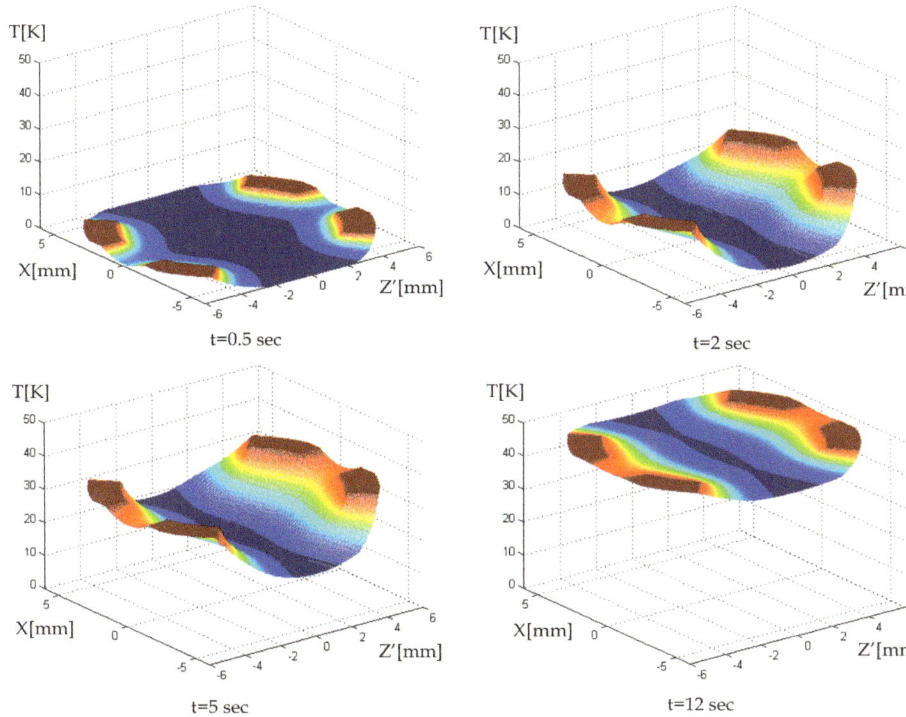

Fig. 13. Distribution of temperature field in character time moments

Static (quasi - static) thermal behavior of PE with enough accuracy is defined by the polynomial of third degree. For defining the dynamic component of instability of frequency we use the methods, which are described in (Taranchuk, Pidchenko, 2002 - 2005). The relative shift of personal frequency of piezoplate fluctuations under the effect of external force $F(\psi)$ applied onto the piezoelement ends of disc shape along its diameter, is calculated as

$$\frac{\Delta f}{f_0} = {}^{f_0}\!\!\diagup_{2R \cdot n} \cdot K_f(\psi) \cdot \Delta F(\psi), \tag{39}$$

where ψ - is the azimuth of force applying $F(\psi)$; f_0 - is the minimal fluctuations frequency; n - is the number of mechanical harmonica; R is the radius of PE; $K_f(\psi)$ is force - frequency coefficient of Rataiskiy.

To find $\Delta F(\psi)$ we take into account the fact, that tensor of elastic deformations of elementary volume of piezoplate is bound with changes of temperature ΔT by a parity $r_{ij} = l_{ij} \cdot \Delta T$, where l_{ij} are the components of tensor of second rank of quartz thermal expansion, and force tension, which originate in this are depicted by generalized law of Gook $t_{ij} = C_{ijkl} \cdot r_{kl} = d_{ij} \cdot \Delta T$, where C_{ijkl} are the components of the tensor of the forth rank of modules of frequency (Cady, 1946).

Realizing the transition $\mathbf{d}' = \mathbf{\Gamma}\ \mathbf{d}\mathbf{\Gamma}^{-1}$ into the plane of piezoplates of the most prevailing one-rotary AT (yxl/β°) and two- rotary SC, FC $(yxbl/\gamma^{\circ}/\beta^{\circ})$ - the cuts of piezoelements and using with a glance of geometric peculiarities of PE the cylindrical coordinate system, we get:

$$\Delta F_r(\psi) = d_{rr}(\psi) \cdot S_e \cdot \Delta T_r(\psi), \tag{40}$$

where $\Delta F_r(\psi)$ are changes of allocated forces, which act in the volume of piezoplate along the radius r; $S_e = R \cdot h \cdot \Delta\psi_e$ is the area of an elementary platform of a face surface of PE, h is the thickness of PE.

Taking into account (40) in parity (39) and carrying out integration on Ψ in the limits from 0 to π and on r in the limits form 0 to R we come to integral from of writing

$$\frac{\Delta f}{f_0} = \frac{f_0}{2n} \cdot h \int_0^{\pi} K_f(\psi) \cdot d_{rr}(\psi) \left[T_R(\psi) - T_0 \right] d\psi, \tag{41}$$

where $T_0 = \frac{1}{h}\int_0^h T_0(h)dh$, $T_R(\psi) = \frac{1}{h}\int_0^h T_R(h,\psi)dh$ - average in the volume temperature in the centre and on the edge of PE.

Received on the basis (44), dynamic components of instability of frequency for two types of quartz - holders are demonstrated on Fig. 14, (a), where curve 1- is the dynamic instability of frequency of quartz holder with four points of fastening of PE (see Fig. 12), curve 2 - for quartz holder with thermal contact on the contour of PE. It can be seed, that in second case there are big deformations of temperature area of PE, which cause therefore much bigger instability of frequency (approximately order).

On Fig. 14, (b) there are settlement (in accordance with (38)) and experimental curves of general shift of frequency of sensor under the thermal effect (from 50°C to 0°C). Good coincidence of curve data proves the adequacy of used mathematical model, which can be used for further research.

The construction of quartz holder influences greatly the general shifts of frequency of PE PRMT. Under this there appear considerable temperature gradients and mechanical tension in piezoplate, which lead to substantial increase of dynamic component of instability. Its maximum values can exceed the frequency shifts, which are caused by quazi-static temperature frequency characteristics (TFC). That is why the construction of PRMT must be optimized with a glance of thermal processes, which take place in quartz holder and piezoelement (Taranchuk, Pidchenko, 2002 - 2005).

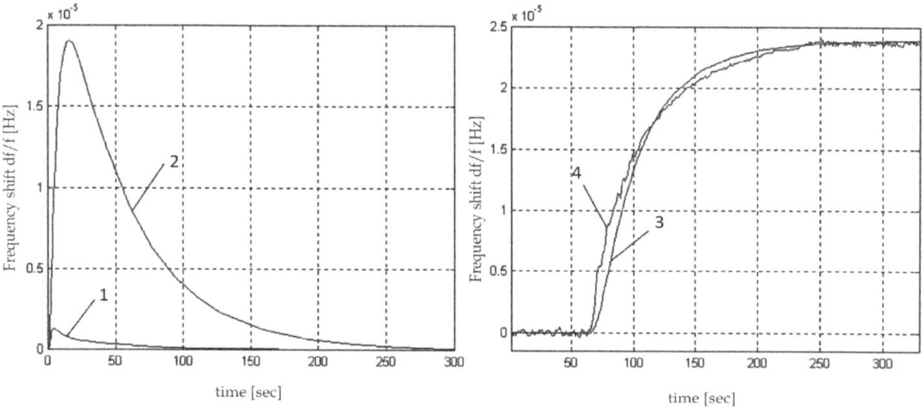

Fig. 14. Dynamic component (a) and frequency shift of piezoelectric resonator (b): curve 1 - is the dynamic instability of frequency of quartz holder with four points of fastening of PE; curve 2 - for quartz holder with thermal contact on the contour of PE; 3 - settlement curve; 4 - experimental curve

8. Implementation of piezoresonant mechanotrons employing modulated interelectrode gap and development research

On the basis of research, the results of which are presented above, the authors worked out the number of sensors of excessive pressure for medical appliances.

On Fig. 15 the PMRT of pressure is presented, in which for modulation of inter electrode gap the metal membrane is used. The case 1 made of polystyrene is connected with metal basis 2. Between top groove of base 2 and union node 11 there is seal ring 12. In a ground part of the basis 2 the node quartz holder 6 which is rigidly bridged by means of a rivet 4 to an elastic element 5 is established. On a working surface quartz holder 6 the flat disk piezoelement 3 AT - cut with the round electrode bridged electrically with a potential conclusion of the sensor 7 is established.

Elastic element 5 through the sealing ring 8, and also the membrane 10 are jammed on a contour by means of a clamping node 11. All nodes of PMRT have hard fixing and do not need applying welding operations for junctions. The size of initial gap between quartz piezoelement 3 and membrane 10 is provided with changes of the flexure of elastic element 5 with the help of screw 9.

In the condition of absence, towards atmosphere, air pressure in the cuff, which is connected by the tube to the nipple part of frame 1, the deformation of membrane 10 is absent, as inner volume of basis 2 is not pressurized and the pressure onto the membrane from both sides is identical. In such condition quartz resonator, which is formed by piezoelement 3, metal membrane 10 and switched into the scheme auto generator PRMT is stimulated on the frequency which corresponds zero excessive pressure. When pressure available exceeds the atmosphere one, the membrane 10 is bended, the gap capacity between unused surface of piezoelement 3 and the surface of central part of membrane goes down: it leads to decreasing the resonance frequency of QR with variable gap (Kolpakov, Pidchenko, Hilchenko, 1998).

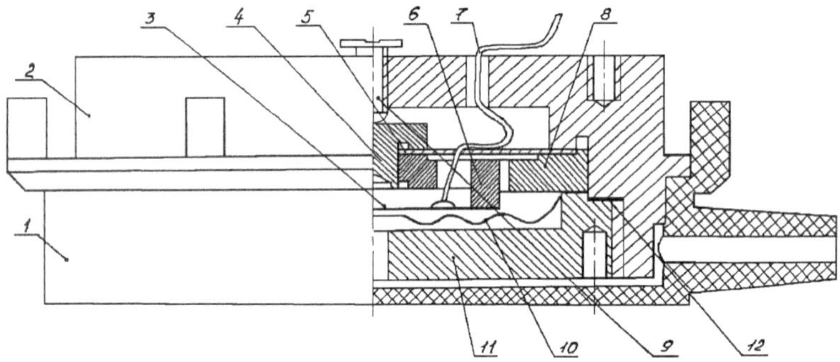

Fig. 15. Construction of PMRT with metal membrane

The basic advantage of this PRMT is complete absence of hysteresis effects, which are inherent in sensors with direct influence onto piezoelement, which substantially increases the reproducibility of graduating characteristics and, as a result, accuracy of measurements.

Typical value of permitted ability of given PRMT with metal membrane comprises 0.05…0.08 mmHg, and main error does not exceed 0.1…0.15%. Highly effective appliance of PMRT with capacity pneumatic control is its use in measuring transducer of pulse air pressure of sphygmographic research (Taranchuk, Pidchenko, et al., 2010).

Pneumatic scheme of PMRT is show Fig. 16, (a) has two channels: K1, which has pneumatic the filter of the bottom frequencies, which consists of pneumatic resistance of filter R_f, pneumatic throttle R_{tr} and under membrane capacity V_{K1}; K2 - is direct channel of action onto membrane. Resulting influence of two channels onto the movement of membrane is equivalent to the operation on given pressure $P = P_{\sim} + P_0$ to the pneumatic filter of the top frequencies. Here P_0 - in constant (slowly changeable) and $P_{\sim}$ - variable (informative) component of pressure.

The frequency of cut PRMT is defined by the formula

$$f_{cut} = \frac{1}{2\pi R_{\Sigma} C_{FC}}, \tag{42}$$

Where $C_{FC} = \frac{V_{K1}}{\Re T}$ is pneumatic capacity of the filtrational chamber V_{K1}; $R_{\Sigma} = R_f + R_{tr}$ is total pneumatic resistance; $V_{K1} = {\pi D_{FC}^2 h_{FC}}/{4}$, D_{FC} and h_{FC} - the volume, height and diameter of filtrational chamber accordingly; $\Re = 287 \frac{Дж}{кг \cdot K}$ is gas constant; T - is the gas temperature (air temperature) .

Resistance of cloth filter is defined as

$$R_F = \frac{\Delta P_F}{\nu \cdot S}, \tag{43}$$

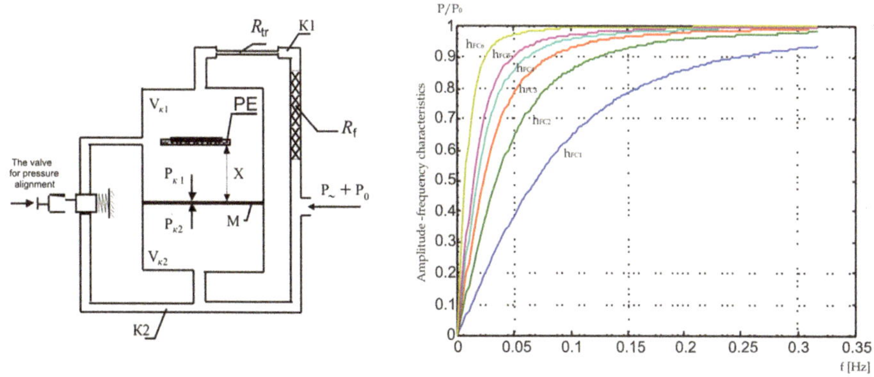

(a) (b)

Fig. 16. (a) Pneumatic scheme of PRMT; (b) The amplitude- frequency characteristics of sensor of pressure for values $l_{tr} = 1$ mm

where ΔP_F is pressure drop on cloth filter; ν - speed of air flow through filter; $S = \pi d_F^2 / 4$ - the area of cross- section of cloth filter opening, d_F is diameter of opening of cloth filter.For the material for cloth filter, type felt: $\Delta P_F = 49$ Pa, $\nu = 0.02$ m/min and for $d_F = 1.5$ mm - $R_F = 8.32 \cdot 10^{10}$ Pa·sec/ m^3.

Resistance of throttle equals

$$R_{tr} = \frac{128 \eta l_{tr}}{\pi d_{tr}^4 \rho}, \qquad (44)$$

where l_{tr}, d_{tr} are the length and diameter of throttle; ρ - the density of air under normal conditions, $\rho = 1.205\,\mathrm{kg/m^3}$; $\eta = A + B \cdot T$ is dynamic coefficient of gases viscosity; A, B are constant coefficients (for air $A = 37.4 \times 10^{-7}$ Pa·sec, $B = 0.506 \times 10^{-7}$ Pa·sec). In normal conditions ($T = 293$K) the dynamic coefficient of air viscosity $\eta = 0.186 \times 10^{-4}$ Pa·sec.

The cut frequency of PRMT s defined by the spectrum of sphygmograph signals - (0.036…60) Hz and comprises approximately $f_{cut} \approx 0.03$ Hz. Having based on chosen cut frequency, taking into account real size $D_{FC} = 40$ mm and received cloth filter $R_F = 8.32 \cdot 10^{10}$ Pa·sec/ m^3, we let's define in conformity with (42) - (44) the constructional parameters of throttle (diameter d_{tr} and the length l_{tr}) and the height of filter chamber h_{FC} (see Table 1, Table 2 and Fig. 16). Analyses of received data shows that substantial influence onto the cut frequency f_{cut} makes the height of filter chamber h_{FC} (see Table 1, Table 2). For $d_{tr} \le 0.1$ mm the resistance of throttle R_{tr} considerably depends on l_{tr} and should be taken into account while calculating PRMT (Fig. 16, (b)), for $d_{tr} = (0.2...1)$ mm and $l_{tr} = (1...10)$ mm the throttle resistance l_{tr} is much smaller that the resistance of cloth filter $R_f >> R_{tr}$ and can not be taken into account while calculating total resistance R_Σ. In such case parameters PRMT are defined as usual by the height of filter camera h_{FC} (see Table 2).

h_{FC}, mm	f_{cut}, Hz			
	l_{tr}, mm under $d_{tr} = 0.1$ mm			
	1	2,5	5	10
1	0,11	0.107	0.093	0.073
2	0.0595	0.0539	0.0465	0.0365
3	0.0397	0.0359	0.031	0.0243
4	0.0298	0.0269	0.0232	0.0182
5	0.0238	0.0215	0.0186	0.0146
10	0.0119	0.0108	0.0093	0.0073

Table 1. Choosing the geometry sizes of elements of PRMT

h_{FC}, mm	1	2	3	4	5	10
f_{cut}, Hz	0.128	0.064	0.0427	0.032	0.0256	0.0127

Table 2. Choosing the height of filter chamber while $d_{tr} = (0.2...1)$ mm and $l_{tr} = (1...10)$ mm

Thus, proposed construction PRMT with successive connection R_f and R_{tr} allows to vary the parameters of throttle in accordance with technological possibilities of production and to realize it without applying special capillary technology (Kolpakov, Dobrova, et al., 1999; Pidchenko, Taranchuk, et al., 2002).

The adantage of given construction of PMRT is in following: the capacity of interelectrode gap is defined only by the amplitude of pulse wave of blood pressure and technological tolerance of membrane mounting, and additional errors of nonparallelism of membrane and PE are excepted by means of removal of: quazi - static component of air pressure. This allows to increase the allowing ability of measuring transformer for sphygmographic research on basis of PMRT while measuring dynamic air pressure in more than three-five times as compared with known devices.

On Fig. 17 it is presented the third type of frequency PMRT of pressure with the use of resonant membrane (RM) , which is developed in paragraph 6 of given chapter.

PMRT with PM works as follows. Under the absence of excessive as to atmosphere pressure of air in the chamber, which is limited by piezoelement 5 and inside surface of cover 1, the deformation of piezoelemet 5, which plays the role of resonating membrane, does not happen. Quartz resonator, connected in the scheme of oscillator PRMT is stimulated on the frequency, which corresponds zero excessive pressure. From the exit of oscillator the information signal is taken, the frequency of which f_0 corresponds the beginning of modulation characteristic of PMRT. Under the pressure in the pressure chamber, which exceeds the atmosphere pressure, there happens the small bending flexure of piezoelement 5, as a result the size of a gap between the free surface of piezoelement and the surface of cylindrical bump would increase, which leads to increasing the frequency $f(P)$. In given PRMT two mechanisms of controlled of frequency of the quartz resonator take place. Capacitor controlled, at the expense of change of size of a gap between a free surface of a piezoelement and a surface of a cylindrical ledge 6, and controlled on the basis of effect tensosensitivity:

$$\Delta_{\Sigma}(p) = \Delta_{C}(p) + \Delta_{\sigma}(p); \tag{45}$$

$$K^{\int} = \frac{\partial f}{f_0 \cdot \partial P} = \frac{N}{2}\int_0^{\pi} K_f(\psi)d\psi, \tag{46}$$

where N is frequency constant; K_f is the coefficient of Rataiskiy; ψ is azimuth of load; f_0 is personal resonance frequency of piezoelement ($f_0 = 10\text{MHz}$); P is applied spread pressure.

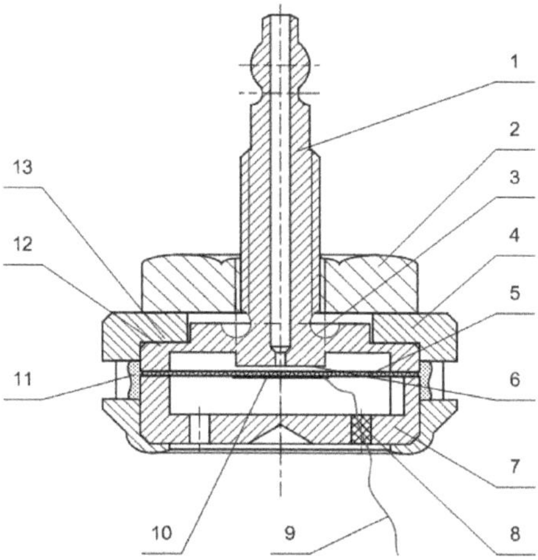

Fig. 17. Construction of pressure sensor with PM:
1- cover; 2 - regulating nut; 3 - annular groove; 4 –ring; 5 - piezoelement; 6 - cylindrical ledge; 7 - basis; 8 - compactor; 9 - potential lead; 10 - mobile electrode; 11- hermetic; 12 - samples; 13 - basis ledge; 14 - groove; 15 - two pairs of diameter - opposite basis ledge

Experimantal researches were held for PMRT with PM with PE of diameter 14 mm and electrode diameter 4 mm under $P \in [0,300]$ mmHg. Static modulation charactetistic of PMRT with PM (see Fig. 18) includes demonstration of both mechanisms of changes the

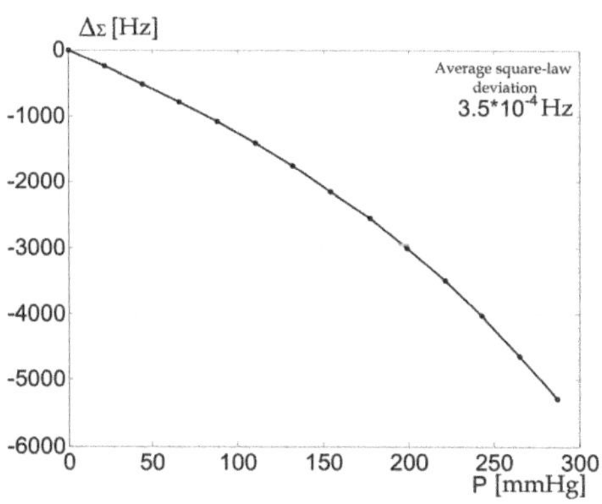

Fig. 18. Summarized modulation characteristic for PMRT with PM

resonance frequency. For estimating the impact of the mechanism of tensosensitivity a separate experiment was held, in which PM of the same size contained two electrodes. Thanks to this the capacity mechanism of frequency was excluded. The analysis of tensosensitivity of mechanisms (see Fig. 19) shows that impact of tensosensitivity of component Δ_σ comprises (2.5…3.5)%. Therefore, the use of model, which takes into account only capacity mechanism of control, for PMRT with PM brings the error, which does not exceed 3.5 %.

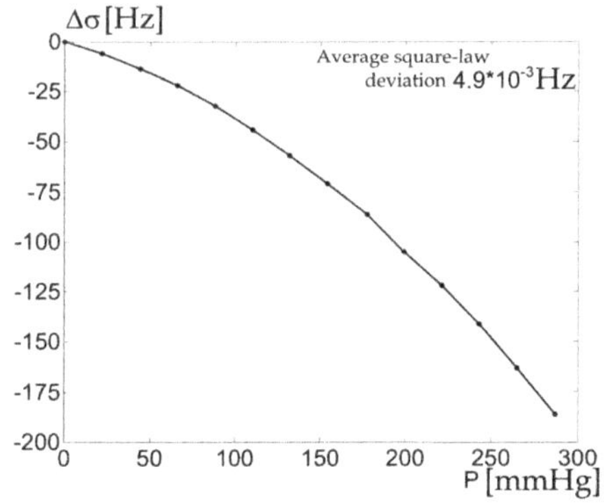

Fig. 19. Tensosensitivity modulation characteristic for PMRT with PM

The temperature error of proposed PMRT is substantially decreased thanks to two peculiarities of its construction.

Firstly, surfaces of a piezoelement and the metal surfaces jamming it, have small roughnesses, therefore the forces of a friction arising because of difference of temperature of Rataiskiy equals zero (see Fig. 20).

This means that in ideal case this component of temperature error equals zero, and practically it is very little.

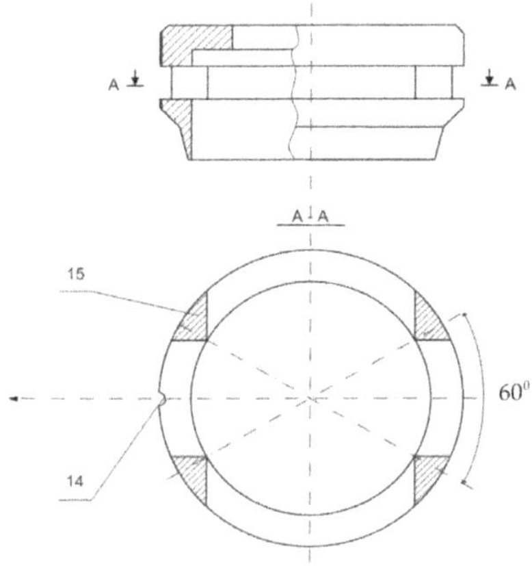

Fig. 20. The construct of quartz holder

The increase of deviation of frequency MC in one and half times can be achieved by increasing the diameter of PE. The results of experimental researches of PMRT with PM ∅18 mm is demonstrated in Fig. 21. As for PMRT with disc flat PE of AT-cut with the diameter 18 mm, the diameter of electrode is 8 mm and thickness $h_{PE} = 169\,\mu m$ which corresponds nominal frequency $f_0 = 10$ MHz, while applying the pressure from 0 to 300 mmHg, the maximal value of banding flexure of piezoelement in order less than its thickness and comprises 16.23 μm. Elements and construction of PMRT with PM are demonstrated in Fig. 22.

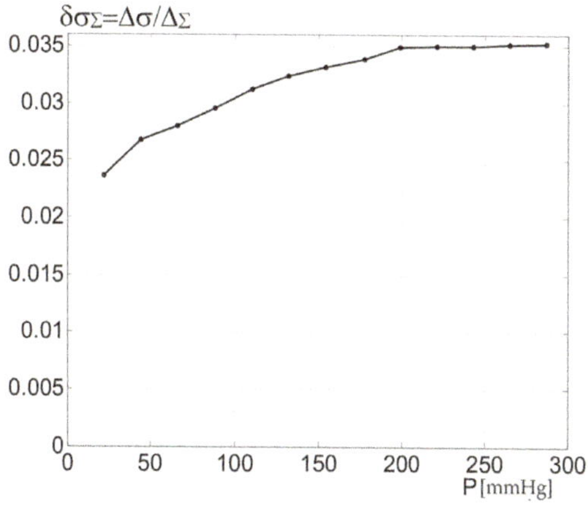

Fig. 21. The relative characteristic of sensitivity

Fig. 22. Elements and construction of PRMT with PM

Typical technical characteristics of sensor of pressure on basis of PMRT with PM:

1. Resolution in the range of applied pressure $P \in [0,300]$:
- for small pressure, less that 10 mmHg.......................................0.008 - 0.01 mmHg;
- for maximum pressure not worse...0.025 mmHg.
2. Full change-over of frequency, not less than ....9.5 kHz.
3. The value of hysteresis after maximum load, not more than ....2 Hz;
4. Basic error of measurement the surplus pressure, not more than.............................0,15%.

Coefficients of approximation of characteristic of transformer of PMRT, calculated in accordance with (13) form: $a_0 = 1938.87\text{Hz}$, $a_1 = 63.95\text{Hz}/\text{mmHg}$ $a_1 = 0.002308\ 1/\text{mmHg}$

Advantages of this type of PMRT with resonant membrane are the exception of effects, which are connected with nonparallelism of membrane and piezoelement, considerable decrease of temperature errors, connected with construct of quartz - holder, simplification of construct of sensor (Pidchenko, Taranchuk, et al., 2003).

9. Application of piezoresonant mechanotrons of surplus air pressure in medical sphygmographic systems to register pulse variation of cardio-vascular human system

On the basis piezoresonat mechanotrons of surplus air pressure the authors developed medical automated multichannel sphygmographic system "BIOTON". This system allows to do computer diagnostics of parameters of human hemodynamics, the registration of local and volume sphygmogrammes for solving the tasks of polycardiography and poly sphygmography. System "BIOTON" differs from the existing POLISPECTR-PWV (Russia), Arteriograph "TensioClinic" (Hungary), Complior (France) by its improving informativeness and quality of reproduction of pulse fluctuations.

Sphygmographic system (see Fig. 23) consists of sensors of pulse oscillation on a basis of PMRT (Sensor 1-n) and the block of microcontroller (MCU). After primary information processing it is transferred into computer on standard interface USB 1.1/2.0.

Each sensor contains measuring transducer of pressure on a base of PMRT, connected into oscillatory system of oscillator. Change of interelectrode gap ПРМТ 1...n under the influence of pressure in a receiver funnel pulse oscillations leads to frequency change on an Oscillator 1...n exit. Further the signal arrives on Frequency multipliers 1...n for increase in information deviation of frequency. On each channel it is realized the method of linearization of graduating characteristics of sensors, for which Synthesizer of direct synthesis (DDS AD 9959) is used, which allows to give the optimal frequency of heterodyning in each channel with the accuracy of less than Hz. From the exit of Mixer 1...n signal of differential frequency comes onto Microcontroller ADUC841, which is used in the regime of period measuring.

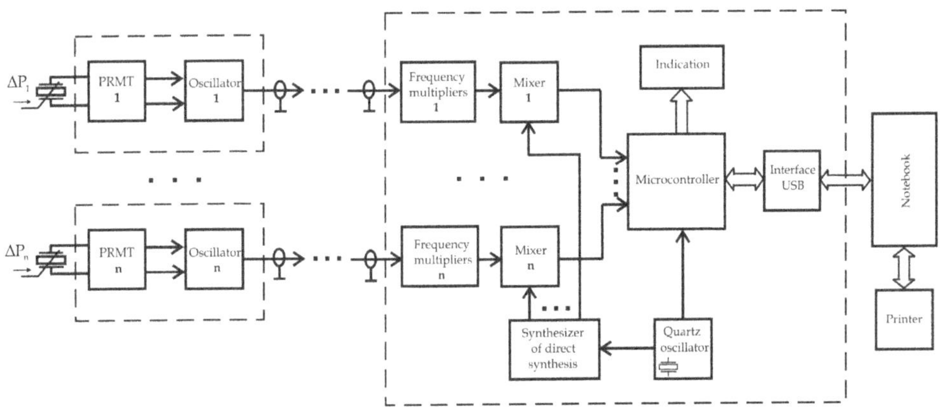

Fig. 23. Structure of sphygmographic system BIOTON

For an exception of additional errors the uniform quartz generator of clock frequency for DDS and MCU is used. Received data in real time are given to standard IBM compatible personal computer (Notebook), which is the part of the system. Further processing is done with the help of worked out special software, which functions under control of OS Windows 2000/XP/Vista/7. The developed software allows to carry out the semi-automatic analysis and data processing, recording of results and creation of databases for research of hemodynamics of cardiovascular system of patients. Possibility contour-time (with allocation of characteristic points) and the spectral analysis sphygmographic curves (see Fig. 24), definitions of an index of augmentation AIx and speeds of distribution pulse waves (PWV) (see Fig. 25), and also allocation of the most typical sites sphygmograms, comparisons two or several pulse oscillation is provided. For elimination of influence of artifacts procedures of a digital filtration and smoothing are used (Taranchuk, Pidchenko, 2008 - 2010).

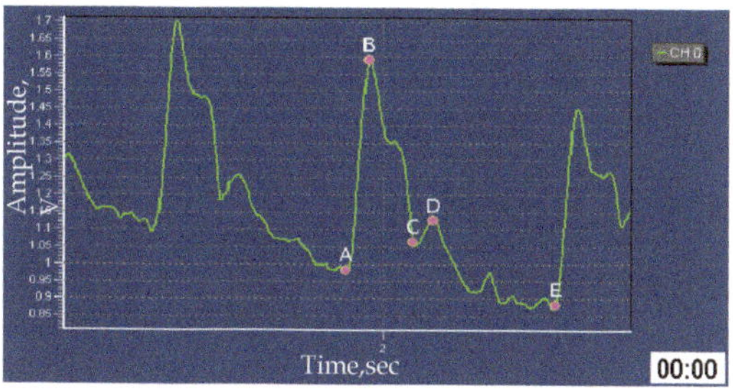

Fig. 24. Contour - time analysis of sphygmogram

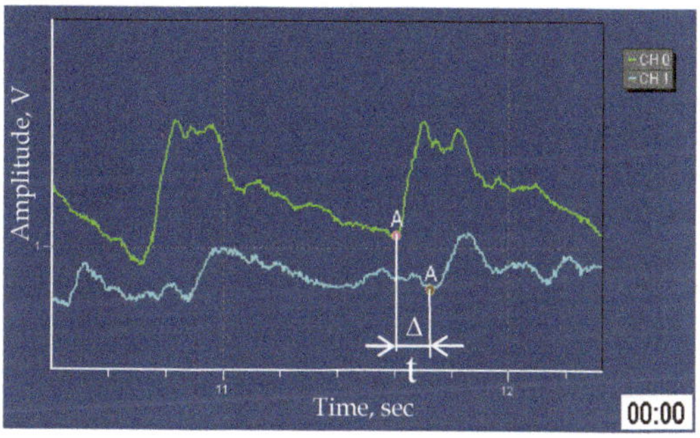

Fig. 25. Defining PWV by signals on carotid artery and femoral artery

System BIOTON is highly effective while definition variability of heart rhythm (VHR) on basis of measured sphygmograph curves of carotid artery during certain time (not less than five minutes). Developed software allows to do statistic, correlation, spectral analysis and form the estimation of analysis results of VHR while doing functional tests (see Fig. 26).

System "BIOTON" is designed for applying in following medicine areas:

1. Cardiology (non- invasive monitoring of hemodynamics, scientific research).
2. Family medicine (primary examination, early diagnosis of atherosclerosis).
3. Nephrology (beginning and progressing of atherosclerosis because of problems with kidneys).
4. Diabetes study (faster process of aging of vessels because of increasing sugar in blood).

5. Obstetrics and gynecology (preeclampsia - endothelial illness, menopause - main factor of risk of atherosclerosis);
6. Children cardiology (often atherosclerosis appears in early age).

Constructive realization of the block of the microcontroller and sensor of surplus pressure on the basis of PRMT are in show Fig. 27.

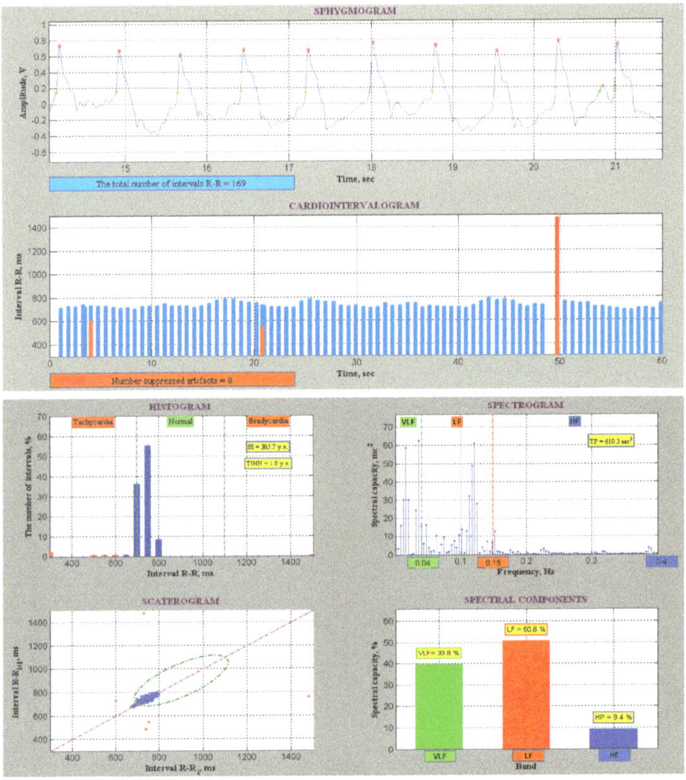

Fig. 26. Results of measurements variability of heart rhythm

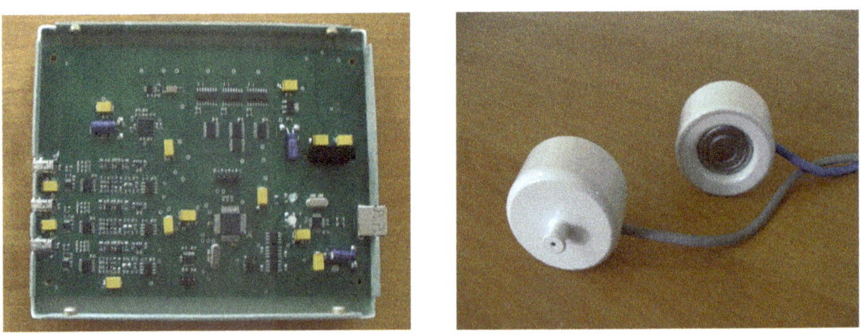

Fig. 27. Constructive realization of the block of the microcontroller and sensor of surplus pressure on the basis of PRMT

10. Conclusion

The result of scientific research presented in present chapter is the creation of new element, named piezoresonance mechanotron, which can be used as basic for constructing wide range of highly informative measuring and diagnostic devices and systems. In combination with linearizer of graduating characteristics, which realize approximation approach, presented constructions of PMRT provide precision measurement of dynamic as well as static pressure on the level of the best world models. They can be used for modernization of existing as well as in the designing the perspective biomedical systems. Wide functional possibilities of PMRT allow, by means of slight changes of its construction, to measure wide range of parameters, such as travel, pressure, acceleration, mass, temperature and others, which lead to micromoving of physical magnitudes. The construct of PMRT permits its realization in microminiature variant on basis of micromembrane element, formed by means of MEMS technology.

11. References

Cady W. (1946). *Piezoelectricity,* New York and London: McGraw-Hill.

Kolpakov F., Pidchenko S., Hilchenko G. (1997). "Pressure Sensor", Patent Russia № 2098783, December 10, 1997.

Kolpakov F., Pidchenko S., Hilchenko G. (1998). "Method for measuring blood pressure", Patent Russia № 2106796, March 20, 1998.

Kolpakov F., Akulinichev A. (1999). Piezoquartz pressure transmitter with an elliptical resonating membrane, *Aerospace technics and technology,* Vol. 12, National Aerospace University (KhAI), ISSN 1727-7337, Kharkov, Ukraine, pp. 103-105.

Kolpakov F., Dobrova V., et al. (1999). Estimation of time diagnostic parameters of blood pressure variation in frequency domain, *Proceeding of the 1999 Finish Signal Processing Symposium,* Oulu, Finland, pp. 189-193.

Kolpakov F., Pidchenko S., Hilchenko G. (1999). "Method of linearization calibration characteristic piezoresonance transmitter with variable interelectrode gap", Patent Russia № 2127496, March 10, 1999.

Kolpakov F., Slavinsky S. (2004). The sphygmography measuring converter "pressure-frequency" dynamic characteristics examination, *Measurement Science Review,* Vol. 4, Sect. 2, pp. 52-58.

Kolpakov F., Pidchenko S., Taranchuk A., et al. (2009). Piezoresonance mehanotron in measurements of parameters warmly - vascular system of the human, *Radioelectronic and computer system,* Vol. 2, National Aerospace University (KhAI), ISSN 1814-4225, Kharkov, Ukraine, pp. 60-70.

Kolpakov F., Pidchenko S. (2011). *Theory and fundamentals implementation of invariant piezoresonance systems,* National Aerospace University (KhAI), ISBN 978-966-662-222-1, Kharkov, Ukraine.

Malov V. (1989). *Piezoresonance sensors,* Energoatomizdat, ISBN 5-283-01507-6, Moscow, Russia.

Pidchenko S., Kolpakov F., Akulinichev A. (2000). Analysis of the characteristics of multi-frequency controlled crystal oscillator, *Measuring and Computing Devices in Technological Processes,* Vol. 3, Khmelnitsky national university, ISSN 2219-9365, Khmelnitsky, Ukraine, pp. 70-75.

Pidchenko S., Taranchuk A., et al. (2002). "Pressure Sensor", Patent Ukraine № 44108 A, January 15, 2002.

Pidchenko S., Taranchuk A., et al. (2003). "Pressure Sensor", Patent Ukraine № 59936 A, September 15, 2003.

Pidchenko S., Taranchuk A. (2004). Identification of the thermal state of the crystal at the stage of oscillation, *Radioelectronic and computer system,* Vol. 3, National Aerospace University (KhAI), ISSN 1814-4225, Kharkov, Ukraine, pp. 36-42.

Pidchenko S., Taranchuk A., Opolska A. (2010). Utilization Features of the Mechanotron for Information Measurement Systems, *Modern Problems of Radio Engineering, Telecommunication and Computer Science: Proceedings of the Xth International Conference TCSET'2010,* ISBN 978-966-553-875-2, February 23-27, Lviv-Slavske, Ukraine, P. 358.

Taranchuk A., Pidchenko S., et al. (2002). The calculation of the temperature field piezoelectric pressure sensor with a modulated interelectrode gap, *Metrology and Computer Engineering: Proceedings of the III International Conference (Metrology – 2002),* Sect. 2, Kharkov, Ukraine, pp. 139-141.

Taranchuk A., Pidchenko S. (2005). Modelling of thermal processes in the piezoresonance sensors with modulated interelectrode gap, *Khmelnitsky State University's bulletin,* Vol. 1, Khmelnitsky national university, ISBN 978-966-330-114-3, Khmelnitsky, Ukraine, pp. 218-222.

Taranchuk A., Pidchenko S., et al. (2008). Sphygmographic channel to study hemodynamic parameters of the cardiovascular system of man, *Khmelnitsky State University's bulletin,* Vol. 3, Khmelnitsky national university, ISBN 978-966-330-114-3, Khmelnitsky, Ukraine, pp. 47-52.

Taranchuk A., Pidchenko S., Opolska A. (2010). Sphygmographic system of high accuracy registration of pulse fluctuations, *Medical and Biological Informatics and Cybernetics: Proceedings of the First Ukrainian Congress,* June 23-26, Kiev, Ukraine, P. 267.

Taranchuk A., Pidchenko S., et al. (2010). "The transmitter with pressure-frequency sensor for sphygmographic research", Patent Ukraine № 53900, October 25, 2010.

13

Approximation and Correction of Measuring Transducer's Characteristics

Janusz Janiczek
Wroclaw University of Technology, Poland

1. Introduction

A very frequent case is that the characteristics of measuring transducers are non-linear and, apart from that, the output signal value may depend on several input values or unwanted influencing factors such as temperature, humidity etc. The need to adjust the form of information about the measured value to human perception possibilities, or to simplify the further processing of measurement information generally requires such adjusting of the slotted line element characteristics so that the resultant characteristics of the slotted line is linear, with assumed error, in the adopted coordinate system. (Bucci et al., 2000; Grzybowski @ Wagner, 1971; Iglesias @ Iglesias, 1988; Patranabis et al., 1988; Patranabis @ Ghosh, 1989)

When the measuring transducer's characteristics is very non-linear, for example in the case of certain transducers for light intensity measurement, a problem with selecting a relevant A/D transducer may occur. All methods of correcting measuring transducer's characteristics cannot eliminate the error related to insufficient A/D transducer resolution. That issue is illustrated by figure 1.

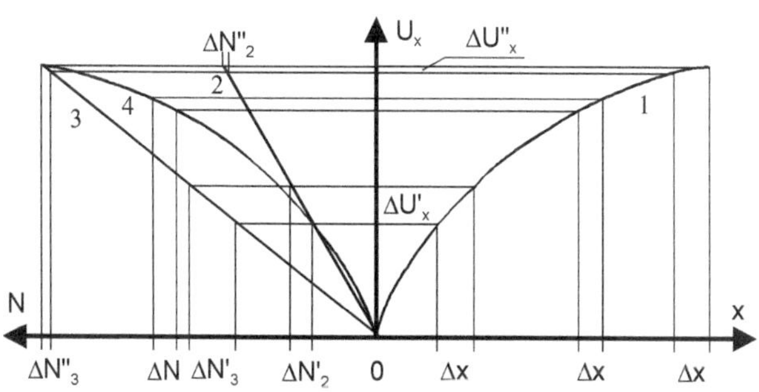

1. measuring transducer's characteristics
2. A/C transducer characteristics of small resolution
3. A/C transducer characteristics of high resolution
4. non-linear A/C transducer characteristics

Fig. 1. Impact of A/D transducer characteristics on resolution error.

At large non-linearity of the measuring transducer's characteristics (curve 1 on figure 3.5) and small inclination of A/D converter characteristics (line 2), the A/D converter resolution is sufficient to process the voltage signal $\Delta U'_x$ to an appropriate digital code $\Delta N'_2$ in the area in which the inclination of measuring transducer's characteristics is high. On the other hand, in the area of small inclination of the measuring transducer's characteristics, the changes in output voltage $\Delta U''_x$ of the transducer are too small to be properly processed by the A/D converter to digital code $\Delta N''_2$. That problem may be partially bypassed using an A/D converter of high resolution (line 3). However, a better solution would be to adjust the processing characteristics of an A/D converter to the measuring transducer's characteristics (curve 4). The A/D converter characteristics should be described by a function reverse to the function describing the measuring transducer's characteristics and then the following linear relation would be obtained $\Delta N = c\ \Delta x$.

That chapter discusses issues related to the approximation of measuring transducer's characteristics and their correction in an analog to digital (A/D) converters The thesis will consider only static characteristics of measuring transducers.

2. Transducers with successive compensation

The block diagram of a transducer with weight compensation is presented in figure 2.1.

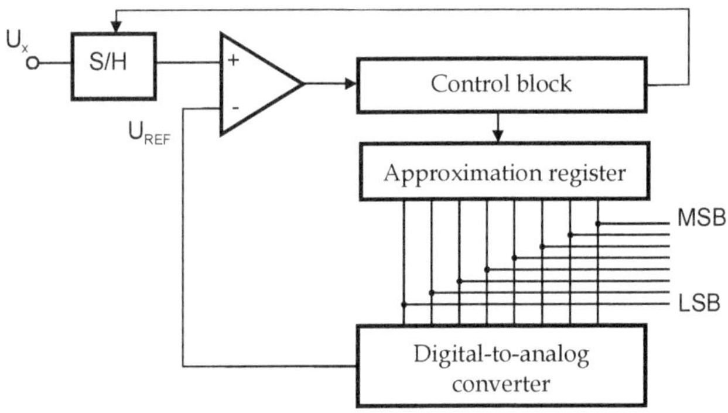

Fig. 2.1. General scheme of transducers with successive compensation.

It consists of a sample-and-hold circuit, a comparator, a digital-to-analog converter, an approximating register and a control circuit. The transducer's operating principle can be found in readily available literature.

In transducers with weight compensation, the only available element, through which the shape of a transducer processing characteristics may be influenced is the reference voltage U_{REF}. For that reason, the change of processing characteristics can be achieved by generating a non-linear reference voltage in relation to the digital output signal N:

$$U_{REF} = a \cdot g(N) \tag{1}$$

where: $g(\cdot)$ - measuring transducer's characteristics approximating function,
a - proportionality coefficient.

The circuit of adjusting transducer processing characteristics with weight compensation by means of reference voltage generation method has been presented in figure 2.2.

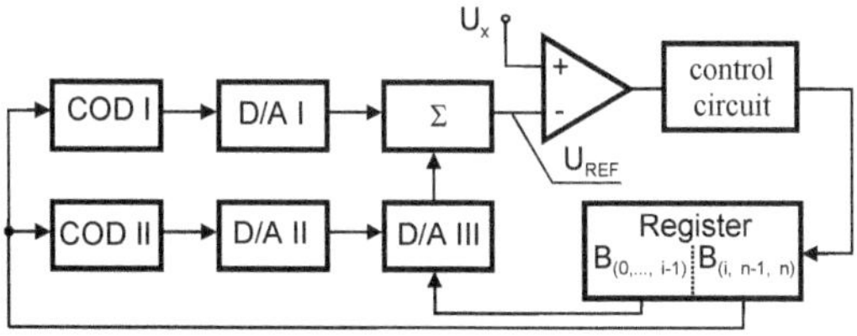

Fig. 2.2. Circuit of adjusting transducer processing characteristics.

Converter circuit consists of a comparator, a control circuit, a digital register and a reference voltage generator containing two coding circuits CODI and CODII, three digital-to-analog converters D/AI, D/AII and D/AIII and an Σ adder. D/A III converter is required to have an input for external reference voltage U^{0III}, thanks to which the multiplication function may be performed:

$$U_{III} = U_{0III} \cdot k_{III} N'' \tag{2}$$

where: k_{III} - processing coefficient of D/A converter III.
U_{III} - output voltage of C/A converter III
N" - digital signal value controlling D/A converter III, dependent on bits $B_{i-1}, \div B_0$

That allows for an adequate change of the Δu value of voltage spike to one bit, and hence the change of inclination of the generated voltage characteristics. The measuring transducer's characteristics should be approximated by a range-linear function with the assumed error, figure 2.3 a), while the range points A_i, A_j, ... should be selected so that their values meet directly, or with the proportionality coefficient, the values N_i, N_j, ... of the A/C transducer output signal.

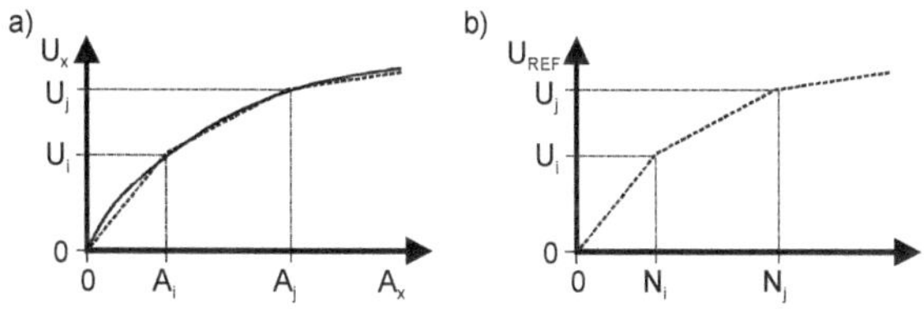

Fig. 2.3. a) Transducer characteristics and its approximation; b) waveform of reference voltage.

More significant bits ($B_n, \div B_i$) of the approximating register, through the coding circuit KODI control the D/A converter I. The CODI circuit, which may be ROM memory, contains a board converting the signal value from the register to such a signal value controlling the D/A converter, as to ensure proper processing characteristics, figure 2.3 b).

The transducer converting process is conducted in the same way as for a linear transducer, figure 2.4, i.e. bits are issued one after another, starting from the most significant. When voltage from the generator exceeds the value of the voltage processed recently, the preset bit is withdrawn and another bit is incorporated. Testing all the more significant bits $B_n, \div B_i$ means that a specified approximation range has been selected, in which the processed input signal is located (status of bits $B_n, \div B_i$ provides the values N_i, N_j). Appropriate inclination of the characteristics required for a given approximation range is received by changing the reference voltage of D/A converter III. The value of that voltage is obtained from D/A

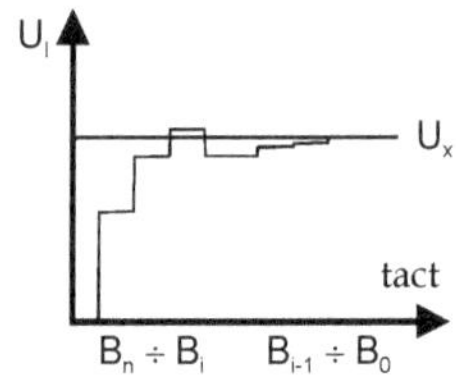

Fig. 2.4. Waveform of voltage generated in the generator circuit.

converter II controlled by bits $B_n, \div B_i$, through the CODII circuit. Hence, after setting a set of bits $B_n, \div B_i$ relevant to the input voltage by the approximating register, further converting process takes place, during which the less significant bits ($B_{i-1}, \div B_0$) control the D/A converter III which already has properly selected inclination of the characteristics.

Beginning of the range voltage values are obtained from D/A converter I and the values of voltages in particular ranges by adding voltage value from D/A converter I with voltage from D/A converter III. Output voltages of converter s C/A I, C/A II and C/A III for i-th range are respectively:

$$U_{Ii} = k_I \ \gamma_I(N') \tag{3}$$

$$U_{IIi} = k_{II} \ \gamma_{II}(N') \tag{4}$$

$$U_{IIIi} = k_{III} \ U_{II} \ k \ (N'') \tag{5}$$

where: γ_I - coefficient value received from cell of memory COD I with address N'.
k_i - coefficient values of transducer processing.
γ_{II} - coefficient value received from cell of memory COD II of address N'.
N' - more significant part of number N of control circuit (bits $B_n, \div B_i$).
N" - less significant part of number N of control circuit (bits $B_{i-1}, \div B_0$).

After summing up voltages from converters D/A in the adder, U_{ERF} voltage is obtained, defined by the relation:

$$U_{REF} = k_I \ \gamma_I(N') + k_{II} \ \gamma_{II}(N') \ k \ N'' \tag{6}$$

The manner of adjusting A/D converter processing characteristics with weight compensation described above requires that the linearized measuring transducer's characteristics is approximated by a range-linear function. Therefore, designing the circuit should be started by determining the approximating function with the assumed error. However, in order not to complicate the shaping circuit, the lengths of approximation ranges should be equal.

The additional errors, which may be introduced by the shaping circuit, are errors mainly from the D/A converter I, especially the linearity error, and the adder. The errors of other transducers have a smaller impact since the values of voltages generated by them are smaller than the voltage from D/A converter I.

The above-mentioned shaping circuit, in a simplified configuration, was completed in a universal temperature meter cooperating with transducers Pt-100, Pt-1000, Ni-100, PtRh-Pt, NiCr-NiAl, Fe-CuNi. The simplified scheme of the meter which allows explanation of its operating principle is presented in figure 2.5. Input circuits containing resistance to voltage transducers and amplifiers have been omitted here.

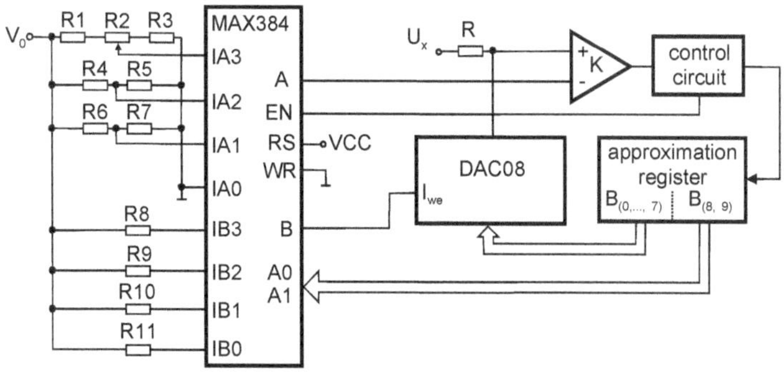

Fig. 2.5. Simplified scheme of a temperature meter.

Because the assumptions require that the absolute error of temperature measurement, submitted by the meter, to be within 1 °C for platinum transducers, 2 °C for nickel transducer and 5 °C for thermocouples, the characteristics of those transducers could be approximated by four sections. Thus, in the place of D/A converter I and the circuit COD I, a voltage generator, containing dividers R1 - R7 and half of analog selector MAX384, has been used. The DAC08 circuit has been used as a C/A III converter. That is a digital-to-analog converter with current output whose inclination of characteristics can be selected by changing the current input I_{IN}. The current can be changed through selecting, in the second part of the analog selector, one of resistors R8 - R11. For each measuring transducer, a separate module containing the selector along with dividers of voltages and current resistors was intended. Module selection was done by specifying the signal for input EN.

Converter circuit operates like an ordinary converter with successive approximation, with addition that in the first four timings (bits B_9 and B_8), the approximation range is determined, where the U_x voltage is located. The voltage from the generator, supplied to K comparator's reversing input, has the beginning of the approximation range value. The value of DAC08 transducer input current is also determined for the selected approximation range. In the other processing timings, the less significant bits are activated (B_7 - B_0) and, through voltage drop on resistor R caused by DAC08 transducer current output, the exact definition of U_x voltage value takes place. The status of bits B_9 - B_0 directly provides the value of measured temperature.

The conducted meter measurements indicated that it meets the parameters adopted in the assumptions concerning ambient temperatures 5 - 35 °C. Processing time was 2.5 ms.

Short converter processing time with successive approximation was used in the meter to measure carbon monoxide concentration. The TGS 2444 measuring transducer manufactured by Figaro USA Inc was used in the meter. In that transducer, the value of carbon monoxide concentration is processed into resistance. The measuring range is within 30 to 1000 ppm. Transducer characteristics has been presented in figure 2.6. It presents relative changes of output resistance, related to nominal resistance. Nominal resistance is given for concentration 100 ppm and ranges from 13.3 kΩ to 133 kΩ. In applied transducer the nominal resistance was 15.2 kΩ.

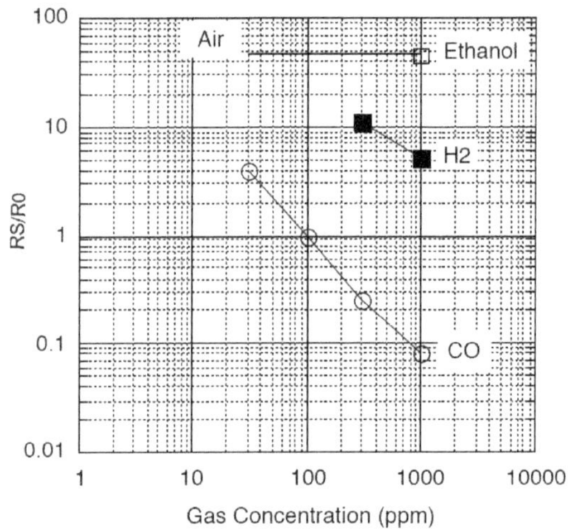

Fig. 2.6. Static characteristics of transducer TGS2442.

Transducer characteristics, presented in figure 2.6, shows a great change of inclination. For 30 ppm, the inclination is -1,52 kΩ/ppm and for 1000 ppm - -1,26 Ω/ppm (TGS 2611). With such high non-linearity of the measuring transducer's characteristics, using an A/C transducer with influenced processing characteristics is beneficial.

Because the meter was expected to be a portable device with battery power supply, a microcontroller MSP430F435 equipped with an LCD field direct control module was used in it. There is an A/C transducer with weight compensation with resolution of 12 bits and processing frequency of up to 200 khz within the internal structure of the micro-controller.

Calculations proved that the resolution of that transducer is too small to receive a proper resolution of measurement for high gas concentrations. Therefore, a circuit capable of adjusting transducer processing characteristics needed to be used.

The value of the digital A/D converter output signal is specified by the relation:

$$N = 4095\frac{U_X}{U_{REF}} \tag{7}$$

The reference voltage U_{REF} is accessible from the outside, which enables manipulation of transducer processing characteristics. The simplified scheme of the meter has been presented in figure 2.7. The device consists of a microcontroller MSP430F435 manufactured by Texas Instruments, an LCD field controlled directly from the microcontroller, a double D/A converter, an RS measuring transducer, highly stable resistors R1, R2 and R3, a heating circuit (elements T2 and R4), a circuit of switching on the measurement resistor (elements T1 and R5) and elements related to power supply and keyboard not shown in the figure.

The measuring transducer requires relevant control, which entail cyclic switching on the Rh heater and in the appropriate phase, switching on the RS measurement resistor, figure 2.8.

Proper operation of a measuring transducer requires that the heating is switched on for 14 ms and, after 981 ms from turning the heating off, the measurement resistor's power supply

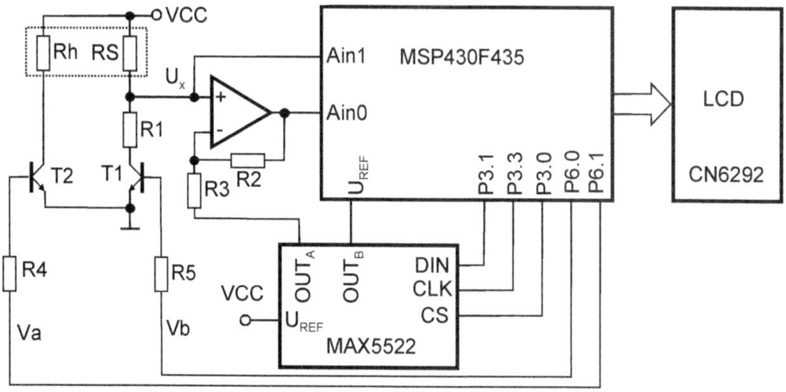

Fig. 2.7. Scheme of A/C transducer characteristics matching circuit.

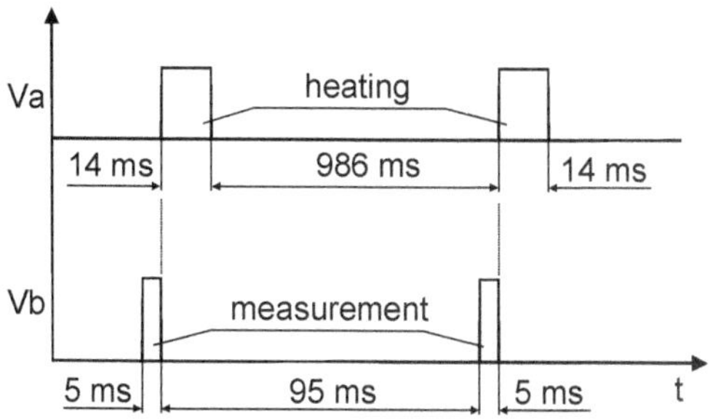

Fig. 2.8. Signals controlling measuring transducer.

is switched on. The measurement resistor activation time is to last 5 ms and the measurement should be taken during that period.

The RS measuring transducer is connected with resistor R1 in a divider circuit, which enables changing the measuring transducer's resistance to a voltage corresponding to it. Output voltage from the divider, after strengthening x6 in the amplifier, is supplied to A/D converter input. VCC voltage value was adopted as 3.3V (voltage supplying the microcontroller). The value of resistor R1 has been selected to obtain the broadest range of voltage changes U_x . Calculations, which have been omitted here, provided the value of 10 kΩ, which provides the range of voltage changes from 0.47 V to 2 V. The chart of relation between gas concentration s and U_x voltage value has been presented in figure 2.9.

The function presented in figure 2.9 is approximated by a range-linear function. With approximation error not exceeding 0.1%, 8 approximating sections were sufficient for approximation. However, the approximating function does not need to be continuous on range endpoints. The values of range points and coefficients k_i of section inclination in particular ranges have been presented in figure 2.10

Inclination coefficients k_i are determined by the relation:

$$k_i = \frac{s_{Pi+1} - s_{Pi}}{U_{Pi+1} - U_{Pi}} \tag{8}$$

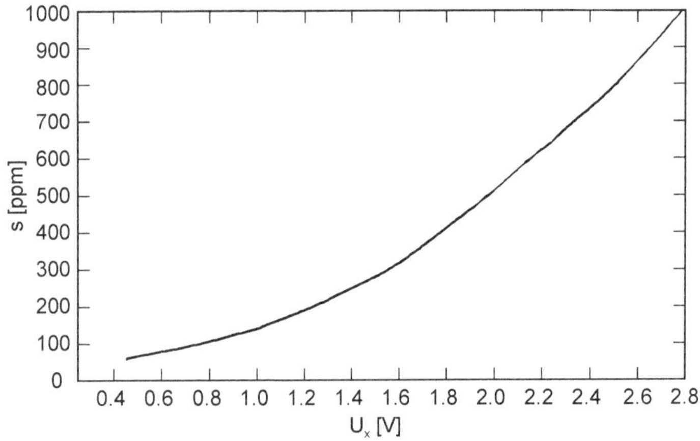

Fig. 2.9. Relation between gas concentration s and U_x voltage value.

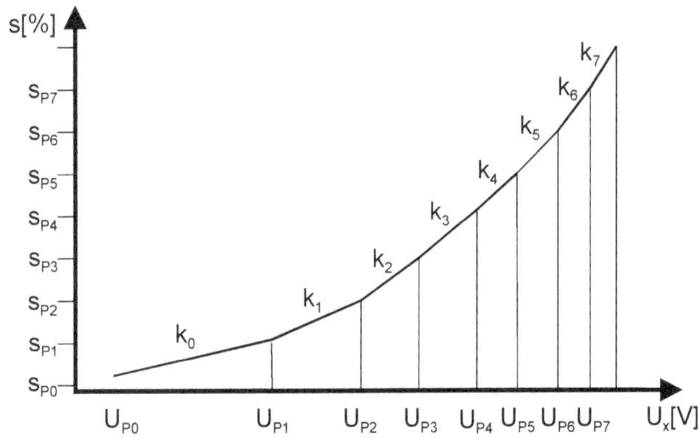

Fig. 2.10. Approximation by range-linear function.

The measurement takes place in two phases. In the first phase, input An1 is active and the initial voltage measurement U_x takes place on the divider output. In that phase, the reference voltage for an A/C transducer is the micro-controller internal voltage. That measurement provides information in which range voltage U_x is located. Then, for a given range, voltage U_{oi} is generated from D/A converter output A for the amplifier. The value of that voltage is selected so that for the value $U_x = U_{Pi}$, the voltage on amplifier output is equal to zero. At the same time, from the transducer output B, a reference voltage U_{REF} is generated for A/D converter. The value of that voltage should allow to obtain a satisfactory measurement resolution in a given range.

The voltage on output B of D/A converter is set proportionally to inclination coefficient value k_i. Such proceeding stems from the fact that the measurement value in i-th range is specified by the relation:

$$s = s_{Pi} + k_i(U_x - U_{Pi}) \tag{9}$$

The value of voltage difference $U_x = U_{Pi}$, after processing in A/D converter, can be specified using the formulation (7):

$$U_x - U_{pi} = \frac{N_{xi} \times U_{REF}}{4095 \times k} \tag{10}$$

where: k - amplifier gain
N_{xi} - value of processing result in i-th range.

Thus:

$$s = s_{pi} + \frac{k_i \times N_{xi} \times U_{REF}}{4095 \times k} = s_{pi} + \Delta s_i \tag{11}$$

If we assume that a linear relation has to be met:

$$\Delta s_i = c\, N_{xi} \tag{12}$$

then the reference voltage should meet the condition:

$$U_{REF} = \frac{4095 \times c \times k}{k_i} \tag{13}$$

Meeting the aforementioned condition may be difficult owing to a limited range of changes in reference voltage. In that particular case, in accordance with catalogue data, the minimum value of that voltage cannot be smaller than 1.4 V, and the maximum may not be greater than VCC - 0.2 V. For that reason, it has been assumed that the reference voltage value should ensure the required measurement resolution:

$$U_{REF} \leq \frac{4095 \times c \times k}{k_i} \tag{14}$$

The final measurement is calculated using the formulation (11). The shaping circuit introduces two kinds of errors. The first is connected with the approximation of the measuring transducer's characteristics by a range-linear function. It stems from the possible finite values of coefficients k_i and threshold voltages U_{Pi}, what is caused by their values being generated by converter D/A. The fact that the approximating sections do not have to be continuous in range points, facilitates approximation.

The second type of errors stems from the accuracy of generating the U_{REF} voltage and errors of A/D converter. D/A converters are 10-bit converters with errors: linearity - 1 LSB, offset - 1 mV, characteristics inclination - 0.5 LSB. Those errors will have the greatest impact at endpoints of the range, where there is greatest inclination of measuring transducer's characteristics. Taking into account the above values, it has been calculated from formulation (11) that the maximum absolute error Δs should not exceed the value of 8 ppm. The conducted simulation research, in which the measuring transducer was replaced with a resistance decade, indicated that the relative measuring error d, (referred to the endpoint of the measuring range) in the whole measuring range did not exceed the value of 0.7%, figure 2.11.

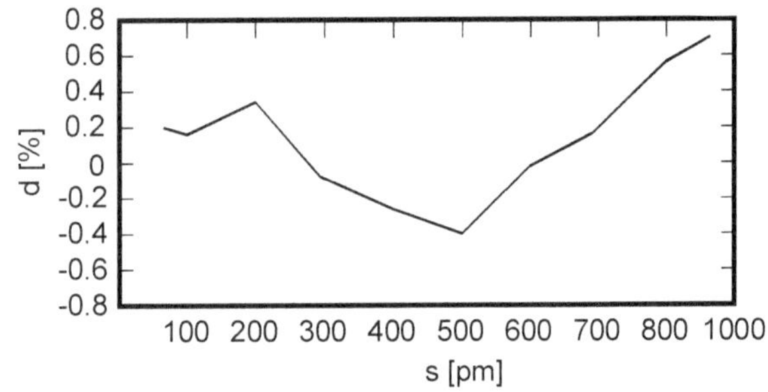

Fig. 2.11. Chart of relative error of carbon monoxide concentration meter.

3. Dual slope converters

Dual slope analog-to-digital converters with are very popular owing to their simple structure, high reliability, high resolution, high processing accuracy and were one of the first to be integrated. They are mainly intended for measuring constant voltages and most often cooperate with different kinds of measuring transducers.

The block diagram of a double integral converter has been presented in figure 3.1 and the waveforms in particular points of the transducer circuit in figure 3.2.

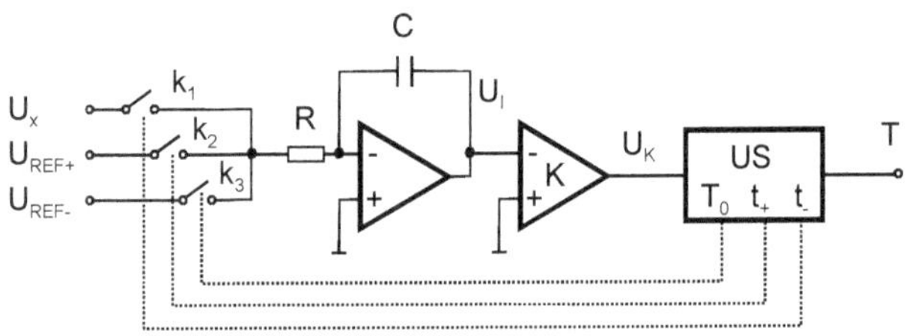

Fig. 3.1. Block diagram of double integral converter.

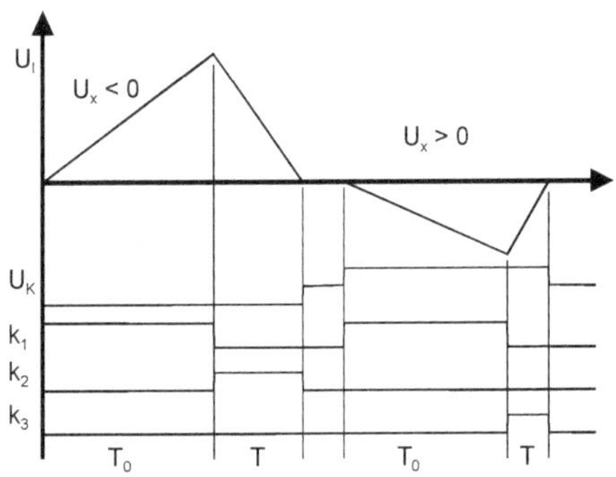

Fig. 3.2. Waveforms in particular points of double integral converter.

The converting process takes place in two phases. In the first phase, which lasts a given time T_0, the processed voltage U_x is connected to the integrator's input through key k_1. Then, after time T_0, voltage U_x is disconnected and reference voltage with polarization opposite to the polarization of voltage U_x is connected to the integrator's input. The level of voltage U_K at the comparator output indicates the polarity of the processed voltage U_x. Reference voltage integration lasts until the moment when voltage U_I at the integrator's output returns to the initial state which will result in activation of comparator K. Time T of reference voltage integration determines the result of processing.

Assuming the ideal integrator and comparator operation, voltage U_I at the integrator's output is defined by the relation:

$$U_I = \frac{1}{\tau}\int_0^{T_0} U_x dt - \frac{1}{\tau}\int_{T_0}^{T_0+T} U_{REF}(t)dt = 0 \tag{15}$$

where: τ = RC – integrator time constant
If voltage U_x is constant over processing, then relation (15) gives the following:

$$U_x = \frac{1}{T_0}\int_0^{T} U_{REF}(t)dt \tag{16}$$

Voltage U_{REF} also has a constant value in a transducer with linear processing and then:

$$U_x = U_{REF}\frac{T}{T_0} \tag{17}$$

Double integral converters typically have a resolution limited to about 13 - 14 bits. It stems from the fact that, with an increasing transducer resolution, the comparator must differentiate between increasingly smaller voltages at the integrator's output. The consequence is the extension of the comparator's response time and its unstable operation. In addition, the presence of noise, interference, hysteresis of the comparator and its insensibility threshold results in the fact that minimum voltage distinguished by the comparator is limited. Increasing transducer resolution, at the same time of voltage integration, would require increasing the clock generator's signal frequency. The above-mentioned factors would lead to transducer's unstable operation.

The need to increase transducer resolution to 16 - 18 bits resulted in preparing a modified double integral transducer circuit whose scheme has been presented in figure 3.3 and the waveforms of signals in selected points of the circuit - in figure 3.4.

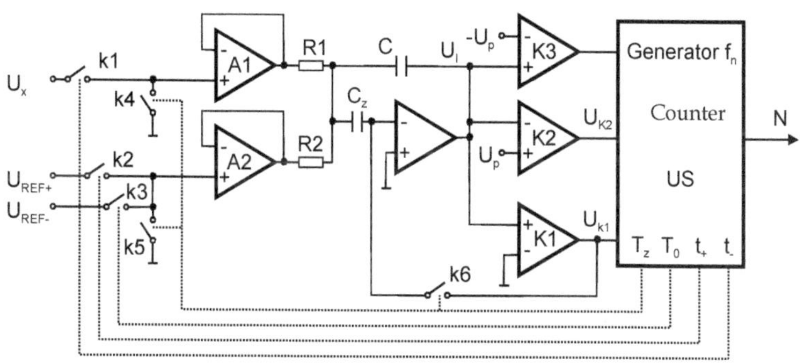

Fig. 3.3. Modified integrating transducer circuit.

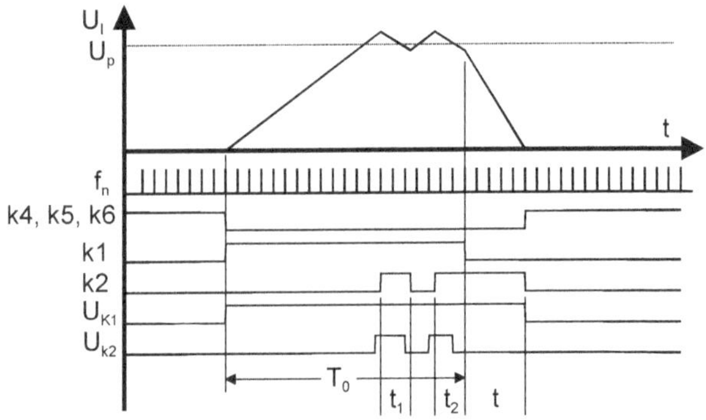

Fig. 3.4. Waveforms of signals in selected points of transducer circuit.

In comparison to the basic circuit of double integral converter, the circuit presented in figure 3.3 has been expanded with two input buffers, two additional comparators and elements allowing cyclical zeroing of the transducer circuit. The control circuit has also been changed.

The modified transducer circuit operates in the following manner: first, zeroing of the set integrator - comparator K1 takes place. That takes place when keys k4, k5 and k6 are closed. Other keys are open. Over that period, the integrator and the comparator K1 are enclosed by the feedback loop and the voltage on the integrator's reversing input reaches the value of that amplifier's offset voltage. Because the inputs of buffering amplifiers are also dense to the mass, the difference of offset voltages of input amplifier and the integrator amplifier is depositing on capacitor C_z. That voltage is remembered on the capacitor at all times throughout the process of converting. As a result, the impact of all amplifiers' offset voltage is disposed.

During integration of the processed input voltage U_x which lasts for T_0, when voltage U_I at integrator's output exceeds the value of voltage U_p, which is the reference voltage for the comparator. Then the comparator activates, which, in turn, causes such action of the control circuit US that, synchronically with impulse f_n of the clock generator switching on takes place - respectively to polarization of the processed voltage - of key k2 or k3 supplying the reference voltage with polarization opposite to the polarization of the processed voltage to the integrator's input. It is summed up together with the processed voltage. Under the influence of the reference voltage, the voltage at the integrator's output reduces its value. If the value of that voltage falls below voltage U_p of the comparator, a change of the comparator's condition takes place and, as a consequence, the control circuit will turn off the reference voltage key again synchronically with the clocking impulse. If the value of the processed voltage is high, that process may be repeated several times. For that converter to operate properly, the absolute value of reference voltage should be greater or at least equal to the absolute value of processed voltage. After the period T_0, the reference voltage is switched on, as in a regular double integral converter.

Because the voltage at the integrator's output returns to the initial state after the period of processing, that proves that the charge supplied from the processed U_x voltage source and the reference U_{REF} voltage source are of equal value. Therefore, an appropriate equality can be written:

$$\frac{1}{R1C}\int_0^{T_0} U_x(t)dt = \frac{1}{R2C}\left(\sum_{i=1}^{k}\int_0^{t_i} U_{REF}(t)dt + \int_0^{t} U_{REF}(t)dt\right) \tag{18}$$

If we assume that R1=R2 and the processed voltage value is constant during integration, the following is obtained:

$$U_x T_0 = \int_0^{T} U_{REF}(t)dt \tag{19}$$

where: T - total time of reference U_{REF} voltage integration.

The processing method has changed in the aforementioned transducer, but function (19) describing the relations occurring in it, is the same as function (16) for dual slope converter.

From the point of view of adjusting processing characteristics of A/D double integral converter, the relation (16) is important, which indicates the process of reference U_{REF} voltage integration. That means that using a generator generating an appropriate waveform of reference voltage will enable obtaining a desired A/D converter characteristics, (Janiczek, 1993).

In the transducer circuit presented in figure 3.5, the set of integrator, comparator, time constant and RC keys remains unchanged. On the other hand, the reference voltage generator - correction voltage, consists of an adder containing an operational amplifier, a D/A converter and EPROM memory. The meter block along with the control block US and the generator block f_n, form a circuit generating signals controlling keys k_1 and k_2. In addition, more significant bits of the meter block are at the same time control signals for address inputs of the EPROM memory. When the second phase of double integral converter operation starts - the phase of reference voltage integration U_o, through closed key k_2, reference voltage U_o is applied to the integrator's input. The voltage is the sum of (with appropriate coefficients specified by resistors R2, R3 and R4), constant voltage U_{REF} and voltage generated by C/A transducer, controlled by signals from the EPROM memory. Signals from the meter, controlling the EPROM memory, in selected time moments t_0, t_1, ... t_n result in sending relevant codes controlling the C/A transducer from the memory, which

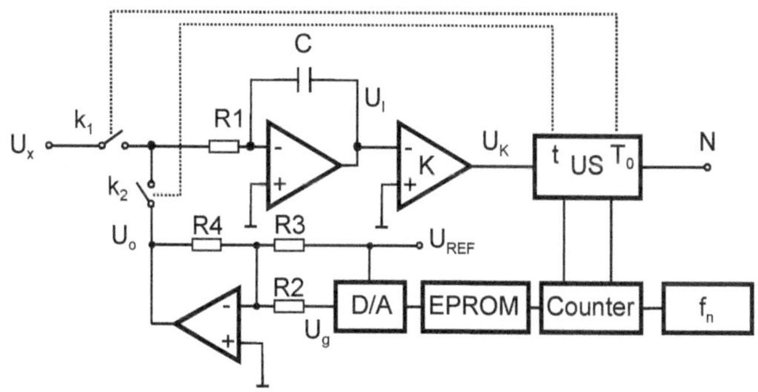

Fig. 3.5. Block diagram of transducer with adjusted characteristics.

results in generating the assumed reference voltage. A sample waveform of reference voltage U_o has been presented in figure 3.6.

Stepped reference voltage, integrated in the second phase of double integral converter operation, gives, as a result, the voltage at the integrator's output described by a range-linear function. The voltage waveform at the integrator's output during processing has been presented in figure 3.7. Time T_0 is the time of the processed voltage integration.

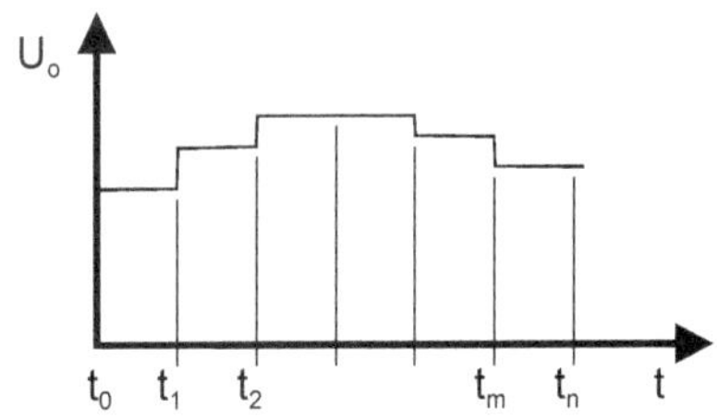

Fig. 3.6. Sample waveform of reference voltage.

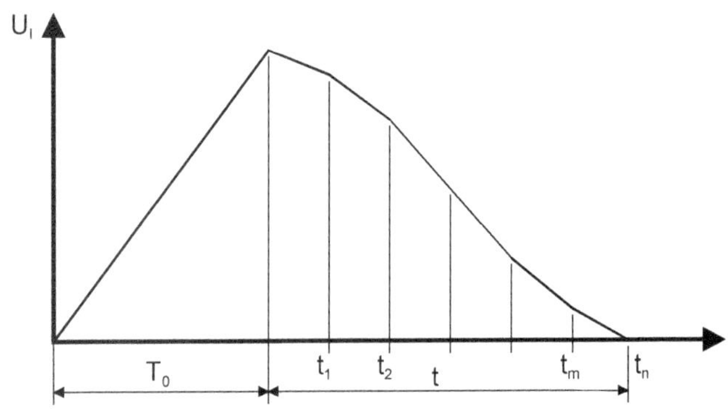

Fig. 3.7. Voltage waveform at integrator's output of double integral converter.

Taking into account that $U_{REF}= U_o$ in the formulation (16), the following is obtained:

$$U_x = \frac{1}{T_0}\left(\sum_{i=0}^{m-1} \int_{t_i}^{t_{i+1}} U_{oi} dt + \int_{t_m}^{t} U_{om} dt \right) \tag{20}$$

where: U_{oi} - reference voltage in the $t_i - t_{i+1}$ time interval
t - moment of completion of reference voltage integration phase.

After integration, the relation is obtained:

$$U_x = \frac{1}{T_0}\left(\sum_{i=0}^{m-1} U_{oi}(t_{i+1} - t_i) + U_{om}(t - t_m) \right) \tag{21}$$

Since time slices $t_i - t_{i+1}$ are the same throughout the entire reference voltage generating time and amount to Δt, and the integration time t of reference voltage ends in moment t_n, the formulation (21) can be presented in the following form:

$$U_x = \frac{1}{T_0}\left(\sum_{i=0}^{m-1} U_{oi} \times \Delta t + U_{om} \times \Delta t_n\right) \quad (22)$$

The formulation (22) is a record of a range-linear function in which Δt determines the range length, U_{oi} - inclination of function in i-th range. Because time slices T_0 and Δt are determined in the transducer by dividing frequency f_n of the clock generator, then, taking account that :

$$T_0 = \frac{k}{f_n} \quad \Delta t = \frac{p}{f_n} \quad t = \frac{N}{f_n} \quad t = m \times \Delta t + \Delta t_n \quad (23)$$

the formulation (22) can be presented in the following form:

$$U_x \times k = p \times \sum_{i=0}^{m-1} U_{oi} + U_{om} \times (N - m \times p) \quad (24)$$

where: N - value of input signal processing result.
Taking into consideration that the reference voltage is equal:

$$U_o = R4 \times \left(\frac{U_{REF}}{R3} + \frac{U_g}{R2}\right) \quad (25)$$

the formulation (24) takes the form:

$$U_x \times k = \frac{R4}{R3} U_{REF}(N - p) + \frac{R4}{R2}\left(p \sum_{i=0}^{m-1} U_{gi} + U_{gm}(N - m \times p)\right) \quad (26)$$

Because, as it seems from the above-mentioned formulations, the function describing the output signal on the integrator in the phase of reference voltage integration is a range-linear function, the measuring transducer's characteristics should also be approximated by a range-linear function:

$$U(x) = \sum_{i=0}^{m-1} \alpha_i \times (x_i - x_{i+1}) + \alpha_m (x - x_m) \quad (27)$$

where: α_i - coefficients of section inclination in i-th range
x_i - range points

Formulations (24) and (27) lead to believe that to obtain a linear relation:

$$N = c\, x \quad (28)$$

the condition must be fulfilled:

$$p \cdot U_{oi} = c \cdot k \cdot \alpha_i \cdot \Delta x \quad (29)$$

where: Δx - range length.

Range-linear function inclination coefficients obtained from the generator of function containing the C/A transducer, have finite values with discretization corresponding to D/A converter's output signal resolution. That fact should be considered when calculating the function approximating the measuring transducer's characteristics. In addition, the number of values that the approximating function coefficients can have is equal to the number of values generated by the D/A converter. In addition, due to simplicity of the shaping circuit's structure, the approximation ranges should have the same length, and, if possible, the number of ranges should be the power of number 2.

The range of changes in coefficient values of approximating sections' inclination is determined by the following relation:

$$\alpha = R4\left(\frac{U_{REF}}{R3} + \frac{U_g}{R2}\right) \tag{30}$$

From formulations (17) and (23), it seems that accuracy of double integral converter depends on the reference voltage stability. Therefore, also in the transducer with adjusted characteristics, the processing accuracy will depend on voltage generation accuracy U_o, with the assumption that the approximation error of the measuring transducer's characteristics by a range-linear function is negligibly small.

The formulation (30) shows that maintaining the assumed coefficient α accuracy depends on the stability of resistors R2, R3, R4 and voltage U_{REF} as well as on the accuracy of generating voltage U_g by D/A converter. The currently available parameters of resistors and sources of reference voltages are such that their impact on the accuracy of coefficients α may be omitted. Thus, the main source of errors will be D/A converter's errors.

The most important D/A converter's errors which should be factored in when determining the generated reference voltage accuracy include: characteristics linearity error, characteristics inclination change error and zero drift. D/A converter characteristics zero drift reduction as well as other analog blocks of slotted line can be performed by means of cyclical zeroing of the slotted line.

The method of adjusting transducer processing characteristics described above is one of the most effective, especially for converters of high non-linearity. An example may be the methane concentration meter with a measuring transducer TGS 22611 E00 manufactured by Figaro USA Inc. (TGS 2611). That is a resistance transducer and it measures methane concentration ranging from 300 ppm to 10000 ppm. Its characteristics, determining the relation of output resistance Rs to reference resistance Ro depending on methane concentration has been presented in figure 3.8. Resistance Ro, defined at concentration 5000 ppm, depending on a specific transducer may vary from 0.68 kΩ to 6.8 kΩ. The transducer used in the described meter, had 1.2 kΩ. Thus, transducer output resistance changes were ranging from 0.9 kΩ to 3.6 kΩ.

The currently available integrated A/D converters do not provide the feature of adjusting their processing characteristics. That is why analog elements that can be found in micro-controller internal structures were used in the structure. The MSP430FG4618 micro-controller manufactured by Texas Instruments was used in the meter for measuring methane concentration. Its structure contains three operational amplifiers, a 12-bit D/A converter, resistors and a set of analog switches. The circuit of dual slope converter, containing those elements, has been presented in figure 3.9.

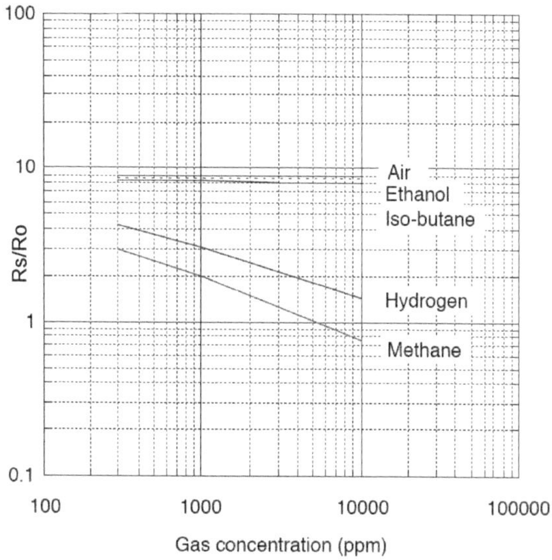

Fig. 3.8. Characteristics of transducer for measuring methane concentration.

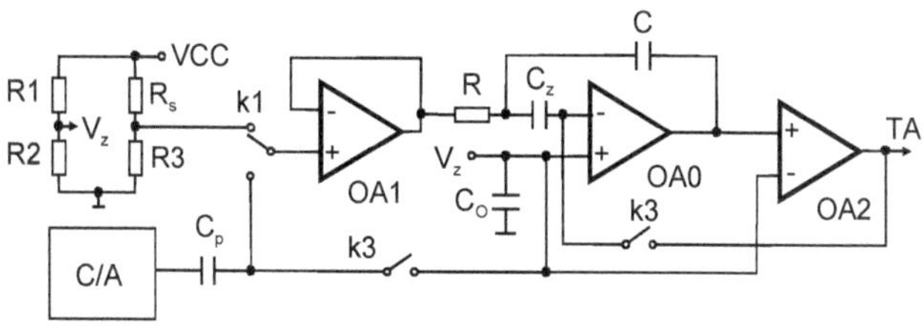

Fig. 3.9. Scheme of dual slope converter.

The integrator with amplifier OA0 and comparator OA2 is enclosed by the zeroing circuit with a capacitor C_z storing offset voltage. Voltage V_z produces artificial mass, in relation to which the input voltage and the reference voltage received from D/A transducer is determined. The capacitor C_p provides an appropriate shift of the output signal level from the D/A converter. Resistors inside the micro-controller have been used as resistor R of the integrator. The measuring transducer is included in the divider circuit, which simplifies the measuring circuit. Since voltage VCC, supplying the divider, is, at the same time, the reference voltage for the D/A converter. Changes in that voltage do not affect the accuracy of measurement. The transducer control circuit has been based on the structure of module Timer A and an appropriate program.

The simplified scheme of the meter circuit has been presented in figure 3.10. It is clear, that almost the entire transducer structure is located inside the micro-controller. Only capacitors and divider resistors are located outside. The diagram has omitted the heater activation circuit and the meter's keyboard.

For selected concentration values, voltage values on bridge diagonal have been calculated and for points obtained in such way, the second-degree approximating polynomial U(s) has been calculated. Its waveform have been specified in figure 3.11.

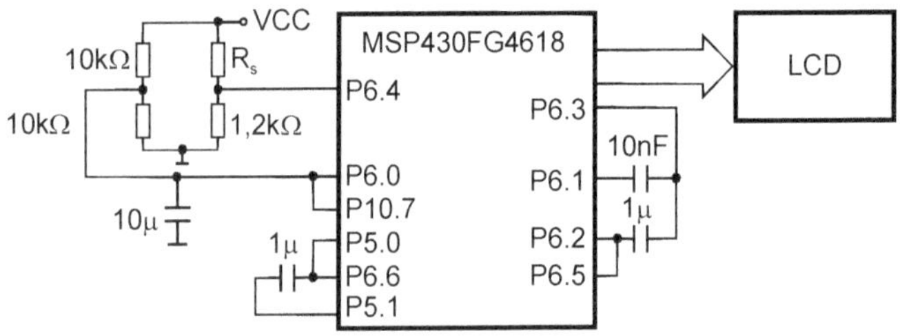

Fig. 3.10. Simplified scheme of meter for measuring CO concentration.

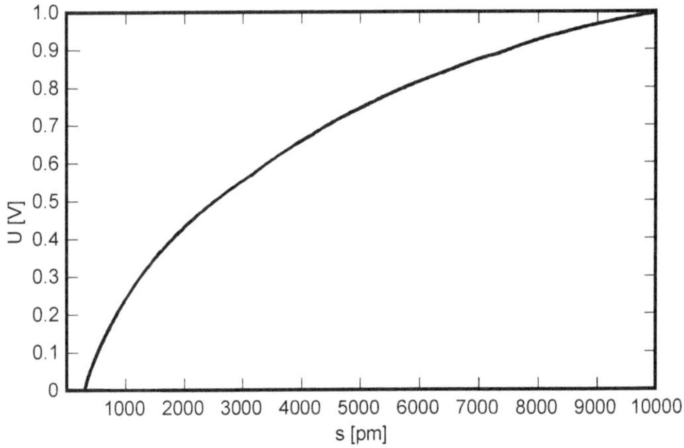

Fig. 3.11. Relation of voltage U on divider output to value of concentration s.

Then, the obtained function U(s) has been approximated by a range-linear function. 10 sections have been obtained. Inclination values of those sections were used to calculate the values entered into the controlling register of the C/A transducer.

Testing of errors produced by the shaping circuit took place in such way that the transducer output resistance, for selected concentration values, was simulated by a resistance decade and the difference between the theoretical concentration value and the value indicated by the meter was defined. The value of relative error d, calculated with respect to the maximum range value, has been shown in figure 3.12.

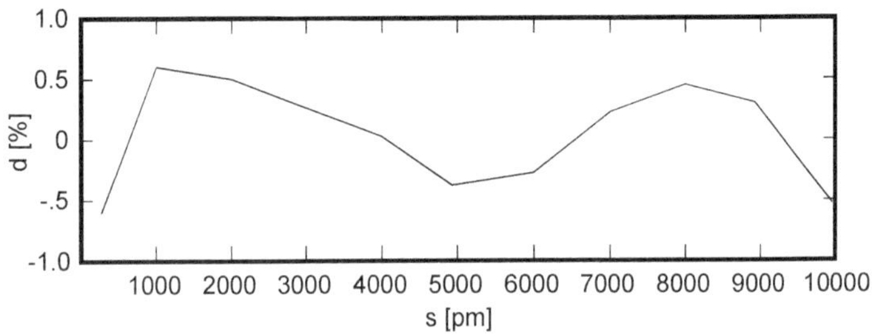

Fig. 3.12. Chart of relative error of carbon monoxide concentration meter.

As it stems from the error chart, its maximum value does not exceed 0.6% and is present at the beginning of the range.

4. Transducer with impulse feedback

Transducer with impulse feedback is also classified into integrating transducers. Its great advantage is that it is a very simple circuit, but it processes signals of just one polarity, although that limitation generally is not disturbing in the case of cooperation with measuring transducers. Especially since it is very well suited for cooperation with measuring resistance transducers. Due to that simplicity and easy control, they are used in measuring devices cooperating with different measuring transducers. Block diagram of a transducer with impulse feedback has been presented in figure 4.1, and voltage waveforms in selected points of the transducer in figure 4.2.

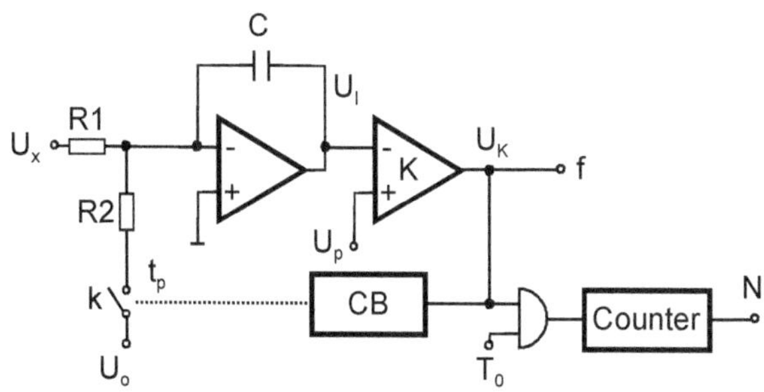

Fig. 4.1. Block diagram of transducer with impulse feedback.

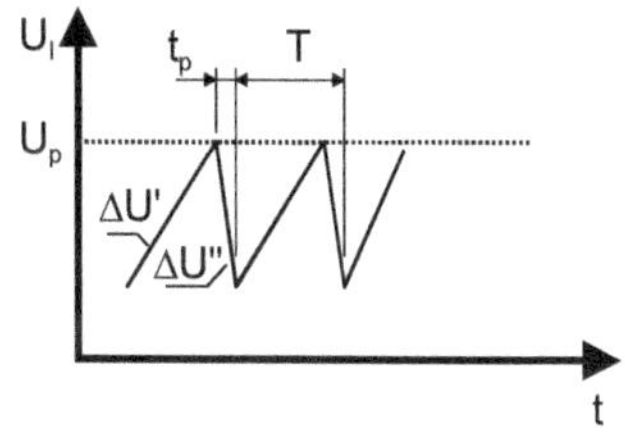

Fig. 4.2. Voltage waveforms in selected transducer points.

The transducer circuit consists of an integrator with operational amplifier, capacitor C and resistors R1, R2, comparator K, control block CB, key k and reference voltage source U_0. The output signal of frequency f being the function of the processed input voltage U_x is obtained at the comparator's output. If numeric values are to be obtained from the transducer, then the meter calculates output signal impulses for a given time T_0.

The circuit operates in the following manner: processed voltage U_x is supplied to the integrator, through resistor R1, at all times. As a result, voltage U_I at integrator's output continues to increase until it reaches the reference level U_p. At that time, the comparator is activated and, as a consequence, control block CB is activated which, for time t_p, closes key k supplying the integrator with reference voltage with polarization opposite to the polarization of input voltage U_x through resistor R2. That results in change of voltage waveform at the integrator's output. After time t_p, key k is turned off and the processing cycle starts over. In the first phase of transducer operation, voltage increase at the integrator's output can be described by the relation:

$$\Delta U'(t) = \frac{1}{R_1 \times C} \int_0^{T-t_p} U_x(t)dt \tag{31}$$

In the second phase of transducer operation, in which reference voltage is added to the integrator, voltage increase at the integrator's output can be described by the following relation:

$$\Delta U''(t) = \frac{1}{R_1 C} \int_{T-t_p}^{T} U_x(t)dt - \frac{1}{R_2 C} \int_{T-t_p}^{T} U_o(t)dt \tag{32}$$

In the steady state, voltage increases at the integrator's output are equal:

$$\Delta U' + \Delta U'' = 0 \tag{33}$$

Thus, summing up the integrals from formulations (31) and (32) the following is obtained:

$$\frac{1}{R_1 \times C} \int_0^{T} U_x(t)dt = \frac{1}{R_2 \times C} \int_0^{t_p} U_o(t)dt \tag{34}$$

If voltages U_x and U_o are constant, then after transforming the relation (34), a formulation determining the frequency of transducer output signal is obtained:

$$f = \frac{1}{T} = \frac{U_x \times R_2}{U_o \times R_1 \times t_p} \tag{35}$$

and the number of impulses calculated during processing time T_0 is:

$$N = T_0 f = \frac{U_x \times R_2 \times T_0}{U_o \times R_1 \times t_p} \tag{36}$$

Formulation (34) shows that in a one processing cycle, the charge supplied to the integrator from voltage U_x source is equal to the charge transferred from the integrator under the influence of reference voltage U_o . Measurement time of processed voltage U_x, namely the time of impulses counting in the meter is T_0 and over the entire period there is balance between the supplied and the transferred charge. Taking that into consideration, the formulation (34) can be presented in the following form:

$$\frac{1}{R_1} \int_0^{T_0} U_x(t)dt = \frac{1}{R_2} \int_0^{Nt_p} U_o dt \tag{37}$$

Assuming that the processed voltage U_x and reference voltage U_o are constant over processing period T_0, then after transforming, formulation (37) takes the form:

$$U_x \frac{T_0}{R_1} = U_o \frac{N \times t_p}{R_2} \tag{38}$$

On the other hand, when the reference voltage U_0 is a variable voltage over the processing period, but it maintains constant value over period t_p, figure 4.3, then that condition can be described by relation (39) arising from formulation (37)

$$U_x \frac{T_0}{R_1} = \frac{N \times t_p}{R_2} \sum_{i=0}^{N-1} U_{oi} \tag{39}$$

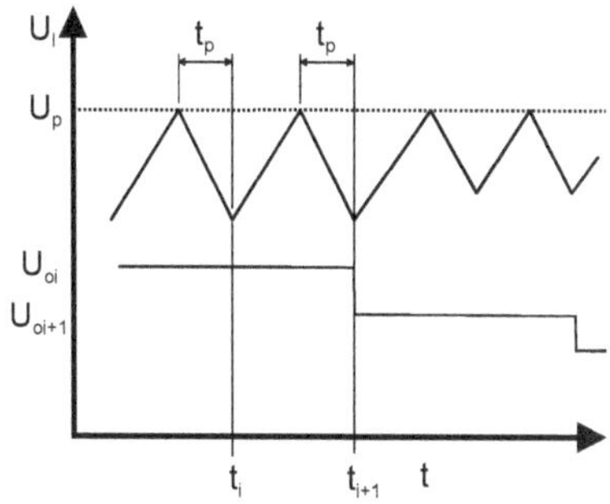

Fig. 4.3. Waveforms at integrator's output at variable voltage U_0.

If voltage U_{oi} is constant over p long ranges, then for the m-th range the relation (39) may be expressed in the form:

$$U_x = \frac{R_1 \times t_p}{R_2 \times T_0} \left(p \sum_{i=0}^{m-1} U_{oi} + U_{om} (N - m \times p) \right) \tag{40}$$

Formulation (40) corresponds to formulation (30) for double integral converter and, as a result, shows that the characteristics of a transducer with impulse feedback can be adjusted in a similar way as it has been presented for a double integral converter. Also, all presented considerations are also valid here with one reservation concerning time t_p.

In the transducer circuit presented in figure 4.1, time t_p is generated by a monostable circuit which does not provide too high accuracy. It may be remedied by introducing modifications in the transducer circuit, presented in figure 4.4. (Janiczek, 1992)

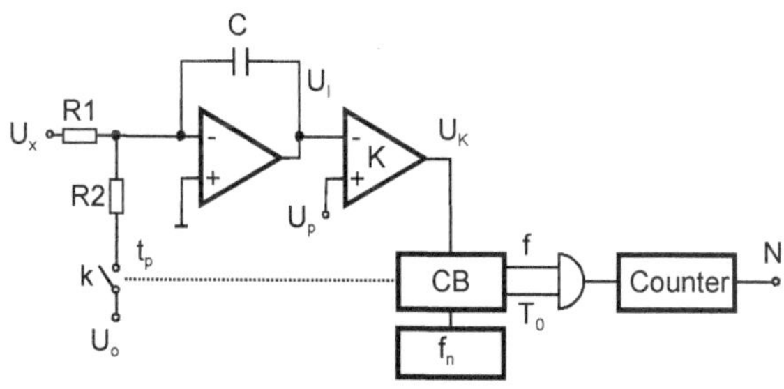

Fig. 4.4. Modified version of transducer circuit with impulse feedback.

The modification of transducer operation involves generation of time t_p – it begins not when the comparator is activated, but synchronically with the first impulse of generator f_n, which comes when the comparator is activated. That is illustrated in figure 4.5.

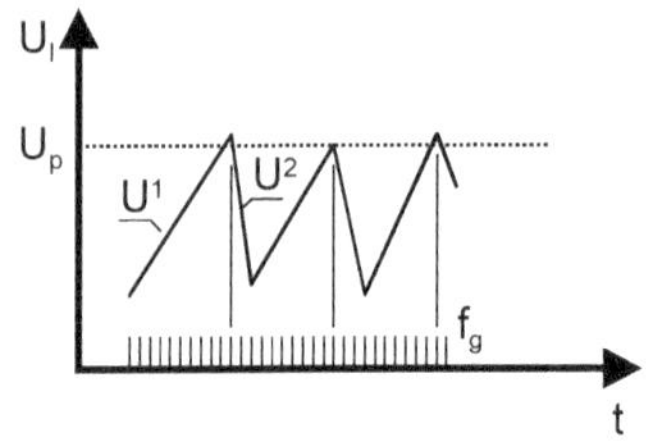

Fig. 4.5. Voltage waveforms in modified transducer circuit.

As a result, time slice t_p may be generated by division of frequency f_n in the divider. That results in a certain deviation of transducer output signal frequency, but after averaging over time interval t_0, that does not affect the processing accuracy. Also, time interval T_0 may be obtained in the same manner. Thus, times t_p and T_0 can be identified as:

$$T_0 = \frac{k}{f_n} \qquad t_p = \frac{q}{f_n} \tag{41}$$

where q, k – frequency f_n division coefficients.
Taking into consideration the relations (41) in formulation (40) the following is obtained:

$$U_x = \frac{R_1 \times q}{R_2 \times k}\left(p\sum_{i=0}^{m-1} U_{oi} + U_{om}\left(N - m \times p\right)\right) \tag{42}$$

Thanks to such solution, the processing accuracy of a transducer depends only on the constancy of relation R1/R2 and accuracy of generating voltage U_o. In addition, there is a possibility to adjust transducer characteristics in the above-mentioned circuit by changing duration time t_p of the feedback impulse.

Assuming, in formulation (37), that both the values of processed voltage U_x and reference voltage U_o are constant, the following is obtained:

$$U_x \frac{T_0}{R_1} = \frac{U_o}{R_2}\sum_{i=0}^{N-1} t_{pi} \tag{43}$$

After considering the relations (43) and transformation, the following formulation is obtained:

$$U_x = \frac{R_1 \times U_o}{R_2 \times k}\sum_{i=0}^{N-1} q_i \tag{44}$$

from which it can be concluded that the characteristics of a transducer with impulse feedback can be adjusted by changing the degree of meter division specifying impulse duration t_p.

Assuming, as previously, that values q_i are constant over p long ranges, the following relation is obtained:

$$U_x = \frac{R_1 \times U_o}{R_2 \times k}\left(p\sum_{i=0}^{m-1} q_i + q_m (N - m \times p)\right) \tag{45}$$

which shows that in such circuit of a transducer with impulse feedback, measuring transducers characteristics approximated by a range-linear function may be adjusted in a very wide range. Moreover, in that transducer, a very high processing resolution and accuracy can be obtained, which practically depends on the offset voltage constancy of integrator's amplifier, the constancy of R1/R2 resistors' relation and the constancy of reference voltage source U_o.

The above-mentioned method of adjusting processing characteristics of transducer with impulse feedback has been applied in the meter for measuring carbon monoxide concentration. A transducer TGS 203 manufactured by Figaro Engineering Inc. (Japan) whose simplified scheme has been presented in figure 4.6. was used in the meter. (Janiczek, 2009)

The A/D converter circuit consists of an integrator, comparator, keys k1 - k5, voltage divider R2, R3 and R4 and reference resistor R1. Rs is the measuring transducer's resistance. Particular phases of transducer operation are controlled by the micro-controller manufactured by Texas Instruments - MSP430F415. Additionally, to improve transducer resistance to changes in temperature and supply voltage, the phase of zeroing the integrator's amplifiers and comparator has been introduced. The zeroing voltage is stored on capacitor C_z. In order for the circuit to be supplied with one voltage, a divider of voltage R2/R3, R4 providing the reference voltage U_z for the integrator has been introduced. That voltage is equal approximately to half the supply voltage. As a result, voltages applied to resistors R1 and Rs, through keys k1 and k2, have opposite polarizations in relation to voltage U_z. The reference voltage U_p for the comparator is obtained from divider R2, R3/R4.

The circuit is zeroed over the T_z time, figure 4.7. Keys k1 and k2, are open and keys k3, k4 and k5 are closed. The integrator's and comparator's amplifiers are enclosed by the feedback loop, and capacitor C_z is charged up to the integrator's amplifier offset voltage. After zeroing the circuit, keys k4 are opened and key k1 is closed for the entire duration of processing and the transducer operates in regular mode of transducer with impulse feedback.

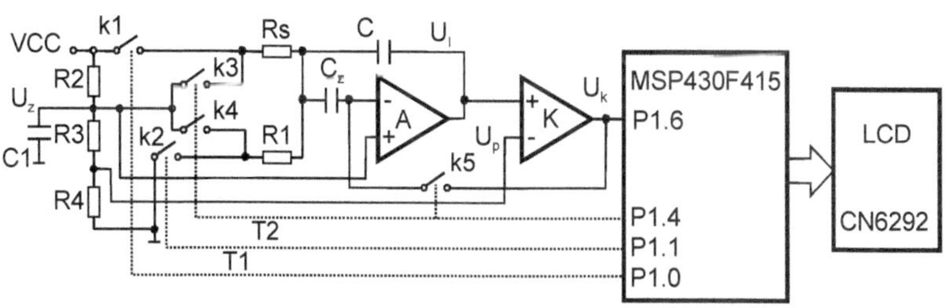

Fig. 4.6. Scheme of carbon monoxide concentration meter.

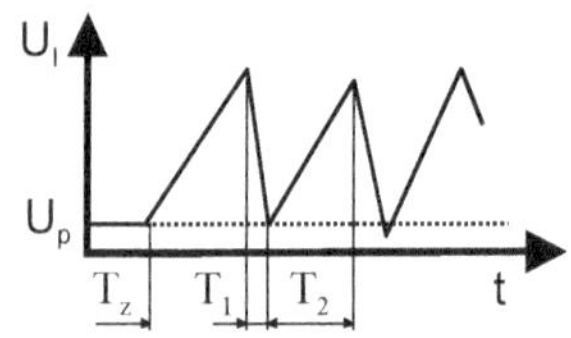

Fig. 4.7. Voltage waveform at integrator's output.

Comparator's output impulses are counted in one of the meter's micro-controllers and a different meter unit generates time slices T_2 of feedback impulse duration. Using the basic relation (38) describing transducer operation and taking account of relation (4.4) describing the measuring transducer's characteristics, the following is obtained:

$$N = \frac{(VCC - U_z) \times R1 \times s^{0.95}}{245.66 \times U_z \times T_2(N)} \times T_0 \tag{46}$$

where: N - number of impulses counted in the meter,
T_0 - processing duration,
U_z - artificial mass voltage.

Figure 4.8 shows the required voltage-frequency transducer characteristics.

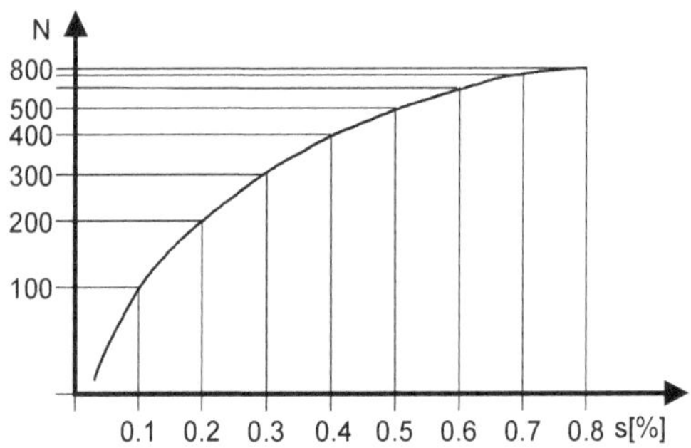

Fig. 4.8. Characteristics of voltage-frequency transducer.

For proper charge balance over the T_2 period, it has been assumed that resistor R1=150 Ω and U_z=VCC/2, hence the formulation (45) takes the form:

$$N = 0.61 \frac{s^{0.95}}{T_2(N)} T_0 \tag{47}$$

In order to obtain a linear relation between the number of counted impulses N and concentration s, the following condition must be fulfilled:

$$T_2(N) = c \cdot s^{0.95} \tag{48}$$

The measuring transducer's characteristics has been approximated with ten sections, which ensured approximation error not greater than 0.3%. Using relation (45), coefficient values determining times T_2 have been calculated.

Conducted tests have indicated that a stable resolution of so constructed A/D converter was reached at the level of 14 bits. The impact of temperature changes in the range 0 - 30° C did not exceed 0.02%. The main factor of that error was changes of R1 resistor values and change of R2/R3, R4 divider's pitch. Change of supply voltage VCC within +/-10% was not noticed. Those parameters significantly exceed the requirements necessary to measure carbon monoxide concentration but they show the capabilities of the measuring circuit.

Conducted meter characteristics tests, in which the measuring transducer was simulated with a variable resistor, indicated that the relative measuring error d (with regard to the range endpoint) did not exceed the value 0.3%, figure 4.9.

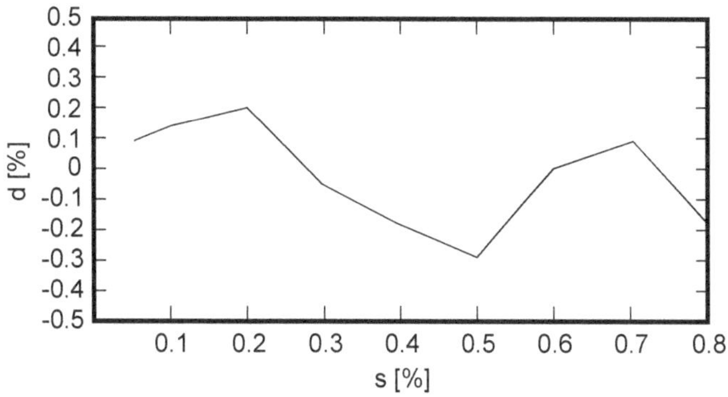

Fig. 4.9. Error chart of transducer for measuring carbon monoxide concentration.

5. Conclusion

This chapter presents a method for digital adjustment of analog-to-digital converter transfer function. If the measuring transducer's characteristics is non-linear, it is necessary to use suitable correction methods. When the measuring transducer's characteristics is very non-linear, for example in the case of certain transducers for light intensity measurement, a problem with selecting a relevant A/D transducer may occur.

A converter with digital adjustments is a useful tool for correcting non-linearity of transducers, especially those with high linearity, in case of which the numerical correction would lead to the lost of resolution. The conducted experiments proved the suitability of the proposed new solution of converter. Such designed analog-to-digital converter achieves a stable resolution and small error of the approximation function.

6. References

Bucci G., Marco Faccio M., Landi C. New ADC with Piecewise Linear Characteristic: Case Study—Implementation of a Smart Humidity Sensor. IEEE Trans. Instr. Measur., Vol. 49, No. 6, Dec. 2000.

Grzybowski W. Wagner F. Nonlinear function from d-a converters. Electronic Engineering, nr 512. 1971.

Iglesias G. E., Iglesias E. A., Linearization of Transducer Signals Using an Analog-to-Digital Converter. IEEE Trans. Instr. Measur., Vol. 37, NO. 1, Mar. 1988.

Janiczek J: Le convertisseur analogique-numerique pour la correction des caracteristiques statiques nonlineaires des capteurs. Measurement Science and Technology. 1992 vol. 3

Janiczek J. Analogue-to-digital converter with digitally controlled transfer function. Measurement Science and Technology. 1993 vol. 4

Janiczek J. Digital adjustment of analog-to-digital converter transfer function. Metrology and Measurement Systems, vol.16, no 3, 2009.

Patranabis D., Ghosh S., Bakshi C. Linearizing Transducer Characteristics. IEEE Trans. Instr. Measur., Vol. 37, No. 1, Mar. 1988.

Patranabis D., Ghosh D. A Novel Software-Based Transducer Linearizer. IEEE Trans. Instr. Measur., Vol. 38, No. 6, Oct. 1989.

Reis G. Nichtlinearen A/D-Umzetzer linearisiert Sensor-Kennlinien. Elektronik No 4, 1980.

TGS 2611 Product Information. Figaro USA Inc.

14

Non Invasive Acoustic Measurements for Faults Detecting in Building Materials and Structures

Barbara De Nicolo, Carlo Piga, Vlad Popescu and Giovanna Concu
University of Cagliari, Engineering Faculty, Italy

1. Introduction

1.1 Backgrounds and generalities

Over the past years both large and small restoration and conservation works on monuments, civil and industrial buildings have become of great interest. As indicators of the historical period in which they were built, all construction works have both their architectural style and the material used in their construction. Indeed, for thousands of years humans built using the same materials (wood, stone, brick, mortar and gypsum) up to the introduction of concrete at the beginning of the 19th century. Although concrete has replaced the old materials used in historical buildings, there still remains the problem of forestalling their deterioration and restoring them, also in the light of the importance of such works from the historical, cultural and economic viewpoints. Problems connected with the restoration of buildings, whether in reinforced concrete, masonry or wood, are quite complex and are essentially linked to the reuse, and thus the redesign, of the existing heritage of buildings. Indeed, cultural, social and economic reasons foster to the desire to lengthen the life of this heritage beyond normal physiological limits and thus its fruition far beyond its useful life. The problems to be addressed vary widely since there are noteworthy differences from one job to another; it is sufficient to consider just the social value of a building of great historical value, which is usually protected by severe restrictions aimed at conserving its artistic and cultural features, or an industrial building the use of which must be completely changed while at the same time maintaining its structural characteristics. It is evident that there is not one single answer to such widely differing situations: each job must be addressed from the cultural, technological and technical standpoints as a special case. The proper management of the rehabilitation of a building implies a knowledge of its real static conditions to be restored, the mechanical, physical and chemical characteristics of the materials of which it is built and the presence and characteristics of defects, anomalies and so on. Of fundamental importance is the diagnosis of materials and structures and many researchers, as well as companies that produce restoration materials, have carried out studies in this field. Methods for structural diagnosis and faults detecting are beginning to appear, albeit in an extremely divergent way, in tenders, rules, guidelines and so on. To exemplify, non-invasive diagnostic techniques are often used to determine whether or not materials compatible with the original structure have been used in restoration works: if not, the building may undergo greater damage than that caused by progressive deterioration. All the factors illustrated above favour the development of new control and diagnostic techniques that produce more and more information on the state of the structure and, indirectly, represent a more precise instrument for use in the planning of restoration works.

In the field of structural faults detection particular importance is given to developments of Non-Destructive Testing measurements techniques (NDT), including automated procedures and information technology to support decision making and evaluation of data. NDT appear to be of great usefulness since, compared to classic laboratory techniques, they are non-invasive, faster and of a general rather than specific nature. These skills have led to the creation, evolution and rapid diffusion of certain diagnostic non-destructive measurements techniques. The main obstacle to the effective, systematic and economical use of NDT in structural diagnostics lies in the gap that exists between the theoretical and interpretative bases and codification of operative modalities, which is to say the almost total lack of standardization procedures; moreover, practical experience has underscored the limits of commercial devices, which have often proved to be incapable of adapting to specific structural problems under investigation and have interfaces that do not allow users to check the validity of data acquired and the accuracy of measurements, thus making the results aleatory and difficult to interpret.

As a major NDT tool, acoustic techniques, based on measurements of the characteristics of acoustic waves propagating through the material, are often used in quality control and faults detection for engineering structures and infrastructures. The studies of acoustic techniques have been focused on medical or materials engineering applications for laboratory testing. The valuable handmade analysis has begun in the first of '90 and nowadays it is performed using considerable approximations to the detriment of results precision. The inaccurate results are also caused by the subjective methodology often used for interpretations. This contributed to form the common opinion that the use of such methodologies in structural diagnostics does not give reliable results. Conversely, the aforesaid techniques could show notable diagnostic properties if used in an appropriate way, that is providing them with innovative techniques, which employ and develop advanced computational tools and testing devices. In the light of this, fuelled by the rapid development of portable personal computers, high-performance computing algorithms and electronic engineering technology, acoustic techniques have evolved dramatically during recent years.

Acoustic material analysis is based on a simple principle of physics: the propagation of any wave will be affected by the medium through which it travels. Thus, changes in measurable parameters associated with the passage of a wave through a material can be correlated with changes in physical properties of the material. Recently, thanks to continuing scientific and technological progress, systematic studies and applications of such methods in many different fields have been developed. The application of methods of automated graphic representation to the so-called *media transparent to vibrations* such as even large metal pieces, has now become routine. Outside the field of engineering, such methods have also been applied to parts of the human body; it is in the techniques of medical analysis such as ultrasonography and computerized axial tomography that some of the most important methods of investigation of structures with acoustics NDT have been developed. The acoustic NDT of materials less transparent to vibrations, such as concrete, mortars, stones, wood, masonry and so on, is still more challenging. For such substantially inhomogeneous media, the development of acoustic testing techniques has been decidedly slower compared to applications on transparent media. This depends mainly on the greater difficulties and the important theoretical and interpretative challenges, not to mention the technological ones represented by the reduced transparency of the materials to vibrations deriving from their high degree of intrinsic lack of homogeneity.

1.2 Methodologies

1.2.1 Sonic and ultrasonic testing methods

Sonic and ultrasonic investigations refer to a complex method for the analysis of materials and the structures of which they are made, based on the study of phenomena connected with the propagation of elastic perturbations inside the materials under study. The signal that penetrates into the material is generated artificially by an external source and acquired by means of a receiver after passing through the medium following appropriate trajectories. From analysis of the processes and parameters connected with the propagation of acoustic perturbations inside the artefact it is possible to collect a great deal of information on the material or structure under study (J. Krautkramer & H. Krautkramer, 1990). This information includes:

- the level of homogeneity of the material concerning several elements or a certain number of samples or only one of these;
- the quality and degree of deterioration of the material;
- the estimate of certain elastomechanical characteristics;
- the identification of possible faults in the material or structure, such as cavities, inclusions and zones having different elastomechanical characteristics;
- the trend in time of different phenomena and related to the stresses the materials undergo.

The intrinsic characteristics of the medium under test intervene decidedly in the several aspects connected with the use of acoustic methods (choice of the most suitable instrumentation, application of the basic principles of the method, criteria for interpreting the results), especially in the choice of the signal frequency to use in the investigation.

In the case of not large homogeneous media, for example, metal pieces, the characteristics of homogeneity favour the use of ultrasonic frequencies above 500KHz. The absence of natural non homogeneities allows the signal to propagate without appreciable reflection, refraction or mode conversion diffusion phenomena, unless these are caused by the presence of possible local anomalies. The analysis of such kinds of media is thus favoured by the negligible attenuation of the signal. This requires the use of very little energy and makes it possible to identify even very small defects by emitting very narrow irradiation bands through strongly directional probes. Substantially, in these cases it is possible to plan the analysis in the smallest details and accurately investigate by closely defined zones from the geometric standpoint.

For investigations in media with a high degree of non homogeneity, for example the stratigraphic analysis of terrains, due to the strong intrinsic heterogeneity of the medium there is a noteworthy diffusion of the signal. This, together with the need to investigate large areas, calls for the use of sources with high emission energy and therefore kept prevalent in low frequencies (usually 5-150Hz), so as to compensate for the high absorption characteristics of the medium and the total loss of signal direction.

When intrinsically inhomogeneous media of not very large dimensions such as those represented by building structures are to be investigated, the problem of revealing perturbations caused by anomalies of great interest arises; however, such anomalies differ slightly in extension, geometry and dimensions from the natural non homogeneities of the medium under study: an emblematic case is that of concrete in which the anomalies may respond to the acoustic input in a way similar to the aggregate. For this category of media, in which we find most of the materials and structures used in civil, industrial and monumental engineering and architecture, it is of fundamental importance to choose correctly the characteristics of the signal employed, attempting a mediation between the characteristics of the ultrasonic signals (enhanced diagnostic precision at the price of greater attenuation and thus a lesser penetrative capacity) and that of the sonic signals (high penetrative capacity at the price of poor definition). This is the case of concrete elements, masonry structures, wood elements and limited volumes of terrain.

Experimental results (Concu et al., 2003a) showed that both sonic and ultrasonic signals reveal the presence of macroscopic anomalies inside a limestone masonry structure, but are differently sensitive to intrinsic structural characteristics and environmental conditions. We performed the experiment on a sample masonry wall of limestone blocks, inside which some metal and wood elements, assumed as anomalies of the structure, were placed in a known position. Moreover, in a central position of the masonry a cavity was excavated designed to hold a rectangular section, but the irregularity of the external blocks and the intrusion of mortar caused the section to become irregular in the executive stage. The propagation velocity of high- and low-frequency signals through the wall thickness was measured: higher velocities are generally associated with a better quality of the material. Velocity data were then elaborated by interpolation and represented in the form of a map of velocity distribution in a generic vertical section of the wall. Fig. 1 shows the velocity maps obtained with the sonic (low-frequency signals) and ultrasonic (high-frequency signals) methods.

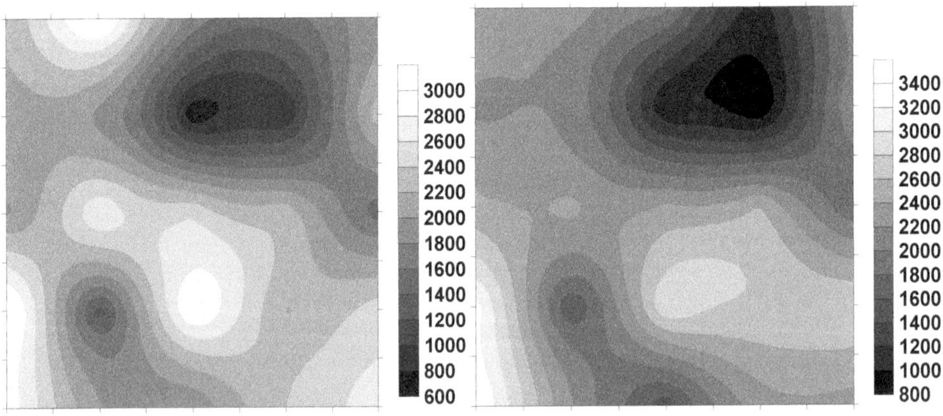

Fig. 1. Maps of velocity [m/s]. Sonic velocity (left) and ultrasonic velocity (right). Extracted from Concu et al., 2003a

The two velocity distributions appear quite similar. Substantially, the two methods show the same velocity range, with maxima and minima located in the same areas. Both maps reveal a well-defined region of minimum velocity in correspondence to the cavity. These results underscore that both methods are useful for revealing the presence of macroscopic anomalies inside stone masonry. The test also made it possible to demonstrate that environmental noises and vibrations impact more on sonic than on ultrasonic method; the latter is instead less practical since it calls for the use of a transducer rather than a hammer as the source, and this requires the precise positioning of source and receiver on the surface of the structure before each signal acquisition.

Independently of the frequency characteristics of the signal employed in the investigation, the parameters associated with the signal penetrating through the medium are the following:

- transit time (or travel time), that is, the time taken by the signal to cover the distance from the source to the receiver inside the material under examination;
- signal propagation velocity, in the sense of the ratio of the distance between source and receiver to transit time;
- signal attenuation characteristics in its passage through the material.

Traditional application of acoustic techniques is based on measurements of the velocity V of acoustic waves propagating through the material. The velocity is obtained from the ratio L/T, where T is the time wave needs to travel along the path of length L. The wave velocity is directly related to structure's elastic parameters, e.g. elastic modulus E, Poisson's number ν and density ρ, thus its analysis provides information crucial for inspections of structures inner conditions.

The propagation velocity, although a significant parameter, limits the investigation at the analysis of propagation times, not taking into consideration important information regarding the way the waves are propagating. For instance, when a wave passing through a specific item encounters any discontinuity, the wave power is certainly attenuated because of scattering phenomena, while the propagation time may not be moved if part of the signal is already able to reach the receiver. Therefore, it would be reasonable to approach the acoustic analysis also in terms of other wave's features changes and not only in terms of propagation times. The higher the intrinsic non-homogeneity level of structures, e.g. masonries, the bigger the advisability of this integrated approach. In fact, as documented by various studies predominantly performed in the geophysics and aeronautics environments, other wave's characteristics such as attenuation, scattering and frequency content, primarily related to the elastic wave power, may allow one to get more and relevant information about the material, because of the reliance of the propagation on the properties of the medium through which waves travel. In fact, different materials absorb or attenuate the wave power at different rates, depending on complex interactive effects of material characteristics, such as density, viscosity, homogeneity. Additionally, waves are reflected by boundaries between dissimilar materials, so that changes in materials structure, e.g. presence of discontinuities or defects, can affect amplitude, direction, and frequency content of scattered signals. Furthermore, all materials behave somehow as low pass filters, attenuating or scattering the higher frequency components of a broadband wave more than the lower. Thus, waves analysis in terms of multiple wave's features can give information on the combined effects of attenuation and scattering as previously described.

The comparative analysis of transit time and amplitude attenuation of acoustic signals carried out on the stone wall previously described confirmed the opportuneness of the joint use of both parameters (Concu et al., 2003b). In this study, we assumed the amplitude attenuation as the ratio of received signal to transmitted signal maximum amplitude. Fig. 2 shows the map of transit time and amplitude attenuation in the generic vertical section of the masonry.

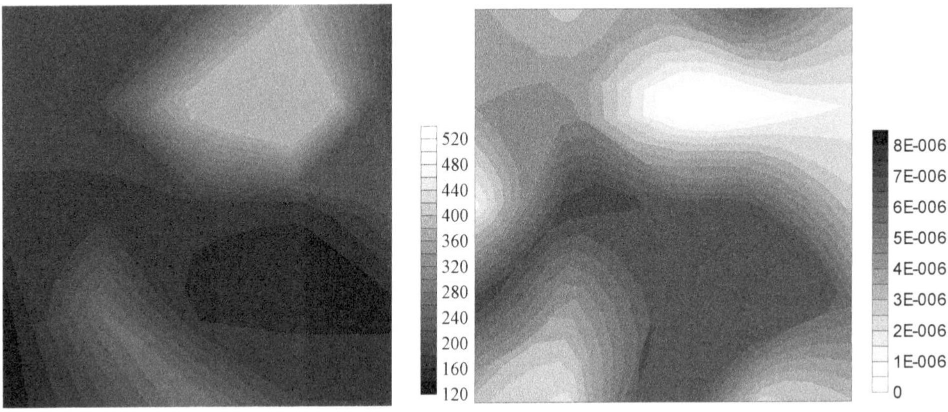

Fig. 2. Map of transit time [μs] (left) and amplitude attenuation (right). Extracted from Concu et al.,2003b

By observing the map we can see that the distribution of transit time shows a clearly defined section with maximum values in association with the cavity inside the wall. This section can be illustrated as a geometric figure of size and position perfectly compatible with the design geometry of the cavity before erection of the wall. Distribution of times in the rest of the map appears on the whole uniform. From this we can deduce that ultrasonic signal transit time (thus velocity) is a parameter capable of identifying macroscopic defects with a good degree of approximation and immediacy, while it does not appear to be sensitive to minor anomalies such as the absence of material between the stone blocks, the presence of mortar joints or small elements of different material. Interpretation of distribution concerning amplitude attenuation is less immediate: we can see a zone of maximum attenuation in correspondence to the cavity, the borders of which are rather blurred. Attenuation values are on the whole rather dispersed. This confirms that amplitude attenuation is a parameter extremely sensitive to all kinds of discontinuities of the material that cause a loss of signal energy and it is therefore quite suitable for use in tests at high definition in small areas. Transit time and amplitude attenuation thus have different diagnostic capabilities and this emphasizes the usefulness of different wave's features integrated use in structural faults diagnoses.

The study of energy characteristics associated with the signal is today addressed also by employing the spectral analysis. By spectral analysis is meant the method based on analysis of the frequency content of the signal travelling through the material of interest. When a signal goes through a medium, the frequency components associated with the input signal are altered since the medium acts as a filter that transmits only a certain frequency band with different degrees of attenuation and phase shifts. It is thus possible to study the effects of the medium's properties on alteration of the input signal by performing a spectral analysis on the signals passing through different materials or different portions of the same material or the same material in different surrounding conditions: a given input signal will emerge with different spectrum frequencies after passing through materials with different characteristics (Priestley, 1981). Substantially, the spectral analysis provides a sort of signature characteristic of the properties of the material travelled through. The material is gone through by a signal of the impulsive kind expressed by means of a function of type A(t), in which the oscillation amplitudes are given as a function of time. The spectrum of the signal offers instead a representation of the same in the frequency domain, in which the signal is expressed by means of a function of type a(f): the amplitudes of the elementary oscillations that make up the impulse are given as a function of the respective frequencies.

1.2.2 Acoustic emission

The acoustic emission technique is based fundamentally on the study of the acoustic signal emitted by the material during deformation, cracking, breaking, collapse and in general during whatever phase causing a release of energy. The single event, that is, the single emission, is thus an impulsive acoustic signal produced by a source within the material following the triggering of any phenomenon capable of releasing energy. The main parameters associated with acoustic emissions and used for the study and application of the method in the field of structural monitoring and faults detection are:

- number of events emitted during the phenomenon - deformation, loading and so on - under observation;
- emission velocity, in the sense of the number of events in a given time interval;
- maximum amplitude of the event;
- duration of the event;
- wave shape of the event;
- frequency spectrum of the event;
- arrival time of the signal at the transducers used for establishing it;
- the so-called "b-value", that is, the slope of the logarithmic curve representing the maximum amplitude of the events as a function of the frequency with which they appear.

These parameters are placed together with the factors that identify the specific issue being tested, such as the stress-deformation load diagrams, microscopic observations and so on, so as to find a relation that makes it possible to arrive at the state of the material starting from observation of the magnitudes associated with its acoustic emission.

The potentialities of the method for the study and monitoring of the behavior of materials, the prediction of their response to different kinds of stresses and the check of defects and anomalies are many. They are based on special phenomena, such as the Kaiser effect, on the application of instruments of mathematical analysis and numerical calculus, and on the comparative study of acoustic parameters and elastomechanical characteristics (Enoki & Kishi, 1991).

The Kaiser effect (Kaiser, 1950) is a phenomenon by which a material under stress emits acoustic signals which are significant only when the level of stress to which the material was previously submitted has been exceeded. In effect, there are emissions even below this level, but the two kinds of events differ greatly: in rapid succession and high energy content the former, associated with the triggering and propagation of new cracks; less frequent and with much lesser amplitudes the latter, associated with the deformation and contact between the surfaces of cracks opened in the previous load cycle. Substantially, the material has a recollection of its history of load and deformation. The uses of this phenomenon are many. It is used to arrive at the maximum stress that a material has undergone, for example the original stress of a rock, estimated by submitting to loading and unloading test samples taken from the rock mass. It is exploited in identifying the breakage surface associated with the material in different load conditions. It can be an indicator of the state of deterioration of the material, for example, of the damage caused to it by loads, since emissions of the second type appear for lower load levels in the most damaged materials.

The possibility of characterizing the source of acoustic emissions arises on using a suitable number of transducers (in any case more than three): it is in fact possible to return to the position of the signal source by recording its arrival times at the different transducers and knowing the value of the characteristic acoustic velocity of the material. By interpolation we find the point inside the material which, for the assigned velocity, satisfies the values of all the times recorded by the transducers. By means of a well-documented mathematical treatment based on Green's functions, on the analysis of tensorial moments and on the calculation of eigenvalue and eigenvector it is possible to identify not only the position but also the volume and spatial orientation of the source inside the medium, thus obtaining its complete geometric characterization (Ohtsu & Ono, 1986).

The comparative study of the acoustic emission parameters and factors of the elastomechanical type makes it possible to obtain many supplementary data on the behavior of the material. Acquisition of the acoustic emission parameters associated with knowledge of the diagrams of load, stress and deformation, as well as the characteristics of the material observed under the microscope before, during and after the phenomena under examination leads to the definition of precise correspondences between acoustic parameter values and certain fundamental data such as:

- level of load, stress and deformation;
- type of defect originating the emission (intra-, inter- and transgranular cracks);
- way of propagation of breaks (tensile or shear) and thus the kind of breakage observed;
- level of creep and the way in which it develops;
- dimension of the damaged zone in the material and so on.

From this naturally derives also the possibility of foreseeing the responses of the material to different kinds of stress, starting from the analysis of its acoustic emission. Substantially, the study of acoustic emissions has great potentialities in the monitoring of complex civil, industrial or natural structures, of single structural elements or laboratory test pieces made up of many different materials such as metals, cement, rock, ceramics, synthetic materials and so on. Interest concerning the use of the acoustic emission technique is proven also by the presence on the market of many kinds of often quite complex instrumental sets capable of acquiring and processing acoustic emission data in real time and proposing diagnostic hypotheses on the materials or structure under investigation by means of comparisons with databases on acoustic parameters. The application of this methodology is of great service in real-time monitoring of the triggering and propagation of breaks in materials in use as well as in the characterization of their behavior and the prediction of their responses to various stresses.

2. Operative procedures

2.1 Direct and indirect transmission

The easier and faster way to get relevant information using sonic and ultrasonic methods is the measurement of the waves propagation velocity V. In fact, the operative procedure to acquire waves velocity and to process data in order to get immediate results is quite simple. In addition, the skills of this parameter are very useful: in fact, from wave's propagation theory it is known that V is dependent on the following material's characteristics: dynamic elastic modulus E_d, Poisson's number ν and density ρ. For a homogeneous isotropic material this function is:

$$V = \left[\frac{E_d}{\rho}\frac{(1-v)}{(1+v)}(1-2v)\right]\frac{1}{2} \tag{1}$$

V is directly related to structures elastic parameters, so that it has been frequently applied for evaluating structures integrity and restoration's effectiveness. Thus the relation between V and E_d, ν, ρ can be exploited for achieving data regarding the structure's health in terms of elastomechanical conditions. In fact, the measurements of V along proper grids of paths leads to the elaboration of maps of velocity, which allow one to define the level of elastomechanical homogeneity of the investigated structure, emphasizing areas where anomalies are located; moreover, the knowledge of V values distribution consents to express a qualitative remark on the mechanical effectiveness of the structure, since it is empirically known that the higher the strength the higher the velocity.

Waves velocity measurements are preferentially carried out applying the Direct Transmission Technique (DTT), in which the wave is transmitted by a transducer (Emitter) through the test object and received by a second transducer (Receiver) on the opposite side. This allows measuring the time T that the wave needs to travel through the object's thickness, from the emitter to the receiver, along a path of length L; the average velocity of the wave is simply obtained from the ratio L/T. The DTT is very effective, since the broad direction of wave propagation is perpendicular to the source surface and the signal travels through the entire thickness of the item. Standards concerning the determination of waves velocity in structures, e.g. Europeans EN 12504-4 (EN 12504-4, 2004) and EN 14579 (EN 14579, 2004), suggest, therefore, the application of this kind of signals transmission.

Nevertheless, there are many kinds of structures, e.g. slabs, retaining walls, piers, in which the DTT cannot be performed, because only one side of the item is accessible. In these cases the Indirect Transmission Technique (ITT), in which both the emitter and the receiver transducers are placed on the same side of the investigated object, or the Semi-direct Transmission Technique (STT), in which transducers are placed on adjacent faces, might be used. Generally speaking, ITT and STT are less effective than the DTT because the amplitude of the received signal is lower, and the pulse propagates in a concrete layer just beneath the surface. These remarks have since now not allowed ITT and STT systematic development, and the scientific literature concerning their use is still quite poor. Despite that, ITT skills of ease to be performed, high potential to evaluate the quality and the characteristics of concrete covering on site, immediacy and low cost, claim to thorough examine its suitability in concrete diagnosis on site, and then to develop studies concerning the standardization of its application. In Annex A of the EN 12504-4 the determination of the pulse velocity via the ITT is illustrated. It is highlighted that there is some uncertainty regarding the exact length of the transmission path, since the areas of contact between the transducers and the item are of significant size. It is therefore suggested to make a series of measurements with the transducers at different distances apart to eliminate this uncertainty. The transmitting transducer shall be placed in contact with the item surface at a fixed point x, and the receiving transducer shall be placed at fixed increments x_n along a chosen line on the surface. The signals transit times recorded should be plotted as points on a graph, showing their relation to the distance separating the transducers. An example of such a plot is shown in Fig. 3, extracted from Annex A. The slope of the best straight line drawn through the points (tan ϕ) shall be measured and recorded as the mean pulse velocity along the chosen line on the concrete surface. Where the points measured and recorded in this way indicate a discontinuity, it is likely that a surface crack or surface layer of inferior quality is present and a velocity measured in such an instance is unreliable.

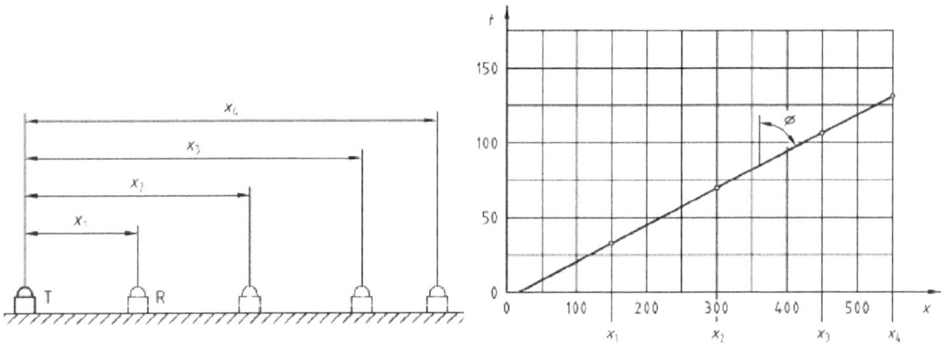

Fig. 3. Example of the determination of pulse velocity by ITT. Extracted from EN 12504-4, 2004 - Annex A

One of the main skills of the ITT is the possibility of cracks depth estimation. An estimate of the depth of a crack visible at the surface can be obtained by measuring the transit times across the crack for two different arrangements of the transducers placed on the surface. One suitable arrangement requires that the transmitting and receiving transducers are placed on opposite sides of the crack and equidistant from it (BS 1881, 1986). Two values of this distance x are chosen, one being twice that of the other, and the transit times corresponding to these are measured. If the first value of x chosen is x_1 and the second value x_2 and the transit times corresponding to these are T_1 and T_2 respectively, then the crack depth h is:

$$h = x_1\sqrt{\frac{4T_1^2 - T_2^2}{T_2^2 + T_1^2}} \quad (2)$$

Equation (2) is derived by assuming that the plane of the crack is perpendicular to the item surface and that the material in the surrounding area of the crack is of reasonably uniform quality. A check may be made to assess whether the crack is lying in a plane perpendicular to the surface by placing both transducers near to the crack and moving one of them far from the crack. If the transit time decreases, this indicates that the crack slopes towards the direction in which the transducer was moved. It is important that the distance x is measured accurately and that a very good coupling is guaranteed between the transducers and the concrete surface. The method is valid provided the crack is not filled with water.

Another method of test uses the indirect method where a discontinuity appears in the graph drawn following the indication of the standards (EN 12504-4, 2004; EN 14579, 2004) as previously explained. In this case, if L is the distance separating the transducers corresponding to which the slope of the line distance-time changes, while T_1 and T_2 are the transit times corresponding to this change, then the crack depth h is:

$$h = \frac{L}{2}\left(\frac{T_2}{T_1} - \frac{T_1}{T_2}\right) \quad (3)$$

As previously stated, the direction in which the maximum energy is propagated is at right angles to the face of the transmitting transducer, so that the DTT is the most effective operative procedure. However, the DTT has some limits too. The major limit consists in describing the wave's characteristics field of the object using for each path only one value of that characteristic, i.e., hypothesizing that the average value is homogeneous along each wave path. This assumption prevents from pinpointing the position of the detected anomaly inside the object. A promising way for overcoming this limit is the use of the tomographic technique, which uses numerical analysis as a real measurement instrument, combining the results of several DTT applications for a sharper and reliable investigation of the object.

2.2 Tomography

One emerging technique for advanced imaging of materials is Acoustic Tomography (AT). AT uses technology invented for the biomedical field to display the interior of engineered structures. The spatial distribution of acoustic velocity and attenuation are imaged and then correlated with properties directly related to physical conditions (Belanger & Cawley, 2009; Rhazi, 2006; Leonard Bond et al., 2000; Kepler et al.,2000; Meglis et al.,2005) . The velocity is determined by the elastic properties and density, while the attenuation is determined by the inelastic property of the medium.

2.2.1 Generalities

Travel time tomography, a type of AT, represents the natural evolution of the DTT: the signals emitted by different sources are acquired by several receivers arranged so as to allow the taking of a large number of measurements of transit time of signals travelling along pathways at different inclinations which intersect each other on flat sections of the structure. This makes it possible to apply an algebraic system whose unknowns are signal velocity at the nodes of a network arranged on the flat section of the medium containing source and receiver. Thus travel time tomography allows determination of velocity distribution on flat sections of the item being investigated. The method's degree of resolution depends on the distance between sources and receivers, on the measurement step, on angular coverage by means of the trajectories of the studied section. Travel time tomography makes it possible to overcome the major limit of the DTT, that is, the impossibility of discriminating an extended alteration along the source-receiver pathway from an anomaly confined to a part of it only, since for each trajectory joining source and receiver the behavior of the element inside the medium is described by means of a single velocity value. In virtue of the thick net of intersecting trajectories, travel time tomography returns velocity distribution on the section under study with high definition, making it possible to localize precisely and in detail any anomalies that may be present.

It has long been believed that attenuation is more suitable than velocity (or travel time) to study the inner properties of materials, because an anomaly has a greater effect on the attenuation of a signal than on the propagation time (Best et al., 1994; Hudson, 1981). In fact, as previously mentioned, wave's characteristics such as attenuation, scattering and frequency content may allow one to get relevant information about the material, because of the reliance of the propagation on the properties of the medium through which waves travel. Different materials absorb or attenuate the wave power at different rates, depending on complex interactive effects of material characteristics, such as density, viscosity, homogeneity. Additionally, waves are reflected by boundaries between dissimilar materials, so that changes in materials structure, e.g. presence of discontinuities or defects, can affect amplitude, direction, and frequency content of scattered signals. In this context special credit has to be given to spectral attenuation tomography, which returns the attenuation coefficient distribution on flat sections of the item being investigated.

The main limitations on the AT widespread diffusion are the longer times of execution compared to the traditional operative procedures, the higher cost of the instrumental sets (arrays of sources and receivers, multichannel acquirers), the complexity of the reconstruction of velocity or attenuation distribution starting from signals acquisition.

As previously stated, acoustic investigation methods exploit the transmission and reflection characteristics of mechanical waves with appropriate frequencies passing through the investigated item. Elastic waves propagate in different manner through different solid materials and cavities, thus enabling fault detection. The waves are in most cases generated by a piezoelectric transducer fed with a voltage pulse. The receivers are accelerometers, appropriate positioned based on the measurement type. The tomography represents an improvement to the classic techniques of wave direct transmission, being able to perform tests also on non-perpendicular wave paths. It is so possible to reconstruct a 2D image of the distribution of the wave propagation parameters (e.g. velocity, attenuation) within the analyzed structure, or in one of its sections. These images allow the identification of variations correlated with defects, malformations, cracks etc. Acoustic tomography implies that a ill posed linear equations system has to be solved, in order to determine the distribution of the chosen wave parameter (e.g. velocity, attenuation) in selected sections of the tested structure, thus highlighting the presence of anomalies (Berryman, 1991). Different inversion algorithms are available for determining this distribution starting from signal transmission and acquisition.

2.2.2 Travel time tomography

An acoustic wave propagating through any object spends a definite time to travel from a point to another of the object. The wave covers the path between the two points spending the time t and propagating with a mean velocity V. When the distance l between the two points reduces to zero a local velocity Vp, and a local slowness s = 1/Vp can be defined for the point p. In mathematical terms this behavior can be expressed by the following equation:

$$\int_{path} \frac{1}{V} dl = \int_{path} s \, dl = t \tag{4}$$

The acoustic behavior of a selected section of the object is then defined when the slowness s(x) is known continuously in every point x of the investigated section. This function can be approximated dividing the section into a grid of n rectangular cells (pixels) in which V is supposed to be constant. The tomographic problem consists in obtaining the slowness of the n pixels starting from the knowledge of m travel times t measured along a series of paths joining couples of transducers located on opposite or adjacent sides of the section. The waves paths depend on the velocity distribution, and their sharp definition is a not easy problem to solve, especially when dealing with structures made of different materials, such as stone masonry, or with a degree of intrinsic non-homogeneity, such as concrete; a valid approximation may be the linear tomography, which considers the paths to be straight. In order to obtain the values of the slowness in the grid, the following equations system, which is the matrix form for equation (4), has to be solved:

$$P\,S = T \tag{5}$$

where :
$T = [t_1, t_2, \ldots t_m]$ is the vector of measured travel times;
$S = [s_1, s_2, \ldots s_n]$ is the vector of slowness;
$P = [l_{11}, l_{12}, \ldots l_{nm}]$ is the coefficients matrix, whose generic element l_{ij} is the length of the ith path in the jth cell.
Thus, the tomographic solution consists in determining the vector S as:

$$S = P_g^{-1}\,T \tag{6}$$

To avoid instability in matrix inversion, the number n of cells must be smaller than the number m of measured travel times. If the inverse of P exists it can be directly evaluated. However, the inverse of P generally does not exist since P is not a square matrix, it is ill conditioned, and it has not full rank. Thus, other methods, such as iterative ones, have to be used to solve the problem.

2.2.3 Spectral attenuation tomography

As previously stated, the propagation velocity, although a significant parameter, limits the investigation at the analysis of propagation times, not taking into consideration important information regarding the way the waves are propagating. When a signal passes through a specific item or structure, certain frequency components of this signal are altered, the item behaving like a filter performing modifications on the frequency components' magnitude and phase. Therefore, it would be reasonable to approach the tomographic problem also in terms of frequency spectrum changes and not only in terms of propagation times. This type of approach is documented by various studies performed in the geophysics environment, and furthermore in the approach of Quan and Harris (Quan & Harris, 1997), based on the observation that the frequency attenuation increments with the frequency of the signal, meaning that the higher frequencies of a signal are more rapidly attenuated than the lower ones. The main advantage of the frequency analysis is the immunity against disturbing factors such as spherical divergence, reflection and transmission effects and coupling of the receiver with the transmitter, which can affect the correct interpretation of the received signals.

For the purpose of estimating attenuation, the process of waves propagation can be assumed as described by linear system theory. If the amplitude spectrum of an incident wave is S(f) and the instrument-medium response is G(f)·H(f), then the received amplitude spectrum R(f) may be, in general, expressed as :

$$R(f) = G(f) \cdot H(f) \cdot S(f) \tag{7}$$

where the factor G(f) includes geometrical spreading, instrument response, source-receiver coupling, radiation patterns, and reflection-transmission coefficients, and the phase accumulation caused by propagation, and H(f) describes the attenuation effect on the amplitude. It can be assumed that the effects included in factor G(f) are not frequency dependent, thus it can be simplified as G(f) = G. In structural diagnosis, the H(f) factor is of greater interest. Experiments indicate that attenuation is usually proportional to frequency (Johnston, 1981), that is, response H(f) may be expressed as:

$$H(f) = \exp\left(-f\int_{path}\alpha_0 dl\right) \tag{8}$$

where the integral is taken along the supposed straight wavepath, and α_0 can be regarded as an intrinsic attenuation coefficient. The tomography's goal is to estimate the medium response H(f), or more specifically, the attenuation coefficient α_0, from knowledge of the input spectrum S(f) and the output spectrum R(f). A direct approach is to solve equation (8) by taking the logarithm and obtaining:

$$\int_{path}\alpha_0 dl = \frac{1}{f}\ln\left[G\frac{S(f)}{R(f)}\right] \tag{9}$$

Equation (9) may be used to estimate the integrated attenuation at each frequency and is called the amplitude decay method. However, as described above, the factor G lumps many complicated processes together, and is very difficult to be determined. Furthermore, the calculation of attenuation based on individual frequencies is not robust because of poor individual signal-to-noise.

To overcome these difficulties, Quan and Harris (Quan & Harris, 1997) developed a statistically based method that estimates the attenuation coefficient α_0 from the spectral centroid downshift over a range of frequencies. An analog relationship to that between signal velocity along the wavepath and travel time connects the attenuation to the difference of the signals' spectrum frequency centroid, the latter being a parameter indicating the center of the signals' distribution in frequency. As mentioned, during wave propagation the higher frequencies are attenuated more rapidly than the lower frequency components, downshifting the centroid towards the lower frequencies. The centroid frequency of the input signal S(f) is defined as:

$$f_S = \frac{\int_0^\infty fS(f)df}{\int_0^\infty S(f)df} \tag{10}$$

and the variance is:

$$\sigma_{S2} = \frac{\int_0^\infty (f-f_S)^2 S(f)df}{\int_0^\infty S(f)df} \tag{11}$$

Similarly, the centroid frequency of the received signal R(f) is:

$$f_R = \frac{\int_0^\infty fR(f)df}{\int_0^\infty R(f)df} \tag{12}$$

and its variance is:

$$\sigma_{R2} = \frac{\int_0^{\infty}(f-f_R)^2 R(f)df}{\int_0^{\infty} R(f)df} \quad (13)$$

where R(f) is given by equation (7). If the factor G is independent on the frequency f, then f_R and σ_R^2 will be independent on G. This is a major advantage of using the spectral centroid and variance rather than the actual amplitudes. For the special case where the incident spectrum S(f) is Gaussian, assuming a linear-dependence model of attenuation to frequency, the attenuation coefficient α_0 can be estimated as follows:

$$\int_{path} \alpha_0 dl = \frac{f_S - f_R}{\sigma_{S2}} \quad (14)$$

where f_S and f_R are the centroid frequency for the source and receiver, respectively, and σ_S is the variance, or bandwidth, of the source signal. The previous relationship states that the attenuation is proportional to the centroid frequency difference which has downshifted from the original source centroid f_S, to the centroid of the received signal f_R. The total amount of centroid frequency downshift depends on the attenuation characteristics along the acoustic path. The tomographic formula relating frequency shift with the attenuation projection is exact only for Gaussian spectra. Yet, similar derivations can also be obtained for other frequency compositions, which implies that the estimates of relative attenuation are not sensitive to small changes in spectrum shapes, and points out the robustness of this method.

It is worth noting that equation (14) is in the same form as (4), with the intrinsic attenuation coefficient α_0 in (14) corresponding to the slowness 1/V in (4). The expression of frequency centroid difference in (14) corresponds to the travel time t in (4). This similarity makes the attenuation tomographic inversion easy to conduct applying the same algorithms developed for travel time tomography, simply replacing 1/V with α_0 and t with $(f_S - f_R)/ \sigma_S^2$.

As stated above, equation (14) is the basic formula for spectral attenuation tomography. It can be written also in a discrete form as:

$$\sum_{j=1}^{N} l_{ij}\, \alpha_{0j} = \frac{f_{Si} - f_{Ri}}{\sigma_{Si}^2} \quad (i= 1,..M) \quad (15)$$

where i represents the ith path, j the jth parameterized cell of the medium, and l_{ij} is the length of the ith path in the jth cell (Fig. 4).

The previous equations system can be written in matrix form as:

$$L\, A = F \quad (16)$$

where :
F = $[F_1, F_2, \ldots F_m]$ is the vector of calculated centroid frequency downshift, in which $F_i = (f_{Si} - f_{Ri})/ \sigma_{Si}^2$ has been assumed;
A = $[\alpha_{01}, \alpha_{02}, \ldots \alpha_{0n}]$ is the vector of attenuation coefficients;
L = $[l_{11}, l_{12}, \ldots l_{nm}]$ is the coefficients matrix, whose generic element l_{ij} is the length of the ith straight path in the jth cell.

The intrinsic attenuation coefficient α_0 is in the unit of [$dBsm^{-1}$].

Moreover, the attenuation coefficient α_0 can be also expressed as:

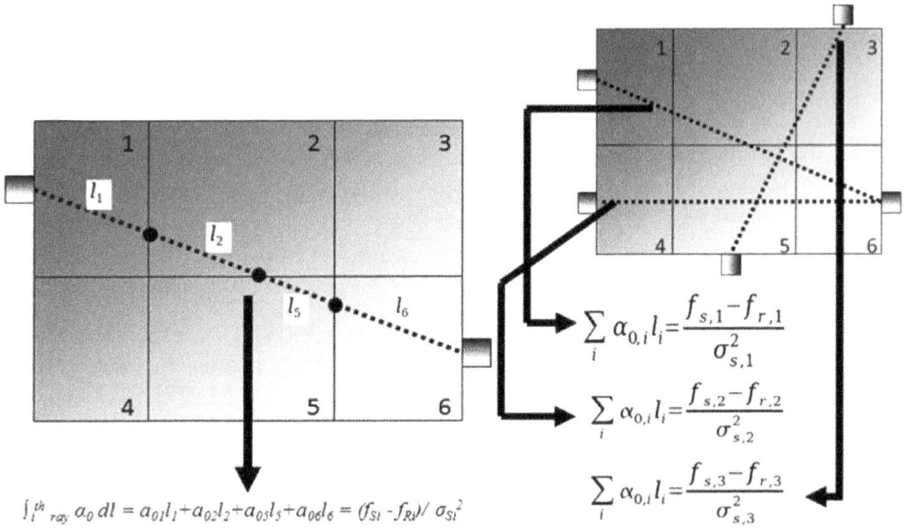

Fig. 4. Example of spectral attenuation tomography system equation

$$\alpha_0 = \frac{\pi}{QV} \tag{17}$$

with Q the quality factor of the material and V the propagation velocity. In wave propagation problems the Q factor is useful for characterizing wave attenuation, being defined as the ratio of the total kinetic energy and energy loss in one vibration cycle (Sheriff & Geldart, 1995; Knopoff, 1964). Geophysicists and seismologists often use the Q factor to study this attenuation in rocks. An infinite Q means that there is no attenuation. This factor is a function of the mineral composition of rocks as well as of their mechanical performances (Ilyas, 2010). Numerous field observations have demonstrated that the quality factor Q appears to be a constant over a large frequency range in the signal bandwidth. This is widely accepted in the geophysics community and is referred to as the constant Q model. Hence, the tomographic inversion can be applied for the case of the spectral attenuation in a similar way as for the arrival times and the propagation velocity. The model obtained for Q is consistent with the one obtained for the velocity; therefore the information regarding the velocity distribution can be used for calculating Q itself.

2.2.4 Resolution algorithms

A common method to obtain the solution of equations system (5) in the least square sense is the Singular Value Decomposition (SVD) (Berryman, 1991; Herman, 1980; Ivansson, 1986). The SVD can be used for computing the pseudo-inverse of the coefficient matrix P. Indeed, this method produces a diagonal matrix D, of the same dimension of P and with non-negative diagonal elements in decreasing order, and two unitary matrices U and V so that $P = UDV^T$. Then, the pseudo-inverse of the matrix P with singular value decomposition is $P^+ = VD^{-1}U^T$ and the solution of equations system (5) can be written as $S = P^+T$. The same result can be assumed for the spectral attenuation tomography, simply considering system (16) instead of system (5). This solution is the minimum norm solution and it is only mathematically suitable. The inverse problem is ill posed and ill conditioned, making the solution sensitive to measurement errors and noise. Regularization methods are needed to treat this ill posedness. It can be shown that the small singular values mainly represents the noise and can be discarded. Truncated SVD (SVDT) may be considered as having a filter, and hence it is less sensitive to high frequency noise in the measurements.

Because of the non linear relationship velocity - travel time and attenuation coefficient - centroid frequency downshift, it is almost impossible to find the solution for systems (5) and (16) by a single step algorithm using a linear approximation. Thus, iterative methods (Gilbert, 1972; Lo & Inderwiesen, 1994) can be used, such as Algebraic Reconstruction Technique (ART) (Gordon, 1974; Gordon et al., 1970) and Simultaneous Iteration Reconstruction Technique (SIRT) (Dines & Lytle, 1979; Lakshiminarayanan & Lent, 1979; Jansen et al., 1991). Both methods need a starting value of velocity or attenuation, and then they modify iteratively this value by minimizing the difference between the measured travel time or centroid frequency downshift and the value calculated in the previous iteration. While ART goes on wavepath after wavepath, SIRT takes into account the effect of all wavepaths crossing each cell. In the n-dimensional space each equation in (5) and (16) represents a hyperplane. When a unique solution exists, the intersection of all the hyperplanes is a single point. A computational procedure to locate the solution consists in starting with an initial solution, denoted by:

$$q^{(0)} = (q_1^{(0)}, q_2^{(0)}, \dots, q_n^{(0)}) \tag{18}$$

where q signifies velocity or attenuation coefficient depending on whether travel time or spectral attenuation tomography is being performed. This initial solution is projected on the hyperplane represented by the first equation in (5) or (16) giving $q^{(1)}$. This value is then projected on the hyperplane represented by the second equation in (5) or (16) to yield $q^{(i-1)}$ and so on. When $q^{(i-1)}$ is projected on the hyperplane represented by the ith equation to yield $q^{(i)}$, the process can be mathematically described by:

$$q^{(i)} = q^{(i-1)} + \frac{b_i - l_i^T q^{i-1}}{l_i^T l_i} l_i \tag{19}$$

where l_i is the ith raw of the matrix L and b_i represents t_i or F_i depending on whether travel time or spectral attenuation tomography is being performed. A single iteration of ART is completed when each row of P has been cycled.

For an over determined problem, m > n, ART does not give a unique solution, but this depends on the starting point. The tomographic system is normally over determined and measurement noise is present. In this case a unique solution does not exist and the solution found by ART will oscillate in the neighborhood of the intersections of the hyperplanes. The SIRT algorithm uses the same equations as in the ART algorithm; the difference is that SIRT modifies the attenuation model taking into account at each iteration the effect of all wave paths crossing each cell. The new value of each cell is the average value of all the computed values for each hyperplane:

$$q_j^{(k+1)} = q_j^{(k)} + N_{jj}^{-1} \sum_{i=1}^{m} \frac{\left(b_i - \sum_{w=1}^{n} l_{iw} q_w^{(k)}\right) l_{ij}}{\sum_{r=1}^{m} l_{ir}^2} \tag{20}$$

Then, using the SIRT algorithm better solutions are usually obtained at the expense of slower convergence.

2.3 Impact echo method

Another development of the classic operative procedures is the impact-echo model, which represents in some way an evolution in the ITT. This method is based on the principle of reflection: a signal propagating in a medium is reflected on encountering an anomaly of any kind inside it (Sansalone & Streett, 1997; Carino, 2001). On striking the medium with a succession of impulses, usually generated by mechanical impact with a hammer, these are reflected from the interface between the medium and the air if the piece is homogeneous or by the defect if the medium presents an interruption in continuity; in this case the reflection caused by the limiting surface is addressed as base echo, while the reflection caused by any imperfections there may be in the medium is addressed simply as echo. By arranging source and receiver on the surface of the medium it is possible to visualize with suitable instrumentation the echoes generated by the reflection of the signal in the medium; this makes it possible to establish the kind of reflection it is, that is, if it is caused by limit surfaces or something else, and to locate the obstacle inside the medium as a function of amplitude and the reciprocal position of the signals corresponding to the echoes. The extraction of information usually takes place in the frequency domain.

2.4 Measurements sets

The basic instrumental set for the performance of sonic and ultrasonic tests is composed of a source, a receiver and a data acquisition and processing unit, often an oscilloscope and PC. The fundamental difference lies in the characteristics of the source: the sonic method calls for the use of signals containing high energy and thus characterized by low frequency for which the excitation of the material is usually performed with the impact of an instrumented hammer; the ultrasonic method requires the use of high-frequency signals since no excessive dissipation of energy inside the material is foreseen, so the signal is introduced by means of a transducer, usually piezoelectric. Fig. 5 shows the schematic of instrumental sets commonly used in the two kinds of tests.

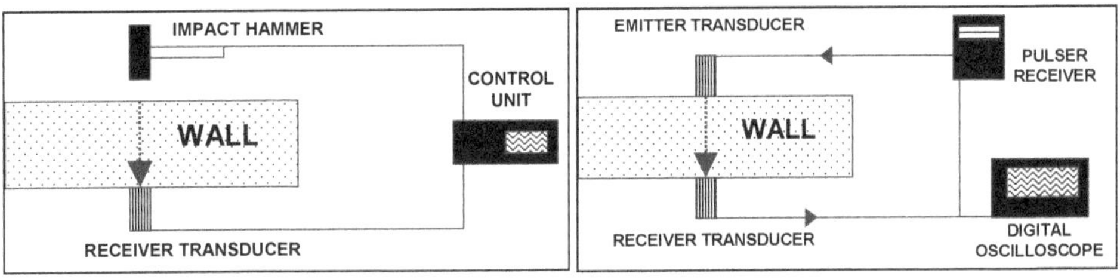

Fig. 5. Instrumentation set up. Sonic Testing (left) and Ultrasonic Testing (right)

When applying the Acoustic Emission procedure, the signal associated to the acoustic emission can be captured by a suitable instrumental setup, usually composed of high-frequency transducers, amplifiers, filters, systems of acquisition and storage such as digital oscilloscopes or acquisition cards, data processing software and so on, as shown in Fig. 6.

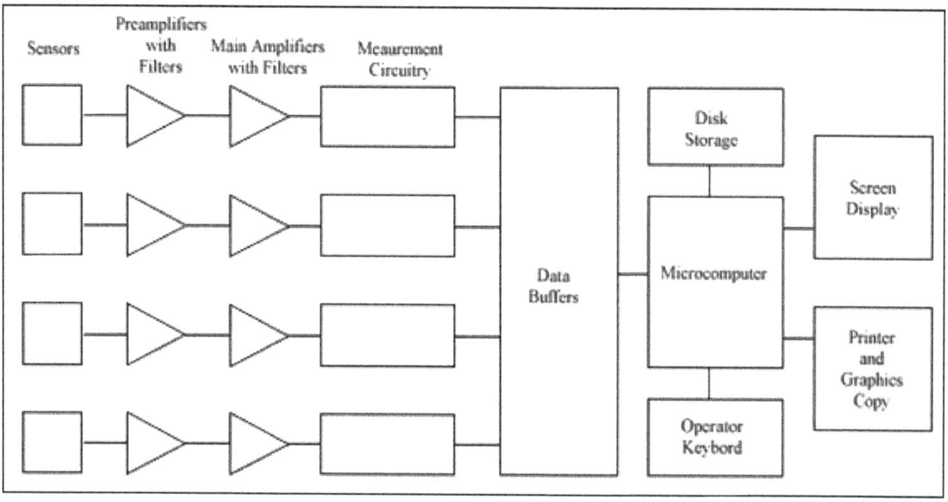

Fig. 6. Schematic diagram of a basic four-channel Acoustic Emission testing system. Extracted from NDT Education Resource Center

Impact-echo testing relies on three basic components, as shown in Fig. 7:

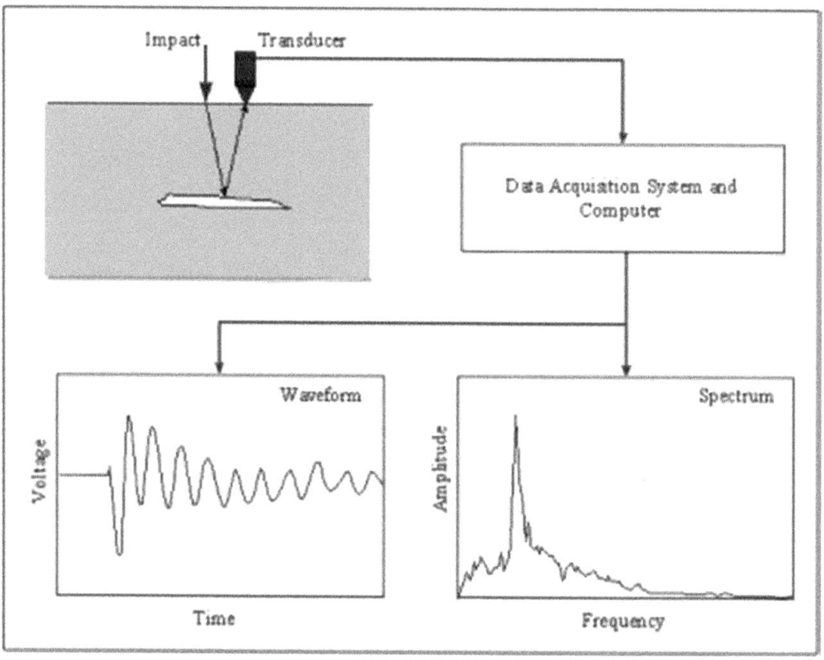

Fig. 7. Schematic diagram of the Impact-Echo test. Extracted from Sansalone & Streett, 1997

- a mechanical impactor capable of producing short-duration impacts, the duration of which can be varied;
- a high-fidelity receiver to measure the surface response;
- a data acquisition-signal analysis system to capture, process, and store the waveforms of surface motion.

3. Case study

In order to deepen the reliability of acoustic methods in buildings faults detecting and materials characterization, an experimental program has been started, experimenting both DTT and AT approaches to a full scale masonry model.

3.1 Materials

The two operative procedures - DTT and AT - have been carried out on a full scale real stone masonry expressly made by the Lab of Structural Engineering Dept. The wall is 0.90 m wide, 0.62 m high and 0.38 m thick, and it is made of Trachite blocks sized 0.20 m × 0.38 m × 0.12 m, settled as shown in Fig. 8 and jointed with cement lime mortar. The block assigned to the central position of the wall was not settled, thus realising a macro-cavity with the same size of the missing block, and assumed as a known anomaly. Mortar joints have been assumed to be 1 cm thick, but since the wall was manually built by a builder, actual dimensions are not so precise.

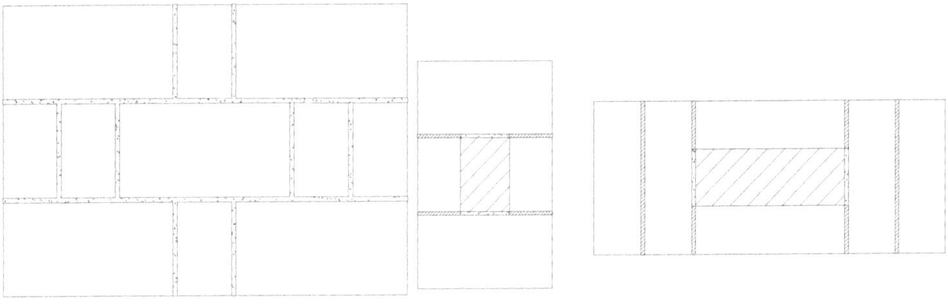

Fig. 8. The full scale real masonry. From left to right: front view, vertical section, horizontal tomographic section

Trachite specimens have been prepared and then tested for the determination of compressive strength and elastic modulus, following the Italian Standards UNI EN 1926, 2000 (UNI EN 1926, 2000) and UNI EN 14580, 2005 (UNI EN 14580, 2005) respectively. Results are shown in Table 1.

Properties	Value (MPa)
Compressive strength	40.5
Static elastic modulus	6100

Table 1. Materials properties

3.2 Methods

3.2.1 Direct transmission technique

As previously stated, when using the DTT the acoustic signal is transmitted through the test object and received by a second transducer on the opposite side of the structure. Changes in received signal provide indications of variations in material continuity. In this case study 220 emitters and 220 receivers have been arranged in a grid of 11×20 nodes in the opposite surfaces of the wall (Fig. 9).

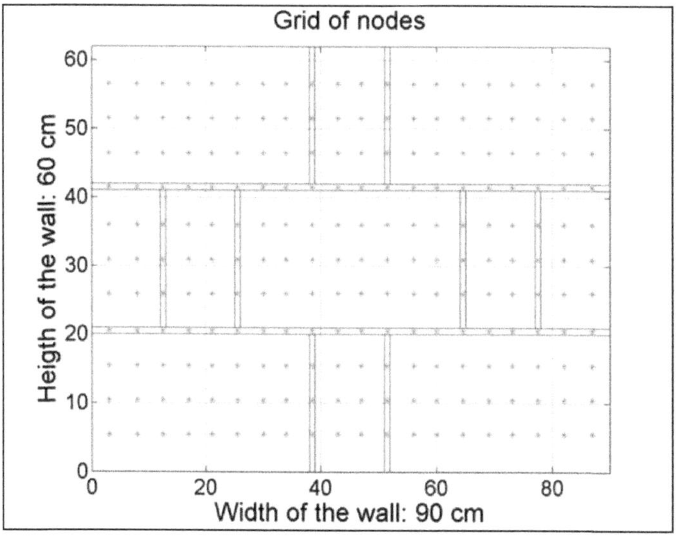

Fig. 9. Grid of test points

At the end of the experimental sessions, 220 signals have been obtained, one for each point of the grid of receivers. From each node of the grid, two parameters - velocity and signal power - have been extracted, in order to investigate the presence of anomalies and obtain some information on the material. The signal power is defined as:

$$P_{Total} = \frac{\int_0^T |x(t)|^2 dt}{T} \tag{21}$$

where T is the time duration of the received signal x(t).

The excitation wave is an acceleration signal. It is a seven and half-cycle tone burst enclosed in a Hanning window, expressed by the following equation (22) and shown in Fig. 10. This waveform has been chosen to reduce the leakage phenomena. The parameter f = 60 kHz is the characteristic frequency of the emitting transducer.

$$y(t) = \frac{1}{2}\sin(2\pi ft)\left[1 - \cos\left(\frac{2}{15}2\pi ft\right)\right] \tag{22}$$

The emitter and receiver are piezoelectric transducers with a frequency of 60 kHz. The emitter is connected to a signal generator PCG10 Velleman Instruments®. Both transducers are connected to a digital oscilloscope interfaced with a laptop. Due to the strong

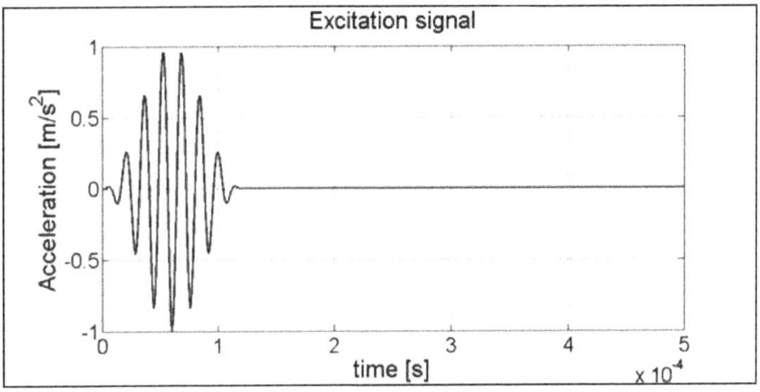

Fig. 10. Excitation signal used in the DTT

attenuation of the transmitted signal, the peak to peak voltage of the generator has been amplified from 13V to 39V using a transformer between the signal generator and the emitter. The transformer has a transformation ratio of 1/3 and a frequency band of (10-200) kHz. Moreover, an amplifier has been connected between the receiver and the oscilloscope. It has gain equal to 100 or 200, a frequency band of (10-200) kHz and input impedance of 1MΩ. The experimental setup is shown in Fig. 11.

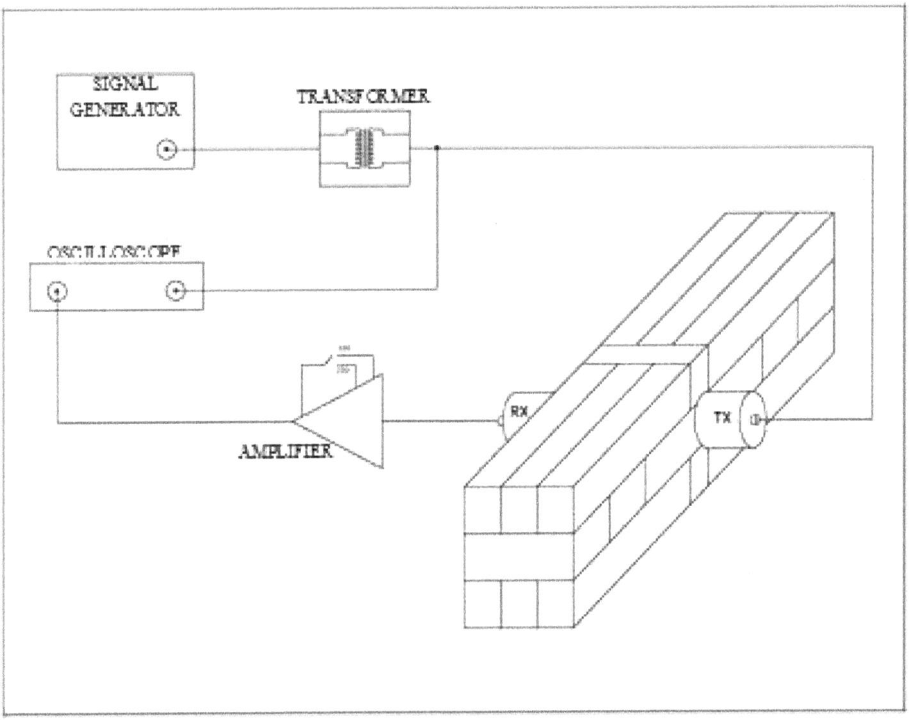

Fig. 11. Experimental setup for DTT application

3.2.2 Acoustic tomography

Aim of the experimental program was also to point out the reliability of both travel time and spectral attenuation AT in detecting building structures faults. With this purpose, the algebraic problem of the tomography inversion has been deepened, the SIRT algorithm has been selected and then numerically developed. After that, the solving algorithm has been implemented in an automated procedure that allows the user to easily obtain a map of the distribution of acoustic parameters (velocity and attenuation) in the selected section of the item. The AT has been applied to a horizontal plane section crossing the wall in order to intercept the central void (Fig. 8). The investigated section was thus 0.90 m wide and 0.38 m thick, and it has been divided in 40 cells 0.09 m × 0.095 m. By using this measurements configuration, the section is crossed by 138 paths, and its coverage by the wave paths is excellent. For each path the quantity $(f_S - f_R)/\ \sigma_S^2$ has been calculated. Thus, systems (5) and (16) consist in 138 equations and 40 unknowns, so as the two equations systems are satisfactorily over determined.

Three kinds of signal have been evaluated in the experimental setup: pulse, sweep and chirp signal. The short voltage pulse p(t), defined as:

$$p(t) = A\,\mathrm{rect}\left(\frac{t}{T}\right) \tag{23}$$

where T is the pulse duration and A the amplitude, is generally preferred for estimating the travel time of elastic waves because it involves high power; on the other hand, when using this signal only a poor control on the signal spectrum is usually achievable, thus making the spectral tomography hardly possible. Because of the huge importance of the signal to noise ratio for FFT spectral analysis, a broadband sweep signal has been preferred as source signal. This signal, expressed by the following:

$$s(t) = A cos\left(2\pi f_0 t + \frac{k}{2} t\right) \tag{24}$$

where T is the pulse duration, f_0 is the lower bound of the frequency bandwidth which increases with $k = 2\pi\Delta f/T$ rate and A is the amplitude, shows a linear relationship between time and frequency. The purpose of using this signal was to extend the frequencies involved in measurements up to 300 kHz, aiming at increasing the analysis resolution and at estimating the attenuation coefficient more accurately in a wider signal band. Indeed, the evaluation of the spectral attenuation coefficient requires a spread spectrum of both transmitted and received frequencies in order to accurately estimate both centroid frequency and variance values. Finally, the feasibility of using a chirp signal, described by the equation:

$$s(t) = A\cos\left(2\pi f_0 t + \frac{k}{2} t^2\right)\,\mathrm{rect}\left(\frac{t}{T}\right) \tag{25}$$

has been evaluated. This kind of signal allowed the performing of received signal spectrum analysis similar to that achievable using the sweep one; moreover, it allowed cross-correlating the received signal with the source one for travel time estimating, so as to perform an effective travel time tomography.

Based on the requirements posed by the mathematical background presented in the previous sections, an innovative measurements system was set-up according to the diagram presented in Fig. 12.

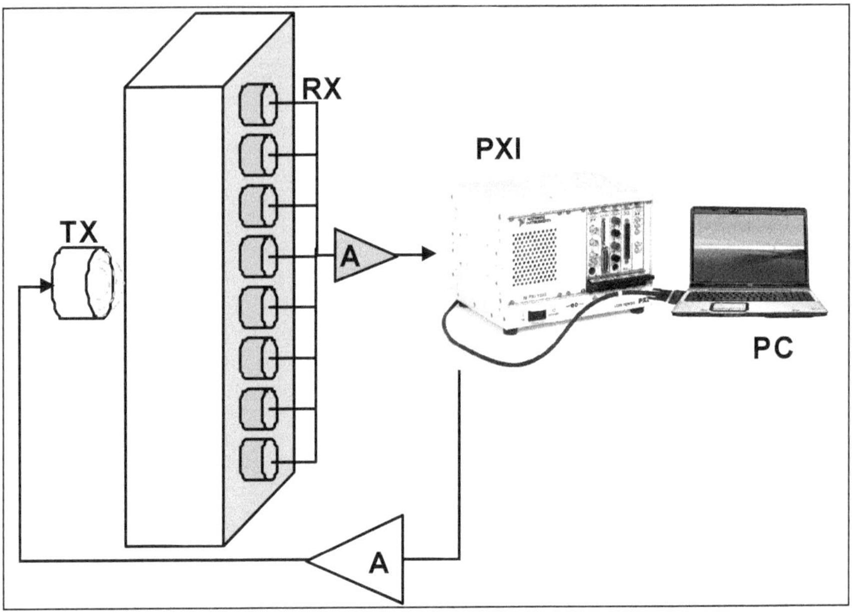

Fig. 12. Set-up of the measurement system

In order to simplify the measurement process, a multi-receiver solution was adopted, using eight accelerometers as receivers for the generated wave. More detailed, the instrumentation setup is composed of the following elements:

- broadband piezoelectric transducer GRW350-D50 Ultran® used as an ultrasonic wave generator. The transducer has diameter of 520 mm, a central frequency of 370 kHz and is fed with a high voltage signal (200 Vpp) for overcoming the high impedance of the analyzed materials;
- eight VS-150-M Vallen® piezoelectric sensors (accelerometers) with a good frequency response in the band of interest (100 - 500 kHz). Each of these sensors is coupled with its own preamplifier with a gain of 40 dB, thus ensuring an elevated sensibility;
- data acquisition system National Instruments® PXI DAQ with two PXIe-6124 boards having a total of 8 analog inputs and 2 analog outputs. The inputs have a sampling rate of 4 MS/s each, at 16-bit resolution. The PXI rack is connected through a PCI-Express card to a laptop computer that commands the entire measuring process.

The interface between the PXI data acquisition rack and the computer is based on the virtual instruments created in the LabView® environment, used both for acquiring the output of the eight reception sensors as for generating the sweep signal for the piezoelectric transducer. This signal is generated on one of the analog outputs of the acquisition board and is subsequently amplified using a custom-build amplifier.

The entire measuring process is encapsulated in a user-friendly application that guides the user by the means of a step-by-step procedure. At the beginning of the procedure, the physical parameters of the structure can be specified, allowing the application to calculate the number and position of the measurement points on the surface of the structure at each step of the measure. Then the application triggers the generation of the signal and acquires the output of the accelerometers for each measurement step. At the end of the cycle, the tomographic algorithm previously mentioned is applied on the entire set of acquired data in order to produce the tomographic map of the analyzed section.

3.3 Results

Results of both DTT and AT operative procedures have been displayed in terms of distribution maps, in order to facilitate data interpretation and anomalies identification. The maps are represented by a 256-levels gray-scale diagram, where the lowest level corresponds to the white color, and the highest level corresponds to the black color. The 256 levels are normalized with respect to the range of the parameter values measured in the represented map.

The map of the distribution of both signal velocity (Fig. 13) and signal power (Fig. 14) has been derived interpolating the data recorded in the grid's nodes (Cannas et al., 2008). Each maps emphasizes areas where anomalies are located, according to feature's specific diagnostic skill. In the center of both the maps the presence of the cavity region can be clearly seen, although it is smaller than expected. The distribution of propagation velocity in the rest of the map seems to be quite uniform, thus confirming that this parameter can easily detect macroscopic anomalies, while it seems to be less affected by minor materials discontinuity such as the mortar joints. The distribution of the signal power, on the other hand, allows to better distinguishing the cavity in terms of both position and extension, and points out that the various interfaces inside the structure affect this feature much more than the velocity.

Fig. 13. Propagation velocity [m/s]: distribution map.

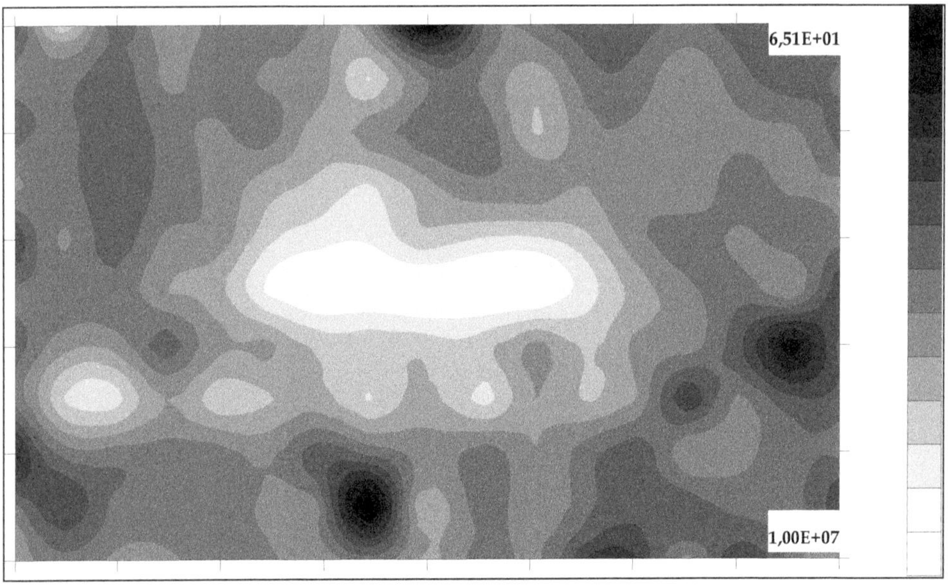

Fig. 14. Signal power [mV]: distribution map.

Figures 15 and 16 show respectively the results of the performed Travel Time and Spectral Attenuation Tomography measurements on the selected horizontal section of the full-scale real masonry (Concu et al., 2010). Showed results of Spectral Attenuation AT have been achieved by using the sweep signal as source.

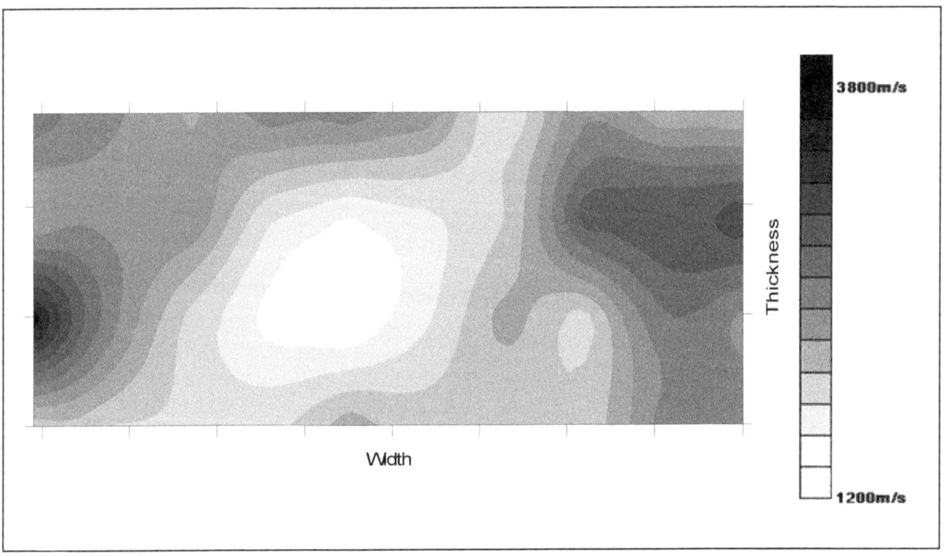

Fig. 15. Travel Time Tomography measurements

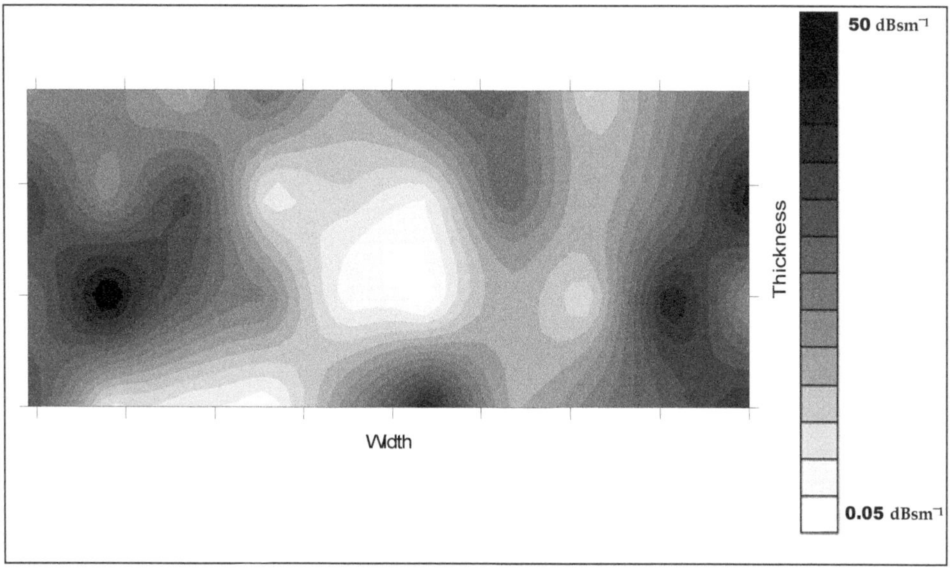

Fig. 16. Spectral Attenuation Tomography measurements

It can be noticed that both maps identify a central region corresponding to the position of the real cavity. The velocity map presents a poor quality in terms of the position and also the extension of the cavity, both parameters being significantly improved by using the attenuation map. It can be also seen that the attenuation map presents a high degree of scattering, confirming the hypothesis that this parameter is very sensible to all discontinuities of a section such as joints, small irregularities, various interfaces. It is worth noting that a tighter grid of measurements would enable a better definition of the cavity's shape, implying longer measuring and computational times.

Results of DTT and AT thus emphasize the main role played by frequency dependent features, such as signal power and spectral attenuation coefficient, when dealing with materials and structures with a quite important degree of non homogeneity. It is worth stressing that the maps derived from DTT do not allow the anomaly to be collocated at its real and correct distance from the wall surface, while the AT maps give a faithful picture of the selected section.

4. Comments on measurements problems

Acoustic measurements are still affected by a quite large amount of uncertainness in terms of stability and reproducibility. This is primarily due to the large number of factors that somehow exert influence on measurements:

- materials (chemical-physical-mechanical conditions);
- signals;
- acoustic parameters;
- instrumental sets;
- data processing methods;

- environmental conditions;
- human factor.

Another big problem concerning acoustic measurements is the gap that exists between the theoretical and interpretative bases and the codification of operative modalities, which is to say the almost total lack of standardization procedures. Actually, only standards concerning the determination of ultrasonic pulse velocity in concrete and natural stones are available (EN 12504-4, 2004; EN 14579, 2004). These standards take into account the problem of measurements stability, repeatability and reproducibility, and address the main factors influencing pulse velocity measurements. It is stated that to obtain some measure of the acoustic velocity that is reproducible and which is essentially a function of the properties of the material (concrete and stone) submitted to testing, it is necessary to take into consideration the different factors that exert an influence on the velocity. This is also essential to establish the correlations existing with the different physical characteristics of the material. The factors that should be carefully considered are the following:

a. moisture content and temperature of the concrete
b. water content of the stone
c. path length
d. shape and size of the specimen
e. cracks, fissures and voids.

- Moisture and temperature of the concrete and water content of the stone
  Moisture content has both chemical and physical effects on ultrasonic pulse velocity. These effects are important especially for establishing correlations with concrete strength. As an example, acoustic velocity can vary significantly between a properly cured standard specimen and a structural element made of the same concrete, because of the influence of curing conditions on cement hydration and the presence of free water in the voids. In the same way, water content has some effects on pulse velocity propagation in stones. Stone humidity, that is to say the presence of water inside the pores, can cause a variation of ultrasonic velocity value up to 50 % with respect to dry specimens or structural components. Concrete temperature effects on ultrasonic pulse velocity should be taken into account only outside the range 10°C - 30°C. Within this range, no significant change in pulse velocity has been experimentally found if corresponding changes in strength or elastic properties do not occur.
- Path length, specimen shape and size
  The path length over which the pulse velocity is measured should be long enough not to be significantly influenced by the heterogeneous nature of the concrete or the stone. For concrete specimen the standard recommends minimum values of the path length depending on aggregates nominal maximum size. The velocity is not generally affected by the variations in the path length, even though the electronic timing devices are likely to give some indications that the velocity can slightly decrease if the length of the path increases. This is because the higher frequency components of the pulse are attenuated more than that lower frequency components and the shape of the onset of the pulse becomes more rounded with increased distance travelled. Thus, the apparent reduction of pulse velocity arises from the difficulty of defining exactly the onset of the pulse and this depends on the particular method used for its definition. The apparent reduction in velocity is normally slight and is within the accuracy of measurements commercial apparatus. Nevertheless, particular care shall be taken when measurements are carried out on important path lengths. The velocity of short vibratory impulses is independent

of the size and shape of specimen in which they travel, unless its least lateral dimension is less than a minimum value. Below this value, the pulse velocity can be reduced appreciably. The amount of this reduction depends primarily on the ratio ultrasonic wavelength to smallest lateral dimension of the specimen, but is insignificant if the ratio is less than unity. Standards give the relationships existing between pulse velocity, transducers frequency and the smallest admissible lateral dimension of the test specimen. If the minimum lateral dimension is less than the wavelength or if the indirect transmission arrangement is used, propagation mode changes and, therefore, measured velocity will be different. This is particularly important in cases where elements of significantly different sizes are being compared.

- Cracks, fissures and voids
  When a wave passing through a specific item encounters any discontinuity, the wave power is most likely attenuated because of scattering phenomena, so that any crack, fissure or void inside the item might obstacle wave propagation. The higher the projection of defects length with respect to transducers size and wavelength, the greater the influence on wave propagation. In this case, the first impulse picked up by the receiving transducer will undergo a diffraction at the edge of the anomaly and this gives a longer path time compared to a transmission taking place in items with no fissures or voids. If part of the signal is already able to reach the receiver - e.g. in cracked elements where the broken sides are kept firmly in contact by compression - then the propagation time may not significantly change with respect to paths with no defects throughout. Transit time is not a decisive parameter also when the crack is filled with a liquid able to transmit wave power. In all these cases, examination of signal attenuation may also provide helpful information. Therefore, it would be reasonable to approach the ultrasonic analysis not only in terms of propagation times, but also in terms of other wave's features changes. The higher the intrinsic non-homogeneity level of structures, e.g. masonries, the greater the advisability of this integrated approach.

The different applications of sonic and ultrasonic investigations have brought to the fore two fundamental issues in the performance of such tests and their repeatability: the influence of pressure exerted on the transducers employed and the effect on acquired data of the application of acoustic coupling agents inserted between the material under examination and the transducers.

Experiments performed on granite specimens in different operative conditions, that is, in presence or absence of the acoustic coupling agent and of pressure on the transducers, allowed us to evaluate the impact of these factors on the characterization of the material and the repeatability of measurements (Concu, 2002). It emerged that the best operating conditions, which ensure excellent measurement repeatability and reliability, are those that call for both the use of the coupling agent placed between the surface of the material and the transducers used for measuring and the application of a constant pressure on the transducers throughout the entire test period. The uncertainty introduced by the variability in pressure and the conditions of the coupling agent (thickness, temperature, viscosity) is in fact converted into a variation of the energy associated with the transmitted signal; this impacts to a greater degree on the wave shape and the signal spectrum rather than on its propagation velocity. From the experiments performed another interesting datum emerged, that is, the influence on the measurement of time between application of the acoustic coupling and data acquisition (Concu & Fais, 2003). Analysis of the values of maximum wave amplitude, amplitude of the maximum spectral peak and velocity relating to ultrasonic signals acquired repeatedly over time, starting from the instant of application of the coupling agent, led to the observation of a behavior of the aforementioned characteristic parameters of the material under examination: as can be seen in Figures 17-19 the three parameters increase progressively for approximately the first sixty minutes from application of the acoustic coupling agent and then become constant. Such behavior is to be taken into account in defining for each tested material the optimal instant for data acquisition so as to reduce the influence of the coupling agent on the test and its repeatability.

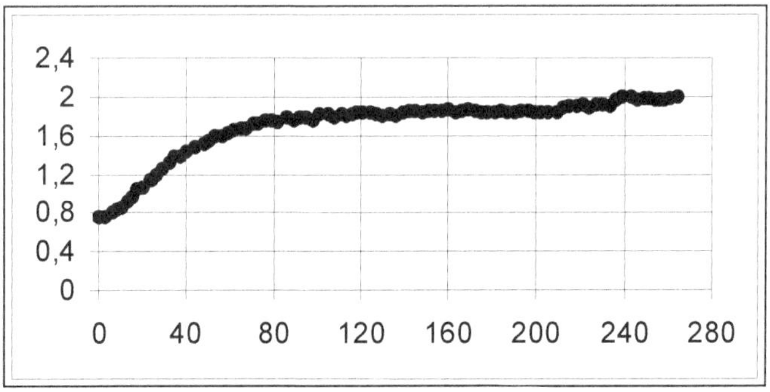

Fig. 17. Maximum wave amplitude [V] vs time [minutes]

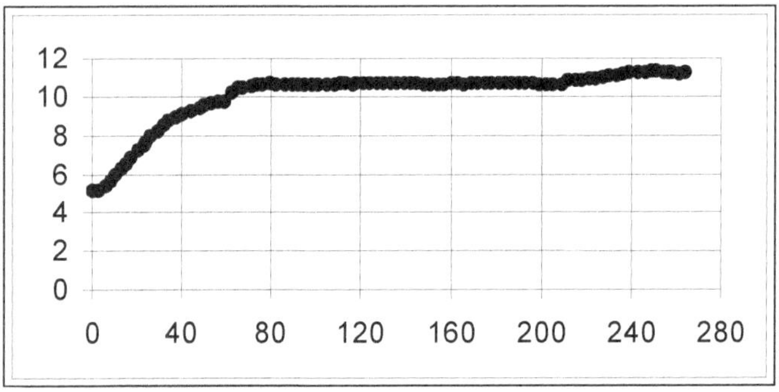

Fig. 18. Amplitude of the maximum spectral peak vs time [minutes]

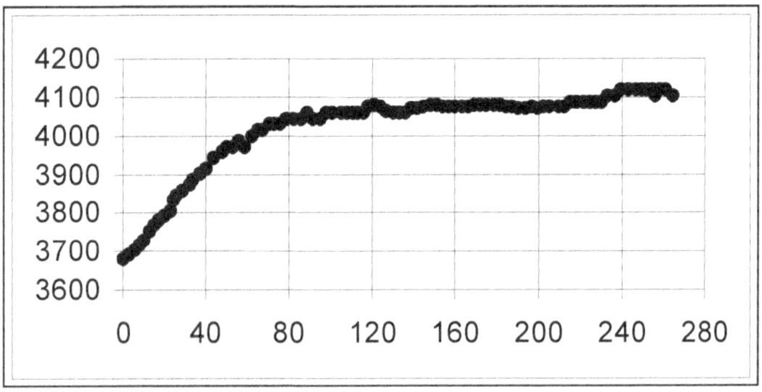

Fig. 19. Propagation velocity [m/s] vs time [minutes]

5. Conclusions

The chapter presented an overview on non invasive acoustic measurements applied for faults detecting in building materials and structures. The state of art has been described, focusing on two main operative procedures: the Direct Transmission Technique and the Acoustic Tomography. Special emphasis has been dedicate to both Travel Time and Spectral Attenuation Tomography, by deepening the aspects of numerical modeling, resolution algorithms, choice of source signal, measuring systems setup.

A case study, reporting the results of an experimental program started with the aim of deepening the reliability of acoustic methods in buildings faults detecting and materials characterization, has been described. Both DTT and AT approaches have been applied to a full scale masonry model with known anomalies inside. Results showed that both DTT and AT can successfully apply to stone masonry characterization. Moreover, results confirmed the suitability of the Spectral Attenuation Tomography - developed according to the model proposed by Quan and Harris, 1997, for seismic surveying – in buildings materials and structures faults detection.

Finally, an outline of the most common problems affecting the acoustic non destructive testing has been illustrated, addressing the main factors influencing the acoustic measurements in terms of stability, repeatability and reproducibility. It has been pointed out that particular attention should be given to chemical-physical-mechanical conditions of the material and to testing modalities such as the pressure exerted on the transducers employed and the effect on acquired data of acoustic coupling agents.

Further researches should deepen various facets, including:

- the potentiality of a novel approach involving the integrate analysis of different features, associated with acoustic waves propagating through the material, acquired both in time and frequency domain;
- the implementation of different inversion algorithms, chosen among the most robust and commonly used, for tomographic measurements, with the aim of highlighting the most suitable one for the specific algebraic problem solution;
- the development of proper measurements setup and automated measuring procedure which allowed problems of measurements stability, repeatability and reproducibility to be minimized.

6. References

Belanger, P.& Cawley, P. (2009). Feasibility of low frequency straight-ray guided wave tomography. *NDT&E International*, Vol.42, No.2, March 2009, pp. 113-119

Berryman, J.G. (1991). *Non linear inversion and tomography*. Lecture notes, Earth Resource Laboratory, MIT

Best, A.I., McCann, C. & Sothcott, J. (1994). The relationships between the velocities, attenuations, and petrophysical properties of reservoir sedimentary rocks. *Geophysical Prospection*, Vol.42, No.2, February 1994, pp. 151-178

BS 1881: Part 203, Recommendations for measurement of the velocity of ultrasonic pulses in concrete, London, 1986

Cannas, B., Cau, F., Concu G. &Usai M. (2008). Features extraction techniques for sonic and ultrasonic NDT on building materials. *Proceedings of 1st International Symposium on Life-Cycle Civil Engineering*, ISBN 978-0-415-46857-2, Varenna, Lake Como, Italy, June, 2008

Carino, N.J. (2001). The Impact-Echo Method: An Overview. *Proceedings of the 2001 Structures Congress & Exposition*, Washington, D.C., May 21-23, 2001

Concu, G. (2002). Il problema del controllo delle prestazioni degli elementi lapidei impiegati nelle costruzioni, PhD. Thesis, University of Cagliari, Italy

Concu G., De Nicolo B. & Mistretta F. (2003a). Comparative evaluation of two non-destructive acoustic techniques applied in limestone masonry diagnosis, *Proceedings of 33rd International Conference and Exhibition Defectoscopy*, ISBN 80-214-2499-0, Ostrava, Czech Republic, November, 2003

Concu G., De Nicolo B., Mistretta F.& Valdés M. (2003b). NDT ultrasonic method for ancient stone masonry diagnosis in Cagliari (Italy) , *Proceedings of Structural Faults and Repair - 2003*, ISBN 0-947644-52-0, London, UK, July, 2003

Concu, G. & Fais S. (2003). In time analysis of a viscous coupling agent effect in ultrasonic measurements. *Proceedings of 3rd International Conference on Non-Destructive Testing of the Hellenic Society for NDT – NDT in Antiquity and Nowadays – Skills-Applications-Innovations*, Chania, Crete, Greece, October, 2003

Concu, G., De Nicolo, B., Piga, C. & Popescu, V. (2010). Measurement system for non-destructive testing using ultrasonic tomography spectral attenuation. *Proceedings of 12th International Conference on Power Electronics and Electrical Engineering*, ISBN 978-1-4244-7020-4, Brasov, Romania, May, 2010

Dines, K.A. & Lytle, R.J. (1979). Computerized geophysical tomography. *Proceedings of IEEE* 67(7), pp. 1065-1073

Enoki, M. & Kishi, T. (1991). *Acoustic Emission. Current Practice and Future Directions*, ASTM, Philadelphia

EN 12504-4 (2004). Testing concrete- Part 4. Determination of ultrasonic pulse velocity

EN 14579. (2004). Natural stone test methods. Determination of sound speed propagation

Gilbert, P. (1972). Iterative methods for the reconstruction of three-dimensional objects from projections. *Journal of Theoretical Biology*, Vol.36, No.1, July 1972, pp. 105-117

Gordon, R., Bender, R. & Herman, G.T. (1970). Algebraic Reconstruction Technique (ART) for three-dimensional electron microscopy and x-ray photography, *Journal of Theoretical Biology*, Vol.29, No.3, December 1970, pp.471-81

Gordon, R. (1974), A tutorial on ART. *Proceedings of IEEE Trans. Nuclear Science*, NS-21, pp. 78-93

Herman, G.T. (1980). *Image Reconstruction from Projection: the Fundamentals of Computerized Tomography*, Academic, New York

Hudson, J.A. (1981). Wave speeds and attenuation of elastic waves in material containing cracks. *Geophysical Journal International*, Vol.64, no.1, January 1981, pp. 133-50

Ilyas, A. (2010). Estimation of Q factor from reflection seismic data, Report GPM 6/09, Department of Exploration Geophysics, Curtin University of Technology, p. 65

Ivansson, S. (1986). Seismic Borehole Tomography - Theory and Computational Methods. *Proceedings of IEEE 74*, pp. 328-338

Jansen, D.P., Hutchins, D.A., Ungar, P.J. & Young R.P. (1991). Acoustic tomography in solids using a bent ray SIRT algorithm. *Nondestructive Testing and Evaluation*, Vol.6, No.3, pp. 131-148

Johnston, D.H. (1981). Attenuation: A state-of-art summary. In Toksöz and Johnston, Eds., *Geophysics reprint series*, Vol.2, Soc. of Expl. Geophys

Kaiser, J. (1950). An Investigation into the Occurrence of Noises in Tensile Tests or a Study of Acoustic Phenomena in Tensile Tests. Ph.D. Thesis, Tech. Hosch. Munchen, Munich, Germany

Kepler, W.F., Leonard Bond, J. & Frangopol, D.M. (2000). Improved assessment of mass concrete dams using acoustic travel time tomography. Part II: application. *Construction and Building Materials*, Vol.14, No.3, April 2000, pp. 147-156

Knopoff, L (1964). The Convection Current Hypothesis. *Reviews of Geophysics*, Vol.2, No.1, pp. 89-122

Krautkramer, J.& Krautkramer, H. (1990). *Ultrasonic testing of materials*, Springer Verlag, New York

Lakshiminarayanan, V.A. & Lent A. (1979). Methods of least squares and SIRT in reconstruction. *Journal of Theoretical Biology*, Vol.76, No.3, February 1979, pp. 267-295

Jansen Leonard Bond, J., Kepler, W.F. & Frangopol, D.M. (2000). Improved assessment of mass concrete dams using acoustic travel time tomography. Part I: theory. *Construction and Building Material*, Vol.14, No.3, April 2000, pp. 133-146

Lo, T.W., Inderwiesen, P. (1994). *Fundamentals of Seismic Tomography*. Soc. of Expl. Geoph., Tulsa

Meglis, I.L., Chow, T., Martin, C.D. & Young, R.P. (2005). Assessing in situ microcrack damage using ultrasonic velocity tomography. *International Journal of Rock Mechanics & Mining Sciences*, Vol.42, No.1, January 2005, pp. 25-34

NDT Education Resource Center. Larson, B. Editor. (2001-2011). The Collaboration for NDT Education. Iowa State University. Available at: <www.ndt-ed.org>

Ohtsu, M. & Ono K. (1986). The generalized theory and source representations of acoustic emission. *J. of Acoustic Emission*, Vol. 5, No. 4, October-December 1986, pp. 124-133

Priestley, M.B. (1981). *Spectral Analysis and Time Series*. vol. I, Academic Press, New York

Quan, Y. & Harris, J.M. (1997). Seismic attenuation tomography using the frequency shift method. *Geophysics*, Vol.62, No.3, May-June 1997, pp. 895-905

Rhazi, J. (2006). Evaluation of concrete structures by the Acoustic Tomography Technique. *Structural Health Monitoring*, Vol.5, no.4, December 2006, pp. 333-342

Sansalone, M. & Streett, W.B. (1997). *Impact-Echo: Nondestructive Evaluation of Concrete and Masonry*. Ithaca, N.Y., Bullbrier Press

Sheriff, R.E. & Geldart L.P. (1995). *Exploration Seismology*. Cambridge Univ. Press, London

UNI EN 1926 (2000). Natural stone test methods. Determination of compressive strength
UNI EN 14580 (2005). Natural stone test methods. Determination of static elastic modulus

15

Planar Microwave Sensors for Complex Permittivity Characterization of Materials and Their Applications

Kashif Saeed[1], Muhammad F. Shafique[2], Matthew B. Byrne[1] and Ian C. Hunter[1]
[1]University of Leeds, School of Electronic and Electrical Engineering,
[2]COMSATS Institute of Information Technology, Department of Electrical Engineering,
[1]United Kingdom
[2]Pakistan

1. Introduction

Accurate measurement of material properties has gained considerable importance over the last decade. The ability to non-destructively monitor specific properties of a material undergoing physical or chemical changes has led to many applications in industry, medicine and pharmaceuticals.

In the food industry, the interest in dielectric properties of agricultural and food materials has been principally for predicting heating rates when the materials are subjected to high frequency electric fields (Oliveira, 2002). Microwave volumetric heating for food preservation and processing applications has been popular since the early 1970s and products such as rubber, wood, paper and many other agricultural products have been studied extensively (Nelson, 1991). An important application of microwaves in the food industry has been the non-invasive determination of moisture content (Kraszewski, 1998) and its effect on the dielectric properties of materials such as granular solids, meats, vegetables and fruits (Mudgett, 1995). Microwave techniques for drying of food products have also been very popular (Schubert, 2005).

The use of microwaves in therapeutic medicine has increased dramatically in the last few years (Golio, 2003). The ability to discriminate between normal and malignant cancerous tissues using microwaves is well reported and has led to the development of many non-invasive techniques for early detection of cancer and harmful tumors in the human body. Moreover, the ability of microwaves for deep tissue heating has resulted in the development of many novel microwave therapeutic treatments. An important and a growing application is the treatment of cardiac arrhytmias using microwave ablation (Rosenbaum, 1993, Greenspon, 2000). Furthermore, novel techniques such as microwave aidedliposuction (Rosen, 2000) and the RF and microwave enhancement of drug absorption are currently being investigated.

In the field of chemistry, dielectric measurements serve as a fundamental tool for the characterisation and evaluation of solvent materials, and have proven to be very convenient for investigating the molecular dynamics of materials. The dielectric analysis of pharmaceutical materials (Benoit, 1996) has increased in importance in recent years and is vital to the understanding and application of microwaves to synthetic organic chemistry (Tierney, 2005). Microwave dielectric heating has been very popular since the early 1980s due to the many benefits it offers over conventional heating (Galema, 1997, Gabriel, 1998). Solvents with higher loss factors lead to higher heating rates than can be achieved conventionally. Moreover, the interaction of microwaves with metal powders is used extensively to create hot-spots which serve as catalysts to accelerate chemical reactions of metals with gases, other inorganic solids, and organic substrates. The reduced reaction times also assist in the synthesis of radiopharmaceuticals (Galema, 1997) and inorganic complexes, and have had a significant impact in drug discovery (Larhed, 2001, Wathey, 2002). Microwaves are also employed during many organic reactions which require dry reaction conditions. In ceramic processing, microwave heating reduces cracking and thermal stress (Michael, 1991), allows quick curing of polymers, and is more economical than conventional heating methods.

In RF and microwave circuit design the dielectric permittivity of substrate plays an important role and requires precise evaluation over a broad range of frequencies. Knowledge of these properties plays a crucial role in the accurate design of multi-layered circuits, Monolithic Microwave Integrated Circuits (MMICs) and Low Temperature Co-fired Ceramic (LTCC) circuits (Edwards, 1991). The following sections aim to provide a general discussion of some well established dielectric measurement techniques, primarily applicable to moderate and high loss liquid samples.

2. Dielectric measurement techniques

The measurement methods relevant for any desired application depends on the nature of the dielectric material to be measured, both physically and electrically, the frequency of interest, and the degree of accuracy required. Being able to design an appropriate holder for the sample and to obtain an adequate model of the circuit for reliable calculations of the permittivity from electrical measurements often proves to be an important challenge.

At low and medium frequency ranges, bridge and resonant circuits have often been used for characterising dielectric materials. At higher frequencies however, transmission line, resonant cavity, and free-space methods are commonly used and have been illustrated in an early review (Altschuler, 1973). In general, dielectric measurement techniques can be categorised as reflection or transmission type, using resonant or non-resonant systems, with open or closed structures.

2.1 Non-resonant methods

In non-resonant methods, the properties of the materials are fundamentally deduced from their impedance and wave velocities therein. When an electromagnetic wave propagates from one material to another, both the characteristic wave impedance and the wave velocity change, resulting in a partial reflection of the wave from the interface between the two materials. Measurements of the reflections from such an interface, and the transmission through it, can provide information for the deduction of permittivity and permeability relationships between the two materials. Non-resonant methods mainly include reflection and reflection/transmission methods. In reflection methods, the properties of the material are deduced from the magnitude and phase measurement of the reflected signals. In reflection/transmission measurements the properties of the material are deduced from measuring the magnitude and phase of both reflected and transmitted signals. To improve confidence in findings it is generally preferred to use results from both reflection and transmission measurements. Non-resonant methods require a means of directing the electromagnetic energy towards a material, and then collecting what is reflected and transmitted through it. For this purpose, any type of transmission line could be used; for instance a coaxial line, a hollow metallic waveguide, a planar transmission line etc.

2.1.1 Waveguide and coaxial transmission line techniques

The use of waveguide and coaxial transmission line cells for complex permittivity measurements was first reported by Nicholson-Ross (Nicolson, 1970) and then Weir (Wier, 1974) who analysed the structure in the time and frequency domains respectively. Following this, Baker-Jarvis (Baker-Jarvis, 1990) produced iterative methods of solutions. The analysis of Nicholson, Ross and Weir led to the development of explicit formulae for the calculation of permittivity and permeability which is commonly referred to as the NRW algorithm. The technique essentially involves filling a section of a transmission line of a certain length with the sample to be measured (figure 1). The change in the propagation constant γ and the characteristic impedance Z_o leads to partial reflections of the wave at the interfaces. The propagation constant is related to the attenuation coefficient a and propagation coefficient γ through the relation $\gamma = \alpha + j\beta$. For a dielectric material the propagation coefficient of a wave within a transmission line is related to the complex permittivity of the filling material through the relation,

$$\gamma = j\sqrt{\frac{\omega^2 \mu_r \varepsilon_r}{c^2} - \left(\frac{2\pi}{\lambda_c}\right)^2} \tag{1}$$

Where ω is the angular frequency, μ_r is the permeability of the material which is equal to 1, c is the speed of light and, λ_c is the cutoff wavelength of the transmission line. In case of a coaxial transmission line supporting TEM propagation the cutoff wavelength is taken to be equal to infinity. $\varepsilon_r = \varepsilon_r^{'} - j\varepsilon_r^{''}$ is the complex permittivity of the material filling the line where, $\varepsilon_r^{'}$ is the dielectric constant and $\varepsilon_r^{''}$ is the dielectric loss of the medium. From equation 1 it becomes possible to extract the complex permittivity of the material filling the transmission line.

$$\varepsilon_r^{'} = \left(\frac{c}{\omega}\right)^2 \left[\left(\frac{2\pi}{\lambda_c}\right)^2 - \alpha^2 + \beta^2\right] \tag{2}$$

$$\varepsilon_r^{''} = \frac{c^2 2\alpha\beta}{\omega^2} \tag{3}$$

The complex propagation coefficient within the section of the line filled with the material is obtained from the S-parameter measurements with a network analyser. Reference planes are

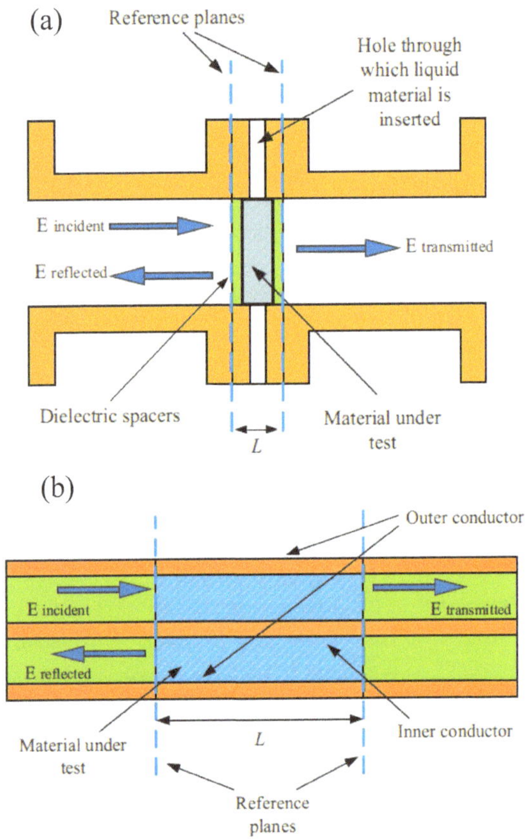

Fig. 1. (a) Waveguide fixture and (b) Coaxial fixture for transmission measurements.

established using well known calibration methods such as the Thru-Reflect-Line (TRL) or the Short-Open-Load-Thru (SOLT) techniques (Bryant, 1993). Owing to the simplicity of this technique, it has been widely adopted and many improvements have been made over the years which allow the permittivity of high and moderately lossy samples to be evaluated with a good degree of accuracy.

This technique is applicable to almost any kind of transmission line but waveguides and coaxial lines are generally preferred at frequencies below 30 GHz. Waveguide transmission cells can be constructed from either rectangular or circular waveguides and the mode of transmission depends on the structure. At frequencies below 2.45 GHz, the use of waveguides is not well suited due to the large volume of the test sample required. Coaxial transmission lines are preferable under these circumstances as they are comparatively small in size and cover a broader bandwidth. It has to be ensured that higher order TE and TM modes do not propagate as they lead to errors in the measured permittivity. For this purpose, standard cell dimensions are used. These are 3.5 mm (0 - 34.5 GHz), 7 mm (0 - 18.2

GHz), and 14 mm (0 - 8.6 GHz). Transmission line cells also suffer from problems with air gaps when the size of the cell becomes too small. The presence of air gaps in the cell leads to significant measurement errors. Over the past 30 years there have been several improvements and variations in the design of transmission line cells each being well suited to a particular class of dielectrics. At frequencies above 30 GHz, the dimensions of cells become too small and one has to resort to using free-space measurement techniques.

2.1.2 Free space transmission techniques

Free-space techniques are also grouped under non-destructive and contactless measuring methods, and are generally employed at higher frequencies (above 10 GHz) (Varadan, 1991). They do not require any special sample preparation, and are particularly suitable for measuring materials at high temperatures and for inhomogeneous dielectrics. They have also been implemented in many industrial applications for continuous monitoring and control. In a typical free-space transmission measurement technique, a sample is placed between a transmitting and a receiving antenna, and the attenuation and phase shift of the signal are measured, from which the dielectric properties of the sample can be determined. Accurate measurement of the permittivity over a wide range of frequencies can be achieved by free-space techniques.

In most systems, the accuracy of the determined permittivity depends mainly on the measuring system and the validity of the model used for calculations. The usual assumptions made with this technique are that a uniform plane wave is normally incident on the flat surface of a homogeneous material, and that the planar sample extends to infinity laterally so that diffraction effects at the edges of the sample can be neglected. Figure 2 shows a typical arrangement of a free-space measurement setup. Multiple reflections, mismatches, and diffraction effects at the edges of the sample are generally considered to be the main sources of errors and have to be accounted for appropriately. To enhance the measurement accuracy, special attention must be paid to the choice of the radiating

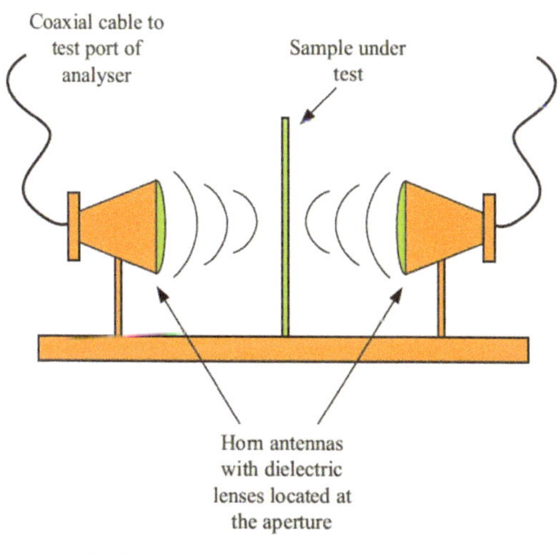

Fig. 2. Typical arrangement of a free space dielectric measurement setup.

elements, the design of the sample holder, and the sample geometry and location between the radiating elements.

2.1.3 Open ended transmission line techniques

Open ended transmission line techniques provide a very convenient and a non-invasive fixture for the evaluation of dielectric permittivity of liquids and semi-solids with no sample preparation. This technique was pioneered by Stuchly and Stuchly (Athey, 1982), who used it to measure the dielectric properties of biological materials.

Over the years, this technique has been improved. In this method, the material to be tested is placed against the cut-off section of the transmission line, and the magnitude and phase of the reflected signal are measured (figure 3). Over the past decade or so several models have been used to relate the permittivity of the material to the reflection coefficient measured at the aperture of the probe. Both waveguide and coaxial transmission lines have been used, although coaxial lines are preferable since they are very broadband and have small dimensions. This technique is well suited to measuring high dielectric constant and high loss samples, and is ideal for characterising lossy solvent materials. It requires calibration reference planes to be obtained at the probe aperture, which are challenging in the case of a coaxial probe. Well known reference materials are often required for calibration.

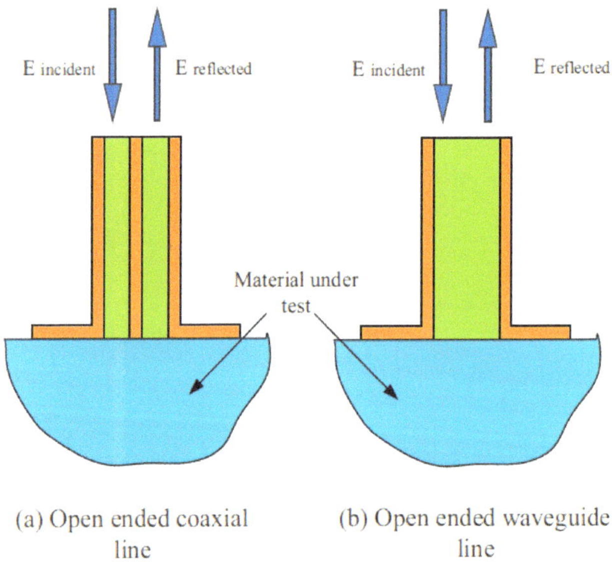

Fig. 3. Open ended coaxial and waveguide transmission lines.

2.1.4 Dielectric waveguide techniques

Dielectric waveguide techniques (Abbas, 2001) have recently been proposed which allow the determination of the dielectric permittivity of low loss planar sheet materials such as, teflon or perspex. The sample to be measured is placed in direct contact between two dielectric waveguides which can be of circular or rectangular geometry (figure 4). The sample can be of arbitrary shape, but its transverse dimensions must be greater than, or in the limiting case equal to, the transverse dimensions of the dielectric waveguide. The system is based on two basic properties of dielectric waveguides. The first property is that the energy of the wave propagating within the waveguide is entirely concentrated inside the waveguide. The other property is that the phase velocity v_φ is equal to the phase velocity of a plane wave in an unbound dielectric medium $v_\varphi = \frac{c}{\sqrt{\varepsilon_r}}$. Suitable expressions which allow determination of the complex permittivity from reflection and transmission measurements can be found in (Abbas, 2001). In order to launch energy into a dielectric waveguide suitable transitions are needed. These are generally made from metallic horn antennas which can be difficult to design.

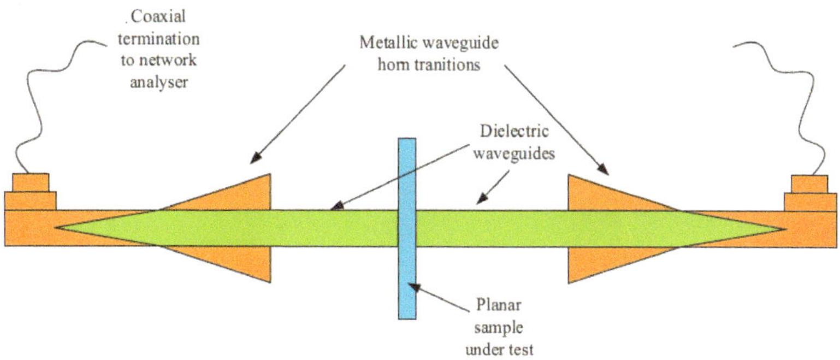

Fig. 4. Dielectric waveguide fixture for the complex permittivity measurement of low dielectric constant materials.

2.1.5 Planar transmission line techniques

Planar transmission lines such as microstrip and coplanar waveguides have long been used as microwave components. They allow ease of fabrication, low cost of manufacture, and compactness, which makes them suitable for industrial applications which use dielectric permittivity measurements. In planar transmission line methods, the material to be tested usually serves as a superstrate or as a substrate, or as a part of either. In the case of solids, the dielectric sample can serve both as the substrate or the superstrate. However, in the case of liquids and semi-solids, it is easier to have the sample as a superstrate. There have been many investigations into the use of planar circuits for complex permittivity measurements of liquids (Stuchly, 1998, Raj, 2001, Facer, 2001, Queffelec, 1994, Hinojosa, 2001, Wadell, 1991, Chen, 2004) Figures 5 and 6 show typical planar cells for dielectric permittivity measurements of liquid materials. The liquid to be measured completely covers the planar circuit and is enclosed inside a low loss container which is fixed firmly or epoxied on top of the board. The enclosure introduces mismatches which can easily be calibrated. The introduction of the liquid dielectric changes the effective permittivity ε_{eff} and the characteristic impedance Z_0 of the line. The dielectric properties are then extracted from the change in effective permittivity using suitable expressions which can be found in literature (Wadell, 1991).

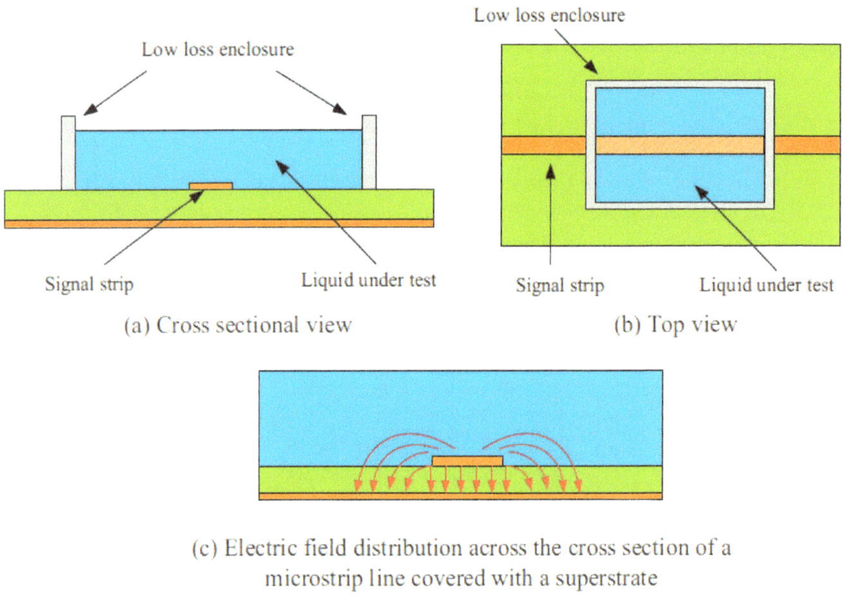

Fig. 5. Microstrip cell for complex permittivity measurement of liquids.

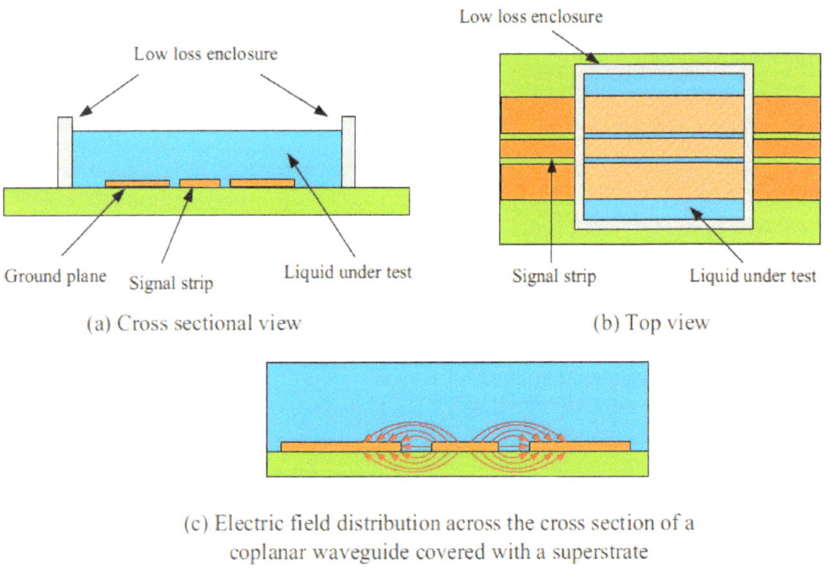

Fig. 6. Coplanar waveguide cell for complex permittivity measurement of liquids.

2.2 Resonant methods

Resonant methods offer the potential of characterising the properties of a material at a single frequency or a discrete set of frequencies with high accuracy in comparison to broadband methods. They can be classified into resonator methods and resonant perturbation methods. *Resonator methods* are those in which the material to be measured serves as a resonator and are only applicable to extremely low loss samples. The different types of resonators used for this purpose can be further classified into dielectric, coaxial surface-wave, and split resonators. Details on these methods can be found in (Chen, 2004). *Resonant perturbation methods* are those in which the sample is introduced into a resonant structure causing a perturbation in the response. The perturbation results in a shift of the resonant frequency and a decrease in the unloaded quality factor of the resonator from which the dielectric properties can be evaluated. The resonant perturbation technique is well suited for low and moderate loss samples. Both reflection and transmission type resonators can be employed for this purpose.

2.2.1 Waveguide cavity resonators

Waveguide cavity resonators are frequently employed for resonant perturbation measurements because of their high quality factors. The resonant cavities are designed in the standard TM (transverse magnetic) or TE (transverse electric) modes of propagation of the electromagnetic fields. The choice of cavity used depends on the particular field distribution of interest. The material to be tested is inserted at a particular location within the cavity where the electric field is at a maximum. Often a rod shaped sample is used, and in the case of liquid and semi-solid materials an appropriate cylindrical enclosure is needed to hold the sample. The insertion of the sample within the cavity causes perturbation to the system which results in a shift in the resonant frequency and a decrease in the unloaded quality factors. This perturbation in the response of the cavity is related to the material properties through cavity perturbation theory which is the most common technique used owing to its simplicity and accuracy. The cavity must be designed for a particular frequency of interest and at lower frequencies the cavity resonators are often large.

Figure 7 shows a rectangular TE_{101} and a cylindrical waveguide TM_{010} cavity operating in the two most widely used modes of propagation for permittivity measurements. Also shown in figure 7 is the magnitude of the electric field distributions for each mode.

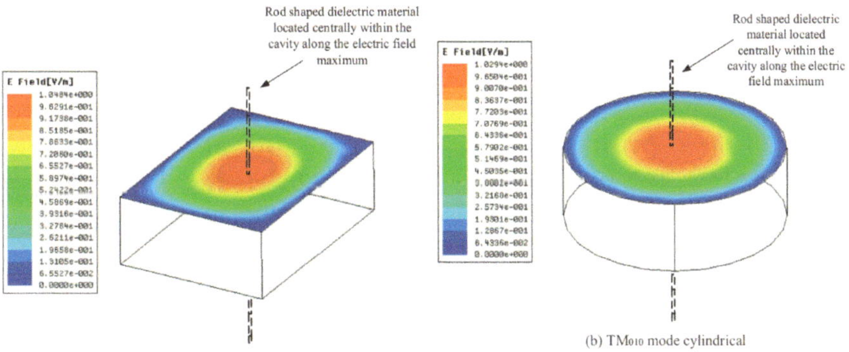

(a) TE_{101} mode rectangular resonant cavity

(b) TM_{010} mode cylindrical resonant cavity

Fig. 7. Waveguide cavity resonators for complex permittivity measurements.

2.2.2 Coaxial cavity resonators

A coaxial cavity shown in figure 8 can also be used for complex permittivity measurements of both moderate and high loss samples. This technique has been demonstrated by Raveendranath and colleagues (Raveendranath, 2000), and can also be analysed using perturbation theory. It offers high quality factors and thus can be used for measuring lossy samples. The coaxial cavity resonator is essentially a straight piece of coaxial transmission line which is coupled to another coaxial line from one end with the other end either shorted or left open. The cavity exhibits resonance at the fundamental with only one electric field maximum and multiple higher order resonances which could also be used for measurements at discrete frequencies. A rod shaped material is inserted within the cavity through a slot cut into the outer conductor, which can slide across the length of the cavity. Since coaxial lines are broadband in nature, resonant measurements can be carried out over a very broad bandwidth.

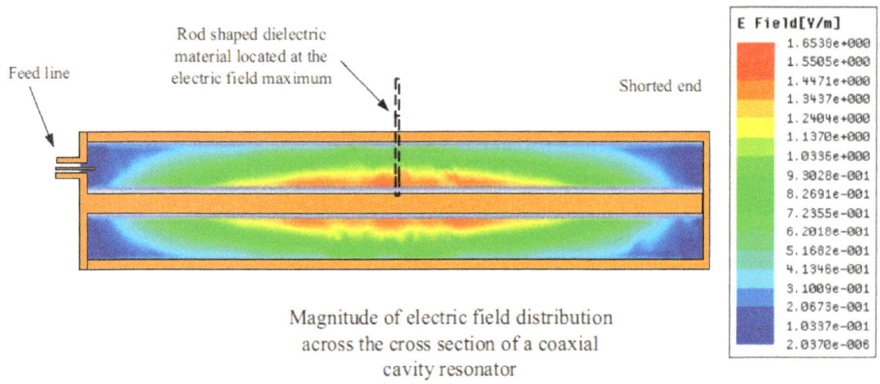

Fig. 8. Coaxial cavity resonators for complex permittivity measurements.

2.2.3 Open ended resonant lines

A very important but not so widely used resonant method of complex permittivity measurement is the resonant open ended line method first demonstrated by Johnson and colleagues (Johnson, 1992). The sensor consists of a long coaxial line with the input feed line located a distance $L_1 = \lambda/4$ away from the shorted end and $L_2 = \lambda/2$ away from the open end section (figure 9). At resonance, the fields are concentrated at the open end aperture of the resonator where the material to be tested is brought into contact. This method is ideally suited for measuring high loss samples and does not require any sample preparation. The sensor is calibrated in air before bringing it in contact with the material under test. The effect of the material in contact with the aperture of the sensor detunes the response.

The response is then tuned back to the original frequency by adjusting length L_1. This change in the length is directly proportional to the permittivity of the material. The deterioration in the quality factor of the resonator after tuning provides information on the loss of the sample.

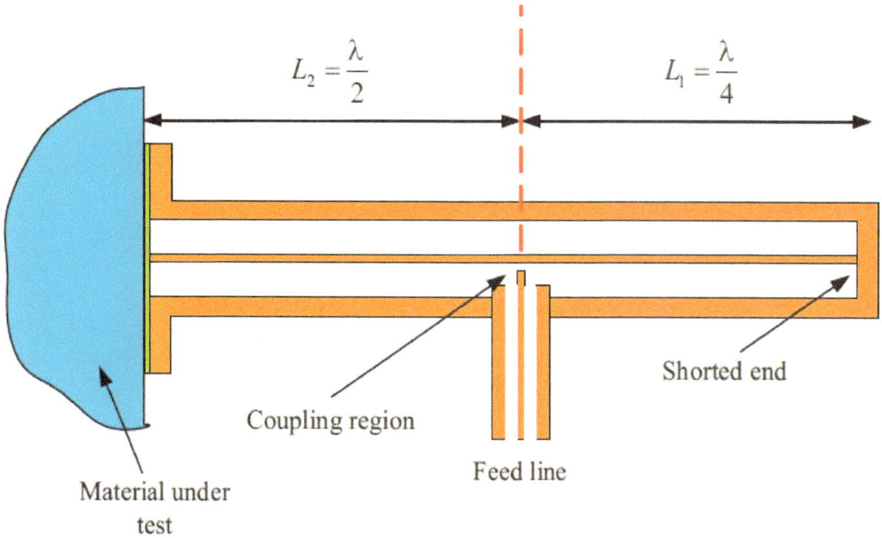

Fig. 9. Open-ended resonant line for complex permittivity measurements.

2.2.4 Planar transmission line resonators

Almost any kind of planar topology can be used for permittivity measurements. Unfortunately, all planar methods suffer from low quality factors (less than 500) and as a result measurements on low to moderate lossy samples have been common. Just as in the case of broadband planar transmission line methods in planar resonator perturbation measurements, the sample under study forms either a substrate or as a superstrate or as a part of either. The perturbation to the system is governed by the properties of the sample and the extent of interaction with the electric fields. Figure 10 shows a planar straight ribbon microstrip resonator fixture proposed by Abdulnour and colleagues (Abdulnour, 1995). This structure is well suited for resonant perturbation measurements on liquid samples. The sample is inserted in a small hole drilled inside the substrate. Because of the higher concentration of the electric fields at the tips of the resonator the sample can be located near the tips. The electric fields interact with the dielectric causing a capacitive perturbation in the response. The advantage of using such a technique is that only a small amount of sample (less than a few microlitres) is required for characterisation. This is highly beneficial, for instance, in the pharmaceutical industry, where accurate characterisation of materials in small quantities is vital. Another well known microstrip resonant measurement fixture applicable to moderately lossy samples was proposed by Boosanovich [Bogosanovich, 2000]. The sensor is essentially a circular disk resonator which is coupled to a coaxial transmission line that extends through the substrate from the bottom. The measurement fixture is shown in figure 11. The sample to be measured serves as a superstrate and the dielectric properties are measured from the shift in resonance and a deterioration in the quality factor of the response. The sample size required for measurements is generally large at low frequencies as the material has to completely cover the patch. Closed forms expression allow the permittivity to be calculated without requiring any reference calibration samples.

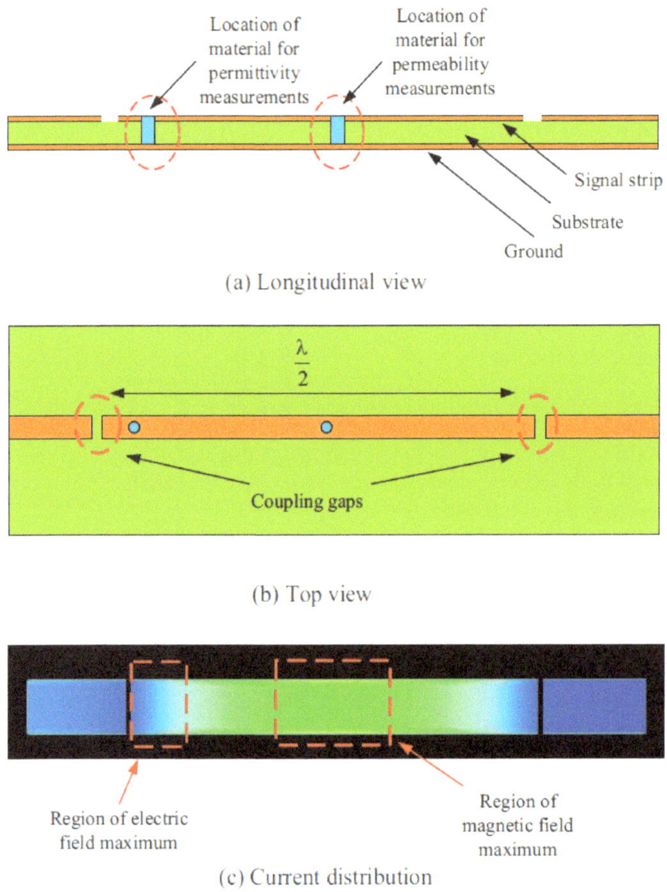

Fig. 10. Microstrip fixture for resonant perturbation measurements on liquid samples.

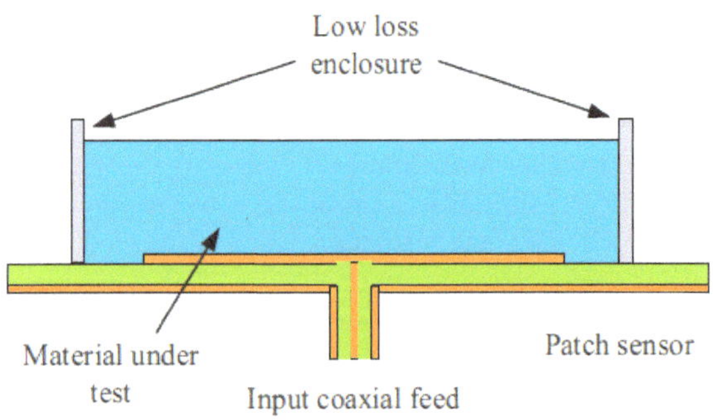

Fig. 11. Patch resonator sensor for complex permittivity measurements.

3. Microstrip resonators

Planar transmission lines have been used extensively in the past to measure electromagnetic properties of materials including thin films, sheet samples and substrate samples. They can offer many attractive features which include compactness, ease of fabrication and, use in a disposable manner. The three most common types of planar transmission lines used widely for materials measurements are stripline, coplanar waveguide and microstrip. Similar to their coaxial and waveguide counterparts, planar circuit methods can also be characterised into broadband and resonant methods. Resonant methods offer higher sensitivity and accuracy and therefore have been the focus of attention. Planar resonator methods are generally applicable to low loss samples such as thin films and substrates. In resonant perturbation methods part of the substrate is replaced by the sample under study. The interaction of the electric field lines with the material under test leads to a perturbation of the system thereby changing its resonant frequency and the unloaded quality factor. Since our aim is primarily to characterise lossy solvent materials using planar technology we will therefore focus our attention on resonant perturbation methods. The use of microstrip resonators for measuring material properties has been investigated by many researchers (Bernard, 1991, Abdulnour, 1995, Bogosanovich, 2002, Fratticcioli, 2004). These methods usually include the material under test (MUT) as a part of the substrate or as an overlay. The complex permittivity of the MUT is related to the resonant frequency f_r and the unloaded-Q factor Q_u, through closed form expressions which have limitations. The techniques can be destructive or non-destructive and in the past have been limited to measuring materials with low or moderate permittivities. In this section we demonstrate how a microstrip perturbation method could be used to characterise common polar and non-polar lab solvents (Saeed, 2007).

3.1 Microstrip transmission line

A microstrip transmission line consists of a strip conductor and a ground plane separated by a dielectric substrate as shown in figure 12(a). w and t are the width and the thickness of the strip conductor respectively. h and ε_d are the thickness and the dielectric constant of the substrate. Since the dielectric constant of the substrate is much higher than that of air the field is concentrated near the substrate. The field distribution on a microstrip line is shown in figure 12(b). The effective permittivity ε_{eff} and the characteristic impedance Z_o of a microstrip line can be determined using the well known closed form expressions found in the literature (Wadell, 1991).

Microstrip resonators such as the ring resonator (figure 12(c)) and the open resonator also sometimes referred to as the straight ribbon resonator (figure 12(d)) form an integral part of many microwave circuits such as oscillators and filters. The ring resonator is a basic resonant structure which offers high unloaded Q factors due to reduced radiative losses. For rings constructed on substrates which have a much higher thickness than that of the strip conductor, we have, at resonance,

$$l = \frac{n\lambda_g}{2} \tag{4}$$

where n is an integer order of resonance and λ_g is the wavelength of the guided structure. To take into account the dispersive nature of a microstrip line the above equation can be rewritten as,

$$l = \frac{nc}{2f\sqrt{\varepsilon_{eff}}} \tag{5}$$

where c is the speed of light, f is the frequency and ε_{eff} is the effective permittivity of the structure. Similarly, the open loop resonator is resonant when its electrical length is half that of the wavelength in the guided structure. The open resonator does however, suffer from increased radiative losses due to the presence of the fringe fields at the ends which result in the inaccurate determination of the resonant frequency. In order to take into account the fringing fields, the length of the resonator is generally taken to be slightly longer.

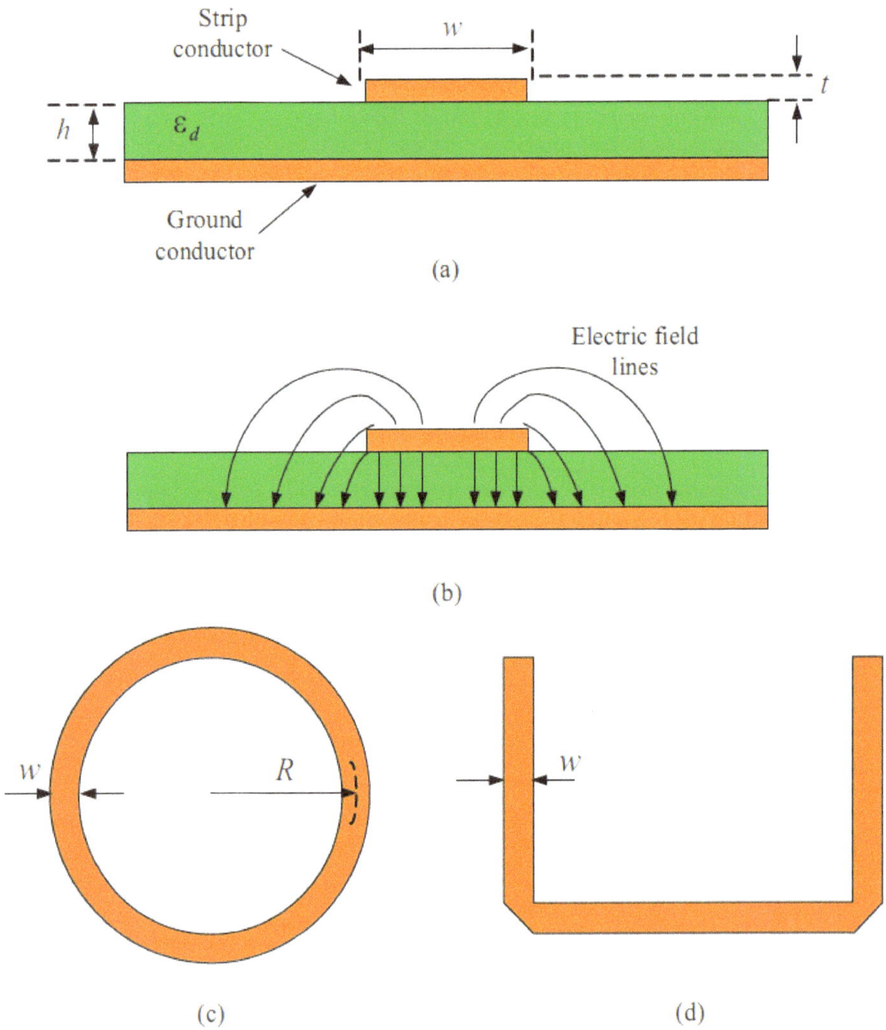

Fig. 12. (a) Geometry of a microstrip; (b) Electric field distribution at the cross section of a microstrip line; (c) Ring resonator; (d) Open loop resonator.

3.2 Microstrip ring resonator for complex permittivity measurements

The earliest use of ring resonators for dispersion measurements was reported by Troughton (Troughton, 1969). Later on it was realised that this simple structure could also be used to characterise the dielectric properties of thin sheet materials. Bernard and his colleagues (Bernard, 1991) used a microstrip ring resonator in a multilayer substrate topology to measure the dielectric constant of thin wafers. This technique was suitable for measuring low dielectric constant and low loss wafers.

The ring resonator could also be used to measure the dielectric properties of solvents. When the material under test is brought in close vicinity to the resonator. The fringing fields interact with the dielectric and energy is coupled into it causing a shift in the resonant frequency f_0. This shift depends on the relative permittivity ε_r' of the solvent and is independent of the dielectric loss ε_r''. The relationship between f_0 and ε_r' can be modelled using a 2nd order polynomial $f_o = a + b_1\varepsilon_r' + b_2\varepsilon_r'^2$ which provides a suitable fit. The coefficients a, b_1 and b_2 can be found by using standard solvents such as methanol, ethanol and 1-propanol of which the dielectric properties are well known. Reducing the perturbation even further gives a linear response and only two calibration media are now required (air and water for which the dielectric permittivity is well known). For demonstration purposes we will use a one port coupled ring resonator as shown in figure 13 (Saeed, 2007). The resonator is modelled as an equivalent *RLC* network where *R* is the resistance, *L* is the inductance and *C* is the capacitance. Figure 14(a) is the simulated response of a square ring resonator designed on an RT/Duroid 5880 substrate with a dielectric constant of 2.2 and a loss tangent of 0.0009. The length of the resonator is 208 mm with the fundamental resonance at 1 GHz. Higher order modes occur at integer multiples of the fundamental frequency at 2 GHz, 3 GHz and so on. Enhanced coupling method with extended input transmission line was used with a coupling gap of 0.4 mm. This is a critically coupled device with a 3dB bandwidth of approximately 10 MHz and a return loss of up to 60 dB. The use of critical coupling allows more accurate measurement of the resonant frequency. The response can be tuned by selectively adding/removing copper.

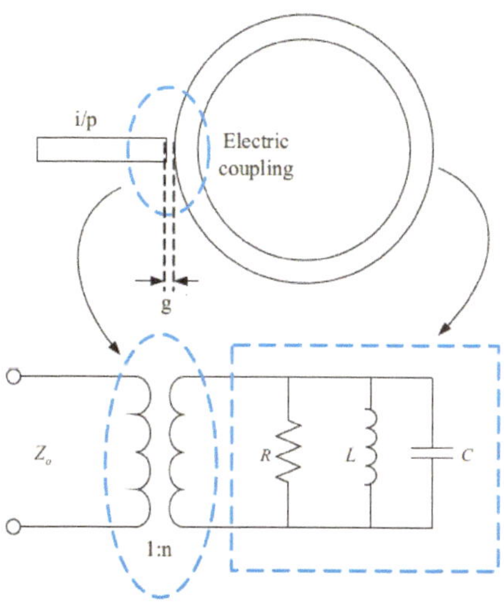

Fig. 13. Ring resonator electrically coupled to a microstrip transmission line through a gap.

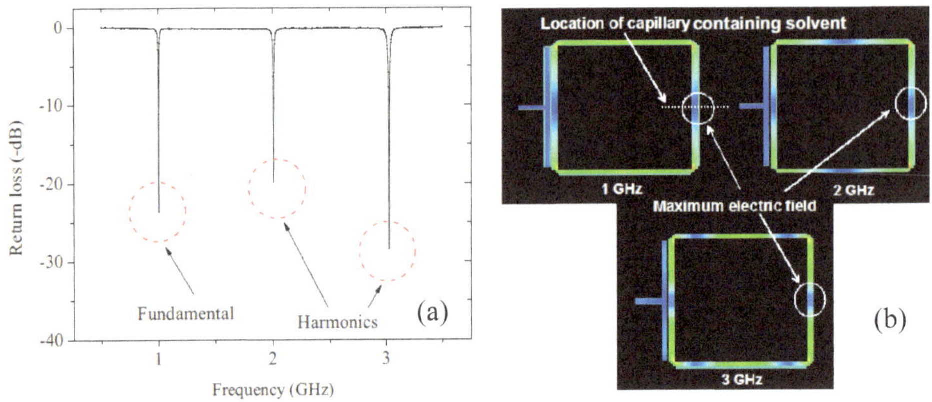

Fig. 14. (a) Simulated return loss performance of ring resonator; (b) Current distribution on ring resonator for first three resonant modes.

Also shown in figure 14(b) is the current distribution on the ring for the first three resonances. It can be seen that the coupling is electric in nature and that several electric field maximums are located along the length of the resonator. For lossy materials only a small portion of the ring can be covered by the material under test. The size and the location of the material governs the perturbation to the system. Liquid samples which can have high dielectric constant and/or high loss can be contained inside a small low loss capillary made of quartz or borosilicate glass and then introduced on the ring. Fixing the capillary on the microstrip in this manner ensures the same amount of perturbation to the system each time the measurement is made. A good location on the ring where the capillary can be located is shown in figure 14(b). Here the electric field maximum is common in all modes.

3.2.1 Location of capillary containing material under test

The capillary containing the material under test can be located either on top of the copper track or inside the substrate. The sensitivity depends on the extent of field penetration inside the material. Figure 15 illustrates two possible locations of the capillary for material measurements. It can be seen that the field lines are more concentrated within the substrate then in air. Therefore it should be obvious that by locating the material inside the substrate (below the copper track) results in higher sensitivity. To do this we can drill a small hole inside the substrate and insert the capillary through it. However, this might not be easily achievable in case of thin substrates. Figure 16 shows the change in the transmission response of a microstrip line when a capillary of 0.25 mm inner radius and 0.3 mm outer radius containing the material under test is placed on top of copper track and inside the substrate. The difference in the transmission responses is shown in figure 17. We can clearly see that higher sensitivity is achieved when the sample is located within the substrate.

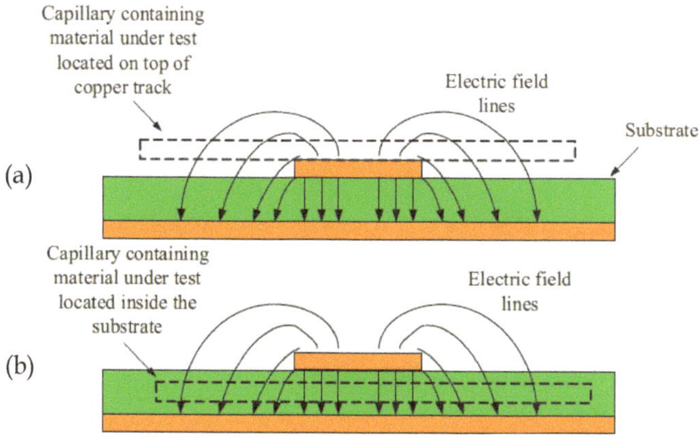

Fig. 15. (a) Capillary located on top of a microstrip line and (b) inside the substrate.

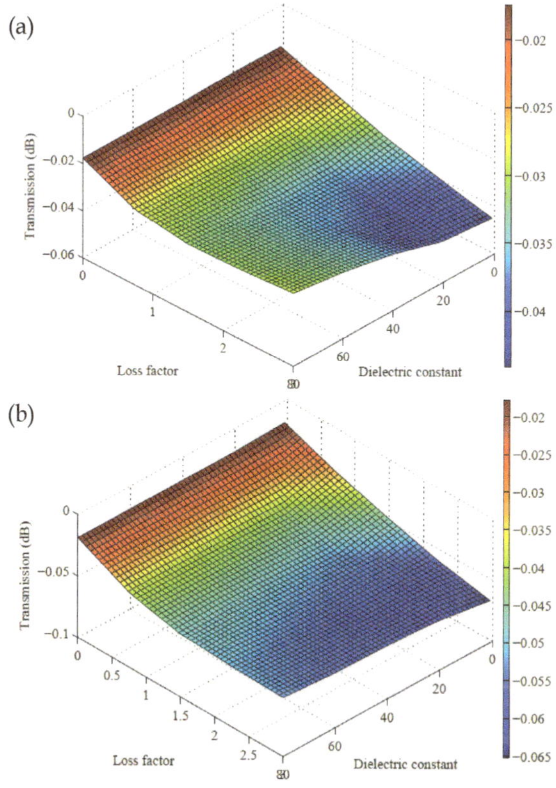

Fig. 16. Transmission response of a microstrip line at 1 GHz with a bororsilicate glass capillary containing material located (a) on top of microstrip line and (b) located inside the substrate. Capillary made of bororsilicate glass with outer radius of 0.3 mm and inner radius of 0.25 mm. RT/Duroid 5880 substrate with $\varepsilon_r' = 2.2$ and $\tan(\delta) = 0.0009$ with thickness $h = 0.787$ mm. Copper thickness $t = 0.035$ mm and width of microstrip line $w = 2.5$ mm.

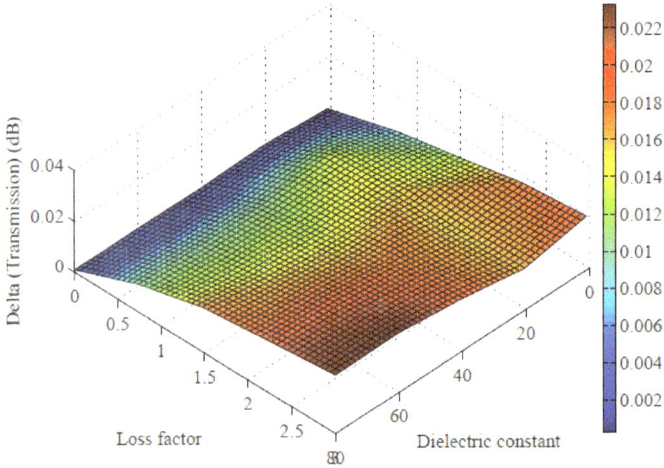

Fig. 17. Difference in the transmission response of a microstrip line when the capillary is located on top of copper track and inside substrate.

3.2.2 Measurements

The solvents used as standards and the unknowns are listed in Table 1 along with literature (Gregory, 2001) and measured values of dielectric permittivity. All samples obtained were of ANALAR standard and have purity greater than 99.5%. The measurements were carried out at 1 GHz, 2 GHz and at 3 GHz. All measurements were performed using an Agilent 8510C network analyzer at room temperature. The analyzer was used to average the data 256 times and the measurements repeated 4 times. Once the dependence of resonant frequency on dielectric permittivity was modelled and fitted using a 2nd order polynomial, the permittivity of the unknown solvents was determined. Four standards namely air, ethanol, methanol and dimethylsulphoxide were used to characterize 1-propanol and ethanediol. Air, 1-propanol and ethanol were then used as standards to characterize xylene, trichloroethylene, butyl acetate,

Solvent	ε_r' (Literature)			ε_r' (Measured)		
	1 GHz	2 GHz	3 GHz	1 GHz	2 GHz	3 GHz
Ethanol	14.14 ± 0.07	8.26 ± 0.05	6.34 ± 0.04	-	-	-
Methanol	30.17 ± 0.08	24.83 ± 0.17	19.71 ± 0.17	-	-	-
Dimethylsulphoxide	45.91 ± 0.07	44.29 ± 0.16	41.84 ± 0.26	-	-	-
1-propanol	6.98 ± 0.04	4.69 ± 0.03	4.14 ± 0.03	7.9 ± 0.16	4.86 ± 0.08	4.26 ± 0.06
Ethanediol	28.49 ± 0.16	17.34 ± 0.19	12.6 ± 0.18	27.8 ± 0.56	17.83 ± 0.3	12.51 ± 0.23
Xylene	-	-	-	2 ± 0.01	1.69 ± 0.31	1.84 ± 0.02
Trichloroethylene	-	-	-	3.08 ± 0.05	2.73 ± 0.218	3.02 ± 0.02
Butyl acetate	-	-	-	4.4 ± 0.06	4.25 ± 0.105	4.21 ± 0.03
Chlorobenzene	-	-	-	5.1 ± 0.01	4.82 ± 0.125	5.18 ± 0.05
Ethyl acetate	-	-	-	5.4 ± 0.03	5.3 ± 0.233	5.67 ± 0.11

Table 1. Measured dielectric permittivity of solvents.

chlorobenzene and ethyl acetate. Figure 18 is the measured and plotted results at 1 GHz, 2 GHz and 3 GHz respectively. The graphs clearly indicate that the resonant frequency is a good fit to the reciprocal of the square root of the dielectric permittivity.

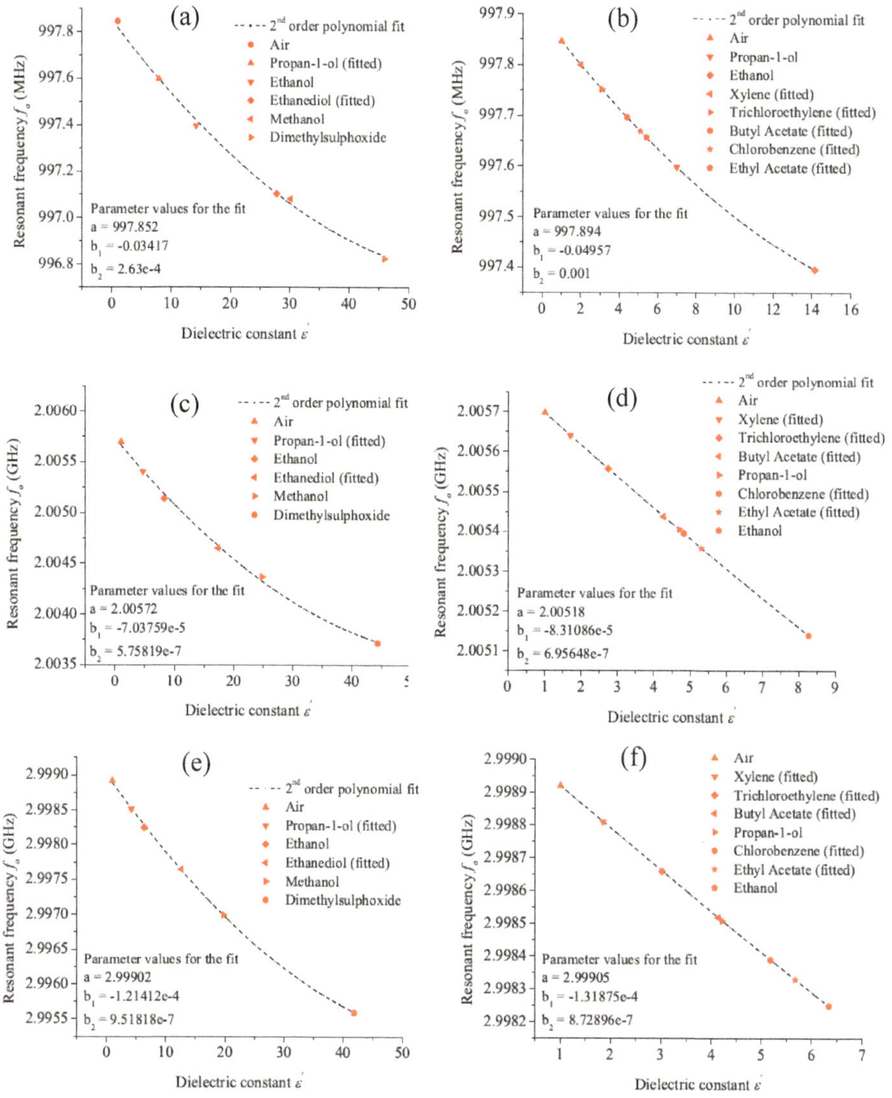

Fig. 18. Resonant frequency f_o versus dielectric permittivity ε_r' for measurement of 1-propanol and ethanediol at (a) 997 MHz (c) 2 GHz (e) 3GHz. Resonant frequency f_o versus dielectric permittivity ε_r' for measurement of xylene, trichloroethylene, butyl acetate, chlorobenzene and ethyl acetate at (b) 997 MHz (d) 2 GHz and (f) 3GHz.

3.3 Microstrip open loop resonator

To demonstrate the use of a microstrip resonator for solution sensing application we use an open loop resonator to characterise glucose-water solutions (Saeed, 2007). The resonator was designed at 1 GHz using RT/Duroid 5880. The length of the resonator is 111.98 mm with a width of 2.4 mm. The resonator is magnetically coupled to a transmission line. The electric field is concentrated at the two tips of the resonator as shown in Figure 19(a). Perturbation can be introduced to the system by placing the capillary across the tips and can be controlled by perturbing a single arm or both.

To demonstrate the high sensitivity of the resonators and the ability to measure small quantities of dissolved impurity, glucose-water solutions have been detected. This could be done either by noticing the change in the resonant frequency of the device or the unloaded-Q or both. The concentration of glucose-water solutions was varied from 0% to 4% (w/v) i.e. 0 g/100 ml to 4 g/100 ml. Figure 19(b) shows the variation in the unloaded-Q and the resonant frequency with varying glucose concentration. It can be seen that as the glucose concentration increases the resonant frequency increases whereas the unloaded-Q decreases. This indicates that the dielectric permittivity decreases and the dielectric loss increases with the addition of glucose. This could be anticipated from dielectric mixing theory (Steeman, 1990). Another conclusion that can be drawn from the graph is that it is simple to relate the concentration to the resonant frequencies and the unloaded-Q as the response is approximately linear. The device shows a decrease of 0.348 in the unloaded-Q factor and an increase of 20 kHz in the resonant frequency for every 1% increase in the concentration of glucose; making this a very sensitive method of detecting impurity concentration. Nevertheless, the sensitivity can be further increased by increasing the perturbation. From the linear relationship the sensitivity $\Delta f/(g/100ml)$ of the device is 25.313kHz/(g/100ml) and in terms of the unloaded-Q $\Delta Qu/(g/100ml)$ is 0.348/(g/100ml).

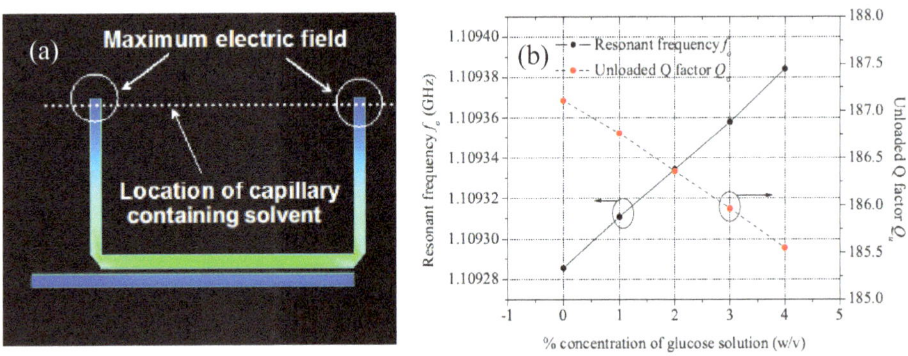

Fig. 19. (a) Current distribution on an open loop resonator at 1 GHz; (b) Resonant frequency f_o and unloaded-Q Q_u for varying concentration levels of glucose solutions.

4. Substrate integrated waveguide cavity resonators

Traditionally, waveguide cavity resonators have been used to characterise materials through resonant perturbation methods. They offer high quality factors, accuracy, and sensitivity and the volume of the material required for characterisation can be as small as a few microlitres. However, waveguide resonators can often be bulky, and in some cases difficult to construct which greatly limits their use. In such circumstances, one has to resort to using planar resonant devices which are generally compact, less expensive and easier to construct. Substrate integrated waveguide resonators are a relatively new class of planar resonant devices which offer higher quality factors in comparison to other planar devices. They are compact and can be constructed easily on a substrate along with the transitions. The devices can be made extremely compact and low loss by fabricating them on a high dielectric constant and low loss substrate. This makes them a suitable candidate for many industrial applications which require measurement of dielectric properties. The compactness of these devices can allow measurements to be made on sample volumes as small as a few nanolitres with better sensitivity. Their high quality factors can allow measurements to be made on a broad range of materials.

The substrate-integrated-waveguide is an artificial waveguide structure that is fabricated on a planar substrate with periodic metallised via-hole cylinders or metallised grooves which serve as sidewalls and solid top and bottom walls of metallization (figure 20). Since their first introduction (Hirokawa, 1998, Deslandes, 2001) a large number of microwave circuits based on this technology have been constructed to demonstrate the validity of the concept in easy integration with planar devices and its significant size reduction. The SIW resonant cavity is an important component in microwave circuits, e.g. filters and oscillators etc., and is constructed in the same way as a conventional waveguide cavity but with smaller dimensions. Due to finite losses in the substrate the quality factor of the cavity is reduced but is still significantly higher than its planar counterparts. If the integrated waveguide is to

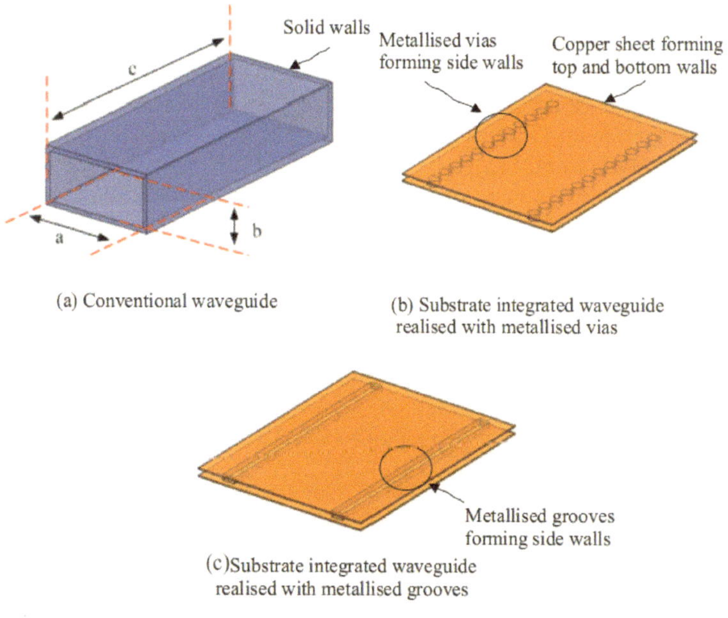

Fig. 20. Construction of a substrate integrated waveguide from metallised grooves and vias on a substrate.

be constructed from metallised grooves (figure 20(c)) the structure can be treated as a rectangular waveguide filled with a dielectric and the dimensions can be calculated easily using equations valid for a conventional rectangular waveguide. If however, the walls of the integrated waveguide are constructed using metallised vias (figure 20(b)) the exact calculations for the dimensions become rather more complex. The exact cut-off frequency of a particular mode in the integrated waveguide then becomes dependent upon the diameter of the vias and the spacing between them. In the past many authors have made use of modelling techniques such as the finite element method and the method of moments to compute the appropriate dimensions of the integrated waveguide and simulate the structures (Deslandes, 2001, Hill, 2001). Cassivi and colleagues (Cassivi, 2002) report on the dispersion characteristics of a substrate integrated rectangular waveguide for the first time and provide approximate equations to estimate the cut-off frequency of the first and second propagating modes. In this section we report on how the substrate integrated waveguide cavity resonator could be used to characterize solvents in small volumes (Saeed, 2008). The sensor is compact and highly sensitive which makes it an ideal candidate to be used for dielectric measurements in the pharmaceutical industry.

4.1 Design of SIW cavity

The dimensions of the TE_{101} SIW resonant cavity constructed from complete metallised walls/grooves can be estimated from $\lambda_0 = \dfrac{2}{\varepsilon_r' \sqrt{a^2 + l^2}}$. For an empty rectangular waveguide cavity resonating at 8 GHz the typical dimensions for the width a, length c and height b would be 23 mm, 32 mm and 10 mm respectively. However, the corresponding substrate integrated waveguide cavity constructed using complete metalised walls and fabricated using RT Duroid 5880 laminate with a thickness of 2.8 mm would have a width of 15 mm and a length of 23 mm. This results in a great reduction in the overall volume of the cavity. Moreover, the electric field is concentrated within a small portion of the cavity. The unloaded Q factor for an empty waveguide cavity is approximately 3500 whereas for a substrate integrated waveguide cavity it is reduced to approximately 700-800 owing primarily due to the losses in the substrate. However, the use of low loss ceramics or polymer substrates will significantly improve the Q. For demonstration purposes an X-band SIW cavity resonator was designed to measure the complex permittivity of solvents. The cavity was a one-port reflection type cavity and the energy was coupled into it by means of a microstrip line [48]. The extent of coupling could be controlled through the use of an offset which makes it possible to tune into the resistive losses of the cavity and achieve return losses in excess of 50 dB. Figure 21(a) is a layout of the cavity resonator. The cavity was formed by complete metallisation of the top, bottom and sidewalls of the substrate. At the intersection of the microstrip and the cavity the metallisation extends towards the microstrip so as to form a complete cavity.

4.2 Location of material under test and sensitivity analysis

As discussed in previous sections for the fundamental TE_{101} resonant mode the electric field is dominant at the center of the cavity and a common practice is to locate the sample at this position for maximum sensitivity. The sample can be inserted either through the broad walls of the cavity or through the side walls (figure 21(b) positions A and B). In case of

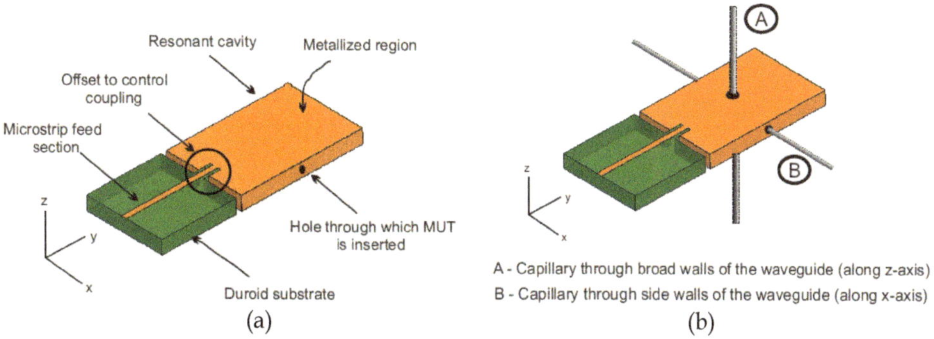

Fig. 21. (a) Layout of substrate integrated waveguide cavity resonator with microstrip feed section and hole within the cavity through which the material under test (MUT) is inserted; (b) Location of capillary containing material under test inside the cavity.

liquid dielectrics the material is contained inside a capillary made of low loss material such as quartz or borosilicate glass and then inserted into the cavity. We choose to insert the capillary through the sidewalls of the cavity through a small hole drilled into the substrate. This gives good mechanical stability and repeatability as the width *a* of the cavity is much larger than the height *b*. Care must be taken to ensure that no air bubbles are created in the portion of the capillary inside the cavity. The capillary does not need to be filled entirely as only the portion of material inside the cavity interacts with the fields. Therefore, the volume of sample required for measurements depends on the inner diameter of the glass capillary and the width a of the cavity.

The perturbation to the system is governed primarily by the location and size of the capillary. As an example we show simulation results for an X-band substrate integrated cavity resonator (8 GHz) made from an RT/duroid substrate of dielectric constant of 2.2, loss tangent of 0.0009 and a thickness of 2.8 mm with a width of 15 mm and a length of 23 mm. The simulation results were obtained using HFSS. Figures 22 and 23 show the fractional change in the resonant frequency $F_s = \frac{f_{ro} - f_{rs}}{f_{rs}}$ and the sensitivity $\frac{dF_s}{d\varepsilon_r'}$ to a lossless dielectric material with varying dielectric constant located within a borosilicate glass capillary (ε_r'=3.4, tan(δ)=0.0015) of various inner diameters (outer diameter = 0.7 mm). f_{ro} is the resonant frequency of the cavity without the sample and f_{rs} is the resonant frequency of the cavity with the sample. From perturbation theory it is anticipated that the fractional change in resonant frequency increases with increasing dielectric constant of material (figure 21). For capillaries with small inner diameters this increment is linear. As the inner diameter of the capillary increases the linearity is no longer valid due to higher perturbation to the system. For even larger inner diameters of the capillary the fractional change in resonant frequency increases very significantly with increasing dielectric constant. As would be expected the sensitivity is higher for low dielectric constant materials for any given diameter of the capillary (figure 22).

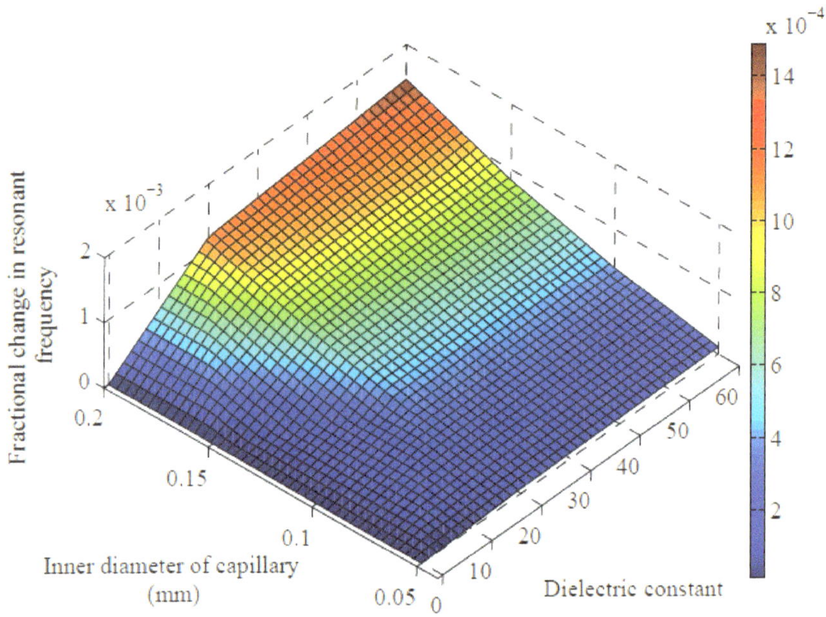

Fig. 22. Fractional change in resonant frequency with varying dielectric constant of an ideal lossless material for different inner diameters of capillary (Outer diameter of capillary = 0.7 mm).

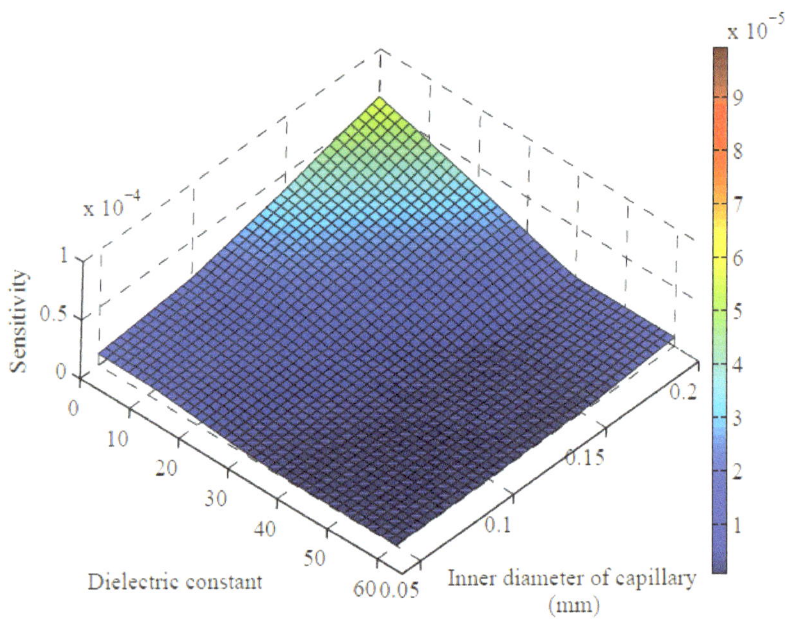

Fig. 23. Sensitivity versus the dielectric constant of material under test with varying inner diameter of capillary (Outer diameter of capillary = 0.7 mm).

For a capillary with an inner diameter of 0.05 mm the volume of material required is 29 nl. The shift in resonant frequency is 0.86 MHz for a dielectric constant change of 60 which is significantly high for samples in nanolitre volumes. This change can be measured easily with a network analyzer. In our case the volume required was less than 0.47 μl as we chose to use a capillary which had an inner diameter of 0.2 mm. The graph given in figure 22 also provides good insight into the range of solvents that can be characterized using a specific size of capillary. It can be concluded that for a capillary of an inner diameter of 0.07 mm this technique is most suitable for measuring solvents with a moderate dielectric constant i.e. $1 \leq \varepsilon_r' \leq 20$ as the sensitivity is high. This technique is also suitable for measuring small dielectric differences in a range of solvents which can be of high dielectric constant. This is achieved by careful adjustment of coupling. The cavity containing the reference material is tuned by varying the coupling offset until the load is critically coupled to the input feed section. The solutions are then compared to the reference material by studying the relative change in the response.

4.3 Perturbation of higher order modes

In order to study the dispersive dielectric properties of solvents over a desired bandwidth it is also possible to observe the perturbation of higher order modes in the resonant cavity. However, this requires the ability to vary the location of the material which can be achieved by drilling holes at various locations along the length of the cavity. To illustrate this we show simulation results on an integrated transmission type cavity formed using the same method discussed in the previous sections but with a longer length of cavity (l = 95 mm). The cavity is loosely coupled to the input and output microstrip feed sections. Figure 24(a) and 24(b) are the reflection and transmission responses of the cavity. The graphs show the perturbation of five resonant modes observed when a cylindrical sample of a dielectric constant $\varepsilon_r' = 20$ is located at the centre of the cavity inserted through the sidewalls. Also shown is the magnitude of the electric field distribution at each resonance. It can be clearly seen that several electric field maximum occur along the length of the cavity. The extent of perturbation of each resonance can be controlled by varying the location of the capillary.

4.4 Fabricated resonant cavity

To study the dielectric properties of solvents an X-band substrate integrated waveguide cavity resonator operating at 8 GHz has been fabricated with complete metallised walls on an RT/Duroid 5880 substrate with a thickness of 2.8 mm. Metallisation of the sidewalls of the cavity was achieved by uniformly depositing a thin layer of silver epoxy. At the intersection of the microstrip and the cavity the metallization extends towards the microstrip by 6.5 mm from either side so as to form a complete cavity. The width and the length of the cavity were optimized using HFSS. The optimized width of the cavity a was 15 mm and the length l was 23 mm. A microstrip to substrate integrated waveguide transition was used and the reflection type cavity was coupled to a microstrip line having a width of 1 mm through an offset length of 3.1 mm. The offset length for critical coupling was optimized to a borosilicate glass capillary located within the substrate through the side walls of cavity (Figure 21(b) position B with capillary along the x-axis). The capillary had an outer diameter of 0.5 mm and an inner diameter of 0.2 mm (0.47 µl volume of material) respectively. The optimised response showed a resonance close to 8 GHz and the electric field distribution in the cavity at resonance is shown in figure 25(a). The fabricated resonator is shown in figure 25(b) and 25(c).

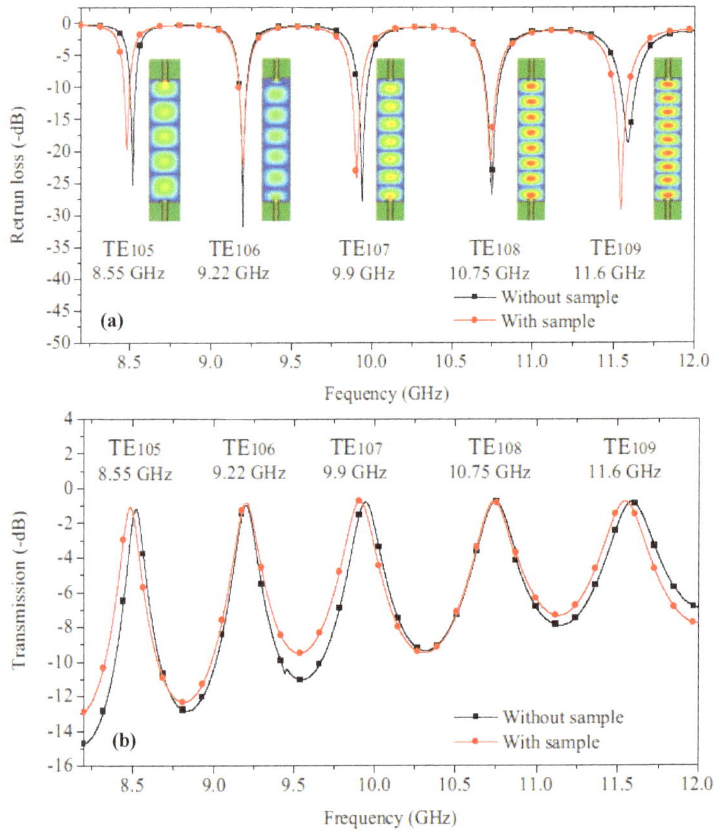

Fig. 24. Simulated (a) Return loss performance and (b) transmission response of integrated waveguide resonant cavity of 95 mm length with sample of dielectric constant $\varepsilon_r' = 20$ located at the center of the cavity.

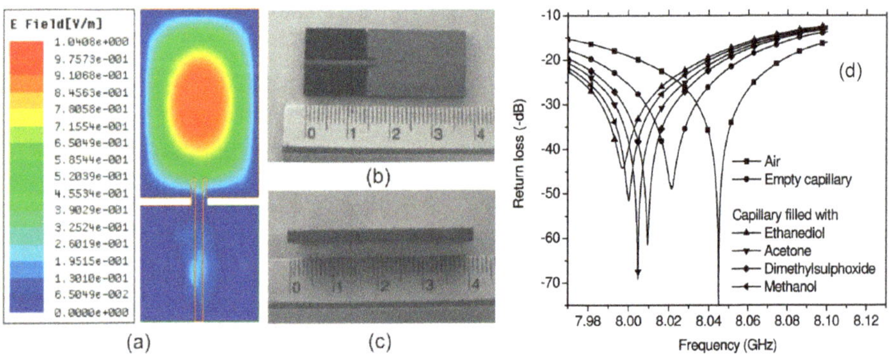

Fig. 25. (a) Magnitude of electric field distribution on the substrate integrated waveguide resonator for the fundamental TE_{101} resonant mode; (b) Top view of fabricated resonator; (c) Side view of fabricated resonator; (d) Measured return loss performance of various solvents.

4.5 Measurements

To illustrate the ability of the compact resonator to discriminate between solvents the capillary was filled with various solvents and their response was measured using an Agilent PNA E8361A at room temperature. The data presented was the result of averaging 256 times and the measurements were repeated 4 times. Figure 25(d) is the measured response for various materials clearly indicating its ability to discriminate among solvents. The resonant frequency shifts and the Q factor changes as various materials are inserted in the capillary. The measured unloaded Q for the empty capillary was approximately 700 and the resonant frequency was close to 8 GHz. As stated in previous discussion the resonator is highly sensitive to moderate dielectric constant materials. Therefore, this technique can be used to characterize and study many alcohol solutions. For demonstration purposes we choose to measure the dielectric properties of isobutanol-isopropanol mixtures. Mixtures with varying volume fractions of dissolved isopropanol were carefully prepared and their response was measured using an Agilent PNA E8361A at room temperature. The shift in the resonant frequency was linearly related to the change in the dielectric constant (figure 26(a)) whereas the change in the 3 dB bandwidth was linearly related to the change in the loss tangent of the mixture (figure 26(b)). The measured return loss performance of the mixtures is shown in figure 26(c). The dielectric properties of the mixtures were measured and compared to those obtained using a TE_{101} transmission type rectangular waveguide cavity and an open ended coaxial probe. A comparison is presented in table 2. From figure 26(a) it can be seen that the resonant frequency decreases as the volume fraction of dissolved isopropanol increases. This is expected as the dielectric constant of the mixture increases with increasing concentration of isopropanol since isopropanol has a higher dielectric constant than that of isobutanol. There is also an increase in the 3 dB bandwidth with increasing volume fraction of dissolved isopropanol due to an increase in loss tangent (figure 26(b)).

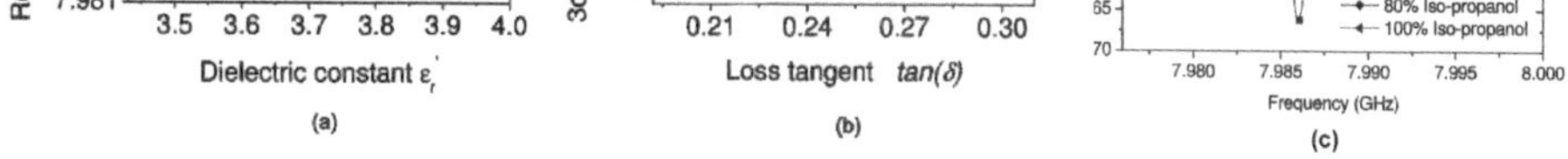

Fig. 26. (a) Measured resonant frequency versus dielectric constant of mixture for increasing volume fraction of dissolved isopropanol (%). (b) Measured 3-dB bandwidth versus loss tangent of mixture for increasing volume fraction of dissolved isopropanol (%). (c) Return-loss performance of isobutanol and isopropanol mixture.

The dielectric data measured using the integrated waveguide technique is in excellent agreement to that obtained using a rectangular waveguide resonant cavity method. For a change in dielectric constant of 0.5 there is a shift of almost 4.4 MHz in the resonant frequency indicating a high sensitivity. Moreover, a change of 2.9 MHz in the 3 dB bandwidth is observed for a change in loss tangent of 0.1. The error in the measured results for both resonant cavity and the integrated waveguide cavity is within ± 0.5 %.

ε_r'			
φ_f	Waveguide cavity	SIW	Coaxial probe
0	3.45 ± 0.04	—	3.18 ± 0.09
0.2	3.56 ± 0.02	3.579 ± 0.02	3.25 ± 0.14
0.4	3.642 ± 0.03	3.667 ± 0.011	3.32 ± 0.1
0.6	3.78 ± 0.023	3.755 ± 0.012	3.39 ± 0.13
0.8	3.814 ± 0.02	3.856 ± 0.017	3.46 ± 0.11
1	3.96 ± 0.04	—	3.52 ± 0.07
$\tan(\delta)$			
φ_f	Waveguide cavity	SIW	Coaxial probe
0	0.197 ± 0.002	—	0.288 ± 0.03
0.2	0.22 ± 0.012	0.218 ± 0.021	0.311 ± 0.03
0.4	0.236 ± 0.03	0.236 ± 0.011	0.326 ± 0.034
0.6	0.26 ± 0.01	0.26 ± 0.015	0.345 ± 0.03
0.8	0.28 ± 0.01	0.27 ± 0.01	0.367 ± 0.035
1	0.3 ± 0.013	—	0.38 ± 0.035

Table 2. Comparison of measured complex permittivity of isobutanol-isopropanol mixtures using a substrate integrated waveguide cavity resonator method, a rectangular waveguide cavity resonator method and an open ended coaxial probe method.

5. Conclusion

In conclusion, we have presented a thorough introduction into the various resonant and non-resonant methods for the characterization of complex permittivity using a range of various transmission line topologies. We have highlighted the advantages and disadvantages of the various techniques and have presented a thorough comparison. Discussion has been focused primarily on the use of planar resonant perturbation methods for dielectric characterization. Two topologies namely the microstrip resonator technique and the substrate integrated waveguide technique have been highlighted as potential candidates for use as compact sensitive sensors for use within the pharmaceutical industry. The discussion has been supported through careful design of sensors and measurements on polar and non-polar common lab solvents for each of the sensors. It has been concluded that the substrate integrated waveguide sensors are highly compact offering high unloaded-Q factors and are capable of performing accurate and sensitive measurements on solvents in volumes of less than a few nano-litres making them ideally for use within the pharmaceutical industry.

6. References

Abbas, Z., Pollard, R., & Kelsall, R. (2001). Complex permittivity measurements at ka band using rectangular dielectric waveguide. *IEEE Transactions on Instrumentation and Measurement*, Vol. 50, No. 5, (October 2001), pp. 1334-1342, ISSN 0018-9456

Abdulnour, J., Akyel, C., & Wu, K. (1985). A generic approach for permittivity measurement of dielectric materials using discontinuity in a rectangular waveguide or a microstrip line. *IEEE Transactions on Microwave Theory and Techniques*, Vol. 43, No. 5, (May 1985), pp. 1060-1066, ISSN 0018-9480

Altschuler, H. (1963). Dielectric constant, In: *Handbook of Microwave Measurements, Vol. 2,* M. Sucher & J. Fox, (Eds.), pp. 530-536, Polytechnic Press, New York (NY), ISBN B-0011007T-C

Athey, T., Stuchly, M., & Stuchly, S. (1982). Measurement of radio frequency permittivity of biological tissues with an open ended coaxial line: Part I. *IEEE Transactions on Microwave Theory and Techniques,* Vol. 30, No. 1, (January 1982), pp. 82-86, ISSN 0018-9480

Baker-Jarvis, J., Vanzura, E., & Kissick, W. (1990). Improved technique for determining complex permittivity with the transmission/reflection method. *IEEE Transactions on Microwave Theory and Techniques,* Vol. 38, No. 8, (August 1990) pp. 1096-1103, ISSN 0018-9480

Benoit, E., Prot, O., Maincent, P., & Bessiere, J. (1996). Applicability of dielectric measurements in the field of pharmaceutical formulation. *Bioelectrochemistry and Bioenergetics,* Vol. 40, No. 2, (August 1996), pp. 175-179, ISBN 0302-4598

Bernard, P., & Gautray, J. (1991). Measurement of dielectric constant using a microstrip ring resonator. *IEEE Transactions on Microwave Theory and Techniques,* Vol. 39, No. 3, (March 1991), pp. 592-595, ISSN 0018-9480

Bogosanovich, M. (2000). Microstrip patch sensor for measurement of the permittivity of homogeneous dielectric materials. *IEEE Transactions on Instrumentation and Measurement,* Vol. 49, No. 5, (October 2000), pp. 1144-1148, ISSN 0018-9456

Bryant, G. (1993). *Principles of microwave measurements,* Peregrinus on behalf of Institution of Electrical Engineers, ISBN 0-83641296-3, London (UK)

Cassivi, Y., Perregrini, L., Arcioni, P., Bressan, M., Wu, K., & Conciauro, G. (2002). Dispersion characteristics of substrate integrated rectangular waveguide. *IEEE Microwave and Wireless Components Letters,* Vol. 12, No. 9, (September 2002), pp. 333-335, ISSN 1531-1309

Chen, L., Ong, C., Neo, C., Varadan V. V, & Varadan, V. K. (2004). *Microwave electronics: Measurement and materials characterization,* John Wiley and Sons, ISBN 0-47084492-2, Chichester (UK)

Deslandes, D., & Wu, K. (2001). Integrated microstrip and rectangular waveguide in planar form. *IEEE Microwave and Wireless Components Letters,* Vol. 11, No. 2, (February 2001), pp. 68-70, ISSN 1531-1309

Deslandes, D., & Wu, K. (2001). Integrated transition of coplanar to rectangular waveguides, *Proceedings of IEEE MTT-S International Microwave Symposium (Vol. 2),* ISBN 0-78036538-0, Phoenix (AZ), May 2001

Edwards, T., & Steer, M. (1991). *Foundations of interconnect and microstrip circuit design (3rd ed.),* John Wiley and Sons, ISBN 0-47160701-0, New York (NY)

Facer, G., Notterman, D., & Sohn, L. (2001). Dielectric spectroscopy for bioanalysis: From 40 Hz to 26.5 GHz in a microfabricated waveguide. *Applied Physics Letters,* Vol. 78, No. 7, (February 2001), pp. 996-998, ISSN 1077-3118

Fratticcioli, E., Dionigi, M., & Sorrentino, R. (2004). A simple low cost measurement system for the complex permittivity characterisation of materials. *IEEE Transactions on Instrumentation and Measurement,* Vol. 53, No. 4, (August 2004), pp. 1071-1077, ISSN 0018-9456

Gabriel, C., Gabriel, S., Grant, E., Halstead, B., & Mingos, D. (1998). Dielectric parameters relevant to microwave dielectric heating. *Chemical Society Reviews*, Vol. 27, No. 3, (May 1998), pp. 213-223, ISSN 0306-0012

Galema, S. (1997). Microwave chemistry. *Chemical Society Reviews*, Vol. 26, No. 3, (June 1997), pp. 233-238, ISSN 0306-0012

Golio, M. (2003). *Microwave and RF product applications*, CRC Press, ISBN 0-84931732-0, London (UK)

Greenspon, A. (2000). Advances in Catheter ablation for the treatment of cardiac arrhytmias, *Proceedings of IEEE MTT-S International Microwave Symposium*, ISBN 0-7803-6435-X, Boston (MA), June 2000

Hill, M., Ziolkowski, R., & Papapolymerou, J. (2001). A high-Q reconfigurable planar EBG cavity resonator. *IEEE Microwave and Wireless Components Letters*, Vol. 11, No. 6, (June 2001), pp. 255-257, ISSN 1531-1309

Hinojosa, J. (2001). S parameter broadband measurements on microstrip and fast extraction of the substrate intrinsic properties. *IEEE Microwave and Wireless Component Letters*, Vol. 11, No. 7, (July 2001), pp. 305-307, ISSN 1531-1309

Hirokawa, J., & Ando, M. (1998). Single-layer feed waveguide consisting of posts for plane TEM wave excitation in parallel plates. *IEEE Transactions on Antennas and Propagation*, Vol. 46, No. 5, (May 1998), pp. 625-630, ISSN 0018-926X

Johnson, R., Green, J., Robinson, M., Preece, A., & Clarke, R. (1992). Resonant open ended coaxial line sensor for measuring complex permittivity. *IEE Proceedings-A*, Vol. 139, No. 5, (September 1992), pp. 261-264, ISSN 0960-7641

Kraszewski, A., Trabelsi, S., & Nelson, S. (1997). Moisture content determination in grain by measuring microwave parameters. *Measurements in Science & Technology*, Vol. 8, No. 8, (August 1997), pp. 857-863, ISSN 0957-0233

Larhed, M., & Hallberg, A. (2001). Microwave-assisted high-speed chemistry: a new technique in drug discovery. *Drug Discovery Today*, Vol. 6, No. 8, (April 2001), pp. 406-416, ISSN 1359-6446

Michael, D., Mingos, P., & Baghurst, D. (1991). Applications of microwave dielectric heating effects to synthetic problems in chemistry. *Chemical Society Reviews*, Vol. 20, No. 1, (March 1991), pp. 1-47, ISSN 0306-0012

Mudgett, R. (1995). Electrical properties of foods, In: *Engineering properties of foods, (2nd ed.)*, M. A. Rao & S. S. H. Rizvi, (Eds.), pp. 389-455, Marcel Dekker, New York (NY), ISBN 0-82475328-3

Nelson, S. (1991). Dielectric properties of agricultural products: measurements and applications. *IEEE Transactions of Electrical Insulation*, Vol. 26, No. 5, (October 1991), pp. 845-869, ISSN 0018-9367

Nicolson, A., & Ross, G. (1970). Measurement of the intrinsic properties of materials by time domain techniques. *IEEE Transactions on Instrumentation and Measurement*, Vol. 19, No. 4, (November 1970), pp. 377-382, ISSN 0018-9456

Oliveira, M., & Franca, A. (2002). Microwave heating of foodstuffs. *Journal of Food engineering*, Vol. 53, No. 4, (August 2002), pp. 347-359, ISSN 0260-8774

Queffelec, P., Gelin, P., Gieraltowski, J., & Loaec, J. (1994). A microstrip device for the broadband simultaneous measurement of complex permeability and permittivity.

IEEE Transactions on Magnetics, Vol. 30, No. 2, (March 1994), pp. 224-231, ISSN 0018-9464

Raj, A., Holmes, W., & Judah, S. (2001). Wide bandwidth measurement of complex permittivity of liquids using coplanar lines. *IEEE Transactions on Instrumentation and Measurement*, Vol. 50, No. 4, (August 2001), pp. 905-909, ISSN 0018-9456

Raveendranath, U., Bijukumar, S., & Matthew, K. (2000). Broadband coaxial cavity resonator for complex permittivity measurements of liquids. *IEEE Transactions on Instrumentation and Measurement*, Vol. 49, No. 6, (December 2000), pp. 1305-1312, ISSN 0018-9456

Rosen, A., Rosen, D., Tuma, G., & Bucky, L. (2000). RF/Microwave aided tumescent liposuction. *IEEE Transactions on Microwave Theory and Techniques*, Vol. 48, No. 11, (November 2000), pp. 1879-1884, ISSN 0018-9480

Rosenbaum, R., Greenspon, A., Hsu, S., Walinsky, P., & Rosen, A. (1993). RF and microwave ablation for the treatment of ventricular tachycardia, *Proceedings of IEEE MTT-S International Microwave Symposium*, ISBN 0-7803-1352-6, Atlanta (GA), June 1993

Saeed, K., Guyette, A., Pollard, R., & Hunter, I. (2007). Microstrip resonator technique for measuring dielectric permittivity of liquid solvents and for solution sensing, *Proceedings of IEEE MTT-S International Microwave Symposium*, ISBN 1-42440688-9, Honolulu (HI), June 2007

Saeed, K., Pollard, R. D., Hunter, I. C. (2008). Substrate integrated waveguide cavity resonators for the complex permittivity characterisation of material, *IEEE Transactions on Microwave Theory and Techniques*, Vol. 56, No. 10, (October 2008), pp. 2340-2347, ISSN 0018-9480

Schubert, H., & Regier, M. (2005). *The microwave processing of foods*, Woodhead, ISBN 1-85573964-X, Cambridge (UK)

Stuchly, S. & Bassey, C. (1998). Microwave coplanar sensors for dielectric measurements. *Measurement Science & Technology*, Vol. 9, No. 8, (August 1998), pp. 1324-1329, ISSN 0957-0233

Tierney, J., & Lidstrom, P. (2005). *Microwave assisted organic synthesis*, Blackwell, ISBN 1-40511560-2, Oxford (UK)

Troughton, P. (1969). Measurement techniques in microstrip. *Electronics Letters*, Vol. 5, No. 2, (January 1969), pp. 25-26, ISSN 0013-5194

Varadan, V. V., Hollinger, R., Ghodgaonkar, D., & Varadan, V. K. (1991). Free-space, broadband measurements of high-temperature, complex dielectric properties at microwave frequencies. *IEEE Transactions on Instrumentation and Measurement*, Vol. 40, No. 5, (October 1991), pp. 842-846, ISSN 0018-9456

Wadell, B. (1991). Microstrip line structures, In: *Transmission line handbook*, B. C. Wadell (Ed.), pp. 93-110, Artech House, ISBN 0-89006436-9, London (UK)

Wadell, B. (1991). *Transmission line design handbook*, Artech House, ISBN 0-9006436-9, London (UK)

Wathey, B., Tierney, J., Lidstrom, P., & Westman, J. (2002). The impact of microwave-assisted organic chemistry on drug discovery. *Drug Discovery Today*, Vol. 7, No. 6, (March 2002), pp. 373-380, ISSN 1359-6446

Weir, W. (1974). Automatic measurement of complex dielectric constant and permeability at microwave frequencies. *Proceedings of the IEEE*, Vol. 62, No. 1, pp. 33-36, (January 1974), ISSN 0018-9219

16

Basics on Radar Cross Section Reduction Measurements of Simple and Complex Targets Using Microwave Absorbers

Marcelo A. S. Miacci[1] and Mirabel C. Rezende[2]
[1]National Institute for Space Research (INPE),
[2]Institute of Aeronautics and Space, Department of Aerospace Science and Technology,
Brazil

1. Introduction

The research of Radar Cross Section (RCS) of simple and complex objects is decisively important to identify targets such as aircraft, missiles, rockets, ships and other objects, with the purpose of improving or rendering difficult their radar visibility in various frequency ranges. The use of RCS measurements of targets have expanded to more than solely military applications in the identification and control processes of defense systems (Burgess, 1988).

Higher the RCS value is, easier it becomes to detect and identify an object. However, when these targets present different geometrical forms and different types of electromagnetic radiation absorber materials (ERAM) on their surfaces, they can become stealthy and practically invisible to radars at determined frequency ranges.

In order to make target identification more precise, it is indispensable to analyze and understand the RCS patterns generated by the targets. These patterns represent the reflection mechanisms in the interaction process of the wave with the target, i.e., the interaction of the wave with the aspects of the target's geometry and the material's physical-chemical characteristics of its surface (Dybdal, 1987).

RCS measurements aim to determine the equivalent effective area of the target when it is impinged by a radar wave. In other words, it is the ratio between the electromagnetic energy irradiated by radar over a target and the energy scattered by it. The scattering measurements can be performed in monostatic condition, whereby the electromagnetic waves reflected by the target are measured in the same direction as the emitting source (radar), or in bistatic condition, when the reflected waves are detected in other directions. The present work describes experimental studies of RCS measurements of targets with simple and complex shapes and the RCS reduction by using ERAM in the microwave frequency range of 5 - 12 GHz (C and X bands). This chapter intends also to give some highlights about the theory involved and some measurement topics concerning target reflectivity, calibration techniques, enhancement methods and some experimental results achieved in this research performed at Materials Division of the Institute of Aeronautics and Space from Brazil.

2. Main definitions

When an electromagnetic wave focuses on an object the energy is spread in all directions. The spatial distribution of energy depends on the target geometry, material composition, the operating frequency and polarization of the incident wave. This distribution of energy is called scattering and the material is often called target (Blake, 1986).

Based on this principle, we can define the radar cross section, or RCS, as a measure of the power that returns to source or reflects in a given direction, normalized in relation to an incident wave power density. The purpose of normalization is to remove the effect of distance and facilitate the description of the cross section independent of the distance between the target and radar (*R*). The RCS is defined as shown in equation 1 (Bhattacharyya & Sengupta, 1991).

$$\sigma = 4\pi \lim_{R\to\infty} R^2 \frac{|\vec{E}^S|^2}{|\vec{E}^I|^2} = 4\pi \lim_{R\to\infty} R^2 \frac{|\vec{H}^S|^2}{|\vec{H}^I|^2} \tag{1}$$

where:
σ: radar cross section of the target (m^2);
$\vec{E}^S$: reflected or scattered electric field (V/m);
$\vec{H}^S$: reflected or scattered magnetic field (A/m);
$\vec{E}^I$: incident electric field (V/m);
$\vec{H}^I$: incident magnetic field (A/m).

The scattered electric and magnetic fields are due to the presence of a target, so, the total field is the sum of the incident and the scattered fields (equation 2):

$$\vec{E}^T = \vec{E}^I + \vec{E}^S \tag{2}$$

The RCS unit is usually given in square meters, or expressed in dB, relative to one square meter (dBm^2 or dBsm) as in equation 3 (Currie, 1989).

$$\sigma(dBm^2) = 10\log_{10}[\sigma(m^2)] \tag{3}$$

These concepts can be applied in the radar equation (Skolnik, 1990), correlating the received power in terms of transmitted power, scattering, distance and antennas gain (equation 4).

$$P_R = \left(\frac{P_T G^2 \sigma \lambda^2}{(4\pi)^3 R^4}\right) \tag{4}$$

where:
P_R: radar received power (W);
P_T: radar transmitted power (W);
R: distance between radar and target (m);
σ: radar cross section of the target (m^2);
G: radar antenna gain (dimensionless);
λ: wavelength (m).

The received power by the antenna from the transmitted radar pulse is directly related with the physical characteristics of the target through the backscattering coefficient. The value of this backscattering coefficient is basically dependent on the following factors (Jenn, 1995):

- the target geometry and surface roughness;
- the target material composition (electrical and/or magnetic properties);
- the wavelength and polarization of the radar; and
- the aspect angle related with the incident wave.

The conceptual definition of RCS includes the fact only one part of the radiated energy reaches the target. The RCS of a target (σ) can be most easily visualized as the product of three factors:

σ = Projected cross section x reflectivity x directivity

The reflectivity is defined as the intercepted radiated (scattered) power by the target. The directivity is given by the ratio of the backscattered power into the radar's direction to the power that would have been backscattered, considering an uniform scattering in all directions (isotropic).

2.1 Scattering matrix

In order to determine the dependence of the RCS in relation to the polarization of the wave, we must consider the relationship between the transmitted *(t)* and the received *(r)* fields, in terms of the horizontal and vertical components of linear polarization of the wave, E_H and E_V, respectively, where the index H denotes horizontal polarization and V the vertical polarization. Then, E_H and E_V can be expressed in terms of the reflectivity of a target illuminated by both polarizations through the proportionality constant a_{ij}, where i denotes the polarization of the transmitter and j the polarization of the receiver (equation 5) (Crispin, 1968) or in matrix notation (equation 6):

$$\begin{aligned} E_H^r &= a_{HH}E_H^t + a_{VH}E_V^t \\ E_V^r &= a_{HV}E_H^t + a_{VV}E_V^t \end{aligned} \tag{5}$$

$$\begin{bmatrix} E_H^r \\ E_V^r \end{bmatrix} = \begin{bmatrix} a_{HH} & a_{VH} \\ a_{HV} & a_{VV} \end{bmatrix} \begin{bmatrix} E_H^t \\ E_V^t \end{bmatrix} \tag{6}$$

The constants a_{ij} shall be considered independent of the distance, but in complex notation due to its phase relationship between the electric field components. For a monostatic radar configuration, a_{HV} and a_{VH} are equal. Therefore, the scattering matrix defines the relationship between amplitude and phase components of the electric fields transmitted and received.

A similar matrix to that one mentioned for linear polarization of the wave can be obtained for elliptical and circular polarizations, using the constants of proportionality describing the left or right hand polarization (Crispin, 1968).

2.2 Frequency regions

When a target has very small physical dimensions compared to the wavelength of the incident wave, it is considered that your analysis is being done in the so called *Rayleigh* region (Blake, 1986). In this region, the shape of the object does not influence in determining their RCS, and for some types of objects the RCS is determined from the volume, instead of considering the dimensions and physical forms. For targets that are comparable in size to the wavelength of incident wave, the RCS varies depending on the frequency and is called the resonant region or *Mie* region.

When the dimensions of the target are large compared with the wavelength of incident wave, the RCS can be determined using the methodology of geometrical optics or by the method of optical physics, and this region is called the optical region. In the following sections, the analysis of some RCS targets of simple and complex geometry is shown in the optical region. Figure 1 is a RCS curve of a sphere as a function of the ratio of the target radius $(a)/$ wavelength (λ), normalized in the optical region. The analysis of this figure shows that there are the distinct regions discussed above. Although it is a perfectly conducting sphere, this behavior is observed for all types of targets (Currie, 1989; Bhattacharyya & Sengupta, 1991).

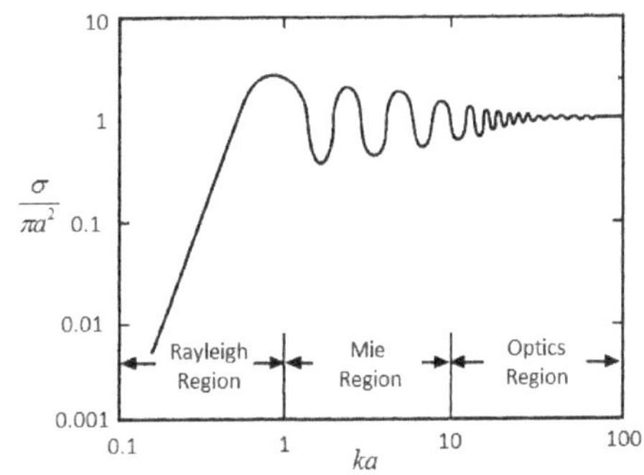

Fig. 1. Normalized RCS of a sphere as a function of ka (where $k=2\pi/\lambda$).

3. Radar cross section reduction measurements

RCS measurements are performed for many reasons; the main one is to verify target detection limits by a radar system to which it is subjected, verifying conformity of the practical models, designed by theoretical equations (Kouyoumjian, 1965).

The measurement as a whole is a complex task due to the many factors that may affect these measurements. Instrumental errors, spurious interferences and reflections are some of contributions to degrade the quality of the data. These problems are compounded when one is interested in reducing the RCS of an object, as in this case, the magnitude of these effects can overshadow the true RCS values.

Many methods have been proposed and used for the measurement of several types of targets. Depending on the size of the targets and radar frequencies used, RCS measurements can be performed on outdoor or indoor ranges, in the last condition, inside anechoic chambers. In measurements, it is important that the radar is illuminated by an electromagnetic wave which is uniform in phase and amplitude. For practical purposes, the maximum tolerance for amplitude variation over a target is 0.5 dB, and the phase should not deviate more than 22.5°. These conditions denote the far-field condition (Balanis, 1982; Jasik & Johnson, 1984), given by:

$$R \geq \frac{2d^2}{\lambda} \tag{7}$$

where:
R is the distance between radar and target;
d is the largest dimension of the target; and
λ is the wavelength of the radar.

This condition ensures good measurements. Errors produced by the instrumentation should neither exceed 0.5 dB nor vary in time in order to avoid instability in the measurement of the RCS patterns. Thus, it is decisive the careful selection of the experimental parameters. In this case, the dynamic range of the system should be at least 40 dB when measuring targets with small RCS. Dynamic range values in the order of 60 dB or higher are preferable when RCS reduction studies are conducted or when absorber materials are used. Another important factor to be taken into account is the target support structure that should not interfere with the incident wave, however in practice such condition isn't always possible.

Analysis of the electromagnetic energy scattering by metallic objects is very important in the understanding of RCS of targets, in the same way that reflections of dielectric and magnetic surfaces are important when studying RCS reduction. Within this context, RCS analysis of targets with simple shapes is fundamental to support the understanding of RCS patterns of targets with complex shapes.

To evaluate the electromagnetic behavior of targets, many methods have been proposed and experimentally used for many decades with well accepted results (Birtcher & Balanis, 1994) and the RCS measurement errors depend on the nature of the target under test, the distance at which this target is being measured and the place of measurements.

Systems are projected in such way that respect the parameters described above. However, they present technical challenges, making adequate the use of experimental procedures that involve new techniques that can minimize errors. When an anechoic chamber is used, the emissions and reflections of spurious radiation are controlled and minimized by using commercial off the shelf ERAM inside the chamber, becoming the background noise levels of this radiation almost null. On the other hand, by using the outdoor range, the measurements are affected by environment variations, therefore needing greater control of the parameters.

Even in measurements performed in indoor range, the RCS patterns of the targets may become impaired due to the occurrence of noise in the system or on account of low backward radiation contribution of the targets with lower RCS.

3.1 Basic instrumentation

The basic instrumentation required for RCS measurements consists of four subsystems that can be controlled by a central station; these subsystems are: positioners and drivers, receiver, transmitter and data acquisition system.

Using a simple setup and placing the transmitter (**TX**) and receiver (**RX**) antennas in the scheme showed in Figure 2, it is possible to measure the RCS of targets in different frequency ranges. The distance between both antennas needs to be tested to eliminate the radiation coupling between them.

In order to have good precision, the transmitter must have means to provide enough power to allow a good signal/noise ratio for the measurement system and it is necessary to take

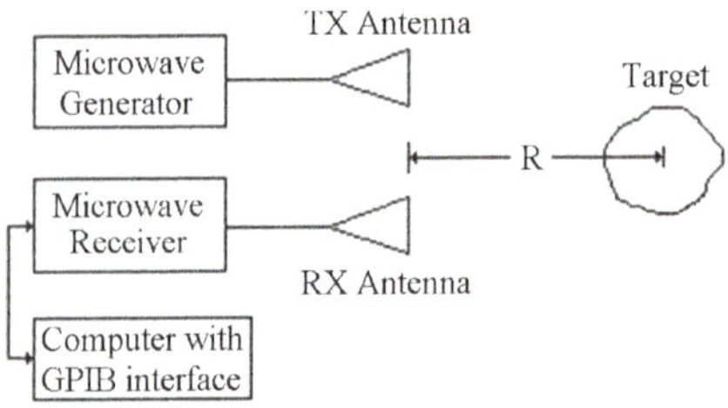

Fig. 2. RCS measurement setup (R is the distance between the antennas and the target).

care with the alignment between the transmission/reception antennas and the target. The use of a laser beam helps in the alignment of the system improving the precision of the measurements. The support column for the target, called as pylon, needs to be recovered with ERAM in order to avoid any possible contribution of reflected waves that prejudices the target characterization.

Figure 3 shows a system used in this study that basically consists of two antennas, one transmitter and one receiver mounted on a tower keeping a certain distance from each other. The target is mounted on a dielectric support and this one is mounted on a positioner which will allow recording the RCS patterns. The transmission is made through a microwave generator that feeds, by a low loss coaxial cable, the antenna input terminals. A spectrum analyzer can be used as a receiver that collects the signals from the receiving antenna. The system has a positioner controller (driver) that controls the rotational speed, azimuth angle and limits. Some equipment is in a control room, where the user monitors the tests through a computer with a GPIB interface.

Figure 3 shows the system composed by:

a. Target under test (square flat plate with 0.2 m side);
b. Pyramidal microwave absorbers;
c. Horn antennas to the 8.2 - 12.4 GHz frequency range;
d. Low loss coaxial cables; and
e. Microwave generator.

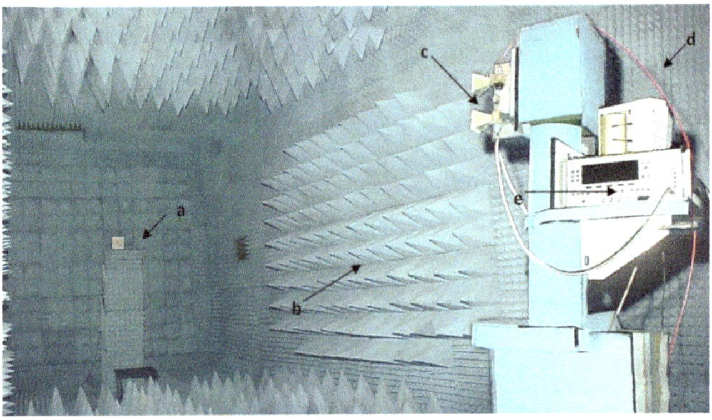

Fig. 3. Anechoic chamber view assembled with the RCS measurement system (courtesy of IFI/DCTA, Brazil).

Figure 4 shows the internal view of the control room (shelter), which is integrated with the instrumentation located inside the anechoic chamber. This figure shows: (a) antenna positioner controllers; (b) spectrum analyzer (receiver) and (c) computer.

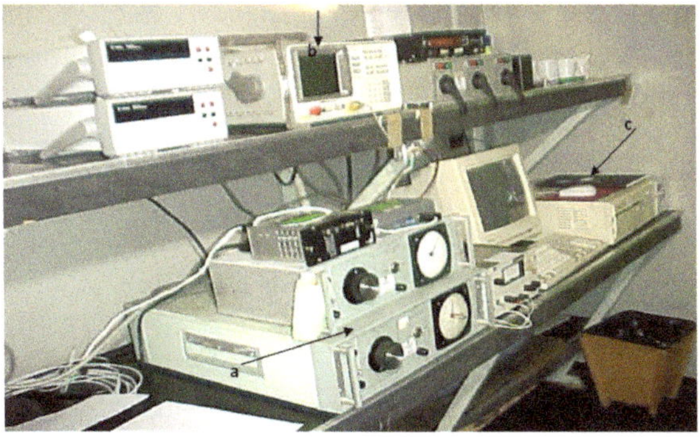

Fig. 4. Internal view of the control room (courtesy of IFI/DCTA, Brazil).

3.2 Amplitude calibration methods

There are two main methods for calibration of systems for RCS measurements: direct and indirect calibrations. The direct calibration involves the use of reference targets with well known values of RCS, such as spheres, flat plates, dihedrals and dielectric lens. The RCS of these targets can be accurately inferred by its physical sizes, to provide a radar return signal that can be compared to the signal of a target under test with unknown value of RCS. In some cases, the reference target may be in a different position relative to the target under test and a modified form of the radar equation is used to estimate the difference, as seen in equation 8:

$$\sigma_u = \left[P_r\left(\frac{R_u}{R_r}\right)^4\left(\frac{F_r}{F_u}\right)^4\right]\sigma_r \tag{8}$$

where:
σ_u: RCS of the target under test (m^2);
P_r: received power ratio between the reference target and target under test (dimensionless);
R_u: distance between target under test and radar (m);
R_r: distance between reference target and radar (m);
F_u: propagation factor of the target under test (dimensionless);
F_r: propagation factor of the reference target (dimensionless); and
σ_r: RCS of the reference target (m^2).

From equation 8, the propagation factors F_r and F_u contain information about the conditions of propagation between the radar and the two targets, including any atmospheric attenuation, scattering or even any undesired multiple reflection, as ground reflection that may be present. If the two factors are the same during measurements, equation 8 is reduced to equation 9:

$$\sigma_u = P_r\left(\frac{R_u}{R_r}\right)^4\sigma_r \tag{9}$$

The accuracy of this calibration method depends on the accuracy of the RCS reference target and the assumption that there are no other targets or significant sources of reflection on the measuring environment and the propagation factor is well known.

The second method of amplitude calibration is known as the indirect calibration method or closure. This method involves precise measurements of the characteristics of the radar system or the transmission and reception system by measuring the received power of a target under test with unknown RCS, and the calculation of RCS is done by using the radar range equation (equation 10) below:

$$\sigma_u = \frac{P_R(4\pi)^3(R_u)^4 L_T L_R}{P_T G^2 \lambda^2 (F_u)^4} \tag{10}$$

where:
σ_u: RCS of the unknown target (m^2);
P_R: received power from the unknown target (W);
L_T: transmitter losses (dimensionless);
L_R: receiver losses (dimensionless);
R_u: distance between radar to unknown target (m);
P_T: transmitted power (W);
G: radar antenna's gain (dimensionless);
λ: wavelength (m), and
F_u: propagation factor (dimensionless).

For this method is considered accurate, the parameters of the radar should be carefully measured, the received power must be precisely determined and the propagation factor should be equal to unity. The errors associated with this technique can occur in the measurements of radar parameters such as received power and also in conditions where the propagation factor is not equal to unity. One advantage of the indirect technique is that the reference target is not required.

3.3 Enhancement measurement techniques (active cancellation)

In order to reduce some error sources present in RCS measurements such as coupling between antennas, reflections in the anechoic chamber, reflection of target support and poor dynamic range, it can be used an active noise cancellation circuit which employs a phase cancelling technique.

A result of works performed at Materials Division/IAE/DCTA in Brazil shows an experimental set-up developed to reduce the error sources presented in indoor RCS measurements in C-band (5.8 to 6.4 GHz). By means of this research it was possible to perform comparisons, with and without the developed system, and observe its influence in measurements of simple objects.

By proposing the implementation of the system, it is expected to allow detection of small RCS targets by increasing sensitivity and dynamic range, so the effective study of RCS reduction can be established. The complete measurement system has a simplified block diagram shown in Figure 5.

The operation principle of the device is based on the detection of spurious signals and the application of the principle of phase cancelling, by generating a signal of equal amplitude

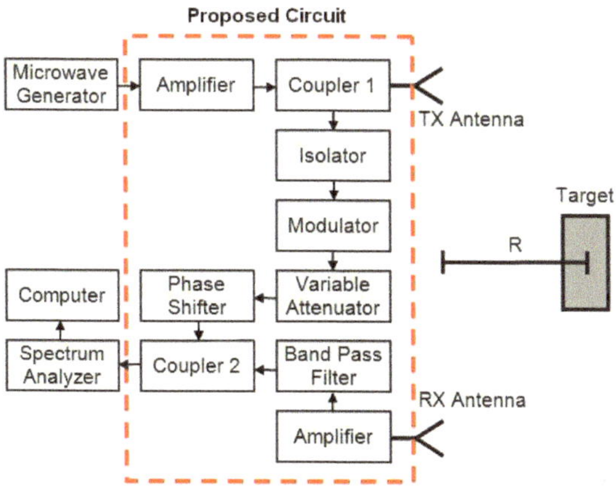

Fig. 5. Block diagram of the measurement system with the active noise cancellation circuit.

and opposite phase. Thus, after operation of the system only the signal of the target under test is expected to be detected. In this case, errors inherent to the RCS measurement systems in an indoor or outdoor range are reduced.

In accordance with Figure 5, operation of the developed device consists of a sample of the signal transmitted by the radar through the directional coupler 1. This signal goes through an isolator, followed by the modulator that performs phase and amplitude modulation (with fine tuning provided by the variable attenuator and phase shifter) and are combined in the directional coupler 2 with the interfering signals. Figure 6 shows the components used, such as: isolator, attenuator, coupler, filter and etc.

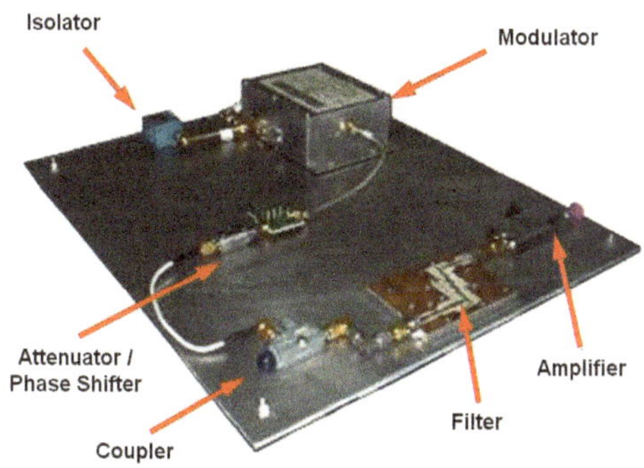

Fig. 6. View of the designed circuit assembled for C-band.

For the evaluation of the system, this circuit was tested by using an anechoic chamber with dimensions of 9.5 m x 4.5 m x 4.5 m available at IFI/DCTA. For this, two corrugated horn antennas were installed, with a transmitter and a receiver and a flat plate measuring 0.3 m x 0.3 m was used as a target under test, at frequencies of 5.9, 6.0, 6.2, and 6.4 GHz. Firstly, the noise level in the chamber was measured with and without the proposed circuit by rotating the azimuth positioner with no targets mounted on that. Some results are shown in Figure 7 and Table 1 presents the results obtained with the circuit for the frequency range.

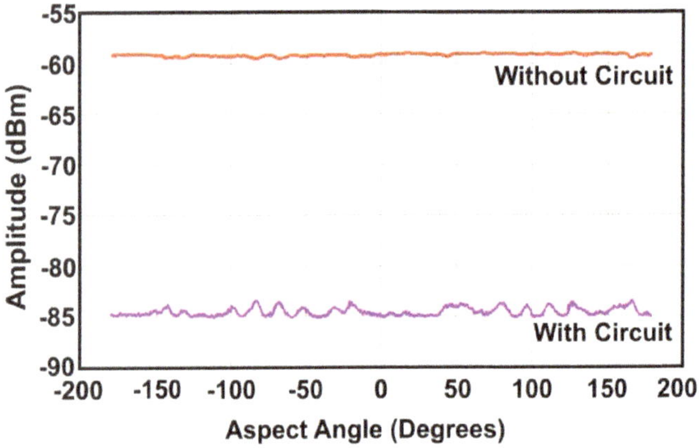

Fig. 7. Noise level of the anechoic chamber, with and without the circuit, at 5.9 GHz.

Frequency (GHz)	Noise level without the circuit (dBm)	Noise level with the circuit (dBm)	Reduction (dB)
5.9	-59	-85	26
6.0	-64	-84	20
6.2	-64	-85	21
6.4	-60	-82	22

Table 1. Anechoic chamber noise level reduction in C-band (deviation: 0.7 dB).

It is observed that due to the reflection from the anechoic chamber, added to the coupling between the antennas, the noise level is relatively high (approximately -59 dBm at 5.9 GHz).

Afterwards, a RCS pattern of the flat plate was measured without the circuit (Figure 8 - left) and it was observed a considerable deformation of the flat plate RCS pattern that can be assessed by foreseen results in literature (Ross, 1966) and experimental measurements (Miacci, 2002). After the circuit's operation, where the interfering signals were reduced, a new RCS pattern of the flat plate was measured and it can be observed as being closer than expected, as seen in Figure 8 (right).

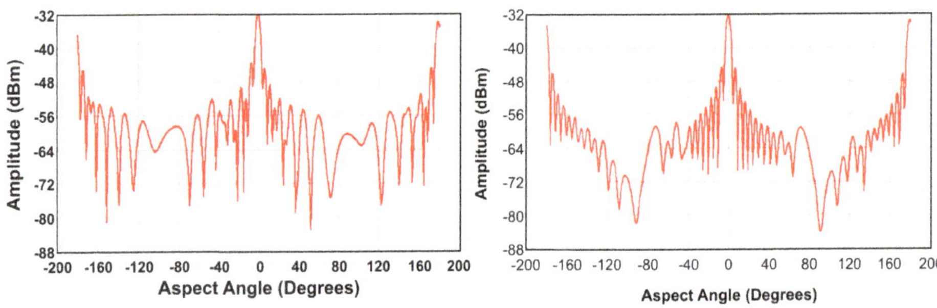

Fig. 8. RCS measurements of the flat plate (0.3 m x 0.3 m) without the circuit (left) and with the circuit (right) at 6.0 GHz.

This last RCS pattern evidences well defined null points and symmetry with a minimal contamination from external signals. It is also observed that the circuit does not influence the maximum peak, hence not interfering in the maximum RCS of the target under test.

Table 2 shows the minimum detectable RCS values and it is shown the increase of the probability of detection of smaller targets, associated to a higher dynamic range of the system, consequently, allowing the study of RCS reduction to be more effective when dealing with targets covered by high attenuation ERAM.

Frequency (GHz)	Detectable RCS without circuit (dBsm)	Detectable RCS with circuit (dBsm)
5.9	-09	-35
6.0	-16	-36
6.2	-15	-36
6.4	-11	-33

Table 2. Minimum detectable RCS values with and without the circuit (deviation: 0.7 dB).

The results denote that the RCS measurements performed with the developed and built circuit improved the RCS patterns with a remarkable reduction of interfering signals in the frequency range. This consideration takes into account the testing environment being almost completely free of spurious signals that reduce precision of measurement systems and can lead to false conclusions in analyses of RCS patterns. These preliminary measurements support other applications in bigger anechoic chambers or even in outdoor measurements.

4. Reflectivity characteristics of simple and complex shapes targets and RCS reduction by using ERAM

In this section it will be treated the reflectivity characteristics of simple and complex targets and discussed the experimental results of the RCS reduction research by using ERAM developed in Brazil.

4.1 Simple shapes targets

The electromagnetic theory allows the RCS calculation of targets that can be mathematically well defined, considering its size and physical forms. In practice this statement refers to objects with simple geometries. The solutions for simple objects are very important for two reasons, firstly, these objects can be considered as reference targets and they can be constructed and used as calibration references to measure complex geometry targets; second, approximations of equations used to determine the RCS of complex geometry objects can be made by detailed knowledge of the behavior of objects of simple geometries and their combinations (Miacci, 2002).

Table 3 shows the equations used for the RCS theoretical calculation for some simple geometry shapes in the optical region (Ruck, 1970; Knott et. al., 1993).

In the next sections is presented some experimental results achieved by the research group in RCS measurements of targets with simple and complex geometries and its RCS reduction when applying ERAM.

Geometry	RCS (m^2)
Flat Plate 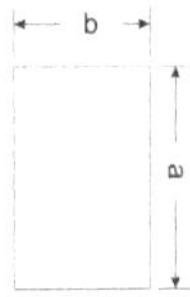	$\sigma = \frac{4\pi a^2 b^2}{\lambda^2}$
Sphere 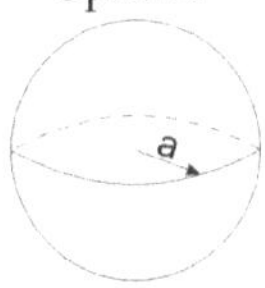	$\sigma = \pi a^2$
Diedral 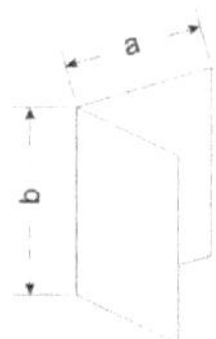	$\sigma = \frac{8\pi a^2 b^2}{\lambda^2}$
Cylinder 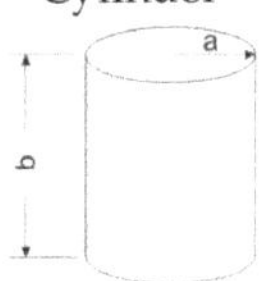	$\sigma = \frac{4\pi a b^2}{\lambda}$

Table 3. RCS of some simple geometries in optical region.

4.1.1 Dielectric lenses (Luneberg reflectors)

In order to validate a RCS measurement setup, firstly, it can be used a certified reference target called Luneberg reflector. A Luneberg reflector (Figure 9 - left) is a sphere constituted of dielectric massive shells. Figure 9 (right) depicts a measured RCS pattern of this target using the setup described in section 3.1, in function of the aspect angles between 0 and 180° in the frequency of 9.375 GHz.

The maximum power observed in this pattern corresponds to the RCS value of 45 m^2, according to the certified value given by the Luneberg reflector's manufacturer (Thomson CSF International Inc.). Using the relationship given by equation 9, it is possible to determine the RCS values of single targets in the same frequency (9.375 GHz).

The almost constant intensity level of the signal, between the aspect angles of -65 and +65°, corresponds to the region with the same scattering level of the reflector, being a characteristic RCS pattern for this target. The obtained RCS pattern shows a good agreement with that one provided by the manufacturer.

Using these data obtained with the Luneberg reflector it is possible to determine the RCS value for other single targets based on equation 9.

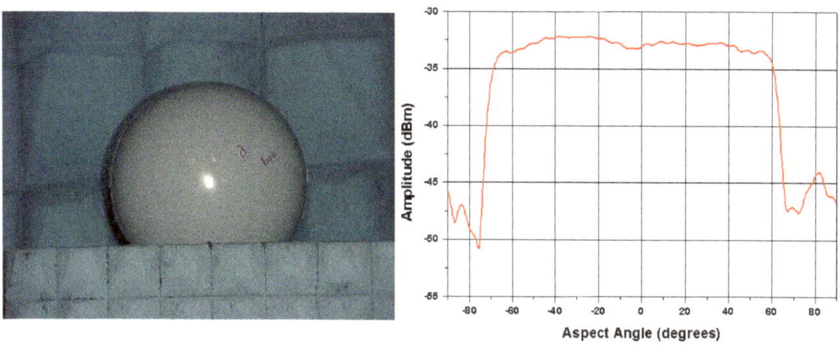

Fig. 9. Luneberg reflector positioned in an anechoic chamber (left) and a measured RCS pattern at 9.375 GHz (right).

4.1.2 Flat plates RCS reduction measurements

The RCS values of a perfect flat rectangular reflector can have its theoretical RCS value calculated as a function of the incident radiation frequency, according to equation depicted in Table 3. The RCS reduction method to evaluate a flat plate requires a double face panel, where one side is used as reflector material (reference) and the other is coated with ERAM. The panel is fixed on a rotating support, which is positioned in front of the receiving and transmitting horns according to the setup described in 3.1. The advantage of this methodology is that it allows the evaluation of the reference and ERAM, by rotating the device from 0 to 360°, evaluating both sides of the panel, one after the other. Figure 10 shows a simplified scheme of the device used in RCS method. With this method, it is not necessary to make two separate measurements, because the RCS pattern of the ERAM is made by rotating the device from 0 to 180° and the reference (metal plate-reflector) is made from 180 to 360°. Thus, it is a self-calibrating measurement (Knott et al., 1993).

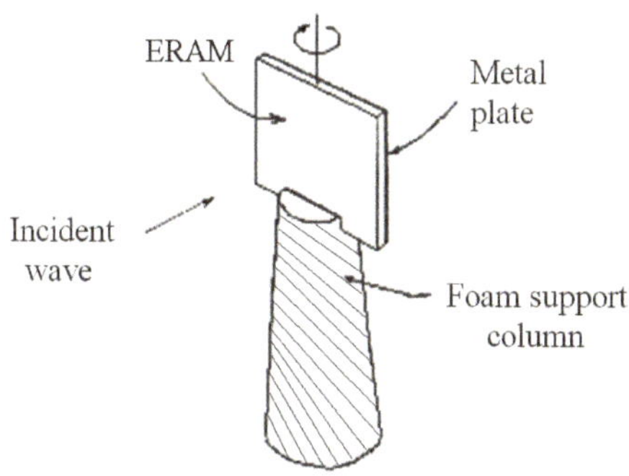

Fig. 10. Device scheme used for the RCS measurements.

The tested ERAM coating was prepared at Materials Division/IAE/DCTA, from Brazil. The ERAM preparation involved the mixture of 60 % (w/w) of a commercial polyurethane matrix loaded with 40 % (w/w) of fillers, being carbonyl iron (20 %) and ferrites of MnZn (10 %), NiZn (5 %) and MgZn (5 %). Physico-chemical characteristics of the fillers and the polyurethane resin as well as the coating preparation procedures were previously described

(Martin, 2002). The ERAM was applied on the aluminum flat panel of 300 mm x 200 mm surface by brushing with thickness of 0.7 to 2.0 mm.

RCS measurements were carried out at the frequency of 8 GHz and the panel (300 mm x 200 mm) was fixed on the rotating support (Figure 11) and rotated from 0 to 360°, at a scanning rate of 0.080 rad/s, characterizing both sides of the panel, i.e., the reference side and the ERAM painted side. From 0 to -180^{0} it is scanned the reference side and from 0 to + 180^{0} the ERAM coated side.

Fig. 11. Flat plate assembled in an anechoic chamber: metal face (left) and ERAM face (right).

Figure 12 shows the RCS pattern of the reference aluminum plate, obtained at 8 GHz, with a rotation of 180°. It is observed a peak at 0° corresponding to values of -25.3 dBm (that results in a RCS value about 32.1 m^{2} using an appropriate reference target), due to the normal incidence of the electromagnetic waves on the reference plate. The position of the plate is a critical point for the success of the measurements, where changes of ± 8°, for example, can change the signal intensity for nearly -35 dBm. This abrupt dropping of the signal for angles different from 0° is due to the flat geometry of the target, scattering the electromagnetic wave impinged on it in different directions of the receiving antenna.

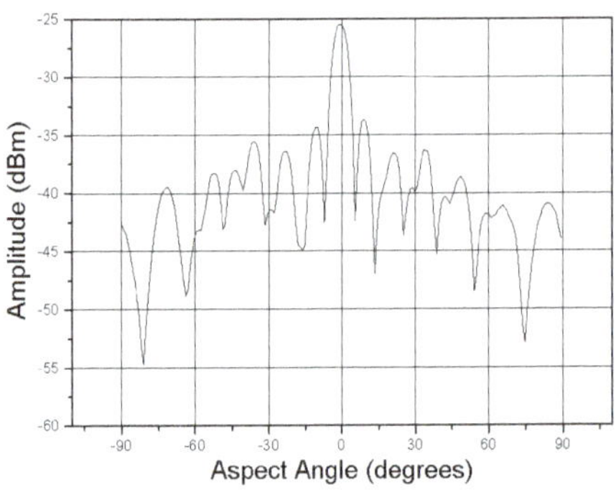

Fig. 12. RCS pattern of an aluminum plate (300 mm x 200 mm), at 8 GHz.

Afterwards, RCS measurements were carried out with the panel having one side coated by ERAM. Figure 13 shows the RCS pattern expressed in dBm and in square meters, both obtained at 8 GHz. In the range of 0 to -180° it is observed a peak at -90°, corresponding to the normal incidence of the radiation on the reference side of the panel.

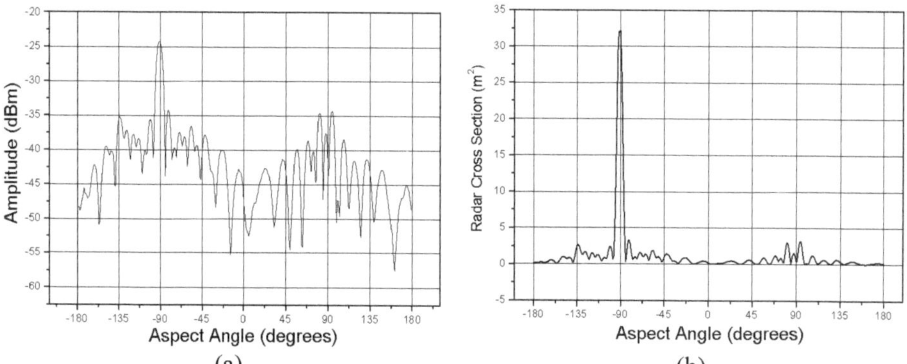

Fig. 13. RCS measurements of the panel (300 mm x 200 mm), at 8 GHz: (a) in dBm and (b) in m^2. Reference side (-180° to 0°) and ERAM coated side (0° to +180°).

The attenuation obtained at the frequency of 8 GHz, presented above, at the incidence angle of 90° is about 10 dB. The measurements showed clearly that the RCS values decrease when the ERAM is applied on the flat plate. The processed ERAM coating is able to attenuate nearly 94 % of the electromagnetic wave, when impinged at 90°, at 8 GHz.

4.1.3 Dihedral and trihedral corner reflectors

Corner reflectors consist in geometries of special interest in electromagnetic scattering problems because they provide a large bistatic or monostatic radar cross section over a broad range of observation and aspect angles. The large echoes from these targets arise from the multiple reflections between two or three mutually orthogonal flat surfaces forming the reflectors. Dihedral corners have been used by many works in RCS reduction studies and also as RCS calibration target (reference target). Trihedral corners yield large backscattering RCS over a wide azimuth and elevation angular ranges and are widely used for external radar calibration.

Both dihedral and trihedral corners are usually present in mechanical structures of ships, aircraft and vehicles, contributing as efficient scattering centers of such targets.

Thus RCS reduction of such geometric corner structures is of great importance in the design and construction of ships and aircraft which will be under surveillance of radar systems. It has been reported that RCS reduction of dihedral corners can be achieved by altering the mutual orthogonality of the flat surfaces. This technique involves changes in original engineering design of the target. Meanwhile, the use of ERAM can also overcome this problem.

This section shows the results involving the RCS reduction of a square dihedral and trihedral corner reflectors, coated with ERAM developed at Materials Division/IAE/DCTA. The characterization was performed using an anechoic chamber in the frequency range of 8 - 12 GHz.

RCS measurements were carried out at the frequencies of 8, 10 e 12 GHz. In each frequency, the dihedral and the trihedral were fixed on the rotating support and rotated from -90° to +90°, at a scanning rate of 0.15 rad/s. In this process two pieces were utilized, one as reference and the other coated with the ERAM (previously described). Figure 14 shows the assembly of a dihedral inside the anechoic chamber used in this work.

Fig. 14. Assembly of a dihedral on the azimuth positioner.

Figure 15 shows the results at 10 GHz of the reference dihedral (-180° to 0°) and this target coated with ERAM (0° to +180°). The dihedral RCS pattern is characterized by multiple reflections of the wave between the orthogonal faces, with two peaks in the aspect angles of -135° and -45°. These peaks are attributed to the wave normal incidence on the flat side of the dihedral. It can be observed at 10 GHz (Figure 15) an attenuation of nearly 13.6 dB and 10 dB at -135° and -45°, respectively in the RCS pattern of the ERAM coated dihedral, in comparison to the reference one. Between +45° and +135° angles it is verified an attenuation of 20 - 24 dB, attributed to the multiple reflections of the wave between the orthogonal faces of the dihedral.

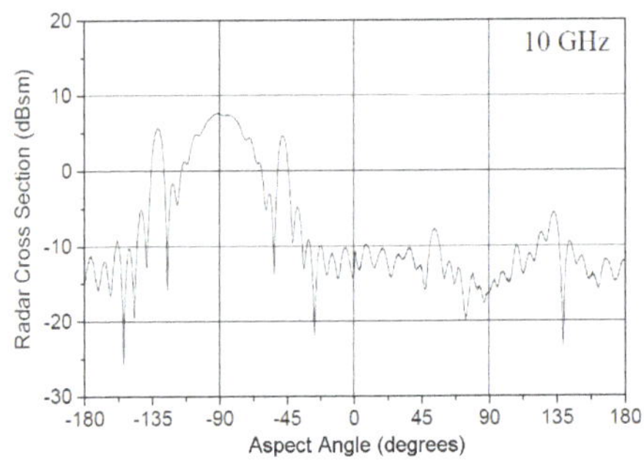

Fig. 15. RCS pattern of a dihedral (120 mm x 120 mm), reference side (-180° to 0°), ERAM coated side (0° to +180°).

Figure 16 depicts the results at 8 GHz of both, the reference trihedral (-180° to 0°) and the ERAM coated one (0° to +180°). The trihedral RCS pattern is characterized by multiple reflections of the wave among the three faces. In a similar way, the resulted RCS pattern is similar to that one verified for dihedrals. At 8 GHz, the RCS pattern shows a reduction of 17 dB for the ERAM coated trihedral in comparison to the reference one, between the aspect angles of +45° and +135°. However, in the angle range of +135° to +180° and 0° to 45° is observed a signal increase related to the reference trihedral.

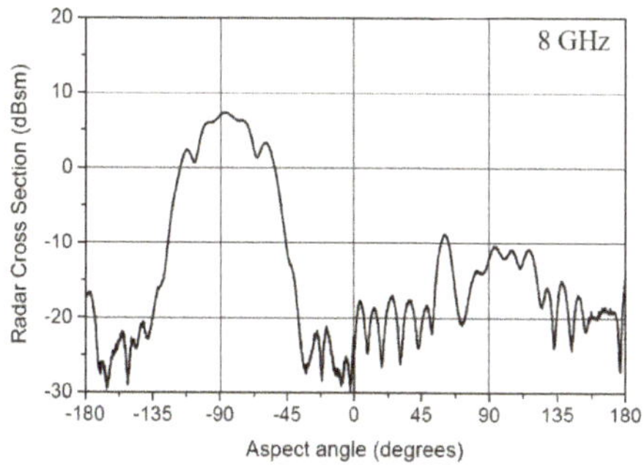

Fig. 16. RCS pattern of a trihedral (120 mm x 120 mm), reference side (-180° to 0°), ERAM coated side (0° to +180°).

4.1.4 Cylinders

The RCS of a metal cylinder has well-known theoretical equations and making use of some considerations, the RCS of a cylinder can be calculated. However, unlike the perfectly conducting sphere, the cylinder is sensitive to wave polarization.

The calculations involving cylinders generally assume that the axial length of the cylinder is large compared to the wavelength. The RCS pattern of a cylinder can be represented by an almost constant value over the entire range of the aspect angle that should, ideally, result in an omnidirectional pattern.

The advantage of using a metal cylinder with a low roughness, as support for testing the reflectivity of ERAM, lies on RCS values nearly constant with the angle of the incident wave. Therefore, the ERAM coated cylinder characterization is more influenced by possible variations of ERAM texture, homogeneity of the absorbing additives distribution and also the absorber thickness. Possible variations of these parameters will result in change of RCS pattern as a function of aspect angle of the incident wave. Figure 17 shows the cylinder mounted on the positioner, located inside an anechoic chamber, uncoated (Figure 17a) and ERAM coated (Figure 17b).

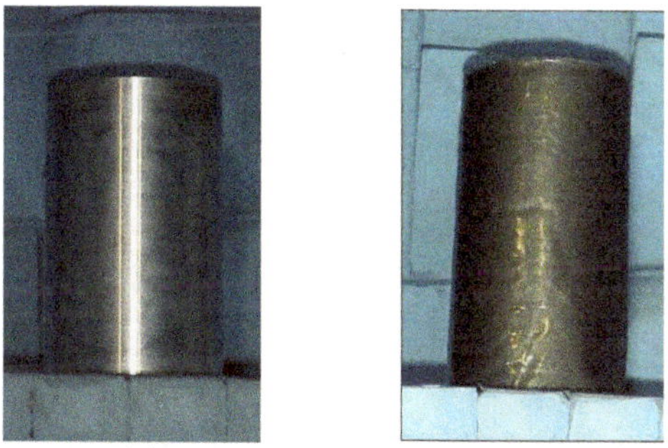

(a) (b)

Fig. 17. Assembly of an uncoated (a) and a coated (b) cylinder inside an anechoic chamber for RCS reduction measurements.

Figure 18 depicts the plotted RCS pattern of a metallic cylinder of 32 cm of length and 15 cm of diameter, rotating the cylinder from +90° to -90° on its axes and keeping the TX and RX antennas in the same position, at 9.375 GHz. The determined RCS value is constant and omnidirectional, equal to 1.61 m^2. The cylinder RCS measurement needs a tight adjusts on the dielectric support to avoid contributions of the reflected waves from this apparatus.

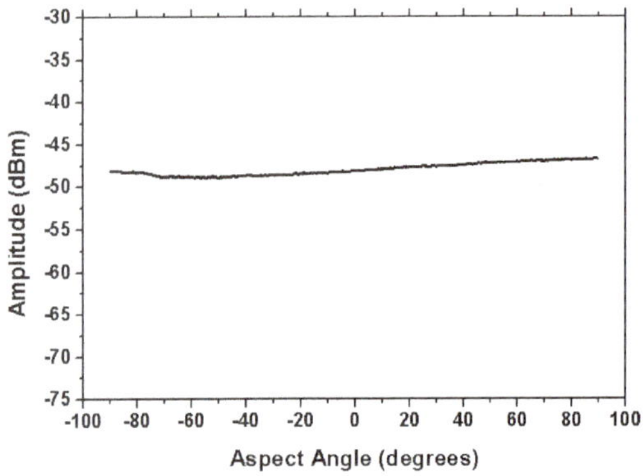

Fig. 18. RCS pattern of a cylinder at 9.375 GHz.

The received power level is almost constant at 9.375 GHz (-49.7 ± 1.0 dBm) rotating the support in 360° around its vertical axe. This curve is typical for this kind of target and it is attributed to the cylinder shape, which contributes only as a line when the wave impinges on it. Afterwards, using the same device and the same cylinder it was obtained the RCS pattern of this target coated with a processed ERAM (Figure 19).

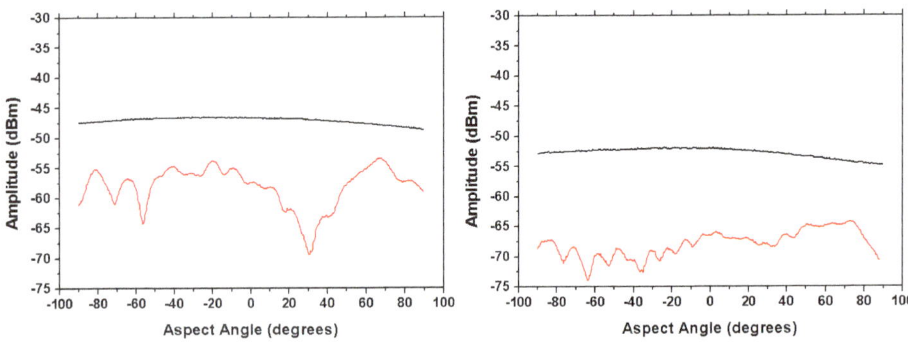

Fig. 19. RCS patterns of cylinder without ERAM (black curve) and coated with ERAM (red curve) at 10 GHz (left) and 12 GHz (right).

Figure 19 shows the RCS pattern of the cylinder at 10 GHz and 12 GHz coated with the ERAM presenting 1.2 mm of thickness loaded with NiZn ferrite and carbon black. In this case, the pattern shows the reflectivity variation in function of the aspect angles, characterizing a RCS reduction of the target. This variation is attributed to the heterogeneity of the processed ERAM, related to both the ferrite and carbon black particles distribution and the absorber thickness.

4.2 Complex shapes targets

The total RCS of a target of complex geometry can be determined by various techniques of different levels of complexity. One approach involves dividing the object into several parts, and these should approach targets of simple geometry. The determination of the RCS of each of these parts, the sum of their individual contributions to the total reflected field, taking into account their respective phase angles and reflectivity, provide the final value of the target RCS of complex geometry.

A complex target has multiple scatters interacting with each other, changing the RCS as a function of aspect angle. For those targets, the peaks of RCS pattern may be many wavelengths apart, but they can add up constructively or destructively by a fraction of degree.

The RCS pattern of a complex target is a function of many variables in the radar system and environmental measures. In addition to the change in aspect angle and shape of the target, there are multipath and atmospheric effects. The main reflection mechanisms of a complex object are: specular reflection, surface waves, edges diffraction and cavities.

In the next section will be treated the RCS studies of complex targets (mainly missiles) done by the Materials Division/IAE/DCTA from Brazil.

4.2.1 Missiles

Missiles possess one of the most fundamental complex geometries to RCS prediction and electromagnetic characterization. These ones present a relatively simple geometry in comparison with more elaborate targets, such as aircraft and ships. However, these shapes allow detailed studies of the scattering phenomenon of the simple geometries that constitute them (Ruck, 1970). In this section, it is shown some RCS results achieved in the characterization of a hypothetical missile and a real missile section, in the microwave range of 5 to 7 GHz. ERAM application on the studied targets is also evaluated.

4.2.1.1 Characterization of a hypothetical missile in indoor measurements in C-band

A prototype was projected and constructed and called 'hypothetical missile'. The purpose is to understand the behavior of an actual complex target, from the well known physical and mechanical characteristics of a prototype. The dimensions of the hypothetical missile correspond to those of a cylinder measuring 0.32 m x 0.15 m and four square flat plates measuring 0.15 m sideways.

The study presented in this section intends to compare the results obtained experimentally with data obtained theoretically, by predictions based on literature (Knott, 1993). An algorithm to simulate the radar signature of complex geometry was developed with the purpose of both estimating the contributions of each simple component of the target and assisting the analyses of the experimental measurements (Miacci, 2006).

The foreseen aspect, by simulation, of the RCS pattern for this type of target is shown in Figure 20. This figure demonstrates the appearance of the peaks relative to the frontal, back and sides contributions of the target. Figure 20 shows also a view inside the anechoic chamber, where the system was assembled for the RCS experimental measurements for the frequency range of 5.8 to 6.4 GHz.

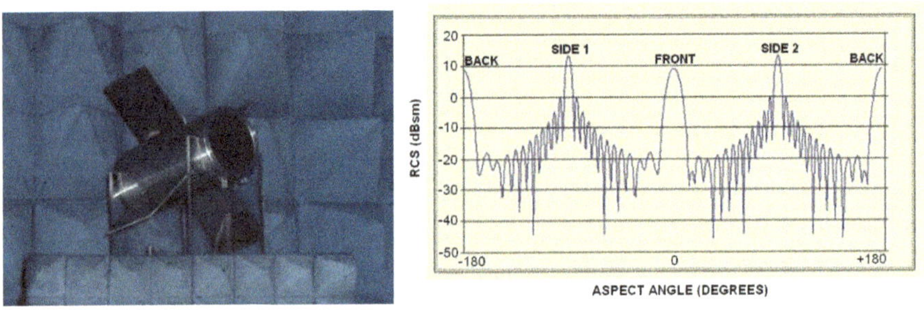

Fig. 20. Simulated RCS pattern for the Hypothetic Missile at 6.0 GHz and the Hypothetic Missile placed in the anechoic chamber.

The same missile was placed in the anechoic chamber and characterized in accordance with the procedures and the RCS patterns obtained at frequencies 6.0 GHz and 6.4 GHz are shown in Figure 21. Comparing Figures 20 and 21 good agreements of the patterns is verified. However, the figures relative to the experimental measurements show less detailing of the secondary lobes. From these patterns, the RCS values can be calculated for the hypothetical missile, as shown in Table 4.

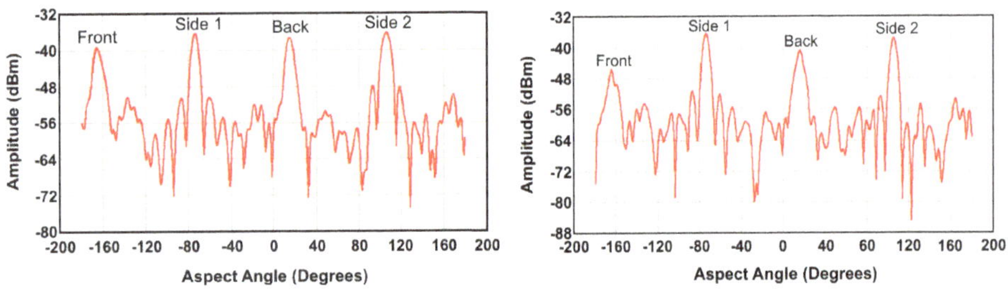

Fig. 21. RCS pattern measured from the Hypothetic Missile at 6.0 GHz (left) and 6.4 GHz (right).

Frequency (GHz)	RCS maximum (dBsm)
5.9	09.2
6.0	11.5
6.2	11.9
6.4	13.0

Table 4. Experimental RCS Values of the Hypothetical Missile (deviation: 0.7 dB).

The analysis of Table 4 shows the maximum RCS variation of the hypothetical missile (9.2 – 13.0 dBsm) as a function of frequency, hence evidencing the sensibility of this measure with small variations on the wavelength of incident radiation. From such measurements it can be affirmed that the system is reliable to accomplish RCS measurements of complex geometry targets.

4.2.1.2 RCS reduction of the hypothetical missile

Reflectivity measurements of the Hypothetical Missile coated with ERAM were performed (Figure 22) and the results were compared with those presented in preceding section. In order to improve the evaluation of the ERAM effect on the Hypothetical Missile, this material was applied in only a portion of this target under test, i.e., a face of the missile was not covered with the absorber. Thus, we applied the ERAM on the front and over one side, while the rear and the other side did not have their surfaces coated (Figure 23).

Fig. 22. View of the Hypothetical Missile with applying of ERAM.

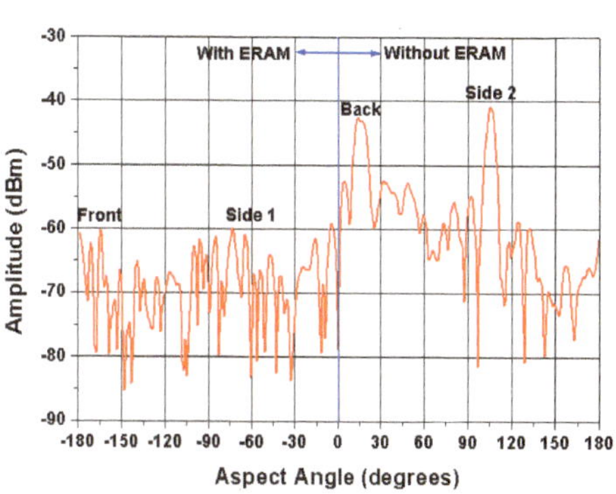

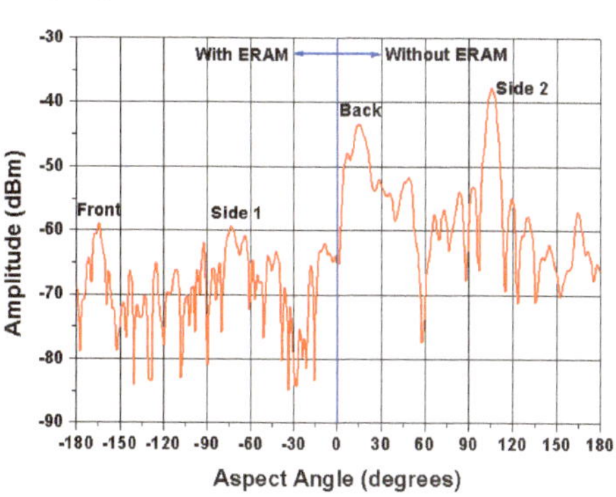

Fig. 23. Reflectivity curve of the Hypothetical Missile partially covered with ERAM, at 6.2 GHz (left) and 6.4 GHz (right).

Table 5 shows the results and compares the RCS reductions achieved.

Frequency (GHz)	RCS Reduction (dB)	Measured RCS (dBsm)
5.9	16	-6.8
6.0	19	-7.5
6.2	19	-7.1
6.4	21	-8.0

Table 5. RCS reduction with the ERAM on the Hypothetical Missile (deviation: 0.7 dB).

The analysis of Table 5 shows that the RCS reduction values (16 - 21 dB) are in accordance with the attenuation values of the used ERAM (~ 20 dB attenuation). This behavior shows that the proposed experimental method is effective for the study of RCS reduction of complex geometries. The main contributions are highlighted in the figures.

4.2.1.3 Characterization of a missile's section in indoor range in C-band

After verifying the reliability of the parameters adopted in the employed methodology, the characterization of the actual complex target was performed. For this, the selected section of the actual missile was placed in the anechoic chamber (Figure 24). The results obtained in the reflectivity measurements are illustrated in Figure 25.

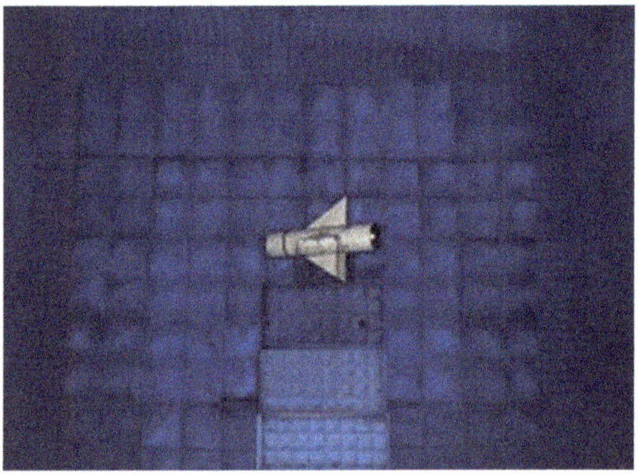

Fig. 24. View of the missile's section positioned in the anechoic chamber.

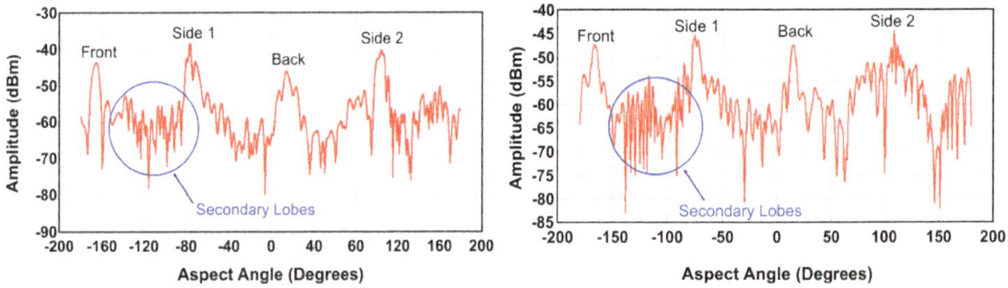

Fig. 25. RCS patterns of the missile's section at 6.0 GHz (left) and 6.4 GHz (right).

Figure 25 represent the obtained curves that are typical of the reflectivity patterns of the missile's section characterized in the anechoic chamber at a frequency range of 5.9 to 6.4 GHz. The comparison of these figures shows that the frequency increase is accompanied by greater detailing of the patterns. This behavior shows a larger contribution of the secondary lobes. These figures reveal also the real complexity of the RCS patterns of complex targets.

The results of these measurements suggest the presence of phenomena commonly seen in experiments performed with this type of geometry. As the missile's section presents similar geometry to that one observed in the hypothetical missile, it is expected that its pattern adjust adequately to those measured for the hypothetical missile.

However, the presences of triangular-shaped flat plates (the missile's warp) and the irregular aspect of the surface contribute with new peaks, increasing the complexity of the pattern. Besides the front part of the missile is not a perfect flat plate, since before this area has the boundary with the tracking compartment, modifying the outline of this target's reflectivity pattern, contributing also to the complexity of the RCS pattern.

Another phenomenon observed is with regard to the non-symmetry of the pattern between the angles of -180 to 0° and 0 to +180°, which reveal peaks of different shapes and amplitudes. This is due to the contributions of physical details of the missile, such as compartments of electronic circuits, conduits where the cables pass through and connectors integrated to the cylindrical body of the target.

Such devices positioned on the missile's body, which at first seem to be of little influence, cause wave scattering on the target's surface, which alter the reflection as a whole, such as diffraction edges and traveling waves. These constructive attributes of the missile are displayed in the measurements and results in alterations in the pattern peaks. These results justify the complexity of the reduction of the target's RCS of complex geometry, as foreseen in literature (Ufimtsev, 1996).

The measured RCS values for these cases can be seen in Table 6.

Frequency (GHz)	Maximum RCS (dBsm)
5.9	7.3
6.0	7.9
6.2	10.2
6.4	10.5

Table 6. Experimental RCS values for the missile's section (deviation: 0.7 dB).

4.2.1.4 RCS reduction of a missile's section

ERAM was used covering selected faces of the missile, i.e., the faces covered and uncovered resulted in a pattern for direct comparison of RCS reduction. Analogously to make with the hypothetical missile, ERAM was applied on the front part and on one side, while the rear and the other side remained uncoated (Figure 26). Figure 27 show the RCS reduction patterns obtained at 6.0 and 6.4 GHz.

Similarly to that observed in the Hypothetical Missile measurements, it is verified the RCS reduction of the faces with ERAM. In a similar way, the application of ERAM promotes attenuation values of the target's RCS close to those determined for the ERAM (~ 20 dB). Table 7 shows the reduction obtained in this case.

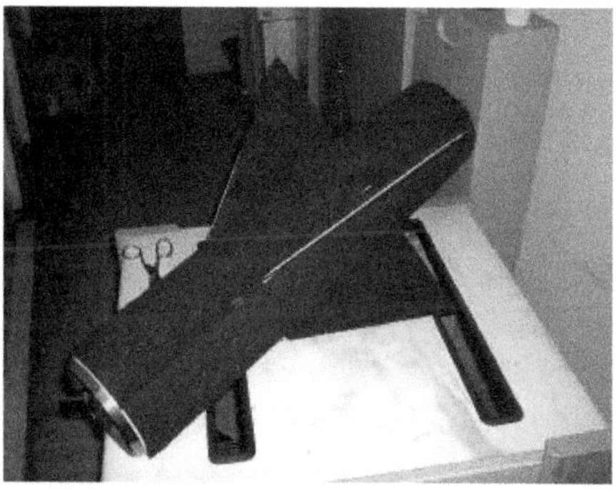

Fig. 26. View of the missile's section covered with ERAM.

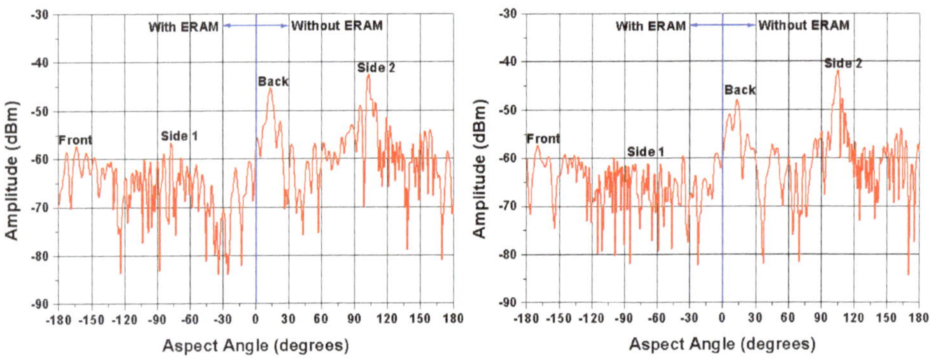

Fig. 27. Reflectivity curve of the missile's section partially covered with ERAM at 6.0 GHz (left) and 6.4 GHz (right).

Frequency (GHz)	RCS Reduction (dB)	Measured RCS (dBsm)
5.9	11.5	-4.2
6.0	14.0	-6.1
6.2	18.5	-8.3
6.4	18.0	-7.5

Table 7. RCS reduction of the missile's section coated with ERAM (deviation: 0.7 dB).

5. Conclusions

This work shows a reliable methodology for RCS measurements of simple and complex targets. Indoor RCS measurements of simple and complex targets in C and X-band are explored and the results show good agreement with the literature. Reflectivity measurements of the studied targets coated with electromagnetic radiation absorbing materials demonstrate that the proposed measurement system is satisfactory for RCS reduction studies.

This work presents also, results of an active cancellation circuit as an enhancement technique for the RCS measurements. The proposed circuit increases the detection probability of smaller targets, associated with a greater dynamic range of the measurement system. Therefore, the proposed circuit makes the study of RCS reduction more effective concerning small targets covered with high attenuation ERAM. An increase of dynamic range of up to 26 dB at 5.9 GHz frequency is observed. Consequently, the minimum detectable RCS values using the proposed circuit increased significantly, as for instance -9.0 dBsm (without the circuit) up to -35 dBsm (with the circuit) at 5.9 GHz.

The recovering of any targets with a determined ERAM developed at Materials Division/IAE/DCTA in Brazil it was possible to compare and to discuss the influence of the ERAM on the observed RCS reduction. The used methodology showed useful for attending aeronautical and telecommunication applications.

For the first time in Brazil, the RCS of missile's sections (hypothetical missile and an actual missile section) was experimentally determined, in C-band in an anechoic chamber.

The evaluated RCS measurement system proved to be effective, with measured values within the expected error range (lower than 1 dB). The obtained results enable to conclude that this methodology is applicable in the characterization of various types of simple and complex geometry targets in indoor and outdoor ranges, attending the far-field condition. In function of the results obtained, employing the presented system, further in-depth studies of RCS reduction is feasible for other military platforms, given its functionality.

6. Acknowledgment

The authors acknowledge FAPESP (São Paulo Research Foundation) (Process 03/00716-4 and 11/03093-4), CNPq (National Council for Scientific and Technological Development) (Process N. 305478/2009-5) for the financial support and also to Dr. Carlos Frederico de Angelis from INPE/Brazil and the Ministry of Defense / Aeronautics Command, for their collaboration and support in this project.

7. References

Balanis, C. A. (1982) *Antenna theory, analysis and design,* Harper & Row, ISBN 0-47160-352-X New York, USA

Bhattacaryya, A. K.; Sengupta, D. L. (1991), *Radar cross section analysis and control,* Artech House, ISBN 0-89006-371-0, Norwood, USA

Birtcher, C. R.; Balanis, C. A. (1994), RCS measurements, transformations and comparisons under cylindrical and plane wave illumination. *IEEE Transactions on Antennas and Propagation,* v. 42, n. 3, p.329-333.

Blake, L. V. (1986), *Radar range performance analysis,* Artech House, ISBN 0-89006-224-2, Norwood, USA

Burgess, L. R., Berlekamp, J. (1988) Understanding Radar Cross-Section Measurements. *Microwaves Systems News and Communications Technology,* p. 54-61.

Crispin, J. W.; Siegel, K. M. (1968), *Methods of radar cross section analysis,* Academic Press, New York, N.Y., USA

Currie, N. C. (1989), *Radar reflectivity measurement: techniques and applications,* Artech House, ISBN 0-89006-345-1, Norwood, USA

Dybdal, R. B. (1987) Radar cross section measurements. *Proceedings of the IEEE,* v. 75, n. 4, p.498-516.

Jasik, H.; Johnson, R.C. (1984) *Antenna engineering handbook,* 2 ed., Mcgraw Hill, New York, USA

Jenn, D. C. (1995), *Radar and laser cross section engineering,* AIAA Education Series, Washington, DC, USA

Knott, E. F.; Schaeffer, J. F.; Tuley, M. T. (1993) *Radar cross section* (2nd ed.), Artech House, ISBN 1-891121-25-1, Norwood, USA.

Kouyoumjian, R. G.; Peters JR. (1965), Range requirements in radar cross section measurements. *Proceedings of the IEEE,* v. 53, p. 920-928.

Martin, I. M.; Dias, J. C.; Rezende, M. C. (2002), Reflectivity in the microwave range of polyurethane coatings loaded with NiZn ferrites. *Journal of Materials Research,* v.1, n.1, p. 1-10

Miacci, M. A. S. (2002), *Experimental Measurements of the Microwave Backscattering of Single Shapes Targets,* (In Portuguese), Dissertation (Master of Science) – *ITA,* Brazil

Miacci, M. A. S. (2006) *Improvement of Missile RCS Measurements by using of Active Cancellation Technique*, (In Portuguese), *Thesis (PhD) – ITA*, Brazil

Ross, R. A. (1966), Radar cross section of rectangular flat plates as function of aspect angle. IEEE *Transactions on Antennas and Propagation*, v. AP-14, p. 329-335.

Ruck, G. T. (1970), *Radar cross section handbook,* Plenum Press, New York, USA.

Skolnik, M. I. (1990), *Radar handbook* (2nd ed.), Mc Graw Hill, ISBN 0-07-057913-X, New York, USA.

Thomson Inc. (1988), *Luneberg Reflectors and Lenses*, Technical Bulletin, September, 8 p, France.

Ufimtsev, P.Y. (1996), Comments on: Diffraction principles and limitations for RCS reduction techniques. *Proceedings of the IEEE*, v. 84, n.12, p.1828-1851.

17

An Intelligent System for Efficient Rigid Film Anticounterfeiting Inspection

Michael Kohlert[1], Christian Kohlert[2] and Andreas König[1]
[1]*Institute of Integrated Sensor Systems, Department of Electrical and Computer Engineering, University of Kaiserslautern, Kaiserslautern,*
[2]*Department of Research and Development, Klöckner Pentaplast GmbH & Co. KG, Heiligenroth, Germany*

1. Introduction

The danger that results from taking ineffective pharmaceutical products or consuming food that does not comply with the requirements, leads to significant health, and life hazard for people. Anticounterfeiting of, e.g., pharmaceutical products, guarantees customers the delivery of original products. The rigid film industry is developing automated systems to detect product piracy. For this aim, an approach to an automated inspection system for rapid and reliable product verification based on optimized insertion of infrared (IR) and ultraviolet (UV) µm-pigments in rigid films has been developed in this work and statistically meaningful samplesets have been extracted. The industrial manufacturing process has been enhanced for optimized insertion of pigments in rigid films with regard to size, type, density, and process step. The pigments are activated with IR or UV light in an encapsulated laboratory system specially developed here. Filter on illumination sources and colour cameras limit the activation and the emission range. Due to optimized film manufacturing and measurement system, the evaluation of µm-pigments can be achieved by a two-stage process of state of the art supervised colour segmentation, followed by a blob analysis. The recognition results of the conceived intelligent engineering system fully meet the industrial specification requirements. The project has advanced from a laboratory, small volume production to a large volume test customer production and evaluation.

2. An intelligent inspection system

Product piracy causes a loss of 600 bn USD world wide every year (Barbieri, 2004). A product copy due to plagiarism is difficult to differentiate from originals. The danger of unknowingly consuming vitally important pharmaceutical products that are either not effective or even poisonous is a stringent societal problem. The financial situation of product markets shows the problem of capital investment in new products when no profit is realized due to illegal copies. To disclose illegal copying of original products the rigid film industry developes reliable application and authentification systems to make product packages and their content safer. As one part of the first stage of the supply chain the rigid film industry is able to achieve this goal for the whole supply chain.

The rigid film industry wants to secure the path from the producer to the customer to avoid replication at low additional costs. The package appearance should not change and the whole system has to be integrated easily in the packaging-process.

Klöckner Pentaplast, Montabaur, Germany, has been looking into ways of marking packaging film in such a way that it can be subsequently recognized by the film manufacturer and can also be used for product protection by the converter.

This work`s aim is to develop measuring systems with costs lower than 2000 € for visual detection of pigments and authentification of the product package. The variable cost of the pigment has to be lower than 0.1 Cent/kg of foil. The reproducibility of equal distribution has to be guaranteed during the production process.

The specifications of the research and development department of Klöckner Pentaplast request the ability to measure the UV- and IR-pigment density in films by using light sources, filters and cameras, and achieve a correct separation of actual pigments and fibres due to contamination.

In this work, the process of calendering is described in section 2.1 (Fig. 1), focusing on the insertion of light emitting pigments for anticounterfeiting.

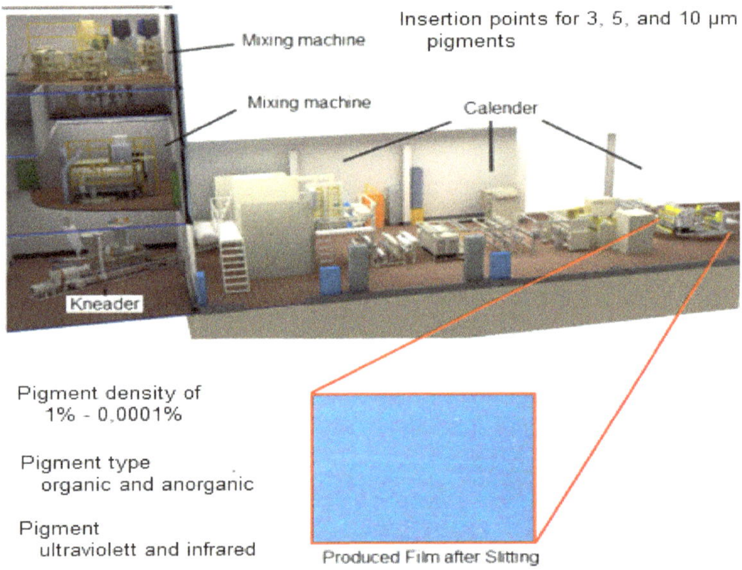

Fig. 1. Insertion points for pigments in the process of rigid film production: 1. mixing machine, 2. kneader, 3. Calendar

A short description on the kinds of pigments, which are used, their activation under IR and UV light sources and related data acquisition will be shown in section 2.2. A hierarchical, two-step classification- approach, that serves to segment the acquired images of pigmented films explained in section 2.3. An overview of the whole anticounterfeiting process is presented in section 2.4, before concluding.

2.1 Basic process and application specific adaptations

The standard production of marked transparent and coloured films can be described in a complete process diagram (Fig. 1). Additional pigments can be inserted in a master batch, which is mixed in two mixing machines with high and low temperature, and the kneader, which plastifies the batch. The insertion of pigments in the calendar where the film becomes formed is inefficient. The produced foil is slittered at the end of the calendaring process for the transport stage. Further changes on the film will be made in a printing firm or the other industrial plants after transportation.

The pigments dedicated to product identification were inserted in the recipe of the mixing machines. Anorganic and organic types have been investigated. Anorganic pigments are FDA approved and do not cause any problems to industrial usage. Thus, polymeric films can be doped with fluorescent pigments (FDA approved, anorganic) for protection of polymeric packages. Standard pigment sizes from manufacturers are 3µm, 5µm, or 10µm.

An inserted pigment density of 0.0001% up to 1% is used in this work. A higher density of more than 0.1% leads to an unintentional coloration of the polymeric film, which was a motivation for an upper and lower boundary and stepwise increment in variation.

The intensity and distribution of the pigments is depending on the thickness of the film. The following permutations have been explored by preliminary work with 300 tests on pigment size, type, density, and film thickness empirically. Starting with the smallest size, smallest density, random type, and smallest film thickness, about 1500 pigmented films were produced with 20 different pigments of three different sizes, 3µm, 5µm, and 10µm. Successful laboratory tests with films of 60µm up to 800µm thickness allow the use of product identification even for multilayered films.

Dedicated measurement laboratory equipment is developed for the detection of the pigments in the films. UV fluorescent tubes (UV-A, UV-B), and light emitting diodes (UV-A, UV-B) activate pigments, and other materials. IR light emitting diodes (880nm - 1050nm), and a laser-diode (980nm) activate the IR pigments used in the film.

The range of settings (Tab. 1) will be completed by achieving best recognition results with measuring devices in chapter 2.2.

Types : = 20	Size : = 3	Density : = 5	Film Thickness : = 5
Organic (natural)	3 µm	0,0001%	200 µm
Anorganic (synthetic)	5 µm	0,001%	300 µm
Ultraviolet (300-370nm)	10 µm	0,01%	400 µm
Infrared (940-980nm)		0,1%	600 µm
		1%	800 µm

Table 1. Ranges of preliminary tested parameters for 1500 films with 20 different pigments

The surface analysis and classification methods show advances for anorganic IR pigments with 5µm, 0.001% density, and 300µm film thickness. The whole tested setting are shown in figure 2.

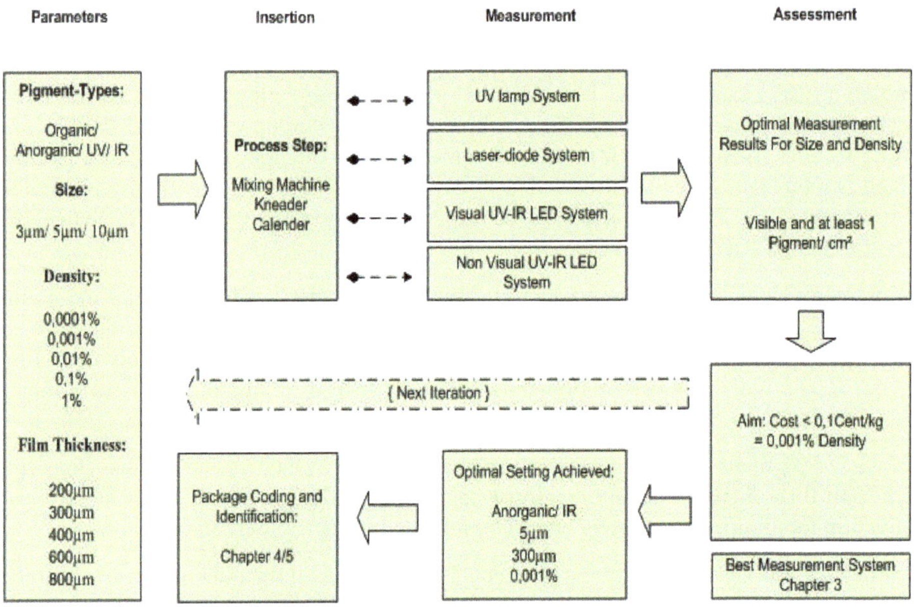

Fig. 2. Tested pigment parameters for 20 types, 3 sizes, 5 densities, and 5 film thicknesses

Four measurement units where deployed. The measurement setup and data acquisition will be explained in chapter 2.2.

2.2 Measurement setup and data acquisition

For the required measurements, four different measuring devices (UV lamp system, visual UV-IR LED system, non visual UV-IR LED system, and laser diode system) were developed in this work.

The settings from chapter 2 have been investigated during the measurement setups and data acquisition with these four units.

The UV illumination is generated by fluorescent lamps and light emitting diodes at activation wavelengths of 300nm to 370nm. Depending on the pigment emission a bandpass- filter is used to detect the pigment wavelength with a colour camera. It is very efficient to use filters for camera and light source to minimize reflections and other influences.

Pigments are activated in UV or IR light and emit in lower or higher wavelengths (Fig. 3 shows an example). These pigments are called up- and down- converters. This feature is used for the following measuring devices.

Blisters are prototypes of pharmaceutical packages that show the produced film in the way they are used later for products. The activation light source for the film is placed in 45° above the object. A camera captures the image in 90°. The coloured image acquisition allows a colour-spaced segmentation as a next step. In this work about 100 blisters were captured.

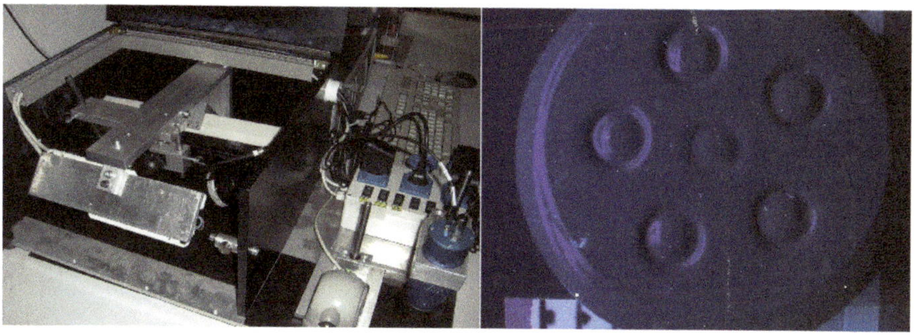

Fig. 3. Measuring device (left) for UV pigments in a blister (right)

The second system is an IR recognition system, which detects pigments with an activation wavelength from 900nm to 1000nm. (Fig. 4)

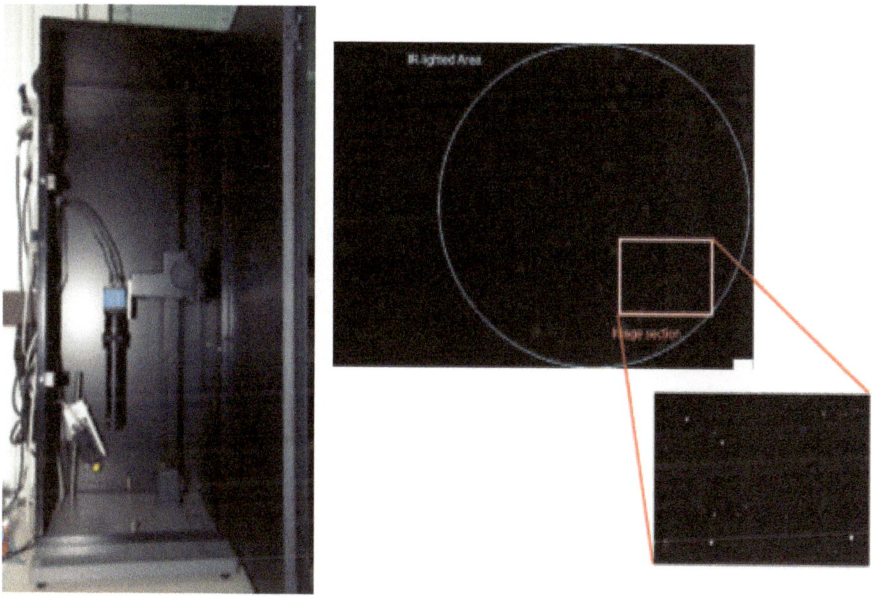

Fig. 4. Image acquisition on 10cm² with a 11-mega-pixel-camera (b/w), image section of 2cm² with a fluorescent IR pigment density of 0,001% and 5µm

A laser diode with an optical spreading device is able to illuminate an area of 10 cm² in an appropriate normalised distribution for image acquisition. The IR lighted area and the selected image section with fluorescent IR pigments are shown in Fig. 4. The illuminated pigments are captured by the black/white camera. For this specific measurement system, 100 testing films have been captured for ensuing analysis.

In Fig. 5 a measurement unit (Visual UV-IR LED System) is shown, which is in development for the recognition of UV, and IR pigments. The hand-held unit allows a fast analysis of rigid film surface at all places in the plant.

Fig. 5. Hand-held unit for internal use for fast recognition of UV, and IR pigments

Different measurement units were tested as described in this chapter. A compact installed laboratory or inline- system is expensive. (Tab. 2)

UV lamp system	Laser-diode System	Visual UV-IR LED System	Non-Visual UV-IR LED System
3000 €	15.000 €	1500 €	1000 €
365nm/ 300nm	975nm	320nm/940nm	365nm/940nm
2 lamps	1 laser-diode	8 diodes/ 8 diodes	12 diodes/ 12 diodes

Table 2. Price range for the different illumination systems

As an alternative the hand-held systems can be used after the calendaring process for fast identification in the plant, at the pharmacy or integrated in a cash terminal. In the textile industry this system can be used for identification of jackets after the production process.

2.3 Two-stage recognition system and results

Based on the optimized rigid films and measurement setups the ensuing evaluation is already feasible by state of the art techniques applied in a two stage approach. The first stage of classification separates the light emitting pigments from the background and the reflections caused by the light source of the measuring device by supervised segmentation. This stage is called here *microclassification*. To separate the pigments from contaminations (fibres) a second classification is carried out on the segmented images, denoted here as *macroclassification*. The origin image is shown in figure 6.

Supervised segmentation of images generates binary images with candidate pigment pixels highlighted. Expert knowledge is integrated during data selection and labelling in the design process. For manually selected relevant sample pixels from 30 specially fabricated films (type: anorganic, size: 5µm, density: 0,001%, not thermoformed), class affiliations were interactively generated. (Fig.7). From the RGB images LAB color-spaced images of the same size were computed, and from the selected pixel coordinates 1200 labeled three dimensional vectors have been acquired, balanced for both classes and randomly split into training and sample sets of 600 vectors each.

Fig. 6. 640x480 pixel RGB image of a rigid film under UV (302nm)

Fig. 7. Manual selection of example pixels representing the two classes reflections/background (left, middle), and pigments (right) for segmentation classifier training

The selected pixels (Fig. 7) define vectors of LAB-colour space for each class to be presented. In a pre-examination the LAB-colour space generated best segmentation results, but other colour spaces can also be used (Hunter Labs [HL], 1996).

The classifier achieves supervised segmentation by assigning a class to each pixel of the RGB image.

For the underlying problem complexity, a non-parametric classifier should be chosen for optimum classification results.

The training vectors of the defined classes separate all samples into different classes by using the L, a, or b data of a sample.

The L, and a data of each vector achieved better results for the Support Vector Machine (Matlab-Toolbox) with 99.83%. Pigment and Background vectors can be distinguished more efficiently. (Fig. 8)

The CorrectRate defines the possibility for becoming classified as a pigment. For efficient use the whole Anticounterfeiting system has to detect 99% correct. Errors in the microclassification will be dominantly compensated in the second stage. The size parameters prevent a wrong assignment (Christianini & Shawe-Taylor, 2000).

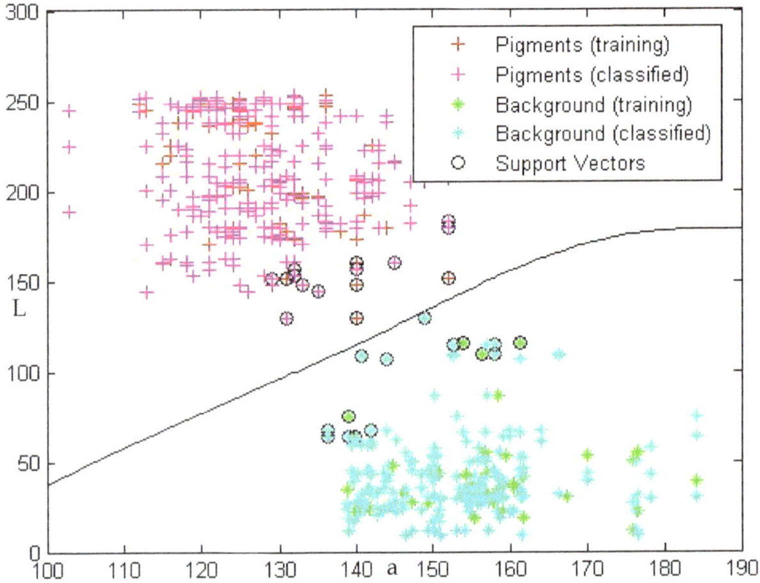

Fig. 8. Scatterplot of 600 training- and 600 testing vectors of L, and a data from L,a,b colour space (SVM - Matlab Toolbox) - CorrectRate = 99.83 %

The a, and b data of each vector showed similar recognition results in comparison with the L, and a data correlation.

Thus, the L, and b data correlates with 99.5%, the L, and a data will be preferred for further classification (99.83%).

The classification results of the training data of 600 vectors (features: L,a,b of colour space L,a,b) and the testing data of 600 vectors are shown in Tab. 3. Four classifiers (kNN, PNN, RNN, and SVM) are known as state of the art methods. These types use different mathematical algorithms for the separation, and classification of sample sets.

	kNN	PNN	RNN	SVM
Classification rate for test set using hold-out approach	k=1 99.5 % k=2 99.5 % k=3 99.5 % k=5 99.5 % k=6 99.5 % k=8 99% k=10 99 %	σ: 0.1 99% σ: 0.03 99% σ: 0.001 60%	99%	99.83 % quadratic Kernel C = 0.1 - 100000 Gamma = 0.001 - 10

Table 3. Microclassification results

Thus, the k-neareast-neighbor (kNN), the reduced-nearest-neighbor (RNN), the probabilistic neural network (PNN), and the support vector machine (SVM) classifier have been applied, and found already to be satisfactorily working (Duda et al., 2001).

The segmented images from the 99.83% correct (SVM) microclassification are used for blob analysis, which has the potential to eliminate residual erroneous pixel classification by the context of pixel cluster shapes. The macroclassification, based on geometrical features of blob analysis, separates light emitting pigments from fibres. (Fig. 9)

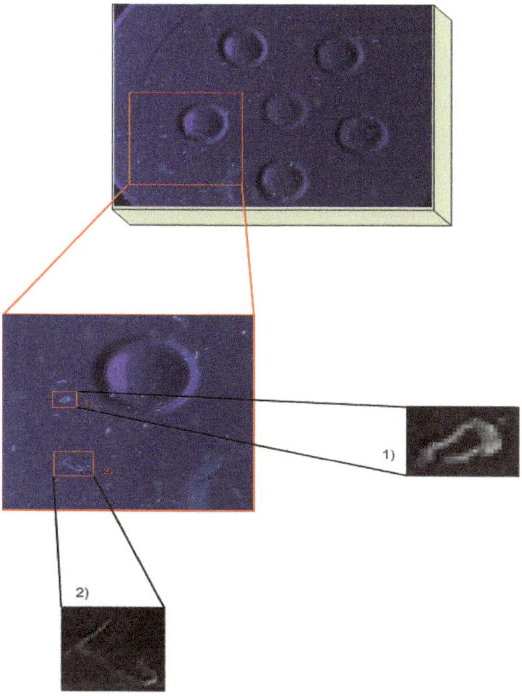

Fig. 9. Separation of fibres from true pigments after segmentation

Fibres originate from contamination or clustered particles (Fig. 10). The geometrical structure of fibres is different from the structure of the pigments.

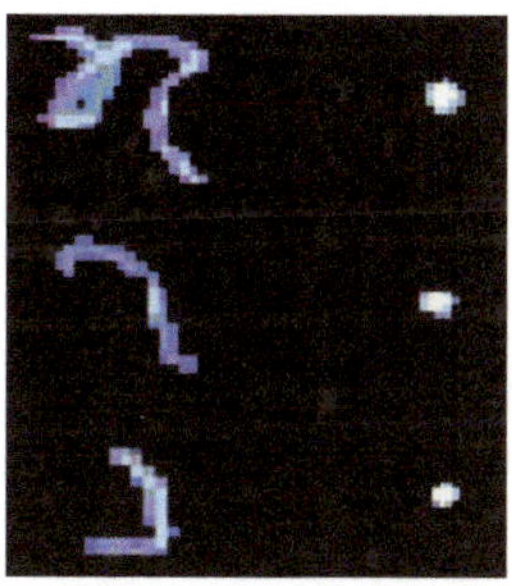

Fig. 10. Subregions containing fibres and pigments for training and test data

From eight blob features, area, perimeter, balance point, compactness, outer circle, inner circle, symmetry, limitative rectangle, the most significant ones (area, rectangle) are automatically chosen by feature selection to distinguish between both classes.

Out of pre-selected 300 vectors for pigments, the training set of 100 feature vectors of pigments, 100 feature vectors of fibres, and the test data consisting of 50 feature vectors of pigments and 50 feature vectors of fibres from the same images were employed. (Tab. 4)

	kNN	PNN	RNN	SVM
Classification rate for test set using hold-out approach	k=1 100 % k=2 100 % k=3 100 % k=5 100 % k=6 100 % k=8 100 % k=10 100 %	σ: 0.1 100 % σ: 0.03 100 % σ: 0.02 72.5 % σ: 0.001 52.5 %	100 %	100 % quadratic Kernel C = 0.1 – 100000 Gamma = 0.001 - 10

Table 4. Classification results for the blob analysis

The result of the k-nearest neighbor classification is an image only consisting of pigments. Classes like background, reflections or fibres are not included any more (Gonzales & Woods, 2008; König et al., 1999; König, 2001).

After detection and elimination of the fibres, the located pigments can serve for a higher-level authentification process in a following step (Fig. 11).

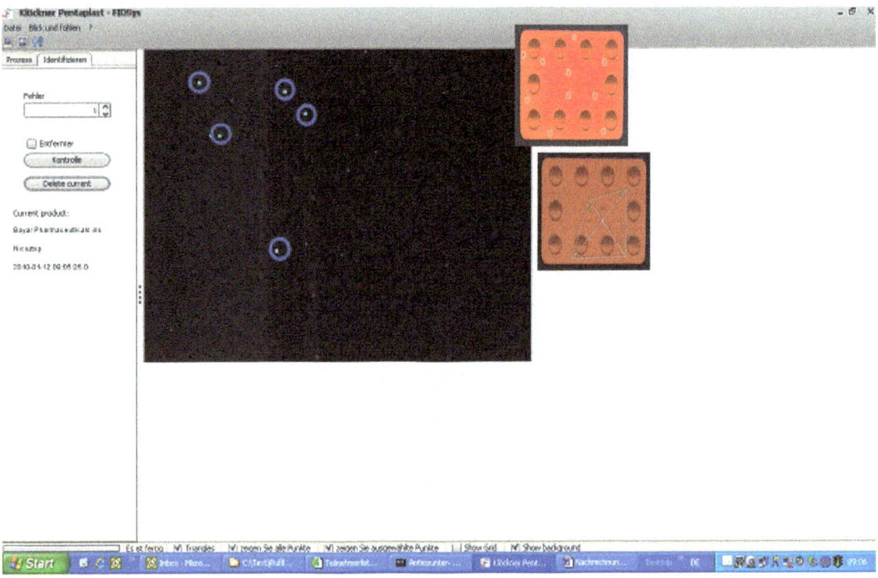

Fig. 11. Selection process (Blue identification pigments in software analysis)

Using the described state-of-the-art method for plane films or blisters as shown in the image, a selection of partial areas allows a defined analysis of completed packages at the end of the production process. This has been experimentally verified on the detection results for three randomly chosen rigid films. The coding process separates package/production id, date, corners and angles to define a unique digital signature for each package. The selected pigments have a geometrical correlation (angles) which is saved in the digital signature of the package code. The pigments are selected randomly, but their coordinates can be recalculated in the verification process (Fig. 12).

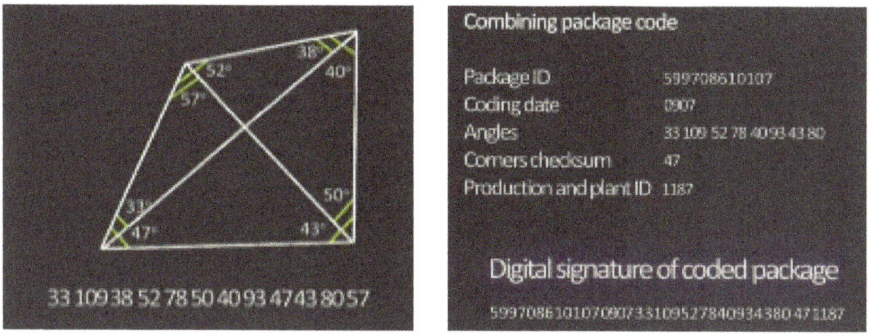

Fig. 12. Coding process

After the polymeric coding a database server saves the scanned angles for further comparison of the packages at the client. The small code allows a fast, brand recognition (Fig. 13).

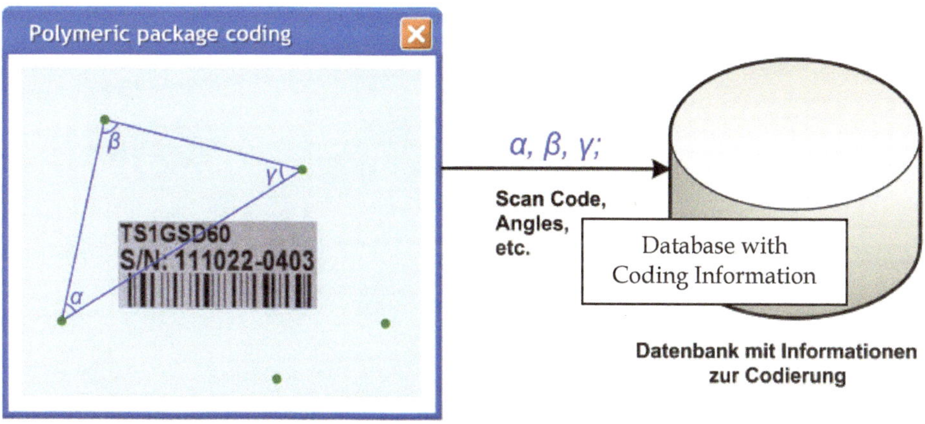

Fig. 13. Polymeric Package Coding

An inline-inspection system (see Fig. 14) is actually in progress for first detection during the calendaring process. The patented technology is available to all users of rigid film from Klockner Pentaplast. The inspection system is not implemented yet. Figure 14 shows an impression of the final system at the end of the development process.

Authentic products can be reliably protected with a fingerprint which is obtained through the incorporation of randomly-distributed pigments in the packaging, which are then made visible (C. Kohlert et al., 2010).

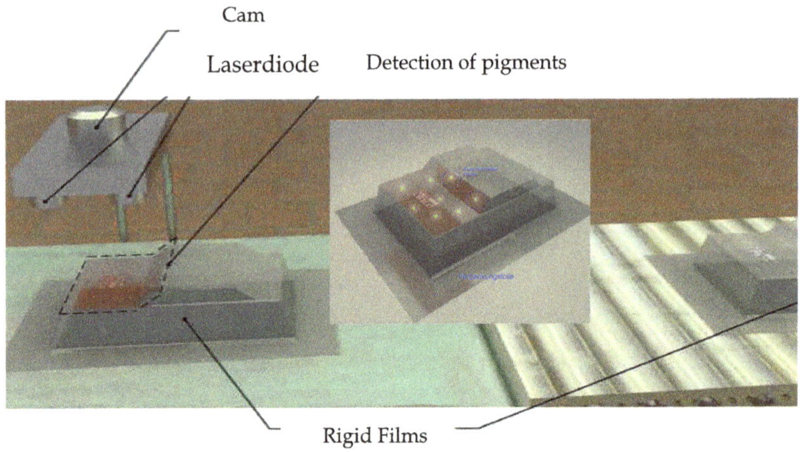

Fig. 14. Inline-Inspection system at the end of the calendaring process – shortly after the thermoforming-/ or the slitting process

2.4 Complete anticounterfeiting system

Using UV or IR pigments for anticounterfeiting is a novel approach in the rigid film industry. The insertion in polymeric recipes allows high level security for the whole supply chain. The complete process of anticounterfeiting rigid film is shown in the next figure (Fig. 15).

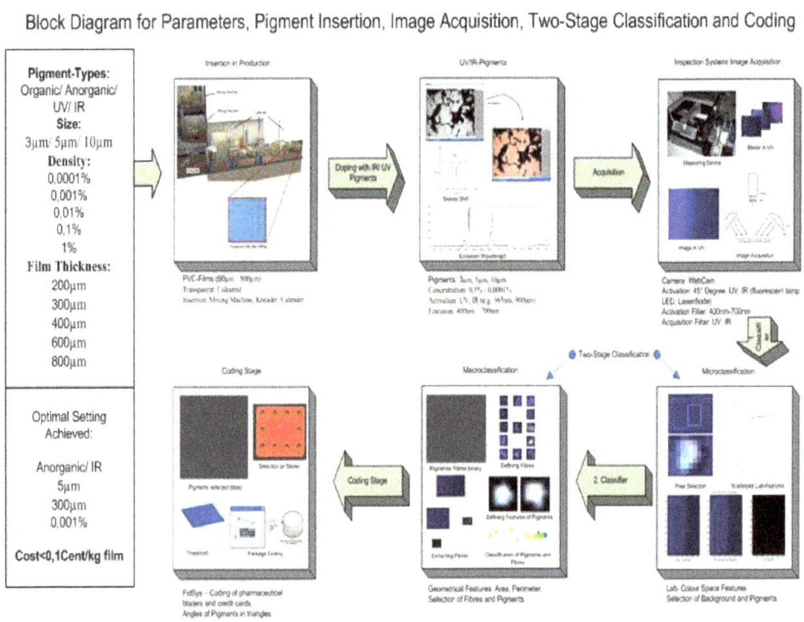

Fig. 15. Processing diagram for pigment insertion, image acquisition, two-stage classification, and coding stage

The insertion of pigments can be done at the mixing machines, the kneader, or the calender. The mixing machines, as the best insertion point, consisting of a mixture of particles (batch) allow a fast disposal of the pigments within the batch.

The choice of the pigment type depends on the measuring device. For optimal results IR pigments are better to distinguish from other batch particles. Most batch particles emit in UV ranges, so the UV emission can be ten times higher than the defined insertion.

A two-stage classification separates pigments from other materials in an image with a correct rate of 100%.

For anticounterfeiting of rigid films a specially developed software authentificates a polymeric package via recording, and reading of random pigment structures correctly. It calculates the geometrical positions of the pigments with 99% accuracy, and checks the apparent authenticity of a film, or a blister within 6 seconds.

3. Conclusion

In this work, a viable and economic implementation of a product authentification system was developed, that is able to recognize, and locate IR and UV pigments in specially fabricated polymeric films. A special feature of this approach is, that film manufacturing and measurement were optimized with regard to recognition accuracy, so that rather basic methods could be employed in the back-end for successful identification, which is also important for aspired low-cost detection devices.

The modification of the polymeric standard process of film production by insertion of pigments was implemented in two of three possible ways. The mixing machine showed best results for optimized distribution. An insertion during the calendaring process causes no optimal results. The kneader is another possibility to insert pigments as fluid, and will be tested in further tasks.

For optimal detection, and separation from fibres, anorganic IR pigments with a density of 0,001% and a size of 5μm showed best results in comparison with UV pigments.

Four different measurement systems, an UV lamp system, a visual UV-IR LED system, a non visual UV-IR LED system, and a laser-diode system were established. The IR LED systems showed best results for film recognition.

A hierarchical two-stage recognition system with state of the art methods for pigment and localization of 100% accuracy has been achieved.

Classification techniques, e.g. Support Vector Machine have been investigated with existing and new data sets to improve system reliability for viable inline-process measuring, and portable inspection systems.

The industrial specification of the company with regard to a homogenous insertion of IR and UV pigments in the calendaring process, and a separation of fibres is fully achieved. In future work, more extensive datasets will be generated, in particular for blob analysis, to ensure wider generality.

The real time behaviour and resource demands of the proposed system have to be regarded carefully, and potentially optimized for hand-held inspection devices (M. Kohlert et al., 2010).

4. References

Barbieri, A. (Sept. 29, 2004). Getting real about fake designer goods, In: *Bankrate.com*, Feb. 26, 2011, Available from http://www.bankrate.com/brm/news/advice/scams/20040929a1.asp

Christianini, N., & Shawe-Taylor, J. (2000). *An Introduction to Support Vector Machines and Other Kernel-based Learning Methods* (1st edition), Cambridge University Press, 978-0521780193, Cambridge, UK

Duda, R.O., Hart, P.E., & Stork, D.G. (2000). *Pattern Classification* (2nd edtion), Wiley-Interscience, 978-0471056690

Gonzales, R. C., & Woods, R. E. (2008). *Digital Image Processing* (3rd edition), Prentice Hall, Upper Saddle River, NJ

Hunter Labs (Aug. 1-15, 1996). Hunter Lab Color Scale. *Insight on Color 8 9*, Feb. 26, 2011, Available from: http://www.hunterlab.com/appnotes/an08_96a2.pdf

König, A., Eberhardt, M., & Wenzel, R. (1999). Image Processing Europe, *QuickCog Self-Learning Recognition System – Exploiting machine learning techniques for transparent and fast industrial recognition system design,* (Sept./Oct. 1999), pp. 10–19

König, A. (2001). Pattern Recognition in Soft Computing Paradigm, *Dimensionality Reduction Techniques for Interactive Visualisation, Exploratory Data Analysis and Classification,* Vol. 2, (2001), pp. 1–37, ISBN 981-02-4491-6

Kohlert, M., Kohlert, C., & König, A. (2010). Lecture Notes in Computer Science, *Automated Anticounterfeiting Inspection Methods for Rigid Films Based on Infrared and Ultraviolet Pigments and Supervised Image Segmentation and Classification*, Volume 6276, (2010),pp. 321-330

Kohlert, C., Schmidt, B., Egenolf, & W., Zistjakova, T. (2010). *Verpackungsfolie für Produktauthentifizierung, Authentifizierungsverfahren und –system*. DE 10 2008 032 781 A1

Permissions

The contributors of this book come from diverse backgrounds, making this book a truly international effort. This book will bring forth new frontiers with its revolutionizing research information and detailed analysis of the nascent developments around the world.

We would like to thank Prof. Md. Zahurul Haq, for lending his expertise to make the book truly unique. He has played a crucial role in the development of this book. Without his invaluable contribution this book wouldn't have been possible. He has made vital efforts to compile up to date information on the varied aspects of this subject to make this book a valuable addition to the collection of many professionals and students.

This book was conceptualized with the vision of imparting up-to-date information and advanced data in this field. To ensure the same, a matchless editorial board was set up. Every individual on the board went through rigorous rounds of assessment to prove their worth. After which they invested a large part of their time researching and compiling the most relevant data for our readers. Conferences and sessions were held from time to time between the editorial board and the contributing authors to present the data in the most comprehensible form. The editorial team has worked tirelessly to provide valuable and valid information to help people across the globe.

Every chapter published in this book has been scrutinized by our experts. Their significance has been extensively debated. The topics covered herein carry significant findings which will fuel the growth of the discipline. They may even be implemented as practical applications or may be referred to as a beginning point for another development.

The editorial board has been involved in producing this book since its inception. They have spent rigorous hours researching and exploring the diverse topics which have resulted in the successful publishing of this book. They have passed on their knowledge of decades through this book. To expedite this challenging task, the publisher supported the team at every step. A small team of assistant editors was also appointed to further simplify the editing procedure and attain best results for the readers.

Our editorial team has been hand-picked from every corner of the world. Their multi-ethnicity adds dynamic inputs to the discussions which result in innovative outcomes. These outcomes are then further discussed with the researchers and contributors who give their valuable feedback and opinion regarding the same. The feedback is then collaborated with the researches and they are edited in a comprehensive manner to aid the understanding of the subject.

Apart from the editorial board, the designing team has also invested a significant amount of their time in understanding the subject and creating the most relevant covers. They scrutinized every image to scout for the most suitable representation of the subject and create an appropriate cover for the book.

The publishing team has been involved in this book since its early stages. They were actively engaged in every process, be it collecting the data, connecting with the contributors or procuring relevant information. The team has been an ardent support to the editorial, designing and production team. Their endless efforts to recruit the best for this project, has resulted in the accomplishment of this book. They are a veteran in the field of academics and their pool of knowledge is as vast as their experience in printing. Their expertise and guidance has proved useful at every step. Their uncompromising quality standards have made this book an exceptional effort. Their encouragement from time to time has been an inspiration for everyone.

The publisher and the editorial board hope that this book will prove to be a valuable piece of knowledge for researchers, students, practitioners and scholars across the globe.

List of Contributors

Md. Zahurul Haq
Bangladesh University of Engineering & Technology (BUET), Bangladesh

Mario Vrazic, Ivan Gasparac and Marinko Kovacic
University of Zagreb, Faculty of Electrical Engineering and Computing, Croatia

André V. Bueno
Federal University of Ceara, UFC, Brazil

José A. Velásquez
Federal University of Technology - Parana, UTFPR, Brazil

Luiz F. Milanez
University of Campinas, UNICAMP, Brazil

Miloš Beković and Anton Hamler
University of Maribor, Faculty of Electrical Engineering and Computer Science, Slovenia

Chih-Jer Lin, Chii Ruey Lin, Shen-Kai Yu, Guo-Xing Liu, Chih-Wei Hung and Hai-Pin Lin
National Taipei University of Technology, Taiwan, R.O.C

Christian Maria Firrone and Teresa Berruti
Department of Mechanics, Polytechnic University of Turin (POLITO), Turin, Italy

Yoshiharu Morimoto and Akihiro Masaya
Moire Institute Inc., Japan

Motoharu Fujigaki
Wakayama University, Japan

Daisuke Asai
Hikari, Japan

Yoshimi Iijima and Hirotaka Sakaue
Japan Aerospace Exploration Agency, Japan

Krzysztof Tomczyk
Cracow University of Technology, Faculty of Electrical and Computer Engineering, Poland

Marek Kciuk and Grzegorz Kłapyta
Silesian University of Technology, Faculty of Electrical Engineering, Department of Mechatronics, Poland

Christian Bräuer-Burchardt, Max Möller, Christoph Munkelt, Matthias Heinze, Peter Kühmstedt and Gunther Notni
Fraunhofer IOF Jena, Jena, Germany

Alla Taranchuk and Sergey Pidchenko
Khmelnitsky National University, Ukraine

Janusz Janiczek
Wroclaw University of Technology, Poland

Barbara De Nicolo, Carlo Piga, Vlad Popescu and Giovanna Concu
University of Cagliari, Engineering Faculty, Italy

Kashif Saeed, Matthew B. Byrne and Ian C. Hunter
University of Leeds, School of Electronic and Electrical Engineering, United Kingdom

Muhammad F. Shafique
COMSATS Institute of Information Technology, Department of Electrical Engineering, Pakistan

Marcelo A. S. Miacci
National Institute for Space Research (INPE), Brazil

Mirabel C. Rezende
Institute of Aeronautics and Space, Department of Aerospace Science and Technology, Brazil

Michael Kohlert and Andreas König
Institute of Integrated Sensor Systems, Department of Electrical and Computer Engineering, University of Kaiserslautern, Kaiserslautern, Germany

Christian Kohlert
Department of Research and Development, Klöckner Pentaplast, GmbH & Co. KG, Heiligenroth, Germany